Introduction to Engineering: Modeling and Problem Solving

Introduction to Engineering:
Modeling and Problem Solving

Jay B. Brockman

University of Notre Dame

John Wiley & Sons, Inc.

EXECUTIVE PUBLISHER	Don Fowley
ASSOCIATE PUBLISHER	Dan Sayre
SENIOR ACQUISITIONS EDITOR	Michael McDonald
EDITORIAL ASSISTANT	Rachael Leblond
EXECUTIVE MARKETING MANAGER	Chris Ruel
SENIOR PRODUCTION EDITOR	Ken Santor
TEXT DESIGNER	Madelyn Lesure
COVER DESIGNER	Jeof Vita
COVER PHOTO	© Curtis Round/Design Pics/Corbis

This book was set in LaTeX by the author and Aptara Corp. and printed and bound by Hamilton Printing Company. The cover was printed by Phoenix Color.

This book is printed on acid free paper. ∞

ISBN 978-0-471-43160-2

Printed in the United States of America

10 9 8 7 6 5 4 3 2 1

*To the memory of my daughter Dylan
and my mother Hope*

Preface

Choosing Engineering

In their 2005 report, *Educating the Engineer of 2020: Adapting Engineering Education to a New Century*, a panel commissioned by the National Academy of Engineering (NAE) in the United States addressed the issues in preparing the next generation of engineers to solve the technical problems facing society in the years ahead [Nat05]. Some of these are old problems that won't go away, such as updating our aging civil infrastructure of roads, bridges, waterways, and the power grid [Ame05]. Other challenges include supplying food, water, energy, and communications to a growing population, providing health care for the increasing number of retirees as people live longer, and maintaining our national defense in the presence of new threats.

Surveys have shown that pre-college students today favor stimulating careers that involve "helping others" [Nat05] [Tay00]. Unfortunately, the data also shows that interest in choosing engineering as a path towards these career goals is declining. As the NAE report notes, there are many factors contributing to this downturn, ranging from preparation for engineering in the K–12 grades, through the compressed and intense course load typically taken by engineering students relative to their college classmates, to the way that the engineering profession portrays itself in society. Among all these factors, however, one of the most important is the students' experience in their first engineering course in college. This course sets the tone for how prospective students view engineering, and plays a pivotal role in their decision to pursue engineering or to switch to another major [Nat05] [PMS03]

Among its suggestions for revitalizing engineering education, *Educating the Engineer of 2020* makes two specific recommendations regarding the first course [Nat05]:

- "Whatever other creative approaches are taken in the 4-year engineering curriculum, the essence of engineering—the iterative process of designing, predicting performance, building and testing—should be taught from the earliest stages of the curriculum, including the first year."
- "Engineering schools [should] introduce interdisciplinary learning in the undergraduate environment, rather than having it as an exclusive feature of the graduate programs."

The main goal in writing this book was to provide a resource for a first course in engineering built upon these two principles.

The Notre Dame Experience

Even before the NAE *Educating the Engineer of 2020* report appeared, a number of engineering programs at colleges and universities had already begun looking at new strategies for attracting and retaining talented students. In the fall of 1998, the new dean of engineering at the University of Notre Dame, Frank Incropera, launched a project to improve the quality of undergraduate education. After an intensive self study, the college came up with two primary recommendations. The first was to enhance the first-year experience for all students considering engineering, and the second was to migrate towards a learning paradigm "focused on developing leadership, teamwork, experiential learning, the use of information technology, student-faculty interaction with industry, and student interaction with faculty (formal and social)." The associate dean for undergraduate studies, Steve Batill, formed a committee with representatives from each of the five engineering departments at Notre Dame to develop a set of learning objectives and a format for the new courses. The committee decided upon three main objectives:

- Understand what engineering is and how it is practiced.
  - societal context
  - relationships and differences between disciplines
  - relationships to mathematics and the sciences
- Develop and apply fundamental engineering skills.
  - problem solving;
  - communications;
  - computer skills.
- Gain practical design experience as part of a multidisciplinary team.

Tom Fuja and I were selected as the course directors. With help from faculty from across the College of Engineering, we came up with four project modules for the course that were designed to present a cross-section of engineering disciplines, as well as to highlight some aspects of real-world engineering practice. Pictured in Figure 1, these initial four projects were as follows:

Launch System   In this project, students used a large slingshot to launch a softball at a target downrange. The key part of the project was for the students to develop a set of models to determine launcher settings, including a model for the energy stored in the spring and a model for the trajectory of the softball. Initially, the models were implemented in spreadsheets, but in later years this was changed to MATLAB.

Bar Code Reader   In this project, students implemented a robot using the LEGO Mindstorms kits that would drive over a barcode and perform an action after reading an encoded command. Prior to using MATLAB in the course, this project served as an introduction to programming and information processing.

pH Controller   Building on the experience of the bar code reader project, the goal of this project was implement a system that would maintain the pH inside of a continuous stirred-tank reactor within specified limits by adjusting the flow of a base solution into the tank. The project introduced mass-balance equations as well as simple control theory. Students also designed and built their own mechanisms for adjusting the flow using the LEGO Mindstorms kits and some additional inexpensive components.

Lightweight Structure   In this project, students used a MATLAB program written by Notre Dame civil engineering professor Dave Kirkner to design a truss structure

Figure 1   Projects from the initial Introduction to Engineering course at Notre Dame, 1999–2000 academic year.

made of K'NEX plastic rods and connectors that would satisfy constraints on load, deflection, and cost.

Writing the Book: Philosophy and Approach

While many of the ideas in this book grew from my participation in developing the Introduction to Engineering courses at Notre Dame, I also had the opportunity to draw upon my own observations, experiences, and biases as a student, teacher, and practicing engineer. In particular there are several key ideas that cut across the book that have grown from my personal background.

Engineering Can Be Fun   Growing up in the 1960s and '70s with the space program a regular fixture on TV, a neighbor who was a ham radio operator, and an ample supply of scrap wood in our new suburban development, engineering seemed like a natural choice for someone who was interested in math and science and who also liked building things, even if I really did not yet know at the time what engineering *was*. While engineering—like any career—does occasionally have some dull and routine aspects, I still find that many parts of the job are as exciting as I'd hoped they would be, and on the best days, it feels as if my job is my hobby! Throughout the text, I've tried to highlight some of the projects and personalities that illuminate this perspective.

Engineering is Inherently Multidisciplinary   This is a view of engineering that I began to see as an undergraduate at Brown University in Providence, Rhode Island, which has a unified engineering department, and where, at least during my years there, all engineering "concentrators" took many of the same core courses during their first three years. As an undergrad, I had summer jobs and internships with an architectural firm in Providence, a civil engineering firm in Boston, and with Philips Electronics in the Netherlands. In each setting, I saw how teams of engineers and others routinely worked together to solve complex design problems, ranging from the Roger Williams Zoo, to the Orange Line of the Boston Subway, to an early form of an electronic newspaper transmitted as a television signal. It was after I graduated, however, and spent several years working for Intel Corporation as a product engineer, that this really hit home.

Modeling is the Key to Making Good Engineering Decisions   Engineering problem solving isn't just a matter of trial-and-error. Good engineering—where "good" means that a solution meets its technical objectives, delivered on time and within budget—depends on the ability to make accurate predictions using a variety of *models*. Models can be as simple as back-of-the-envelope calculations or as complex as detailed computer simulations. Engineers use *theoretical* models that are based on laws of nature, *empirical* models based on experimental data, as well as combinations of the two.

Many first-year engineering courses have introduced projects that stress creativity and teamwork—both very important aspects of the engineering process. Nevertheless, the bulk of course hours in the undergraduate engineering curriculum still focuses on core topics from the natural and physical sciences and mathematics, and the main reason for this is to equip students with the tools needed to make decisions backed up by quantitative analysis. I believe that an introduction to modeling should thus be a critical aspect of the first engineering course.

Engineering is More than Applied Math and Science   A common misconception about engineering held in some quarters of college campuses is that engineering is "just" applied math and science. To a degree, engineering programs themselves hold some responsibility for this. In 1968, Herbert Simon of Carnegie Mellon University delivered a series of lectures at MIT which were published the following year as the monograph *The Sciences of the Artificial* [Sim96], where he discussed his ideas for properly educating engineers and designers.

What are the topics of an engineering curriculum distinct from the natural sciences? Simon asserted that the list should include at least the following: evaluation of designs, the formal logic of design, the search for alternatives, theory of structure and design organization, and representation of design problems. Simon's ideas made a very strong impression on me as a graduate student, and after 15 years of teaching engineering, they do even more so today. In many respects, this book is a attempt to present an introduction to engineering written at a level appropriate for first-year college students, following the principles that Herb Simon outlined in 1968.

Organization of the Book

Generally, this book relies only on material from the natural sciences and mathematics that students would have seen in high school. It uses algebra, geometry, and some trigonometry, but no calculus. The physics involves only a few basic laws, such as conservation of mass and energy, Hooke's Law, and Ohm's Law, and reviews these as necessary. Using these topics as a foundation, the book focuses on the engineering method, specifically how engineers solve open-ended problems using models to support their decisions.

This book is organized into three parts. **Part I: The Engineering Mindset**, discusses how we represent and solve engineering problems. It has three chapters:

- **Chapter 1: Engineering and Society** presents an engineering world view. It describes the "engineering method" as something related to, yet different from the "scientific method." It gives examples of how engineering products—along with the teams that make these products and the society that uses them—have many interconnected parts and describes how engineers and scientists use the notion of a *system* to make sense of this network. The chapter then gives an overview of the major engineering disciplines, and shows how they relate to each other. Finally, the chapter concludes with a discussion of engineering and computing, and describes how computing has become a vital part of the engineering method.

- **Chapter 2: Organization and Representation of Engineering Systems** describes ways that engineers represent systems and illustrates techniques that help make sense of the network of ideas that surround most engineering problems. The chapter introduces a type of network diagram called a *concept map* that we use extensively throughout the text to sketch a wide variety of scenarios. This chapter concludes with a detailed example of representing the problem of providing water for rural communities in developing nations, a problem that affects over a billion people worldwide.

- **Chapter 3: Learning and Problem Solving** presents tips and techniques for studying engineering, as well as a methodology for attacking engineering problems ranging from short homework assignments to long-term projects. This chapter contains a catalog of *heuristics* or "rules-of-thumb" that will help students get "unstuck" when solving problems.

A key part of the engineering method is collecting and analyzing information to make technical decisions. To do this, engineers use *models* that approximate real-world situations. **Part II: Model-Based Design** describes some of the different kinds of mathematical models that engineers use.

- **Chapter 4: Laws of Nature and Theoretical Models** illustrates the evolution of a set of theoretical models by tracing some of the history behind the Ideal Gas Law, the Law of Conservation of Energy, Hooke's Law, and Conservation of Mass. The chapter then shows how early ideas behind these laws led to the development of the steam engine and later to the internal combustion engine. Further it describes how improvements in engine design, in turn, led to new and better models. We conclude the chapter with a detailed example using theoretical models of force, pressure, work, and efficiency to complete a detailed design of a hand pump, first introduced in Chapter 2.

- **Chapter 5: Data Analysis and Empirical Models** introduces some of the mathematical and graphical tools that scientists and engineers use to analyze data when building and using models. We begin the chapter by considering two theories three hundred years apart—Boyle's Law for gasses and Moore's Law for integrated circuit manufacturing—and use these to demonstrate ways of testing how well a theory fits the data. Next, we look at building and using empirical models, which involves using experimental data even when there's no formal theory for how a system works. Following this, we introduce techniques for quantifying the uncertainty in experimental data using statistics and probability. Finally, we present a graphical method for visualizing tradeoffs between design options, and give a detailed example of using this technique to determine settings for a large slingshot to launch a softball at a target downrange.

- **Chapter 6: Modeling Interrelationships in Systems: Lightweight Structures** builds on the material from Chapters 4 and 5 to use models to analyze and design more complex systems. Specifically, it examines the design of a type of lightweight structure called a *truss*, which is made up of many individual elements working together. Further, analyzing a truss requires looking at it from several perspectives at the same time, such as using Newton's Laws to examine the balance of forces, or considering the properties of engineering materials to determine how it deforms. Mathematically, this leads to a system of equations that must be solved simultaneously.

- **Chapter 7: Modeling Interrelationships in Systems: Digital Integrated Circuits** is conceptually similar to Chapter 6 in that it looks at a system with many parts and multiple perspectives, but in a different domain. Specifically, this chapter looks at the analysis and design of simple digital integrated circuits as a network of switches that has both logical and electrical behaviors.

- **Chapter 8: Modeling Change in Systems** looks at how engineers use models to predict how a system will change over time. We use a technique that marches through changes in the system in small, but finite steps that do not depend on any background in calculus, and that can be easily implemented in a spreadsheet or simple computer program. After looking at how we can simulate the trajectory of a ball launched from a slingshot, the main example in this chapter develops a model for estimating when we will run out of liquid petroleum.

The computer has become a critical tool for modeling. Some of these techniques are explored in **Part III: Engineering Problem Solving with MATLAB**. *MATLAB* is an integrated environment for technical computing that combines a powerful programming language with hundreds of predefined tools and commands for applications including graphics, statistics and data analysis, and simulation. MATLAB is widely used by scientists and engineers in both academia and industry. This part of the book systematically introduces MATLAB, adding features as needed to implement the models from Part II, and can be read concurrently with those chapters.

- **Chapter 9: Getting Started with MATLAB** is a gentle introduction to MATLAB, showing how to use it as a calculator to perform arithmetic, how to use variables to substitute numbers into formulas, and how to save work in a file called a *script*. The chapter provides examples of scripts that implement models from Chapters 3 and 4.

- **Chapter 10: Vector Operations in MATLAB** introduces a way to represent a list of data using a MATLAB feature called a *vector*. It also shows how to do arithmetic with vectors—say, adding the values of two vectors together— as well as how to produce plots of one vector versus another. This chapter complements Chapter 5, using MATLAB to write scripts for analyzing and plotting experimental data.

- **Chapter 11: Matrix Operations in MATLAB** describes how to work with tables of rows and columns of data in MATLAB called *matrices*. After introducing basic matrix notation, this chapter shows how to use operations on matrices to construct tables that display values for the output of a model with respect to combinations of values of two input variables. Next, the chapter presents a variety of ways of plotting such three-dimensional data and uses these techniques to look at tradeoffs in the pump design from Chapter 4, using methods from Chapter 5. Finally, this chapter provides an introduction to matrix arithmetic and linear algebra in MATLAB and shows how to use MATLAB to solve systems of linear equations, using examples from structural analysis in Chapter 6 and circuit analysis in Chapter 7.

- **Chapter 12: Introduction to Algorithms and Programming in MATLAB** opens by explaining what an *algorithm* is, and showing ways to represent them graphically using *flowcharts*, as well as with text descriptions called *pseudocode*. Next, this chapter introduces the basic MATLAB language features for writing scripts that implement algorithms, such as functions, "if" statements, and loops. Finally, the chapter concludes with detailed examples of MATLAB scripts for simulating systems that change over time, building on the material in Chapter 8.

Acknowledgements

First of all, I'd like to thank three people at Notre Dame, without whom this project would never have gotten off the ground. Frank Incropera, Dean of the College of Engineering from 1998 through 2006, initiated the Introduction to Engineering (EG 111/112) course sequence and has provided strong support and great advice all along. Frank also introduced me to John Wiley & Sons as a publisher. Tom Fuja, chairman of the Department of Electrical Engineering and Steve Batill, chairman of the Department of Aerospace and Mechanical Engineering—whom I would have loved to have had as co-authors had their schedules permitted it—helped write the original course notes from which the book began. Tom and I worked together as the course co-directors for EG 111/112 for its first few years, and he wrote some of the original material on symbols, signals, and computing that is now part of Chapter 6. Steve helped formulate many of the ideas on engineering design and modeling. The "SolderBaat" example in Chapter 3 came from his senior design course and he supervised the team of ND undergraduates that designed and characterized the original slingshot project in Chapter 5. The material in Chapter 5 on trade studies also grew from research in the area of multidisciplinary design optimization, under a grant from NASA Langley Research Center, in a collaboration between me, Steve, and John Renaud.

Many thanks to Joanne Van Voorhis of the Sloan Career Cornerstone (careercornerstone.org), who graciously allowed me to use the material that she wrote describing the various engineering disciplines in Chapter 1.

The hand pump examples in Chapters 2 and 4 are based on Steve Silliman's efforts over many years working with rural communities in Haiti and Benin to develop safe drinking water supplies. Steve is currently the Associate Dean for Undergraduate Studies in the College of Engineering and has worked with the Introduction to Engineering program at Notre Dame since its inception. I'm also deeply grateful to Steve for his careful technical review of the pump models, and for clearing up some of my misunderstandings on how pumps work, in addition to providing the original example and access to pump hardware.

Ed Maginn of the Department of Chemical and Biomolecular Engineering at Notre Dame contributed the example of estimating how much CO_2 a typical car produces in Chapter 3. He also introduced me to the work of Donald Woods of McMasters University and Phil Wankat of Purdue University, which formed the basis for the problem solving framework used throughout the text.

The material in Chapter 6 on lightweight structures began with a truss design project using K'NEX (www.knex.com) kits that was originally developed by Dave Kirkner of the Department of Civil Engineering and Geological Science at Notre Dame and Billie Spencer, who is currently with the Department of Civil Engineering at the University of Illinois, Urbana-Champaign. Steve Batill and Dave Kirkner wrote the original draft of the course notes on which the chapter is based.

Many of the end-of-chapter problems as well as solutions were developed by Ramzi Bualuan of the Department of Computer Science and Engineering at Notre Dame, who also runs the Introduction to Engineering summer program for high school students, together with ND engineering undergraduate students Maggie Merkel (Civil Engineering, 2008), Peter Nistler (Computer Engineering, 2008), and

John Souder (Chemical Engineering, 2008). Lynnwood Brown of WikiRing Consultants worked with me in developing the online problem set database.

Rumit Pancholi, a graduate student in the Creative Writing program at Notre Dame, provided tremendous assistance in editing the manuscript.

In addition to the people above who participated in conceiving and writing the book, there were many others from Notre Dame and other institutions who provided examples, reviews, suggestions, and invaluable feedback. Thanks to all, and I apologize to anyone that I may have left out.

Boston Museum of Science: Ioannis Miaoulis

California Institute of Technology /NASA Jet Propulsion Laboratory: Erik Antonsson, Gary Block, Fehmi Cirak (currently University of Cambridge), Andy Downard, David Politzer, Paul Springer, Thomas Sterling (currently Louisiana State University), Ed Upchurch, Roy Williams

California State Polytechnic University, Pomona: Francelina Neto

University of Connecticut: Zbigniew Bzymek

Cray, Inc.: Kristi Maschhoff

IBM: John Cohn, Peter Hofstee

Intel: Shekhar Borkar

Iowa State University: Martha Selby

University of Maryland: Leigh Abts

University of Melbourne: Ray Dagastine, Jamie Evans, Roger Hadgraft, Rao Kotagiri, Andrew Ooi, David Shallcross, Harald Sondergaard, Tony Wirth

University of Michigan: Jason Daida

University of Missouri, Columbia: Craig Kluever

Norwich University: Paul Tartaglia

University of Notre Dame: László Barabási (currently Northeastern University), Gary Bernstein, Joanne Birdsell, Kevin Bowyer, Joan Brennecke, Pat Dunn, Pat Flynn, Natalie Gedde, Alex Hahn, Bob Howland, Jeff Kantor, Lloyd Ketchum, Tracy Kijewski-Correa, Peter Kogge, Craig Lent, Marya Lieberman, Mark McCready, Leo McWilliams, Kerry Meyers, Al Miller, Dave O'Connor, Sam Paolucci, Wolfgang Porod, Joe Powers, Keith Rigby, Matthias Scheutz (currently Indiana University), Mihir Sen, Greg Snider, Mike Stanisic, Bill Strieder, Aaron Striegel, Flint Thomas, John Uhran, Mitch Wayne

Ohio University: Daniel Gulino

Portland State University: Lemmy Meekisho, Hormoz Zareh

Rice University: Jim Tour

Rose-Hulman Institute of Technology: Patricia Brackin

University of Southern California: Jean-Michel Maarek

Stanford University: Larry Leifer

Tennessee State University: Hamid Hamidzadeh

Tufts University: Chris Rogers

University of Virginia: Michael Fowler

Virginia Tech: Michael Gregg

Louisiana Tech University: Kelly Crittenden

University of Queensland: S. J. Witty

Zyvex Corporation: Robert Freitas

I would also like to thank Cray Inc. and the Defense Advanced Research Projects Agency (DARPA) for their support of my research activities during the years that I wrote this book while visiting at Caltech. I would also like to thank the National Science Foundation for its continued support through both research and curriculum development grants that have helped make this work possible.

The team at Wiley has been fantastic, and I look forward to continuing to work with them in the future. In particular, I'd like to thank Rachael Leblond, Mike Mc-Donald, Chris Ruel, Ken Santor, Dan Sayre, Gladys Soto, and Joe Hayton, who took on the project and guided it through the first few years.

Finally, I thank my wife Jean, kids Abby and Sam, brothers Rob and Reed, and my father Mel for their love and support through more than five years of writing.

Jay Brockman
South Bend, Indiana
January, 2008

Contents

PART 1

THE ENGINEERING MINDSET

Engineering and Society

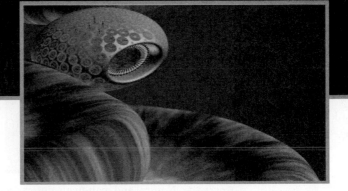

LEARNING OBJECTIVES

- to articulate a view of our environment as containing both naturally occurring and human-made or *artificial* things and to discuss the role of engineers in developing and producing these artificial things to meet human needs and desires;
- to describe what is meant by a *system* and to discuss examples of systems, including the engineering working environment;
- to discuss some of the opportunities and challenges facing engineers over the next decade;
- to describe the focus of some of the major undergraduate engineering disciplines and to list some of the professional settings in which engineers with these degrees are employed;
- to discuss the role of computing and information processing in engineering practice.

1.1   INTRODUCTION

In his 1968 text *The Sciences of the Artificial* [Sim96], Herbert Simon observed that the world is filled with two kinds of things: natural things and man-made or *artificial* things. Natural things are shaped by the processes of nature, by wind, rain, and sunlight, and by the rumblings of earth's molten core. Historically, the role of the natural sciences is to investigate these natural phenomena. When geologists, physicists, or chemists analyze natural things, they generally seek answers to questions such as:

- How does this thing work?
- What is it made of?
- How did it get to be that way?

By comparison, the man-made things on our planet—and elsewhere in our solar system—are all designed and built with a *purpose*: to satisfy mankind's complex needs and desires. The list of basic human needs includes food, water, shelter, sanitation, communication, and the development and entertainment of our minds. For many of us, the air we breathe daily is artificially heated or cooled and filtered, and the sights

Figure 1.1   Natural and man-made desert scenes: Delicate Arch, Arches National Park, Utah and the Strip, Las Vegas, Nevada. (Delicate Arch photo courtesy of the US National Park Service.)

and sounds we take in are filled with signs and symbols. Simon uses the term *artificial* without the negative connotations sometimes attached to it, simply to mean "made by human work or art." Similarly, he uses the term *artifact* to describe the products of these artificial processes. While many artifacts are hard goods, they can also assume other forms, such as processes, events, or services. From the time we formed our first communities, our survival has depended upon our ability to produce artifacts, and every known civilization has left an imprint on the natural world through the artifacts it has made.

The main business of engineering is to apply technology in concert with natural phenomena to develop these things that we need or want. Whereas the natural sciences traditionally seek to discover how things *are*, engineering focuses on the question,

- What form *should* we give to this thing so that it will effectively serve its purpose?

Engineering is a profession as diverse as society's needs. Engineers work in every conceivable business setting, from large corporations and factories to small start-up companies and consulting firms. Engineers work at construction sites and farms, on off-shore oil platforms, and in space. They serve in the military and teach in universities. People with degrees in engineering have brought their unique training to bear in other professions as well, including medicine, law, business management, and the arts. Engineers help shape our world using tools ranging from heavy equipment to computer software, and develop new tools as needed. They gather and generate information and make critical decisions using this information: will a structure fail under a given load? Can a proposed computer system for a data center be air cooled or does it need to be liquid cooled in order to function reliably? Engineers in university, government, and corporate research labs create new knowledge that can be applied to produce new or improved products, processes, and services. Finally, engineers both study and practice in a wide range of specialties or disciplines. The best-known of these—mechanical engineering, civil engineering, electrical engineering, and chemical engineering—are very broad fields with many sub-specialties.

While there is a lot of diversity in engineering, the main point of this book is to provide a common foundation that engineers of all disciplines share. In short, this book is an introduction to viewing the world through the eyes of an engineer, looking at how engineers apply science and technology to solve problems facing society.

1.2   THE ENGINEERING METHOD

The word "engineer" derives from the Latin *ingenium*, which refers to one's native genius, one's ability to design or create things. Since engineers apply technology to develop the things that we need, they require a solid background in the sciences and mathematics. Science and mathematics, however, are only part of an engineer's technical training. As the science curriculum teaches the scientific method, the engineering curriculum must also teach an *engineering method*.

Most engineering problems are *open-ended*, in that they don't have a single solution. Just because a problem has more than one possible solution, however, doesn't necessarily make it easier to solve; in fact, having multiple acceptable solutions often makes solving the problem more difficult. Most new engineering students find open-ended problems incredibly frustrating. One reason for this is that most mathematics and science training in high school typically stresses coming up with the "right" answer to a problem, and further, fosters an expectation that this "right" answer will also have a simple and elegant form. Second, in open-ended problems, the problem statement frequently doesn't provide enough information to apply a familiar technique, such as solving an equation. To circumvent this difficulty, it's often necessary to make assumptions, and knowing what assumptions to make and determining if they're reasonable often comes only with experience. To complement their study of the natural sciences and traditional mathematics, engineers and other students of the "artificial sciences" need also learn skills such as:

- how to represent a design problem
- how to make assumptions
- how to generate possible ideas for designs
- how to effectively conduct a search for a solution
- how to plan and schedule activities
- how to make efficient use of resources
- how to organize the components and activities of a team design project

1.2.1   Science, Mathematics, and Engineering

Engineering, science, and mathematics have grown side by side, and each has benefitted from developments in the other. At the core, however, their objectives and methods are different. Richard Feynman, who taught freshman physics at Caltech and received a Nobel Prize in 1965 for his contributions to the understanding of subatomic particles, compared the process of scientific discovery to trying to figure out the rules of chess through observation. The set of rules the scientist tries to deduce are the fundamental "laws" of chess. After watching a few games, he or she may formulate a "law" of conservation of bishops, stating that each player has two bishops, one that's always confined to red squares and one that's always confined to black squares. Then one day, the scientist walks in on a game in progress and sees two white bishops on black squares. What happened, and what the scientist didn't see, was that one bishop was captured, a pawn crossed for queening, and it turned into a bishop on the same color as the other bishop. Rather than despair, the scientist changes the law and keeps watching intently in hopes of finding a mechanism that helps explain the change. To extend Feynman's analogy, a mathematician would develop a precise language for describing a chess board and pieces, and for describing the rules for how the pieces move. Without such a language, scientists would find it extremely difficult to accurately discuss the laws that they've discovered. By analogy, an engineer would use the laws that the scientist came up with, described in the language of the mathematician, to formulate a strategy for winning a game of chess. This simple example illustrates some of the fundamental challenges of engineering: the scientists may not have figured out all the rules of the game, and even if they have, it's still really difficult to win at chess against a good opponent.[1]

As professionals, scientists work extensively in engineering and engineers, in turn, contribute greatly to basic science. Many college engineering students chose their majors because they enjoyed their math and science courses in high school, while the careers of many scientists began with tinkering. In high school, David Politzer's favorite class was shop. His skill at designing and making things led to a job in a physics lab as an undergraduate, building equipment for experiments to measure the decay of positrons, the "antimatter" counterpart of electrons. This project eventually led him to question *why* it was important to make these measurements, which drove him to want to learn more about the theory behind the experiments. David emphasizes that invention and discovery are equally important and challenging pursuits, although in his own career as a physicist, he's received more attention for discovery. In 2004, he won the Nobel Prize in physics for his contributions toward understanding the forces that bind the smallest known objects—called quarks—together. While his research is still largely theoretical, today David teaches the applied and experimental side of the freshman physics course at Caltech. "Engineers create a lot of neat [stuff]," he says, and his favorite part of the course is designing and presenting dramatic and sometimes explosive demonstrations of theory in practice.

1.2.2 Ingenuity: From Lifting Weights to Microelectronics

Shekhar Borkar was born in Mumbai, India and studied physics at the University of Bombay as an undergraduate. He earned a Master's degree in Electrical Engineering from the University of Notre Dame and started work at Intel, where he is now Director of the Microprocessor Technology Lab. "We engineers don't know anything about quantum physics," he jokes, "the real physicists are a lot smarter." Still, despite their "ignorance" and maybe even because of it, engineers have found extremely clever ways to skirt some of the laws of nature. They can't actually break or violate the laws, but through creative readings, engineers have managed to accomplish things that might seem impossible through stricter readings of "the law." As an example, Shekhar describes how Intel and other semiconductor manufacturers have found ways to project patterns onto chips using masks so fine that waves of light won't "fit" through the holes without becoming extremely distorted. Scientists including Grimaldi, Newton, Gregory, and Huygens first observed the distortion of waves through narrow slits in the 1600s, while Young, Fresnel, and Lord Rayleigh developed modern theories in the 1800s. Today's chip designers can't prevent these distortions,

---

[1]This is not to say that the scientist and mathematician couldn't also bring their skill to bear to play a winning game!

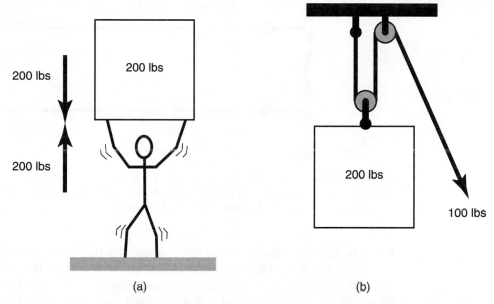

Figure 1.2   Newton's Third Law says that "for every action, there is an equal and opposite reaction." But we can still find ingenious ways to lift a heavy weight with a small applied force that don't violate this law.

but by exploiting properties of how light waves combine, engineers have discovered ways of distorting the masks themselves to project light in the desired pattern.

One of the first engineering problems mankind must have faced was how to lift an object heavier than a person can carry. While the 20th century produced some truly amazing structures, we still marvel at the ancient ones—such as the pyramids in Egypt and Mexico or the monoliths of Stonehenge—because we wonder how they could have been built without modern tools. While the discovery of Newton's Laws of Motion in the 1600s gave us a framework for reasoning about this problem, people clearly invented ways to lift massive stones and other objects thousands of years earlier. According to Newton's Laws, if gravity is bearing down on an object with a force of 200 lbs, an equal and opposite force of 200 lbs must be pushing up on it if the object is at rest. Figure 1.2(a) illustrates one consequence of these laws: if a person is to hold a 200 lb block overhead, he or she must be able to apply a force of 200 lbs. This, of course, is more than most people could raise overhead unassisted, and a strict and conservative reading of Newton's Laws might suggest that it isn't possible for a weaker person to do so.

With a bit of ingenuity, however, it's possible to build a machine that enables people to lift a heavy block, applying only their own weight, and still not violate Newton's Laws. One particularly clever machine is the block-and-tackle. As shown in Figure 1.2(b), a block-and-tackle consists of a set of pulley blocks and a rope or cord. It is believed to have been invented by Archimedes around 200 B.C.—nearly 2000 years before Newton—but Newton's Laws help us explain how it works. As a first step toward understanding the operation of a block-and-tackle, consider pulling on a rope attached to a wall with a force of 100 lbs, as shown in Figure 1.3. According to Newton's Third Law—for every action there is an equal and opposite reaction—the wall must be pulling back on the rope with a force of 100 lbs. Next, suppose that we could somehow cut into the rope and measure the force that one end of the rope

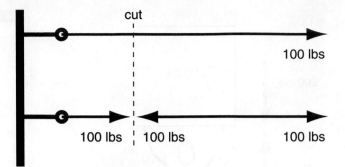

Figure 1.3   If we could splice into a rope with a force of 100 lbs applied to one end and measure the internal forces, by Newton's Third Law, the fibers on either side of the cut would exert an equal and opposite force of 100 lbs on each other.

exerts on the other. Both of these forces would also be 100 lbs according to the Third Law, since one side of the cut must balance the force exerted by the person pulling, and the other end must balance the force exerted by the wall. In fact, no matter where we cut into the rope nor how many times, the forces applied to the ends of the rope sections will always be 100 lbs.

Now, consider the analysis of the block-and-tackle in Figure 1.4. In this configuration, a 200 lb weight is hung from pulley A and a 100 lb force is applied to the free end of the rope. If we made three cuts through the rope along the dotted line, the forces on the ends of the sections would all be 100 lbs. In particular, each end of the section looped around pulley A would be pulled upwards with a force of 100 lbs. Thus, there would be a total force of 200 lbs lifting the weight, which is equal and opposite to the force of gravity, and so the weight would be suspended.

Note that a block-and-tackle doesn't violate any laws of nature; in fact, our analysis depended on Newton's Laws. Rather, this simple example illustrates a typical engineering solution in which a creative application of a law, coupled with an indirect attack on the problem, accomplishes something that may initially seem impossible.

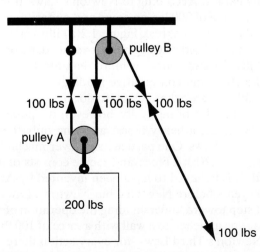

Figure 1.4   Because of the equal-and-opposite internal forces in the rope, each end of the segment around pulley A exerts an upward force of 100 lbs, for a total of 200 lbs.

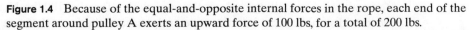

1.2.3 Engineering Models

As the block-and-tackle example helped illustrate, sometimes engineering solutions have preceded the scientific theories that explain how and why they work. Oftentimes, however, success requires knowledge of the relevant theory: NASA, for example, could never have landed a probe on an asteroid or gathered dust from the tail of a comet without a deep understanding of Newton's Laws regarding the gravitational forces between bodies. Engineers apply scientific theories by constructing *models* of their designs. A model is an approximation of a real system; when actions are performed on the model, it will respond in a manner similar to the real system. As illustrated in Figure 1.5, models can have many different forms, ranging from physical prototypes, to simple pencil-and-paper calculations, to computer simulations.

The ability to use models to describe physical processes and phenomena is a core skill for all branches of engineering, and this is the main reason why science courses make up a large portion of the engineering curriculum. Even within your engineering courses, you may expect that a sizable part of your workload will be devoted to learning about models for different phenomena. Structural Analysis, Circuit Analysis, Thermodynamics, Computer Architecture, Fluid Mechanics—all of these courses typically revolve around a fundamental set of models that engineers of various disciplines use to gather information and make decisions. Oftentimes in engineering courses, you'll encounter topics that are also covered in your basic science courses.

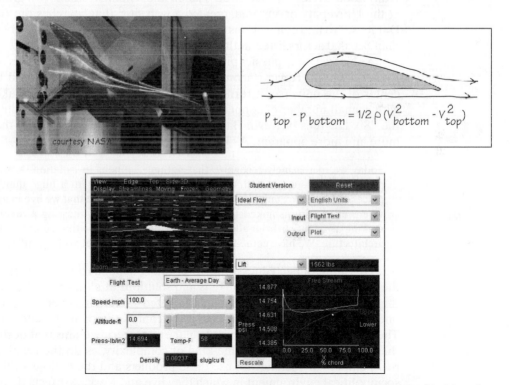

Figure 1.5 Different forms of models for estimating the lift force on the wing of an aircraft. Clockwise from top left: Small-scale prototype model of an airplane for wind tunnel. Rough sketch and calculation based on Bernoulli's equation. Foilsim II, a web-based simulator for students developed by NASA [BES+99]. Images of wind tunnel and Foilsim II courtesy of NASA.

Usually, however, the engineering courses will describe the same "laws" that you cover in your science courses in a different light, with the goal of using the laws to build models as part of an engineering method.

1.3 NETWORKS AND SYSTEMS

1.3.1 Everything is Connected to Everything

Late one night in 1993, Craig Fass, Brian Turtle, and Mike Ginelli were watching television at Albright College when a commercial for Kevin Bacon's movie *The Air Up There* came on. It occurred to them that Bacon had been in so many different types of films that it should be possible to connect him to many different actors with a sequence of links through movies [FTG96]. Leonardo DiCaprio, for example, has a "Bacon number" of 2 links: he co-starred in the 1993 movie *This Boy's Life* with Robert De Niro, who co-starred with Kevin Bacon in the 1996 movie *Sleepers*. Elizabeth Taylor—who starred in the 1939 classic *Gone with the Wind*—also has a Bacon number of 2: she appeared in the 1979 movie *Winter Kills* with Eli Wallach, who had an uncredited role in the 2003 movie *Mystic River*, co-starring Kevin Bacon. After the three students appeared with Bacon on the *Jon Stewart Show* in 1994, their game *The Six Degrees of Kevin Bacon* spread rapidly through college campuses and Hollywood circles. In 1996 Brett Tjaden and Glenn Wasson, two graduate students at the University of Virginia, created a web site[2] that searches the Internet Movie Database[3] to find connections. Surprisingly, the average Bacon number for the more than 800,000 actors listed in the database is only 2.96.

As Notre Dame physics professor Albert-László Barabási points out in his book *Linked* [Bar03], Kevin Bacon is no more the center of Hollywood than he is the center of the universe. It turns out that almost any actor can be linked to another with a small number of connections through the vast web of relationships. Further, the "Six Degrees of Kevin Bacon" is but one example of a reality that's becoming more and more apparent. As Barabási writes,

> Today we increasingly recognize that nothing happens in isolation. Most events and phenomena are connected, caused by, and interacting with a huge number of other pieces of a complex universal puzzle. We have come to see that we live in a small world, where everything is linked to everything else. We are witnessing a revolution in the making as scientists from all different disciplines discover that complexity has a strict architecture. We have come to grasp the importance of networks [Bar03].

This interconnectedness, that "everything depends on everything" characterizes the shape of engineering problems and solutions at all levels. When we look under the hood of a car, or at a bridge, an integrated circuit, or a chemical processing plant, one of the first things that strikes us is the complexity of all the interconnected parts. The connections among the people in the engineering teams that designed and built these artifacts exhibit the same kind of complexity, as do the relationships among the members of society who use these products and the regions of the natural and sociopolitical environment in which they live and work. Artifacts themselves form a kind of link between the needs and desires of a society and its technical capabilities,

[2]www.oracleofbacon.com.
[3]www.imdb.com.

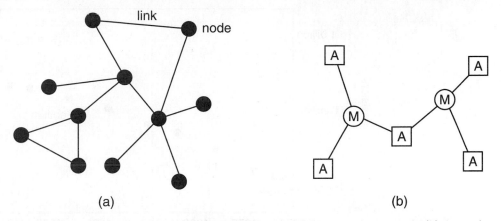

Figure 1.6   (a) A graph composed of nodes and links is a diagram of a network. (b) A graph of the Internet Movie Database can represent both movies and actors by nodes (labeled M and A), with links representing appearances of actors in movies.

including the ingenuity, tools, and materials at its disposal. As archaeologists know, from the early farmers of Mesopotamia to the casino builders in the Nevada desert, we can learn a lot about a society by studying its connections to the artifacts it produces.

Diagrams help us visualize complex situations, and we typically sketch networks using a type of diagram that mathematicians call a *graph*, as illustrated in Figure 1.6(a). A graph consists of a set of *nodes* typically drawn as circles or boxes, connected by *edges* or *links* drawn as lines. The nodes in such a network diagram could represent many possible things: people, places, events, pieces of hardware, or simply ideas. In a graph of the network of movies and actors, for example, both movies and actors could be drawn as nodes, while the links represent the shared relationship of actors appearing in movies, as shown in Figure 1.6(b). In general, a link represents something *shared*. This could be a physical connection between objects, such as a wire connecting two components in an electrical circuit or a pin connecting two beams in a truss structure. It could also be a flow of information such as a conversation between two people. Finally, a link could represent an environmental resource—such as a water supply, an oil field, or the atmosphere—shared by the population of a planet becoming more and more crowded.

1.3.2   A Web of Innovation

The web of connections surrounding any design project poses some of the greatest challenges for engineers, but also some of the greatest opportunities. In this section, we take a look at how an expanding network of innovation has produced the tremendous growth in computing and communications technology we witness today.

Moore's Law   For all its complex behavior, deep inside, a microprocessor is a network of tiny electrical switches called *transistors*. In 1965, Gordon Moore, one of the founders of Intel, suggested that the number of transistors that could be integrated onto a single working chip would double every two years. As Figure 1.7 shows, this prediction has held for more than 30 years. The most fascinating aspect of "Moore's Law" —as this prediction has been called—is that it's not so much a "law" but a

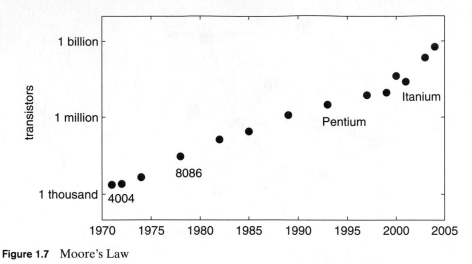

Figure 1.7   Moore's Law

rallying point for an entire industry that has motivated a diverse group of engineers and scientists to innovate at an astonishing pace. Each point in Figure 1.7 represents a complete cycle that includes conceiving the idea for a new product, then designing, manufacturing, marketing, and shipping it every couple of years. While Moore was at the helm of Intel for much of this time, he did not single-handedly drive this trend, nor is the story behind it one of Intel's technology alone. Rather, the 35-year progression is the result of a vast and expanding network of ideas and individuals that encompasses far more than the technology of making tiny switches out of silicon.

In addition to advances in semiconductor technology, sustaining exponential growth in processor complexity for over three decades required advances in computer systems design, software, applications and other areas. Figure 1.8 shows just part of the web of technologies that surrounds the transistors. Innovations in a technology at one end of a link help drive innovations in the technology at the other end. There's no "chicken" and no "egg" in this network; the entire network functions together, where change in one area leads to change in another. It's interesting to note that some companies that have direct links with Intel in shrinking transistors also have significant other businesses, such as Nikon, which provides photographic patterning technology. Other companies have a more distant relationship, but still play an important role in the growth-through-innovation process.

From Supercomputing to Video Games: Applications and Computing Technology   Kristi Maschhoff did well in her math and science classes in high school, but according to her, she wouldn't have considered engineering as a major in college had she not been given a strong "push" by her parents. In the early 1980s, engineering was not a popular choice for women in college. She studied electrical engineering at the University of New Mexico—and played a lot of soccer—but was not highly motivated or excited by her classes. A summer internship at Kirtland Air Force Base after her sophomore year, followed by a part-time engineering aide position at the local electrical utility company (PNM) during her junior and senior years, helped spark a new interest in mathematical modeling and simulation, in particular the algorithms used to solve the large system of equations required. Kirtland provided her an introduction to computer simulation of physical systems—in this case the

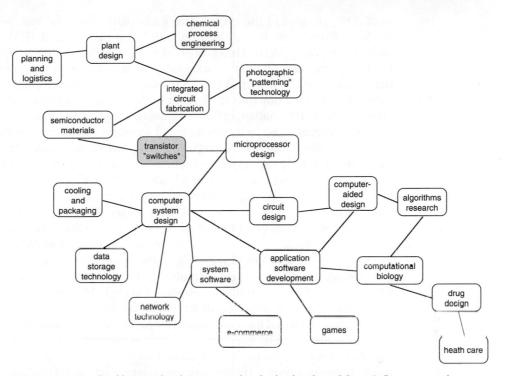

Figure 1.8   A web of innovation in many technologies has kept Moore's Law on track.

scientists were modeling lasers—and Kristi had the unique experience to run simulations on an early Cray computer system. From this experience she chose to study Applied Mathematics in graduate school.

Today she is a Technical Project Leader for Cray, Inc., a company that since the 1970s has designed and manufactured some of the most powerful supercomputers in the world. Most of Cray's business today focuses on providing computer systems for the massive calculations needed for national security and defense, scientific research, weather forecasting and climate modeling. Another emerging supercomputer application is modeling the mechanics of molecules, such as how protein molecules fold—a key concept to understanding and hopefully finding cures for genetically based diseases such as Alzheimers's, Parkinson's, or Mad Cow disease. Kristi describes the symbiotic relationship that these and other scientific applications of great importance have had with the computer and microelectronics industries. Advances in microelectronic technology have made larger and faster computer systems possible, which in turn allows scientists to include more complexity, and in turn better accuracy into their models, which can mean the difference in predicting both the path and the intensity of a hurricane before it makes landfall. On the other hand, the needs of climatologists and other scientists have helped drive the demand for more powerful computers, which in turn has helped sustain Moore's Law.

While the needs of science and commerce continue to spur growth in computer and information technology, one of the biggest drivers today is personal electronics. From its roots in the 1880s, making equipment for tabulating census data using punched cards, IBM Corporation has continued to produce some of the most advanced computing systems for business and research. While the company still focuses on "business machines," today, video games are also serious business for IBM, and as of 2007, all three of the major video game platforms—the Microsoft XBOX 360, the Sony Playstation 3, and the Nintendo Wii —make use of IBM microprocessor technology [Valb] [Vala] [How]. In order to be successful in the personal electronics market, game chips require an exquisite balance of low power consumption, high performance, and low cost. While IBM's experience in developing mainframe computers has trickled down to the benefit of "embedded" microprocessors, its recent efforts to meet the challenging constraints of the games market have spawned new ideas that percolate up to its business and scientific computer systems.

John Cohn is one of the engineers who has helped make IBM's technology available for a wide variety of products, including video games. In high school, John was a nerd, "as geeky as geeky gets," according to him, and a few decades later, he still is. With his frizzy gray hair and beard, John looks the part, and volunteers much of his spare time introducing K-12 students to engineering with his "Jolts and Volts" presentations. Growing up in Houston during the 1960s, home of NASA's manned spaceflight center, John was inspired—as many of us engineers of a certain age were—by the quest to land people on the moon. During high school, he tinkered with electronics at home, and then attended MIT to study Electrical Engineering. Later, John returned to graduate school at Carnegie Mellon, where he specialized in computer-aided design (CAD) of integrated circuits.

At IBM, John and his team are developing CAD tools and design standards that will enable engineers to share building blocks in novel ways to create unique products, and then simulate their designs to ensure that they will work properly before sending them to manufacturing. In this way, companies such as video game makers, server designers, and networking companies can all use common components to design completely different chips to meet their specific needs while adhering to extremely tight schedules and budgets.

Globalization As the expanding world population pushes people into closer proximity, innovations in communications technology strengthen the connections between us. When the Americas were being settled in the sixteenth and seventeenth centuries, it would take weeks or months for news in the colonies to reach Europe. Today, through wireless communications and network technologies such as the Internet and the World Wide Web, people in all parts of the globe can exchange vast quantities of information nearly instantaneously. This revolution in information technology tremendously impacts how our artificial systems—social and political, as well as technological—operate.

With expanded communications through the Internet, and increasing computerization of engineering and business processes, developing nations—in particular India and China—have become important partners in the global technical economy. Moreover, as a National Academy of Engineering study notes, engineers from these countries are "willing and able" to perform technical jobs at much lower wages than their counterparts in industrialized nations [Nat04]. These nations have invested heavily in education, and today China graduates three times as many engineers per year as the United States [Nat04].

In the 1970s, small teams of engineers who worked in adjacent offices designed most IC chips, which were then fabricated, packaged and tested at facilities on the same site or nearby. By the mid-1980s, it became common for parts of the manufacturing process to be located overseas. Today, all aspects of the engineering process, from design and development through manufacturing and technical support, are routinely shared by teams that span the globe. The Indian cities of Bangalore and Hyderabad have become major development centers for computer and communications hardware and software, and as of 2006, Intel, Microsoft, and Cisco have invested or plan to invest more than $1 billion each to open engineering facilities in India.

This trend toward global business practices will continue, and through technology, will become more efficient. According to Intel's Shekhar Borkar, the only real obstacle to the productivity of global engineering teams is that "people still need to sleep." Because of time differences, for example, a project meeting may be scheduled at 9 PM California time, 1 PM Manila time, 7 AM Jerusalem time, and 10:30 AM Bangalore time. Shekhar believes that over the next decade, there will be a strong focus on quality-of-life issues, so that engineers can better balance their work and home lives. In particular, broadband communication networks, together with improved teleconference and project collaboration software, will enable more people to work from home—or wherever they happen to be—without having to regularly commute to the office. Shekhar notes that he became especially aware of this while sending and receiving email on a flight between California and Frankfurt, Germany as he was crossing the polar ice cap. As of 2006, there's already strong evidence that people are becoming comfortable collaborating online with others across the globe, as demonstrated by the success of chat rooms and blogs, gaming communities, and international sales over eBay.

1.3.3 Systems

Making Designs Manageable The span of the web attached to a given engineering problem can be truly overwhelming, and engineers and scientists expend much of their intellectual effort in trying to make sense of it. The most powerful tool that they use to organize the vast network of information is the notion of a *system*. According to the Oxford English Dictionary, a system is "a set or assemblage of things connected, associated, or interdependent, so as to form a complex unity." Thinking in terms of systems enables us to draw a ring around a portion of a network and consider that portion as a single entity. As illustrated in Figure 1.9, the ring is called the *boundary*

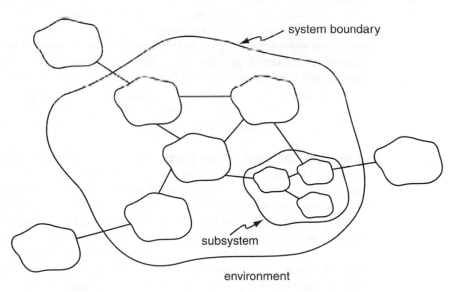

Figure 1.9 A system.

of the system and the region outside the boundary is called the *environment*. The components inside of a system may themselves be systems, or *subsystems*. For example, we might think of a portable MP3 player as a system containing a motherboard and a hard drive, and the hard drive as a subsystem containing a motor, platter, and other parts. This pattern may continue through many layers, into the motor and its components, and so on.

We commonly find this arrangement in all types of engineering designs. At first glance, there doesn't seem to be any physical reason why systems *have* to be organized this way. For example, it's conceivable that we could find a way to build an automobile that isn't arranged into distinct subsystems—you just step on the gas, and via some complex, jumbled mechanism, the car moves. In fact, to people who are unfamiliar with an automobile's subsystems, when they look under the hood of a car it may appear to be organized just that way. But experience tells us to look for a systematic organization and if we don't see one immediately, we sense that we are missing something.

Systems predominate in engineering designs for both a psychological and a physical reason. The psychological reason is that the classic "boxes-in-boxes" organization of engineering systems directly reflects the way people solve problems, which in turn reflects the way that we organize and retrieve information in our minds. As we will see in Chapter 2, people can only juggle so much information, and breaking a complex problem down into subproblems, or composing a design out of subsystems, enables us to hide details and focus on a smaller number of concepts at a time. Further, it provides a way for a team of people to work simultaneously on different parts of a design. In short, we design systems as such in order to make them intellectually manageable, and we see the imprint of this feature of human problem solving in nearly every artifact produced.

The physical justification for systems is that designs organized this way tend to be more robust. This is true for manmade systems as well as naturally occurring ones; we find the "boxes-in-boxes" organization in living organisms as well, with distinct circulatory, digestive, musculoskeletal, and other subsystems. Simon explains this scenario using a parable of two watchmakers, one who builds watches by packing all of the parts into the case at once, and the other who has found a way of organizing the design into stable subsystems—perhaps using a few more parts—that can be combined together later [Sim96]. Whenever the doorbell rings, each watchmaker gets up to answer it, and the piece of the design that he's currently working on springs apart. While the first watchmaker must restart assembly of the watch from the beginning, the second only loses one subassembly, and over time, he'll successfully complete many more watches than the first. Further, the watches built with stable subsystems will be easier to diagnose and repair when they break down.

More Than the Sum of the Parts The overall behavior of a system is typically more than just the sum of the behaviors of its parts. We sometimes say that the behavior of the system *emerges* when the components function together. An example of such a system is a soccer team in which the individual "components" are skilled athletes. In the context of the system, they play different roles, such as goalkeeper, fullback, midfielder, or forward. As a team, their objectives are to score goals and to defend against the opponent's offense, and they do so in coordinated patterns or plays. It would be impossible to discern this behavior, however, by looking at the athletes individually; it only emerges when the team works together as a system. We observe the same effect in an engineered system such as an airplane. Two of the important behaviors of an airplane are structural integrity—that it can withstand the forces imposed upon it without breaking up, and aerodynamics—that it can fly. The responsibility for providing these behaviors, however, is distributed among each of the major structural components of an airplane including the fuselage, wings, stabilizer, rudder, and landing gear. Conversely, changing the design of one of these structural components would affect both the structural integrity and the aerodynamics of the entire system.

Boundaries and Interfaces   A system interacts with its environment via the links that cross the boundary. Sometimes, it's straightforward to decide where to draw the boundary of a system, while other times it's less obvious. It clearly makes sense, for example, to locate the boundary of a biological system at the wall of a cell or at the skin of a person. With computer software, on the other hand, it can be difficult to determine where the boundaries lie; they aren't necessarily restricted to the shell of a computer's case, since some programs may depend on the system being connected to the Internet.

When things go wrong in the design of a complex system, it's often because of a misunderstanding at the boundary. There are many ways to break down a design problem, and some breakdowns are easier to work with than others. In general, a good breakdown is one in which each of the components can be designed independently, or at least as close to independently as possible. That way, it's possible to make decisions regarding the design of one part without worrying about how that decision will affect another part. As one example, let's consider the problem of designing an automobile. If we wanted to divide this task among three designers, one possible breakdown would be to have one designer work on the portion forward of the windshield, another person work on the passenger compartment, and the third person work on everything behind the passenger compartment, as shown in Figure 1.10. Breaking the design problem down in this manner would likely lead to problems. The reason for this is that the design of each of the components depends very strongly on the design of each of the others. Both physically and conceptually, many natural connections between the pieces are severed by this partitioning, and it'd be very difficult to keep the interfaces between the parts consistent if they were developed independently.

A better breakdown would split the design into subsystems—such as chassis, powertrain, and electrical—as shown in Figure 1.11. The subsystems still depend on each other, but the connections between them are less complex and more easily managed, physically as well as conceptually. Nonetheless, the subsystem designed by one team member is part of someone else's environment. The decisions each team member makes can impact the designs of the others and the most difficult part of a large team project is negotiatating tradeoffs in order to improve the system as a whole.

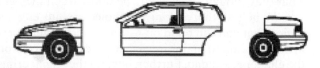

Figure 1.10   A breakdown of an automobile design problem that would be difficult to manage.

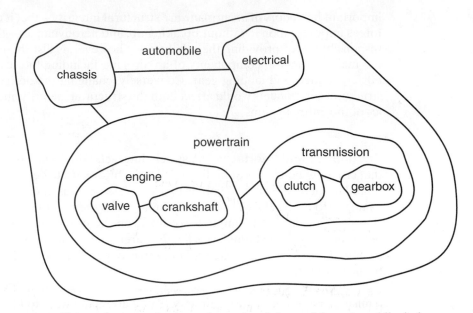

Figure 1.11 Using subsystems leads to a better breakdown of the automobile design problem.

1.4 ENGINEERING DISCIPLINES AND MAJORS

1.4.1 Introduction

Viewing a complex organization in terms of systems and subsystems makes it easier to manage, as well as easier to see the big picture. Inside of each subsystem, however, there is a lot of detail that requires specialized knowledge to comprehend. Looking again at an automobile, for example, a single person might easily learn how the power train connects to the electrical system and the chassis, but could spend an entire career trying to master the knowledge required to design each of these on her or his own. Because no one can be an expert at everything, the study and practice of engineering has evolved into a set of *majors* and *disciplines*. The main purpose of this section is to provide an overview of the most common engineering disciplines.

Before doing so, however, it's important to emphasize that engineering is inherently *multidisciplinary*. Engineers commonly work in teams of people from many different backgrounds. Engineers don't need to know the inner workings of each discipline, but they should understand the interfaces between them. Also, engineers need to be prepared to adapt; an informal survey of the author's friends and colleagues—mostly 20–25 years out of college—showed that most have switched areas more than once, and few are currently working in the area that they majored in, or are doing what they thought they would be doing when they graduated.

We believe that making the choice to study engineering—and developing a foundation that you can build on—is much more important than the choice of which field in engineering to major in. Society *needs* engineers of all disciplines in order to face the challenges ahead. Further, beyond the engineering profession, society would also benefit from more people in business and public policy who see the world from an engineer's perspective.

1.4.2 Overview of Engineering Disciplines

Because engineering is such a diverse and dynamic field constantly adapting to meet society's needs, it's impossible to summarize each area in detail and still stay relevant. Fortunately, there are excellent resources available on the WWW that provide regular updates on the state of the profession, as well as educational opportunities. In particular, as a starting point, we refer readers to the *Sloan Career Cornerstone Center* (www.careercornerstone.org), a non-profit resource center for those exploring career paths in science, technology, engineering, mathematics, computing, and medicine supported by the Alfred P. Sloan Foundation. Much of the material on individual disciplines that follows is derived, with permission, from their materials. The Career Cornerstone web site contains extensive information, including descriptions of and links to degree programs, professional societies, "day in the life" scenarios, and employment data. Another excellent web site, *TryEngineering* (www.tryengineering.org), is geared toward pre-college students and includes tips on preparation for engineering, as well as life profiles, lesson plans, and games.

Aerospace Engineering Aerospace engineers create machines, from airplanes that weigh over half a million pounds to spacecraft that travel over 17,000 miles an hour. They design, develop, and test aircraft, spacecraft, and missiles and supervise the manufacture of these products. Aerospace engineers who work with aircraft are called aeronautical engineers, and those working specifically with spacecraft are astronautical engineers.

Aerospace engineers develop new technologies for use in aviation, defense systems, and space exploration, often specializing in areas such as structural design, guidance, navigation and control, instrumentation and communication, or production methods. They often use computer-aided design (CAD) software, robotics, and lasers and advanced electronic optics. Some specialize in a particular type of aerospace product, such as commercial transports, military fighter jets, helicopters, spacecraft, or missiles and rockets. Other aerospace engineers are experts in aerodynamics, thermodynamics, celestial mechanics, propulsion, acoustics, or guidance and control systems.

Aerospace engineers typically work in the aerospace product and parts industry, although their skills are becoming increasingly valuable in other fields. For example, in the motor vehicles manufacturing industry, aerospace engineers design vehicles that have lower air resistance and, thus, increased fuel efficiency.

Agricultural Engineering Agricultural engineers combine engineering principles with biological and agricultural sciences. They work to develop equipment, systems, and processes that improve how the world's food supply is produced and distributed. They're involved in problem solving and analyzing current systems with expertise in improving the current process. They often have to look beyond a specific challenge, such as a machine or storage solution, and consider a larger system, and how improvements or changes would affect the whole.

Agricultural engineers often work in teams and their duties involve analysis of current methods and equipment applied to the production, packing, and delivery of food products. They might work in a group with other engineers, or those outside of engineering, to solve problems related to systems, processes, and machines. They may be involved in designing a water irrigation system, or in determining alternative uses for agricultural byproducts. They may participate in legal or financial consulting regarding agricultural processes, equipment, or issues.

Some agricultural engineers focus on machinery and design equipment important in agriculture and construction. Such engineers might have a special interest in crop handling, hydraulic power, or the growth of specific crops, and work for machine manufacturing firms. Other agricultural engineers design buildings or other structures used for livestock, storage of grains, or experimental growing facilities. Still other agricultural engineers focus on developing systems for food processing, such as drying processes, distillation, or long-term storage.

The type of job agricultural engineers have often determines whether they work inside or outside. However, most work inside a majority of the time. Some agricultural engineers whose tasks require visits to farms, animal operations, or seed manufacturers may find that they travel frequently. Many agricultural engineers find that working directly with growers, for example, provides immediate job satisfaction as it allows them to interact with people their work affects.

Architectural Engineering Architectural engineers apply engineering principles to the construction, planning, and design of buildings and other structures. They often work with other engineers and with architects, who focus on function layout or aesthetics of building projects. Architectural engineering often encompasses elements of other engineering disciplines, including mechanical, electrical, fire protection, and others. The architectural engineers are responsible for the different systems within a building, structure, or complex. Architectural engineers focus on several areas, including:

- the structural integrity of buildings,
- the design and analysis of heating, ventilating, and air conditioning systems,
- efficiency and design of plumbing, fire protection and electrical systems,
- acoustic and lighting planning, and
- energy conservation issues.

Most architectural engineers work in the construction industry or related areas. Others choose to work at non-profit organizations or firms. They spend most of their time in offices, consulting with clients and working with other engineers and architects. In addition, they often visit construction sites to review the progress of projects. Architectural engineers also work in different geographic locations based on the site of a construction project.

Bioengineering By combining biology and medicine with engineering, biomedical engineers develop devices and procedures that solve medical and health-related problems. Many do research, along with life scientists, chemists, and medical scientists, to develop and evaluate systems and products for use in the fields of biology and health, such as artificial organs, prostheses (artificial devices that replace missing body parts), instrumentation, medical information systems, and health management and care delivery systems.

Bioengineers design devices important for various medical procedures, such as computers that analyze blood or laser systems that perform corrective eye surgery. They develop artificial organs, imaging systems such as magnetic resonance, ultrasound, and x-ray, and devices for automating insulin injections or controlling body functions. Most engineers in this field require a sound background in one of the basic engineering specialties, such as mechanical or electronics engineering, in addition to specialized biomedical training. Some specialties within bioengineering or biomedical engineering include biomaterials, biomechanics, medical imaging, rehabilitation engineering, and orthopedic engineering.

Approximately 40 percent of biomedical engineers work for companies that manufacture products, primarily in the pharmaceutical, medicine manufacturing, and medical instruments and supplies industries; many others work for hospitals. Some also work for government agencies or as independent consultants.

Chemical Engineering Chemical engineers work in manufacturing, pharmaceuticals, healthcare, design and construction, pulp and paper, petrochemicals, food processing, specialty chemicals, polymers, biotechnology, and environmental health and safety industries, among others. Within these industries, chemical engineers rely on their knowledge of mathematics and science, particularly chemistry, to overcome technical problems safely and economically. And, of course, they draw upon and apply their engineering knowledge to solve any technical challenges they encounter. Their expertise also applies to law, education, publishing, finance, and medicine, as well as many other fields that require technical training.

Specifically, chemical engineers improve food processing techniques and methods of producing fertilizers, to increase the quantity and quality of available food. They also construct the synthetic fibers that make our clothes more comfortable and water-resistant; they develop methods to mass-produce drugs, making them more affordable; and they create safer, more efficient methods of refining petroleum products, making energy and chemical sources more productive and cost-effective. They also develop solutions to environmental problems such as pollution control and remediation.

Civil Engineering From the pyramids of Egypt to the space station Freedom, civil engineers have always faced the challenges of the future, advancing civilization and building our quality of life. Today, the world is undergoing vast changes: the technological revolution, population growth, environmental concerns, and more. All create unique challenges for civil engineers of every specialty. The next few decades will be most creative, demanding, and rewarding for civil engineers.

Today, civil engineers are in the forefront of technology. They are users of sophisticated high-tech products, applying the latest concepts in computer-aided design (CAD) during design, construction, project scheduling, and cost control. Civil engineering is about community service, development, and improvement—the planning, design, construction, and operation of facilities essential to modern life, ranging from transit systems to offshore structures to space satellites. Civil engineers are problem solvers, meeting the challenges of pollution, traffic congestion, drinking water and energy needs, urban redevelopment, and community planning. Our future as a nation will be closely tied to space, energy, the environment, and our ability to interact with and compete in the global economy. Civil engineers will perform a vital role in linking these themes and improving quality of life for the 21st century. As the technological revolution expands, as the world's population increases, and as environmental concerns mount, civil engineers' skills will become increasingly essential.

Civil Engineering branches into seven major divisions of engineering: Structural, Environmental, Geotechnical, Water Resources, Transportation, Construction, and Urban Planning. In practice, these aren't always hard and fixed categories, but they offer a helpful way to review a very diverse and dynamic field.

Computer Engineering, Computer Science, and Software Engineering Computer technology and information processing have become an important part of all engineering disciplines. Computer Engineering, Computer Science, and Software Engineering are three closely related fields that focus on developing the technology that many other disciplines depend upon. While there's significant overlap among them, they are separate fields that we summarize below.

Computer engineers analyze, design, and evaluate computer systems, both hardware and software. They might work on system such as a flexible manufacturing system or a "smart" device or instrument. Computer engineers often find themselves focusing on problems or challenges that result in new state-of-the-art products that integrate computer capabilities. They work on the interface between different pieces of hardware and strive to provide new capabilities to existing and new systems or products. The work of a computer engineer is grounded in the hardware—from circuits to architecture—but also focuses on operating systems and software. Computer engineers must understand logic design, microprocessor system design, computer architecture, and computer interfacing, and continually focus on system requirements and design. While software engineers primarily focus on creating the software systems individuals and businesses use, computer engineers, too, may design and develop some software applications.

Computer scientists impact society through their work in many areas. Because computer technology is embedded in so many products, services, and systems, computer scientists work in almost every industry. Design of next-generation computer systems, computer networking, biomedical information systems, gaming systems, search engines, web browsers, and computerized package distribution systems are all examples of projects a computer scientist might work on. Some computer scientists focus on improving software reliability, network security, and information retrieval systems, or even work as consultants to a financial services company.

Computer software engineers apply the principles and techniques of computer science, engineering, and mathematical analysis to the design, development, testing, and evaluation of the software and systems that enable computers to perform their many applications. Software engineers working in applications or systems development analyze users' needs and design, construct, test, and maintain computer applications software or systems. Software engineers can be involved in the design and development of many types of software, including software for operating systems and network distribution, and compilers, which convert programs for execution on a computer. In programming, or coding, software engineers instruct a computer, line by line, how to perform a function. They also solve technical problems that arise. Software engineers must possess strong programming skills, but are more concerned with developing algorithms and analyzing and solving programming problems than with actually writing code.

Electrical Engineering Electrical and electronics engineers conduct research, and design, develop, test, and oversee the development of electronic systems and the manufacture of electrical and electronic equipment and devices. From the global positioning system that continuously provides the location of a vehicle to giant electric power generators, electrical and electronics engineers are responsible for a wide range of technologies. Electrical engineering has many subfields, some of the most common of which we outline below.

Telecommunications is a prime growth area for electrical/electronics engineers. This includes developing services for wired and wireless networks for homes and businesses, as well as satellite, microwave, and fiber networks that form the backbone of the civil and military communications infrastructure.

Power engineers deal with energy generation by a variety of methods, such as turbine, hydro, fuel cell, solar, geothermal, and wind. They also deal with electrical power distribution from source to consumer and within factories, offices, hospitals, and laboratories, and they design electric motors and batteries. In industry, power engineers work wherever electrical energy is used to manufacture or produce an end product. They are needed to design electrical distribution systems and instrumentation and control systems for the safe, effective, and efficient operation of the production facilities.

The **computer industry** serves many sectors, and electrical engineers play a major role. Electrical engineering has strong connections to computer engineering and at many universities, the computer engineering and electrical engineering programs co-exist within the same department.

The chief enabling technology at the heart of the electronic components booming computer industry is **semiconductor technology**, in particular the development and manufacture of integrated circuits. As integrated circuits companies search for faster and more powerful chips, they seek engineers to investigate new materials and improved packaging—engineers who can handle the challenge of competitive pressure and ever-shorter development time. Manufacturers of microprocessors and memory chips, for example, continuously improve existing products and introduce new ones to beat the competition and meet customers' expectations of ever-higher performance. Semiconductor products include not just digital ICs but also analog chips, mixed-signal (analog and digital) integrated circuits, and radio-frequency (RF) integrated circuits, as well as power devices.

Environmental Engineering   Using the principles of biology and chemistry, environmental engineers develop solutions to environmental problems. They are involved in water and air pollution control, recycling, waste disposal, and public health issues. Environmental engineers conduct hazardous-waste management studies in which they evaluate the significance of the hazard, offer analysis on treatment and containment, and develop regulations to prevent mishaps. Environmental engineers are concerned with local and worldwide environmental issues. They study and attempt to minimize the effects of acid rain, global warming, automobile emissions, and ozone depletion. They also work to protect wildlife; many environmental engineers work as consultants, helping their clients to comply with regulations and to clean up hazardous sites.

Environmental engineers' job duties include collecting soil or groundwater samples and testing them for contamination; designing municipal sewage and industrial wastewater systems; analyzing scientific data; researching controversial projects; and performing quality control checks. Some environmental engineers work in legal or financial consulting regarding environmental processes or issues.

Industrial Engineering   Industrial engineers determine the most effective ways to use the basic factors of production—people, machines, materials, information, and energy—to make a product or to provide a service. They are the bridge between management goals and operational performance. They are more interested in increasing productivity through the management of people, methods of business organization, and technology than are engineers in other specialties, who generally work more with products or processes. Although most industrial engineers work in manufacturing industries, some work in consulting services, health care, and communications.

To solve organizational, production, and related problems effectively, industrial engineers carefully study the product and its requirements, use mathematical methods such as operations research to meet those requirements, and design manufacturing and information systems. They develop management control systems to aid in financial planning and cost analysis and design production planning and control systems to coordinate activities and ensure product quality. They also design or improve systems for the physical distribution of goods and services. Industrial engineers determine which plant location has the best combination of raw materials availability, transportation facilities, and costs. Industrial engineers use computers for simulations and to control various activities and devices, such as assembly lines and robots. They also develop wage and salary administration systems and job evaluation programs. Many industrial engineers move into management positions because the work is closely related.

Manufacturing Engineering   Manufacturing engineers are involved with the process of manufacturing from planning to packaging of the finished product. They work with tools such as robots, programmable and numerical controllers, and vision systems to fine-tune assembly, packaging, and shipping facilities. They examine flow and the process of manufacturing, looking for ways to streamline production, improve turnaround, and reduce costs. Often, a manufacturing engineer works with a prototype, usually created electronically with computers, to plan the final manufacturing process. In a globally competitive marketplace, the manufacturing engineer's job is to determine methods and systems to produce a product in an efficient, cost-effective way to provide a marketing edge for the final product.

Materials Engineering   Materials engineering is a field of engineering that encompasses the spectrum of materials types and how to use them in manufacturing. Materials span the range: metals, ceramics, polymers (plastics), semiconductors, and combinations of materials called composites. We live in a world both dependent upon and limited by materials. Everything we see and use is made of materials: cars, airplanes, computers, refrigerators, microwave ovens, TVs, dishes, silverware, athletic equipment of all types, and even biomedical devices such as replacement joints and limbs. All of these require materials specifically tailored for their application. Specific properties are required that result from carefully selecting the materials and from controlling the manufacturing processes used to convert the basic materials into the final engineered product. Exciting new product developments frequently are possible only through new materials and/or processing.

New materials technologies developed through engineering and science will continue to make startling changes in our lives in the future, and people in materials science and engineering will continue to be key in these changes and advances. These engineers deal with the science and technology of producing materials that have properties and shapes suitable for practical use.

Activities of materials engineers range from primary materials production, including recycling, through the design and development of new materials, to the reliable and economical manufacturing of the final product. Such activities are common in industries such as aerospace, transportation, electronics, energy conversion, and biomedical systems. The future will bring ever-increasing challenges and opportunities for new materials and better processing. Materials are evolving faster today than at any time in history. New and improved materials are an "underpinning technology"—one which can stimulate innovation and product improvement. High-quality products result from improved processing and more emphasis will be placed on reclaiming and recycling. For these many reasons, most surveys name the materials field as one of the careers with excellent future opportunities.

Mechanical Engineering Mechanical engineering is one of the largest, broadest, and oldest engineering disciplines. Mechanical engineers use the principles of energy, materials, and mechanics to design and manufacture machines and devices of all types. They create the processes and systems that drive technology and industry.

The key characteristics of the profession are its breadth, flexibility, and individuality. Individual choices of engineers determine their career paths. Mechanics, energy and heat, mathematics, engineering sciences, design and manufacturing form the foundation of mechanical engineering. Mechanics includes fluids, ranging from still water to hypersonic gases flowing around a space vehicle; it involves the motion of anything from a particle to a machine or complex structure.

Mechanical engineers research, develop, design, manufacture, and test tools, engines, machines, and other mechanical devices. They work on power-producing machines such as electric generators, internal combustion engines, and steam and gas turbines, as well as power-using machines such as refrigeration and air-conditioning equipment, machine tools, material handling systems, elevators and escalators, industrial production equipment, and robots used in manufacturing. Mechanical engineers also design tools other engineers need for their work. Mechanical engineers work in production operations in manufacturing or agriculture, maintenance, or technical sales; many are administrators or managers.

Mining Engineering Mining and geological engineers, including mining safety engineers, find, extract, and prepare coal, metals, and minerals for use by manufacturing industries and utilities. They design open-pit and underground mines, supervise the construction of mine shafts and tunnels in underground operations, and devise methods for transporting minerals to processing plants. Mining engineers are responsible for the safe, economical, and environmentally sound operation of mines.

Some mining engineers work with geologists and metallurgical engineers to locate and appraise new ore deposits. Others develop new mining equipment or direct mineral-processing operations that separate minerals from the dirt, rock, and other materials with which they are mixed.

Mining engineers frequently specialize in the mining of one mineral or metal, such as coal or gold. With increased emphasis on protecting the environment, many mining engineers work to solve problems related to land reclamation and water and air pollution. Mining safety engineers use their knowledge of mine design and practices to ensure the safety of workers and to comply with state and federal safety regulations. They inspect walls and roof surfaces, monitor air quality, and examine mining equipment for compliance with safety practices.

Nuclear Engineering Nuclear engineers research and develop the processes, instruments, and systems for national laboratories, private industry, and universities that derive benefits from nuclear energy and radiation for society. They devise ways to use radioactive materials in manufacturing, agriculture, medicine, power generation, and many other ways.

Many nuclear engineers design, develop, monitor, and operate nuclear plants used to generate power. They may work on the nuclear fuel cycle—the production, handling, and use of nuclear fuel and the safe disposal of waste produced by the generation of nuclear energy. Others research the production of fusion energy. Some specialize in the development of power sources for spacecraft that use radioactive materials. Others develop and maintain the nuclear imaging technology used to diagnose and treat medical problems.

Petroleum Engineering   Petroleum engineers search the world for reservoirs containing oil or natural gas. Once they discover these resources, petroleum engineers work with geologists and other specialists to understand the geologic formation and properties of the rock containing the reservoir, determine the drilling methods to use, and monitor drilling and production operations. They design equipment and processes to achieve the maximum profitable recovery of oil and gas. Petroleum engineers rely heavily on computer models to simulate reservoir performance using different recovery techniques. They also use computer models for simulations of the effects of various drilling options.

Only a small proportion of oil and gas in a reservoir will flow out under natural forces; therefore, petroleum engineers develop and use various enhanced recovery methods. These include injecting water, chemicals, gases, or steam into an oil reservoir to force out more of the oil, and computer-controlled drilling or fracturing to connect a larger area of a reservoir to a single well. Because even the best techniques used today recover only a portion of the oil and gas in a reservoir, petroleum engineers research and develop technology and methods to increase recovery and lower the cost of drilling and production operations.

Other Engineering Degree Areas   In addition to the main engineering fields covered within this site, there are many accredited engineering programs in other areas. These include:

- Ceramic Engineering
- Construction Engineering
- Drafting and Design
- Engineering (General)
- Engineering Management
- Engineering Mechanics
- Engineering Physics/Engineering Science
- Forest Engineering
- Geological Engineering
- Metallurgical Engineering
- Naval Architecture and Marine Engineering
- Ocean Engineering
- Plastics Engineering
- Surveying Engineering
- Welding Engineering

1.4.3 Professional Organizations

Professional organizations and associations provide a wide range of resources for planning and navigating a career in engineering. These groups can play a key role in your development and keep you abreast of what's happening in your industry. Associations promote the interests of their members and provide a network of contacts that can help you find jobs and move your career forward. They can offer a variety of services including job referral services, continuing education courses, insurance, travel benefits, periodicals, and meeting and conference opportunities. Many professional societies also have student chapters. Student engineers are encouraged to join their local chapter and participate in programs and activities to help network with other students and professional engineers.

Appendix C provides a partial list of professional associations serving engineers. A broader list of professional associations is available at the Sloan Career Cornerstone website (http://www.careercornerstone.org/assoc.htm).

1.4.4 Innovation at the Interfaces Between Disciplines

Over the period 2001–04, a committee organized by the National Academy of Engineering (NAE), with members from academia, industry, and the government, addressed the question, "What should engineering be like in 2020?" The results were published in two reports: *The Engineer of 2020* [Nat04] and *Educating the Engineer of 2020* [Nat05]. The reports envision a complex future with opportunities from emerging breakthrough technologies, and challenges in supporting a growing and shifting world population in the face of constrained natural resources and an aging man-made infrastructure. The NAE reports state that "because of the increasing complexity and scale of systems-based engineering problems, there is a growing need to pursue collaborations with multidisciplinary teams of experts across multiple fields [Nat04]." Further, the breakthrough technologies will typically lie at the interface between disciplines, drawing on innovations from each to create new possibilities. To illustrate, we take a brief look at two "hot" areas, *nanotechnology* and *biotechnology*.

Nanotechnology and Molecular Engineering   Nanotechnology is the science and engineering of developing components and systems at a molecular scale. Over the next few decades, molecularly engineered materials, structures, and systems will become prevalent in nearly all branches of engineering, and we'll see applications of nanotechnology in products as diverse as paints and coatings, clothing, electrical circuits, artificial organs, computer displays, and even molecular-scale robots or nanobots. Figure 1.12 shows the relative sizes of some micro and nano structures, from the width of a human hair down to the diameter of a hydrogen atom.

Nanotechnology will extend the revolution of fabricating fantastically small systems with integrated electrical and mechanical components that began at the micro scale. An example of a "microelectromechanical system" used in consumer products today is Texas Instruments' Digital Light Processing (DLP) technology. DLP combines computer circuitry and an array of micro-mirrors to produce images for large-screen televisions and projectors, as illustrated in Figure 1.13 [You93]. A DLP chip contains an array of up to 2 million pixels; each pixel consists of a computer memory cell and a micro-mirror. Each micro-mirror switches back and forth up to thousands of times per second, controlling the brightness of the pixel in an image. Creating this novel technology required the combined efforts of experts in electronics, mechanical structures, materials, and information processing.

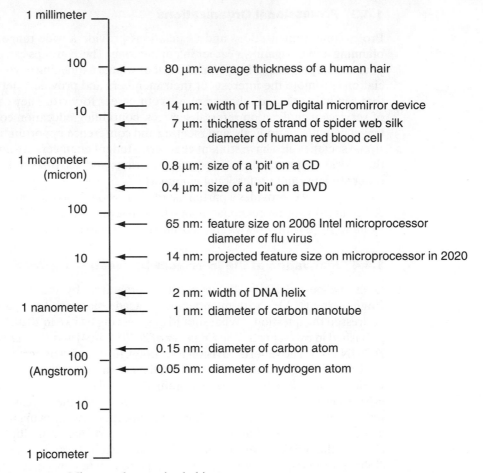

1 millimeter

100 ← 80 μm: average thickness of a human hair

10 ← 14 μm: width of TI DLP digital micromirror device
← 7 μm: thickness of strand of spider web silk
diameter of human red blood cell

1 micrometer ← 0.8 μm: size of a 'pit' on a CD
(micron) ← 0.4 μm: size of a 'pit' on a DVD

100 ← 65 nm: feature size on 2006 Intel microprocessor
diameter of flu virus

10 ← 14 nm: projected feature size on microprocessor in 2020

← 2 nm: width of DNA helix
1 nanometer ← 1 nm: diameter of carbon nanotube

100 ← 0.15 nm: diameter of carbon atom
(Angstrom) ← 0.05 nm: diameter of hydrogen atom

10

1 picometer

Figure 1.12   Micro- and nano-sized objects.

Nanotechnology promises to build microelectromechanical structures a thousand times smaller than a DLP micro-mirror. Much current research explores ways of using carbon atoms as basic building blocks. A carbon nanotube is a carbon molecule in the shape of a cylinder. Carbon nanotubes have extraordinary mechanical and electrical properties. They're one of the strongest and most flexible materials we know, with a strength-to-weight ratio 500 times greater than that of aluminum. They also conduct electricity one million times better than copper. Nanotubes have been added to plastics to create high-strength electrically conductive composites. One application of this already in production is coating automotive gas lines to dissipate static electrical charge that can lead to explosions [Bau02]. Engineers envision myriad other applications of carbon nanotubes, ranging from body armor and space suits, to wiring for integrated circuits.

By combining nanotubes with other molecules, researchers envision building tiny components such as gears, electrical probes, and even vehicles, as shown in Figure 1.14. In 2005, a research team led by Jim Tour, Professor of Chemistry, Mechanical Engineering, Materials Science, and Computer Science, at Rice University, demonstrated the world's first single-molecule nanocar, and a year later, added a motor to it [Tou] [SOZ+05] [SMS+06]. The vehicles measure just 4 × 3 nanometers—20,000 of

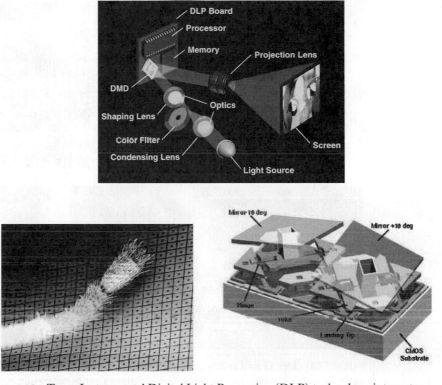

Figure 1.13 Texas Instruments' Digital Light Processing (DLP) technology integrates a complete microelectromechanical system onto a chip. Top, a DLP chip as part of a system for processing color images for a large-screen television. Bottom left, an array of DLP micro-mirrors compared to an ant's leg. Bottom right, micro-mirror detail. (Images courtesy of Texas Instruments.)

them could be parked side-by-side across the diameter of a human hair—and consist of an organic molecule chassis with four independently rotating axles. The wheels are spherical molecules of carbon, hydrogen and boron. The motor is a paddlewheel—developed by Ben Feringa at the University of Groningen in the Netherlands—that rotates when struck by light. Tour and his colleagues hope to build on this work by designing nanotrucks that can ferry atoms and molecules in non-living systems much as hemoglobin transports oxygen in living cells.

Bioengineering and Biotechnology   "Steve Austin, astronaut, a man barely alive. Gentlemen, we can rebuild him . . . we have the technology." So began each episode of the 1970s TV series *The Six Million Dollar Man*. Three decades later, the U.S. is spending nearly 20 *billion* dollars per year in bioengineering and biotechnology research. *Bioengineering* is an umbrella field that integrates engineering with any biological process or system. This includes biomedical engineering, which focuses on applications of engineering in health care, as well as *biotechnology*, which uses biological materials or living organisms to develop a wide variety of products and processes. Biotechnology provides the means to control innate characteristics of living things, and thus raises many moral and ethical questions regarding the essence of life, what makes each organism unique, what's appropriate to use or modify, and what isn't. As a result, research and development teams in the field must include

Figure 1.14 Top: Gears and probe tips made from carbon nanotubes [HGJD97]. Images courtesy NASA Ames Research Center. Bottom: Computer model of a nanocar [SOZ+05] (left) and a fleet of actual nanocars on a gold surface, photographed through a scanning tunneling microscope (right). Images courtesy of Rice University.

collaborations not only with a broad group of technical specialists, but also with public policy makers and ethicists.

People have found ways to use and modify living organisms to serve a wide variety of human needs for thousands of years. Hieroglyphics show that the ancient Egyptians knew how to use yeast microbes to leaven bread and ferment alcoholic beverages. The earliest farmers knew that they could select plants with certain favorable characteristics—such as resistance to disease—and cull others, in order to improve their crops from season to season. Later, they learned to cross-breed plants and animals to produce new generations with hybrid features from both parents. The turning point in the history of biotechnology, however, came in 1953 when James Watson and Francis Crick discovered the structure of DNA, the molecule found in every cell that contains the instructions for how living things develop. Today, scientists and engineers have learned how to modify the structure of a DNA molecule, effectively programming an organism to develop in a certain way. Many believe that this technology has the ability radically to transform the way we develop all types of products, from food to computers, as well as the way we treat disease. Advances in computing technology will likely enable us to process genetic information and develop new, individually customized drugs.

Intriguing possibilities lie at the interface between bioengineering and nanotechnology. Nanoscale robots or nanobots may be used for a variety of jobs, such as repairing torn tissues, unclogging arteries or transporting drugs to specific locations in the body [Nat04]. Figure 1.15 shows an artist's conception of a microbivore, a robotic white blood cell that could search for microbiological pathogens in the bloodstream and destroy them [FJ05].

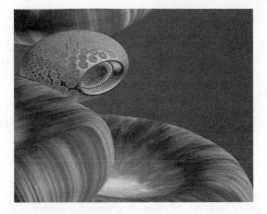

Figure 1.15 Artist's concept of a microbivore, an artificial white blood cell. Image courtesy of Zyvex Corp.

1.5 ENGINEERING AND COMPUTING

Like all professions today, engineers rely heavily on computers as tools for managing and processing information. This includes communication, searching for information, and producing documentation. It's fair to say that on the whole, engineers have a more intimate relationship with computers than professionals in most other fields. In this section, we look at two aspects of that relationship: first, learning to use a computer in your development as a logical thinker, and second, the use of computers for processing numeric data or "number crunching."

1.5.1 Programming and Logical Thinking

From creating a system for procuring and distributing clean drinking water, to developing a fabrication line for manufacturing integrated circuits, to formulating a test strategy for a new aircraft, engineers of all disciplines develop "programs" for solving problems. One of the most important communication skills an engineer must have is the ability to describe programs clearly and unambiguously.

Fundamentally, computers are machines that automatically follow a prescribed set of rules, written by engineers, to perform calculations. Just as factories contain manufacturing processes, computers contain *computational processes*. The "stuff" that computational processes operate on is called *data*, from the Latin term for something "given" or "factual." The rules and procedures that govern the operation of computational processes are called *computer programs*, written in precise, specialized languages called *programming languages*.

Engineers have always been early adopters of new technology, and as the digital computer rose to prominence in the second half of the twentieth century, engineers were among the first professionals to find ways to use the computer to make their work more productive. In the early days, this meant that they *had* to learn to program computers themselves if they wanted to use them. Today, however, with the advent of the software industry, most engineers use computer-based tools developed by others, rather than hammering them out for themselves. With the notable exception of the many engineers whose job is to produce computer software, programming for most engineers has gone the way of the manual transmission in cars: it's there for those who need or enjoy the control it provides, but easy-to-use, more efficient options are available.

It's a curious fact that as computers integrated into schools and homes during the 1990s through the growth of the Internet and availability of cheap hardware and software—to the extent that virtually all students entering college today have used them extensively—students today are more intimidated by them than they were a generation ago. Engineering students in general are no exception to this, and many eye computer programming with suspicion, perhaps because of the "hacker and geek" reputation that has surrounded it. Andy Downard, who earned his Bachelor's degree in Chemical Engineering as well as an MBA at Notre Dame before pursuing a Ph.D. at Caltech, sheds some light on the issue. According to Andy,

> In high school I liked math and science, but I had not had any exposure to computer programming. I almost did not major in engineering because I was so intimidated by programming. I foolishly considered how much experience my friends majoring in Computer Science had on me and thought that I would never catch up .... Thankfully, I stuck with Chemical Engineering and learned that programming is very logical, and similar across different languages. The working knowledge that I quickly developed of Excel, MATLAB, Mathematica, and C enabled me to literally write up programs that solved problems in five minutes that would have taken me five hours or more by hand. I worked with technology commercialization for a year after I graduated and I actually rarely wrote my own programs. Most of what I did was easily handled by Excel or MATLAB, which I think is common for most engineering jobs. Now I am happy to be back in graduate school where I have just learned FORTRAN and I am using it to design better medical diagnostics.

There was no small number of engineering educators who advised us *against* integrating computer programming into this text, as it might scare students away from engineering at a time when attracting and retaining students is a serious goal for the engineering profession. And as with Andy Downard's experience, most practicing engineers do *not* need to be able to program computers today in order to reap their benefits. While we recognize and agree with *some* of the concerns of our colleagues, we also believe that some exposure to computer programming is an important pursuit in the intellectual development of an engineer. Just as reading and writing essays and technical memos helps us learn to formulate and express arguments clearly, reading and writing computer programs teaches us to describe complex procedures logically and precisely. While people may take liberties in interpreting the steps in a program, a computer will do exactly what a program tells it to do, without trying to decide what the author meant. As every experienced computer programmer knows, the real system that you're troubleshooting when you debug a computer program is your own thought process and your ability to clearly state a set of instructions.[4]

1.5.2   Number Crunching

As technical decision-makers, engineers perform calculations, and a computer—in one form or another—is simply *the* tool for this job. While traditional number crunching is just one of many tasks for which computers are used today, this is after all, the reason why they were invented. Even the word "calculate" pays homage to one of the very first tools that humans devised for working with numbers: the pebble, or *calculus* in Latin. Over the span of the twentieth century, the substrate of a computer has come full circle back to a small rock, in this case a chip of silicon. The word "computer" itself referred exclusively to a person who performed mathematical

---

[4]See Pirsig, Robert M. *Zen and the Art of Motorcycle Maintenance* (New York: HarperTorch, 2006) page 417.

calculations until 1897, when the January 22 issue of the journal *Engineering* first used it to describe a mechanical device for doing the same.

When we think of engineers, computers, and number crunching, the image that often pops into mind is long lists of data and complex calculations scrolling by on a screen. Most of the calculations engineers perform, however, despite the stereotype, are actually quite simple. Even so, the computer still makes these tasks much more manageable. Below, we look at some of these most common tasks and where a computer fits in.

Arithmetic The most common calculations an engineer—like everyone else—performs is simple arithmetic: the "big four" functions of addition, subtraction, multiplication, and division, as well as the *transcendental* or "scientific functions" such as square roots, trigonometric functions, and exponentiation and logarithms. Because arithmetic is both tedious and routine, it was an early target for computing machines. A few milestones in the history of calculating machines are shown in Figure 1.16.

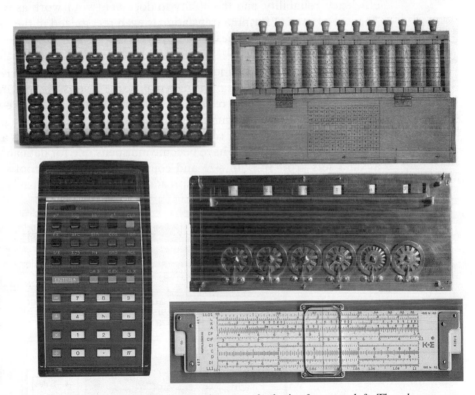

Figure 1.16 Calculating machines over the ages, clockwise from top left: The *abacus* was developed in ancient times and is still in use today. John Napier, who invented logarithms in the early 1600s, developed a calculating device called *Napier's Bones* for multiplication and division (photo copyright Science Source/Photo Researchers, Inc.). Blaise Pascal invented a machine called the *Pascaline* in the mid-1600s that used a system of gears to implement carries in addition (photo copyright J-L Charmet/Photo Researchers, Inc.). The *slide rule*, a descendant of Napier's Bones, was used extensively by scientists and engineers before the invention of the electronic calculator. The Hewlett-Packard *HP-35*—so named because it had 35 keys—was the first pocket-sized scientific calculator; it sold for $395 when it was introduced in 1972 (photo copyright David G. Hicks, Museum of HP Calculators).

Substituting Numbers into Formulas   The most common mathematical scenario in solving engineering problems is to first express a "word problem" as a set of equations using symbols for variables and constants, then solve the equations algebraically to obtain the result as a formula, and finally substitute in values for the symbols to get a numeric result. This final substitution of numbers into formulas only requires arithmetic, but even with a calculator, it's still a tedious process prone to errors.

One of the main differences between the digital computers invented during the twentieth century—which from now on we'll just call "computers"—and the earlier calculating machines is that computers were conceived to process *any* kind of information that could be encoded as strings of ones and zeros. This includes not only numbers, but also text, images, sound, and more. The fact that computers can operate on any kind of information has tremendous implications, but returning to number crunching and the principle of focusing on the simplest features that have the greatest impact, we note that computers let us express and manipulate formulas as well as numeric values. This means we can type in formulas as text and let the computer handle the substitution of values as well as the calculation and the printing of the results. This is a great advantage over simple calculating machines in terms of efficiency, reliability, and the ability to document your work as well as the answer. The importance of formula translation is even recognized in the name of one of the first programming languages: FORTRAN.

Data Analysis and Plotting   Another very common number-crunching activity is analyzing collections of data. The data may be the results of experiments, simulations, or financial data. The goal of any kind of data analysis is to make sense out of a set of numbers. The main techniques for doing so are arranging the numbers in tables, plotting them as graphs, or reducing them using statistics such as averages.

Computer-based tools have become indispensable for data analysis, particularly as data sets grow in both size and complexity. One such tool is the *spreadsheet*, illustrated in Figure 1.17. The first spreadsheet program was VisiCalc, released in 1979

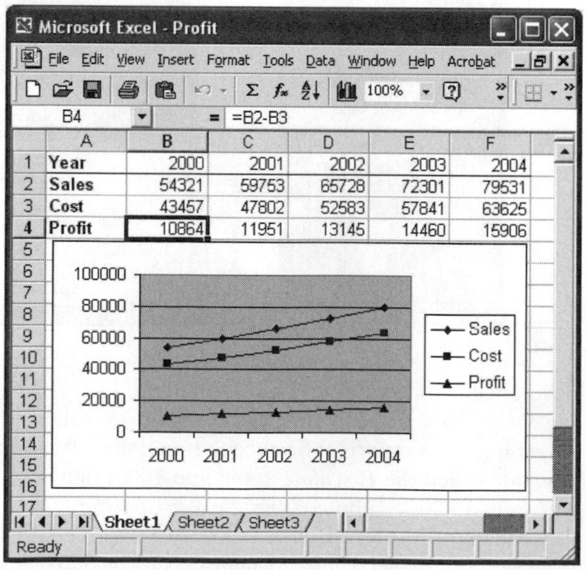

Figure 1.17   A spreadsheet program operates on data organized in rows and columns.

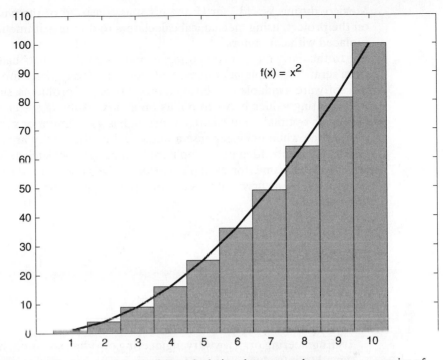

Figure 1.18   A numerical method for calculating the area under a curve as a series of rectangles.

for the Apple II computer. Spreadsheet software was *the* reason many businesses bought their first personal computers in the early 1980s and remains, along with communication, word processing, and presentation software, among the most widely used business software applications.

Numerical Methods   Still another common number crunching task is using *numerical methods* to solve mathematical problems. In general, a numerical method is a precise sequence of steps that may be carried out either by hand or implemented in a computer program, as an alternative to solving the problem symbolically using algebra or calculus. A typical example of a numerical method is finding the area under a curve, as illustrated in Figure 1.18. In this example, suppose we want to find the area under the curve described by the function $f(x) = x^2$, where x ranges from 0 to 10. Using calculus, we could obtain an exact symbolic solution of this problem as $x^3/3$ for any given value of x. A numerical method for solving this problem would be to approximate the area as a series of 10 rectangles, where the width of each rectangle is 1 and the height of each rectangle is the value of x^2 at each of the 10 points x equals 1 through 10. A key point of the numerical method is that it didn't require knowledge of calculus to obtain an exact solution; instead, it just required the ability to calculate x^2 to obtain a close approximation. This is the goal of all numerical methods: to reduce a complex mathematical problem to a series of arithmetic operations that don't require advanced mathematical skills to apply. In the days before computers, teams of people performed numerical methods in "assembly lines," where one person performed additions, another squared a number, and so on, each passing a partial result to the next person in line. In developing the first atomic bomb at Los Alamos, New

Mexico, during World War II, these teams consisted of the wives of the scientists on the project, using mechanical calculators to do the arithmetic, before they were replaced with computers.

In the early days of computing, engineers and scientists had to write programs from scratch to implement numerical methods. Today, there are libraries of numerical software available for solving a broad range of problems, such as finding areas, interpolating values between points on a curve, solving systems of equations, and finding an optimal point where a function has a minimum or maximum value. Most of the time when engineers use a numerical method, they aren't even aware of it, because it's embedded in a computer-aided design tool such as a simulator. A high-realism video game, for example, performs the same type of calculation as finding the area under a curve millions of times per second in order to simulate the motion of an object.

PROBLEMS

1. **Open-Ended Problems**

   What is an "open-ended" problem?

2. **TryEngineering.org**

   TryEngineering.org (www.tryengineering.org) has some online video games designed to introduce students to aspects of the engineering method. Play some of these games and review them in terms of which were the most interesting and why.

3. **Exploring Interrelationships**

   Section 1.3.2 and Figure 1.8 discuss how the miniaturization of integrated circuits over the last few decades—Moore's Law—involved a network of interrelated activities among many different people, businesses, and applications. Develop a presentation or brief report that tells a similar story focused on a different problem area. The description doesn't have to be heavily researched; the important thing is to focus on the complexity of connections, and where engineering fits into the picture. Web searches can be a very good way to get snapshots of some specific businesses or people involved in the process, similar to the brief bios used in the text. Some example areas of focus include:

   - Public transportation systems
   - Fast food
   - Personal music/multimedia systems
   - Clothing

4. **Research Activities at Your School**

   In addition to teaching undergraduate courses, engineering professors typically also engage in research. Find out about a research project involving engineering faculty at your school and answer the following questions:

   (a) What is the title of the project?

   (b) In a few sentences, what are the goals of the project?

   (c) How is the project funded?

   (d) How many people are involved in the project, including faculty and students? Is more than one school involved?

(e) Are undergraduate students involved in the project? If so, give an example of a contribution that an undergraduate has made. What background did the student have before joining the project?

5. **Roles of Engineering Disciplines**
Using the descriptions in Section 1.4 as a basis, discuss the roles that engineers from two or more different disciplines might play in addressing the following issues:

- providing an adequate energy supply for household use or transportation
- preventing or containing an epidemic such as influenza
- producing a motion picture

6. **Interview a Senior**
Interview an engineering student from your school who is graduating this year, asking at least the following questions:

- What was his or her major?
- What were his or her favorite classes?
- What were his or her most difficult classes? Most difficult year?
- What was the most interesting project that he or she worked on?
- What suggestions does he or she have to help you be successful in your major?
- What are his or her plans for after graduation?

7. **Interview an Engineer**
Section 1.4.1 makes the statement that engineers need to be prepared to adapt, and that few of the author's friends and colleagues 20–25 years out of college are doing what they thought they would be doing when they graduated. Interview an engineer (perhaps arranged through local industry or an alumni organization) who graduated at least 10 years ago and report on what you've learned.

8. **Engineering Professional Societies**
The Sloan Career Cornerstone Center provides links to the major engineering professional societies and organizations
(http://www.careercornerstone.org/engineering/engprofassn.htm).
Choose one that is of interest to you and answer the following questions:

(a) What are the main objectives of this organization?
(b) What are the main benefits that it offers to its members?
(c) Does the organization have a special student membership? If so, what benefits and activities does it specifically offer for students? Is there a student branch of this organization on your campus?

Organization
and Representation
of Engineering
Systems

LEARNING OBJECTIVES

- to define the attributes of an engineering design problem and to describe the relationship among the purpose, environment, and form of an artifact;
- to describe a simple *cognitive model* for how the human mind stores, organizes, and retrieves information;
- to draw *concept maps* consisting of circle-and-arrow diagrams that illustrate ideas and the relationships between them; to construct concept maps for a variety of simple situations;
- to define a *hierarchy* and to discuss why hierarchies are important in representing ideas; also, to successfully use hierarchical concept maps to illustrate the organization of parts in a system.

2.1 WHAT WE THINK ABOUT HOW WE THINK

Of all the factors that influence the shape of an engineering system, the most important is the human mind itself. Since engineering systems are the products of human thought processes, systems are organized in the manner in which people conceive them. In this section, we look at some of the theories of how the human mind stores, accesses, and processes information—the "ingenuity" behind engineering. Researchers in the fields of *cognitive psychology* and related areas including computer science have actively studied human problem solving over the past several decades, but the question of "how we think" dates back to the ancient Greek philosophers.

It may seem unusual for a section on human cognition to appear in an intro-ductory engineering textbook, but for several reasons we believe this is important material for an engineer to understand. In the late 1990s and early 2000s, the National Research Council (NRC), working on behalf of the National Academy of Science and the National Academy of Engineering, conducted a multi-year study on how people learn and on best practices for teaching. This study produced two influential books, *How People Learn: Brain, Mind, Experience, and School* [BBC00] and *How Students Learn: History, Mathematics, and Science in the Classroom* [DB05]. An over-riding principle in the NRC study is that having an awareness of how you learn, and being able to monitor your own understanding of material, both play critical roles in your development as an independent learner. The authors of the study term this knowledge about your own thought processes *metacognition*; its prefix "meta" means "after, along with, or beyond." The years you spend on a formal education cannot adequately equip you with all the knowledge you'll need in the future. As the NRC study states,

> You are the owners and operators of your own brain, but it came without an instruction book. We need to learn how we learn. [DB05]

Just as a driver can achieve better mileage and performance from an automobile by knowing minimally about how an engine and transmission work, a skilled prob-lem solver can, similarly, achieve better mileage and performance from his or her cognitive machinery by understanding how *that* works!

2.1.1 Example: Doing Math in Your Head

Let's begin our introduction to human problem solving by considering a problem involving single-digit arithmetic.

$$2 \times 7 = ?$$
$$5 \times 4 = ?$$
$$2 + 0 = ?$$
$$9 + 6 = ?$$

Most adults will produce the correct answers to problems such as those above very quickly—in less than a second. In fact, we're so proficient at solving these kinds of problems that we couldn't even explain how we do it; we simply know the answer.

Now, let's look at the problem of multiplying a four-digit number by a two-digit number.

$$4132 \times 57 = ?$$

Given a pencil and paper, most adults could solve this problem in under a minute, but could not easily solve this problem in their heads in any given time. Why is this? We don't lack a strategy for solving the problem, since we could easily explain the basic approach of multiplying through by a single digit, shifting, adding, and so on. Similarly, we don't lack basic factual information; all that's required is the ability to perform single-digit arithmetic, which we can do very easily. Rather, the obstacle to solving this problem is our memory: specifically, our inability to keep track of all the partial results and access them when we need them.

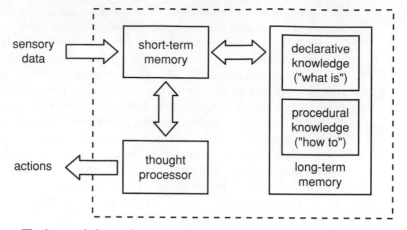

Figure 2.1 The human information processing system.

2.1.2 A Model for Cognitive Processing

Current theory in cognitive psychology models the human problem-solving system with three main components, illustrated in Figure 2.1:

- **long-term memory**, the extremely large store for facts we accumulate over a lifetime. This includes *declarative* or "what is" knowledge, as well as *procedural* or "how to" knowledge;
- **short-term memory**, a much smaller store for bits of information we're currently working with;
- a **thought processor**, which operates on facts in short-term memory.

According to this model, short-term memory temporarily stores information gathered through our senses (hearing, seeing, etc.) before passing it on to long-term memory. Research has shown that if information from our short-term memory doesn't successfully transfer to long-term memory within a few seconds, then we forget the information. Further, all thought processing occurs in information our short-term memory stores, and concepts our long-term memory stores must transfer back to short-term memory before we can work with them. Short-term memory, then, is the bottleneck in this system.

Short-Term Memory   Two key questions in understanding the function of short-term memory are:

- How much and what kind of information can short-term memory store?
- How do we access information in long-term memory to transfer it to short-term memory?

In 1956, George Miller of Harvard wrote an article in *The Psychological Review*, entitled "The Magical Number Seven, Plus or Minus Two: Some Limits on Our Capacity for Processing Information," which states that the number of distinct items of information a person can hold and recall from short-term memory is seven, plus or minus two. The following example illustrates this effect. To illustrate, try the following simple experiment: look at the sequence of letters below for about five seconds and try to memorize them, starting from the beginning. Then, try to write down the sequence on paper from your memory.

w t o r c p r h u d o e c n

If you're like most people, you could probably recall and write down approximately seven letters, plus or minus two. Given this model, we can better understand why multiplying a four-digit number by a two-digit number in your head is so difficult. If we count all of the independent pieces of information we need to solve the problem—products of individual digits, when to carry a number to the next column, partial sums, etc.—we find more than seven pieces. Our short-term memory becomes saturated and we're unable to process all the information.

Miller also considered the question of what constitutes an "item" in memory. He determined that we organize information into "chunks"—a term still commonly used—and we combine small chunks of information to form larger chunks. Miller found that the capacity of short-term memory is seven chunks, regardless of what the chunks represent. To test this with another simple experiment, spend five seconds memorizing the following list of words, and then try to write them down from memory:

thud pour crew no gold pine gear book

Most people will be able to reproduce between five and all seven of the words, and can easily remember the first four. Note, however, that the letters in the first four words—"thud," "pour," "crew," "no,"—are the same fourteen letters in the earlier example, rearranged. By chunking the letters into words, we're able to represent more than seven letters in our short-term memory, but we're still limited to seven unrelated words, plus or minus two. Now consider the following list of ten words and try the same experiment of memorizing them for five seconds, then writing them down:

worth pound cure old dog new tricks time saves nine

Most readers will recognize this list as a combination of phrases from the sayings, "an ounce of prevention is worth a pound of cure," "you can't teach an old dog new tricks," and "a stitch in time saves nine." Chunking the words into three phrases as such, most people will have little difficulty recalling all ten words, which contain a total of 42 letters. Moreover, the first three words, "worth pound cure," is another reordering of the 14 letters from the first example, which is now very easy to remember. Further, if people do make mistakes in recalling this list, they wouldn't miss individual letters, but would drop or even add entire words.

Long-Term Memory and Schemas   From Miller's experiments and related work, we've come to see knowledge as being transferred between long-term memory and short-term memory in chunks. But what makes up chunks of knowledge, and how are they organized and accessed in long-term memory?

Today, it's widely considered that knowledge in long-term memory is organized as a collection of concepts linked through relationships, called *schemas*. Under this model, chunks of knowledge are collections of related ideas in the network tied to some central concept. *Facts* are very small chunks of knowledge that relate one concept to another through a single relationship. Recalling a fact from long-term memory involves traversing relationships from concept to concept until we locate

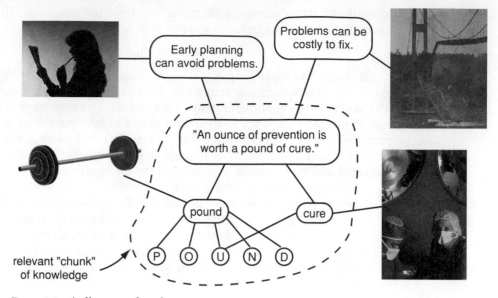

Figure 2.2 A diagram of a schema.

the needed information. Learning involves storing a fact in long-term memory by forming links between a new piece of information and existing ones. Many kinds of relationships link ideas, and you can understand the richness of relationships by letting your mind wander in "free association." Figure 2.2 illustrates a schema centered upon the concept "an ounce of prevention is worth a pound of cure." Several concepts are linked to this, including the concepts behind the meaning of this maxim, namely, that "early planning can avoid problems" and that "problems can be costly to fix." Linked to this latter concept may be concepts of classic engineering disasters, such as the failure of the Tacoma Narrows Bridge. For the problem of recalling a list of letters we described in the previous section, however, the important concepts are simply the words that comprise the phrase and the letters that comprise the words, and other concepts related to these are irrelevant. In order to take advantage of using the phrase as a tool for recalling the letters, you would need to locate the chunk of knowledge enclosed by the dotted line within the vast network of concepts in your long-term memory, and then bring it into your short-term memory for processing.

2.1.3 "How To" Knowledge and Problem Solving

Up to this point, we've focused on the representation of knowledge used in answering a question of the form "what is." This type of knowledge is often called *declarative knowledge*, because it declares facts about concepts. A second type of knowledge called *procedural knowledge* addresses questions about "how to" do something, and is the basis for all types of human problem solving. Nobody knows exactly how the human mind stores and uses procedural knowledge. Cognitive psychologists and computer scientists working in the field of artificial intelligence, however, have developed models that simulate how people use procedural knowledge to solve problems

[And80] [And83]. In this section, we consider a popular approach to representing procedural knowledge and look at how small chunks of procedural knowledge can chain together to solve complex problems.

Using Rules to Represent Chunks of Procedural Knowledge   A common approach to modeling procedural knowledge is to represent small chunks of procedural information as condition-action pairs as *rules* of the form:

> IF   some condition exists,
> THEN   perform some action.

The "condition" part of a rule examines facts that may be present or absent within declarative knowledge. The "action" part of a rule adds new facts to declarative knowledge when the condition is true, and may also cause a person to take physical action in his or her environment.

When the "thought processor" activates a given rule, it checks the condition against the set of facts in memory. If the condition is true—if it matches a fact or set of facts—then we say that the rule *fires* and the action is performed. As an example, consider the following rule that encodes a chunk of knowledge about "how to" determine if object X is a bird:

> IF   X is an animal
>   and X can fly
>   and X has feathers
> THEN   X is a bird

Sometimes, more than one rule applies to a particular situation. For example, consider the following two rules:

> IF   the goal is to stay dry
>   and it is raining
> THEN   carry an umbrella

> IF   the goal is to stay dry
>   and it's raining
> THEN   wear a raincoat

In this case, there's no stated reason to prefer one approach to staying dry over the other, and either rule could fire. In other words, there's more than one right answer to solving the problem of staying dry when it's raining. On the other hand, suppose that we "relearned" the first rule as:

> IF   the goal is to stay dry
>   and it's raining
>   and you have a free hand
> THEN   carry an umbrella

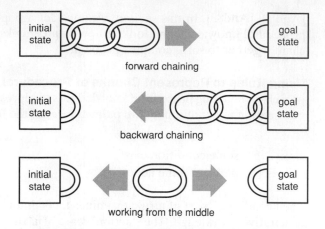

Figure 2.3 Strategies for problem solving. Adding links to the chain corresponds to finding productions that narrow the gap between the initial state and the goal.

In this case, if you don't have a free hand, only the second rule, whose action is to wear a raincoat, would fire. Many problems have more than one possible solution and our brains are able to keep track of these options.

Linking Chains of Rules to Solve Complex Problems   Sometimes, we're fortunate enough to get a perfect match on a single rule that completely solves a problem for us. Most of the time, however, we find that no direct solution exists, so instead we have to creatively piece together a solution as a series of steps between our starting point and our goal. Three different strategies for linking up a complete solution are working forward from the starting point toward the goal, called *forward chaining*, working backward from the goal, or *backward chaining*, or working *from the middle*, illustrated in Figure 2.3.

One example of chaining together a solution is booking flights between two cities when no direct flight between them exists. Suppose, for example, that we want to book a flight between South Bend, Indiana and Monterey, California. Since there are no direct flights between South Bend and Monterey, it's necessary to find a route composed of flight segments through other airports. A flight route map, as in Figure 2.4, shows many possible solutions.

- Working forward, we'd first book a flight from South Bend to a hub that serves it, such as Chicago or Cincinnati, and then continue to book segments until we reach Monterey.
- Working backward, we'd first book a flight to Monterey from one of its hubs, such as San Francisco or Los Angeles, and then continue to book flights with successively earlier departures until we reach South Bend.
- Working from the middle, we'd first book a direct flight that covers most of the distance, such as between Chicago and San Francisco, and then add flights to either end.

A classic problem from Maier [Mai31] provides another example of linking pieces together to form a solution to a complex problem. Suppose that you're in a room, pictured in Figure 2.5, and find two strings hanging from the ceiling and are asked to tie the ends of the strings together. While they're long enough to be tied

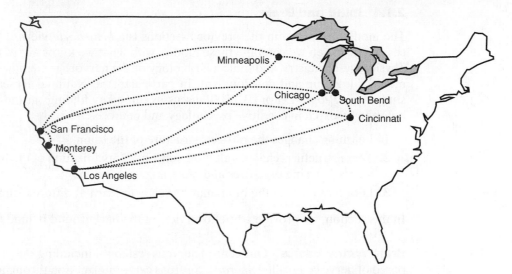

Figure 2.4   A flight route map is a diagram of the problem space for flying from an originating airport (the initial state) to a destination airport (the goal state) [Sky05].

together at some height, they're also too short by several feet for you to simply grab the end of one and reach the other. The room also contains a collection of objects, including a chair and a pair of pliers, which you may use in solving the problem. How can you tie the strings together?

One solution is to stand on the chair and tie the pliers to one of the strings to make a pendulum, swing it, and then move the chair under the other string, catch the pendulum, and finally tie the strings together. How would forward- and backward-chaining strategies apply to finding this solution? In a forward chaining approach, you might consider what's initially available, such as the pliers, and look for ways to use them, perhaps by tying them to a string. In the backward-chaining approach, you might picture the two strings tied and think of ways to bring them together: if you can't reach both strings simultaneously, then maybe there's a way to have one string come to you. Either approach may lead to a valid solution, and people often find it useful to switch strategies when stuck on a difficult problem.

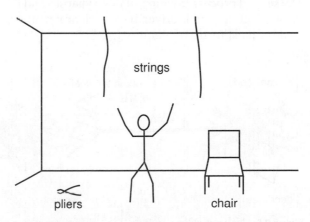

Figure 2.5   The string problem.

2.1.4 Mind and Brain

The model presented in the previous sections is a *behavioral* view of how we process ideas. When we combine these models with what we know about the anatomy, physiology, chemistry, and molecular biology of the nervous system, we understand more of learning and development. In particular, the National Research Council study *How People Learn* [BBC00] discusses three key findings at the convergence between research in cognitive psychology and neuroscience:

1. Learning changes the physical structure of the brain.
2. These structural changes alter the functional organization of the brain; in other words, learning organizes and reorganizes the brain.
3. Different parts of the brain may be ready to learn at different times.

In this section, we provide a brief overview of the background behind these points.

Brain Wiring Basics   Our central nervous systems—including the brain—is composed of nerve cells called *neurons*. Neurons carry information throughout the body, communicating with cells in sensory organs, muscles, and other neurons. Figure 2.6 illustrates the structure of a neuron. The cell body is covered with two types of specialized projections. *Dendrites* carry information into the cell, whereas *axons* carry information out of the cell. One neuron passes information to another by releasing chemicals called *neurotransmitters* across a tiny gap called a *synapse*. The release of neurotransmitters across a single synapse will either excite or inhibit activity in the receiving cell. Each neuron integrates the effects of all of its inputs across each of its synapses to determine its output.

The formation of synapses determine the basic wiring pattern of our brain. There are approximately 200 billion neurons in the human brain; over a lifetime, the number of neurons doesn't change, but the number and connectivity of synapses does change. Two basic processes make these connections, either by overproduction of synapses and then "pruning" them, or by creation of new synapses. The first process, overproduction and pruning, is especially important in early development. For example, David Hubel and Torsten Wiesel received the 1981 Nobel Prize for discovering that the area of the brain that controls sight, called the visual cortex, is shaped by this process, and that infants reduce the number of synapses in the visual cortex during the first few months of life as they accumulate visual experiences [Hub88]. The second process, creation of new synapses and the growth of dendrites, continues throughout life and is driven by experience; neuroscientists believe that this process governs most forms of memory [BBC00].

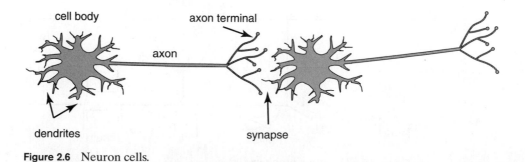

Figure 2.6   Neuron cells.

Learning Adds Connections in the Wiring   The experiments with rats that William Greenough and his colleagues conducted at the University of Illinois showed that learning causes increased development of synapses [BIA+90] [BBC00]. These experiments compared four groups of rats. One group was taught to traverse an obstacle course over a month of training, a task that required learning but little exercise. A second group was given mandatory exercise using a treadmill. A third group was given voluntary exercise on a wheel, and a fourth control group was given no exercise at all. At the end of the study, the two "exercise" groups had a higher density of blood vessels in their brains than either the "obstacle course" or "no exercise" groups. The number of synapses per neuron, however, was highest in the "obstacle course" group, suggesting that learning produced synapses, while exercise alone did not.

Functional Regions of the Brain and Effects of Instruction   Non-invasive techniques such as *positron emission tomography*, or PET scanning, and functional *magnetic resonance imaging* or MRI, as well as surgeries, enable researchers to study regions of the human brain that are active during various activities. These experiments determine that memory operations are distributed over different parts of the brain. The prefrontal cortex of the brain, located just behind your forehead as shown in Figure 2.7, plays a role in short-term, working memory [Bea97]. Separate parts of the brain control declarative memory of facts and events, and procedural memory for how to perform tasks. These two processes involve the hippocampus and the neostriatum, respectively, as shown in Figure 2.8.

While specific regions of the brain are typically associated with particular functions, parts of the brain can be "trained" to perform different functions through instruction. In all deaf people, for example, parts of the brain that normally process auditory information become organized to process visual information [BBC00]. Furthermore, there are different patterns of electrical activity in the brains of persons who communicate using sign language from those who do not [FC86] [BBC00]. To further illustrate that instruction can change the organization of brain function, persons who have had strokes or portions of their brains removed have been able to regain lost functionality after instruction and long periods of practice [BBC00].

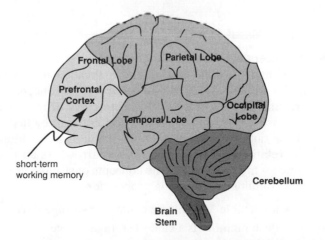

Figure 2.7   The prefrontal cortex of the brain plays a role in short-term, working memory.

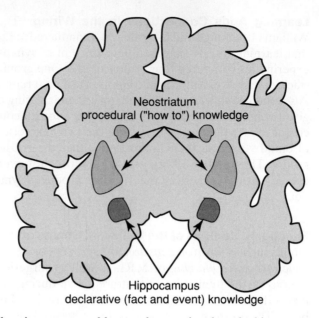

Figure 2.8 Declarative memory of facts and events involves the hippocampus. Procedural memory for "how to" perform tasks involves the neostriatum.

2.2 CONCEPT MAPS

Diagrams that pictorially represent chunks of information are crucial tools for problem solving. In the same manner that writing down partial results helps us solve multiplication problems we otherwise would have difficulty doing in our heads, diagrams alleviate the bottleneck of our short-term memories by making the information we need readily accessible. There are many types of diagrams and other kinds of representations of problems used throughout engineering, and learning how to use and select (or devise) an appropriate representation for a problem—so you can understand a problem and then solve it—is a critical engineering skill. In this section, we'll introduce a very basic type of diagram, called a *concept map*, and show how to use one to organize information.

2.2.1 What Is a Concept Map?

A concept map is a network diagram or graph, where the nodes correspond to concepts and the edges correspond to relationships between concepts. Labels on the nodes and edges indicate the names of the concepts and relationships. Propositions are statements formed by connecting two or more concepts with relationships. We "pronounce" propositions in a concept map by first stating the name of the object at the tail of the arrow, then the name of the relationship, and finally the name of the relationship at the head of the arrow. As an example, Figure 2.9 illustrates a concept map focused on the concepts of storage devices for a portable MP3 player, that contains the following propositions:

- a portable MP3 player contains a storage device
- flash memory is a kind of storage device
- a disk drive is a kind of storage device

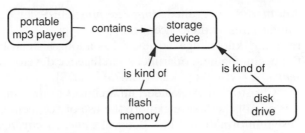

Figure 2.9 A concept map focused on storage devices in a portable MP3 player.

Various kinds of "circle and arrow diagrams" appear regularly in many different contexts. Computer-aided design (CAD) tools commonly use graphs as part of their user interfaces, as the examples in Figure 2.10 illustrate. The reason for this is simple: they create a clear impression of a situation and they tend to "stick" in people's minds. Concept maps are especially powerful because they help establish patterns for situations that can be applied over and over again in solving related problems, and we'll use concept maps repeatedly throughout the text for this purpose. Beyond their value for presenting ideas, the act of constructing a concept map, piece by piece, can have a tremendous impact in a student's—or teacher's—understanding of a body of knowledge. In fact, research has shown that a person's ability to draw a concept map for a body of knowledge indicates their level of understanding of the material, and that test questions based on drawing concept maps—which require students to pull ideas together—are much better evaluation tools than short answer tests that require little more than rote memorization and a shallow understanding of concepts.

Joseph Novak and his colleagues at Cornell University introduced the use of concept maps as a formal teaching and learning tool in the course of their research in understanding changes in children's knowledge of science [Nov91] [ANH78].

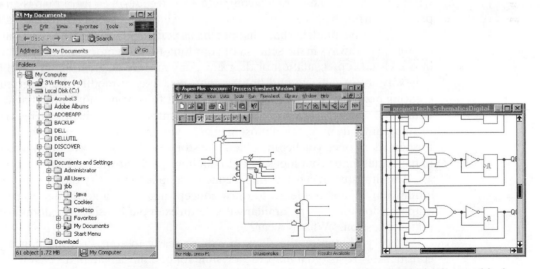

Figure 2.10 Examples of engineering information represented as graphs in the graphical user interfaces of computer tools. The nodes and edges have different shapes, but the representations are still graphs; left, file browser from Microsoft Windows; center, chemical process flow diagram from Aspen Plus; right, digital circuit schematic from Electric.

Another system, called *Mind Mapping* [Buz95], has also been widely adopted, with similar aims as the concept-mapping approach. The use of such diagrams for representing knowledge, however, has a long history. In the late 1950s, researchers in the beginning field of artificial intelligence developed a very similar notation called *semantic networks* [Ric58] [Sim73] [Qui78].

Christopher Alexander, an architect at the University of California, Berkeley, was one of the first to explore the use of "concept map-like" diagrams to organize design information. In his book, *A Pattern Language* [AIS77], Alexander provides a catalog of 253 interrelated patterns for the design of buildings and communities, encompassing details ranging from the design of window panes and the selection of chairs, to the distribution of population centers within a region. Since the publication of *A Pattern Language*, other design communities have experimented with the approach, and a book on design patterns for software development with a similar organization has attracted much interest [GHJV95].

2.2.2   How to Build a Good Concept Map

As we discussed in the previous section, concept maps are a valuable tool for teaching and learning because:

1. drawing a concept map helps you brainstorm and organize ideas, thus facilitating a deeper knowledge of material;
2. monitoring your ability to draw a concept map for a body of knowledge tests how well you understand it;
3. a well-constructed concept map is an effective tool for presenting ideas to others.

Given these preliminaries, we now turn our attention to how to draw a concept map and how to recognize a good one. First of all, concept maps are neither right nor wrong, nor are they ever really complete. Constructing a concept map is an iterative process, starting with a preliminary sketch that undergoes many refinements. In terms of our cognitive model, what's happening as you draw a concept map is that you are exploring pathways in the schema of your long-term memory and transcribing them into a visual representation or diagram. As you stare at the diagram, you discover new ways of organizing and linking concepts. As you modify the diagram, you're creating new knowledge that gets written back to your memory, and that stimulates your mind to search new pathways for related information, which you, in turn, add to the diagram. The process is almost magical—you can't stop your mind from searching for new ideas once you begin! When you stop drawing, you have not only a diagram in hand that maps your ideas, but also an improved knowledge base in your mind.

The first step in constructing a concept map is to identify the domain by posing a question. When you learn to draw concept maps, it's a good idea to practice by mapping domains you're familiar with. As an example, consider the domain centered on the question, "What is a car?"

Once you identify the domain, you're ready to start drawing. Since you will considerably revise your map, use a system for building a map that's easy to change. A whiteboard, if you have one, is a better choice than paper and ink. Another approach is to write concepts on PostIt notes and stick them to a large sheet of wrapping paper, where you can move them around easily. A better way is to use a drawing or presentation program on your computer, or even one of several programs specifically for drawing concept maps.

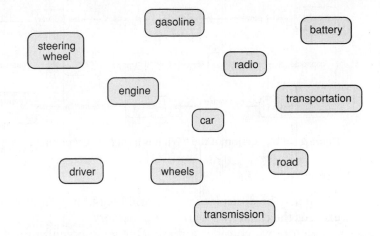

Figure 2.11   Initial set of concepts for a concept map focused on "What is a car?"

With drawing tools in place, now begin listing concepts. Start with the focus of the domain; in this case, a "car," and continue to add ten or so concepts as quickly as they come to mind. Each concept should consist of a word or a few words at most; if it takes more words than this to express an idea, this indicates you have several concepts lumped together that could either be broken down into more concepts, or represented by a single overriding concept. Figure 2.11 illustrates an initial set of concepts.

After you form the initial set of concepts, the next step is to link them with relationships. As with concepts, try to limit the labels on relationships to at most a few words. Figure 2.12 shows a first version of the concept map with relationships. This map is a start, but isn't very effective in deepening our understanding of cars. While each of the supporting concepts relates to the central concept of "car," there's no indication of how they relate to each other. By exploring these relationships, we achieve a better understanding of what a car "is" and how it works. In addition, in terms of our cognitive model, this concept map arranges concepts into one big chunk,

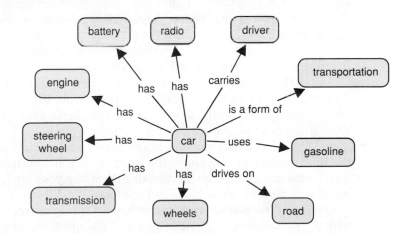

Figure 2.12   A version of the "What is a car?" concept map that ties all the concepts into one big chunk.

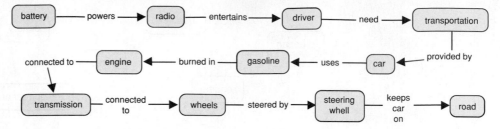

Figure 2.13 A version of the "What is a car?" concept map arranges concepts in a line.

which may be difficult to recall from long-term into short-term memory because it exceeds the capacity of short-term memory.

Before we consider a reasonably good organization for the "What is a car?" concept map, we'll take a look at another problematic one, illustrated in Figure 2.13. In this case, the concepts are arranged in a line. Each concept, except for the ends, connects to two other concepts; however, the map's problem is that in order to explain the relationship between a concept near the beginning of the line and one near the end of the line, it's necessary to follow a rather circuitous path. As an experiment, study Figure 2.13 for 20 seconds and determine how much you can recall from memory. In terms of the cognitive model, the concepts are still arranged as one big chunk with 11 concepts and 10 relationships for a total of 21 items—far more than the capacity of short-term memory.

Constructing a good—specifically a *memorable*—concept map involves striking a balance between the map shapes shown in Figures 2.12 and 2.13, so that the map can be easily broken down into manageable chunks. According to our cognitive model, a manageable size is approximately five to nine items, which includes both concepts and relationships. To do this, we find concepts that bind together small groups of concepts. Typically, this involves *adding* more concepts to the map. Figure 2.14 shows our version of the "car" concept map organized this way. We've added three new concepts to the map, "passenger environment," "drive train," and "electrical system;" each serves to organize a small chunk of knowledge. We've also introduced a *plural relationship*, drawn as an arrow that splits into multiple arrows. This plural relationship groups relationships of the same type together and effectively increases the capacity of a chunk. Dotted lines enclose each of five suggested chunks in the graph. The central chunk consists of the concept of "car," its three subsystems, and its purpose—"transportation." Peripheral chunks linked to the central chunk further describe a car's subsystems and its purpose.

Even though the map in Figure 2.14 has more concepts than the maps in Figures 2.12 and 2.13, it's easier to recall information from the former. Repeat the earlier experiment by studying this map for 20 seconds, then try to reproduce it from memory. Even if you can't reproduce the map verbatim, you should be able to *reconstruct* the main ideas behind the chunks of information and the relationships between them— the most important concept of the exercise. If you can, then you've *meaningfully learned* the concepts in the map, linking them into your internal schema rather than merely *memorizing* a rote list of isolated facts. The differences you may find between the map you reconstruct from memory and the original include associating "radio" with the electrical system rather than the "passenger environment," or even adding concepts that don't appear in the original map. These differences help you understand

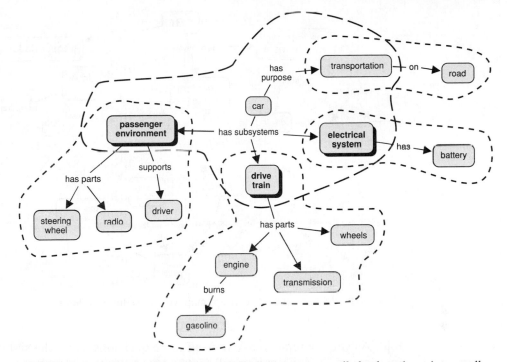

Figure 2.14   A version of the "What is a car?" concept map easily broken down into smaller chunks and enclosed by dotted lines. The three highlighted concepts, additions to the map, organize the chunks.

what a car is. Recall that we discussed the same phenomenon in the context of trying to memorize lists of words or phrases in Section 2.1.2.

The final step in drawing a concept map is to look for additional *cross-links* between concepts, both within chunks and between chunks in different parts of the map. This step provides significant insight toward understanding a domain. For example, the radio—a part of the passenger environment—is also tied to the electrical system and powered by the battery. Similarly, we may ask if there are connections between the electrical system and the drive train. After some reflection, you may have the insight that some cars have electric motors, but there are also connections between the drive train and the electrical system in cars with gasoline engines. The question is, how do you discover these additional connections? One approach we strongly recommend is using a combination of brainstorming and research. One relevant and useful form of research is an Internet search engine. For example, if you search the terms "car engine battery" or "car engine battery gasoline," you'll find references to many web pages that explain concepts such as an alternator, which uses power from the engine to charge the batteries, or spark plugs, which use an electrical spark to ignite gasoline in the engine, or a timing chain, which synchronizes the rotation of the engine crankshaft to when the spark plugs fire. Cross-links can also be used to explain the inner workings within chunks. For example, within a drive train, the clutch couples the engine to the transmission. Figure 2.15 illustrates a version of the "car" concept map with cross-links added. One of the effects of adding cross-links is that the map becomes more densely interconnected and the earlier chunks become less visually apparent. To help clarify the central concepts, they appear in a larger

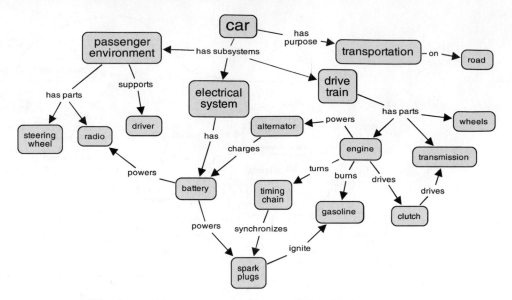

Figure 2.15 "What is a car?" concept map with cross-links added.

font. As you continue to ask yourself more questions about the system and search for answers, the concept map will continue to grow outward from the central concept of "car," as you add new concepts and refine relationships and, most importantly, expand your knowledge.

2.2.3 Hierarchies

In learning to draw concept maps, we discovered that some organizations reflect a more meaningful understanding than other organizations, and are thus easier to recall. In terms of our cognitive model, the best organizations let us transfer manageably-sized chunks of information between our long-term and short-term memory as we navigate a body of knowledge. We determined that a central concept linked to a few supporting concepts is a particularly effective organization. These concepts themselves were then linked to a few supporting concepts, and so on; such organizations are termed *hierarchies*. Hierarchical organization is one of the most powerful concepts in the design and analysis of systems, and the ability to organize information or decompose problems hierarchically is one of the most important skills for an engineering student to develop.

A map of a hierarchical organization is a graph with a characteristic shape, called a *tree*, illustrated in two forms in Figure 2.16. The graph on the left has a "spider" or "snowflake" shape, while the graph on the right has the "classic tree" shape. The common terminology for describing trees is a wonder of mixed metaphors. A tree has a single node, called the *root*, which is its initial point of access. In the spider shape, the root is in the middle and in the classic—upside down—tree shape, the root is at the top. Each node in the tree is either an *internal* node or a *leaf* node. Each internal node has one or more *child* nodes connected to it by edges, and is called the *parent* of its children. A *leaf* node has no children. In the spider shape, children radiate outward from their parents; in the classic tree shape, children are connected below their parents.

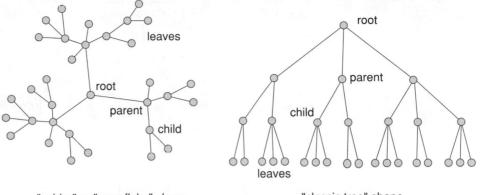

"spider" or "snowflake" shape   "classic tree" shape

Figure 2.16   The characteristic shape of a map of hierarchical organization is a graph called a *tree*.

As we move along a tree from the root toward its leaves, we funnel from more general concepts to more specific, where each parent is in some sense a generalization of its children. Concepts closer to the root are often called *high-level concepts*, while those further from the root are called *low-level concepts*—consistent with the drawing of a classical tree.

Parts Hierarchies   A *parts hierarchy* is a hierarchical concept map that illustrates how a complex object is composed of a set of parts, which may themselves be composed of a set of smaller parts. A relationship labeled "contains," "has," or "has-part" directed from parent to child expresses the notion that a child concept is a part of a parent. Equivalently, we could use a "is-part-of" relationship directed from the child to the parent. We've already encountered the "has-part" relationship in the "What is a car?" concept maps in Figures 2.12 through 2.15, where, for example, a power train "has-parts" engine and transmission. Note that concepts in a parts hierarchy may represent either physical objects or abstract ideas. Just as a physical object such as a bicycle has handlebars and wheels, a story has a plot and a setting, and a typical engineering homework problem has an unknown quantity and assumptions. Figure 2.17 shows a parts hierarchy of the United States, from the country level to the city level.

In a strict hierarchy, as represented by a tree, each node has exactly one parent, except for the root, which has none. Oftentimes, however, it's convenient to think of one concept as part of two different things. In a car, for example, a starter motor may be thought of as part of both the power train and the electrical system. Figure 2.18 shows a concept map that illustrates this scenario. We will still refer to such a map as a parts hierarchy, but the graph no longer qualifies as a tree because the "starter-motor" node has two arrows pointing toward it and hence *two* parents.

Class Hierarchies   An important method people use to help understand and manage ideas is organizing concepts into classes. A *class* may be defined as a set of concepts that share a common set of attributes or properties. A *class hierarchy*, also called a *taxonomy*, is a hierarchy of classes, where more general classes of things are successively broken down into more specific classes of things. In a taxonomy, child classes *inherit* attributes from their parents, which means that the children possess all the attributes of their parents, yet have additional attributes of their own.

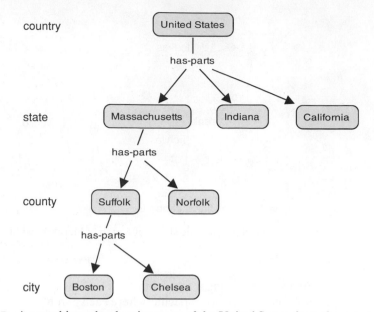

Figure 2.17   A parts hierarchy showing parts of the United States, from the country level down to the city level.

Figure 2.19 illustrates two ways of expressing inheritance relationships. A particularly famous class hierarchy is the taxonomy of all living things developed by the Swedish biologist Carolus Linnaeus in the mid-1700s. Linnaeus's taxonomy is illustrated in part in Figure 2.20.

In the graph in Figure 2.20, the edges labeled "is-a" represent the inheritance relationships. Thus, an animal is a living thing, a dog is a canine, and through inheritance of characteristics, a dog is also an animal as well as a living thing. Classes higher up in the hierarchy are considered more *abstract*. The Oxford English Dictionary defines *abstract* as

abstract *adj*, withdrawn or separated from matter, from material embodiment, from practice, or from particular examples.

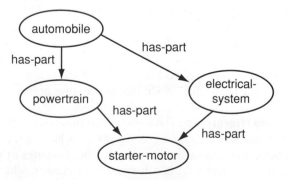

Figure 2.18   A starter motor is part of both the power train and electrical system of an automobile. This graph is not a tree because the "starter-motor" node has two parents, but it still represents a hierarchy.

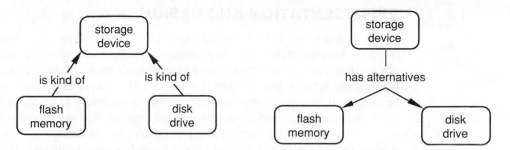

Figure 2.19 Ways of expressing that flash memory and disk drives both belong to the class of storage devices.

Classes lower in the hierarchy are more *concrete*, or having more substance. An *instance* is a fully concrete realization of a particular class. For example, "living thing" is a very abstract description of the particular instance "Carolus Linnaeus," whereas "human" is a more concrete classification.

In Linnaeus's taxonomy, as well as in most taxonomies used in various branches of natural science, the classification scheme is based on the *form* of things, such as what they look like or how they're composed. By comparison, when we consider organizing man-made things into categories, we generally find it more useful to classify them by *purpose*. Hardware stores, for example, aren't organized into aisles of "things made of metal" and "things made of plastic;" instead, they're organized into departments for plumbing supplies, electrical supplies, tools, or fasteners.

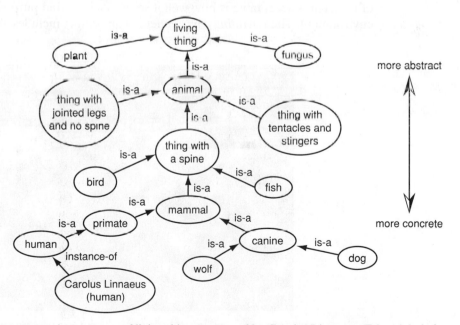

Figure 2.20 A taxonomy of living things proposed by Carolus Linnaeus. Edges labeled "is-a" denote inheritance relationships between classes of things.

2.3 REPRESENTATION AND DESIGN

Figure 2.21 depicts René Magritte's 1929 painting *The Treason of Images*, a picture of a form we clearly recognize as a tobacco pipe. Written below the image is a French phrase that translates to "this is not a pipe." In this paradox, Magritte asks the viewer to question what makes a "pipe" a "pipe." The surrealist painter would perhaps be surprised to learn that pragmatic engineers routinely ask similar questions, and furthermore, that there's a straightforward engineering way to answer them.

To understand the engineering perspective in answering this question, we first recognize that artifacts have not only a form, but also a *purpose*. If we understand the purpose of a pipe is "to smoke tobacco," then in order to accept it as such, we'd need to draw tobacco smoke coming from this pipe. While Magritte's goal was to indicate that because we couldn't use the *image* of the pipe for smoking, there's also no way to determine whether the particular object the image depicts could function as a pipe, either. For example, we can't determine whether the stem is hollow. Thus, as engineers, we have no reason to quibble with Magritte's assertion.

2.3.1 Purpose, Environment, and Form

As Magritte's pipe helped us illustrate, all artifacts have both a form and a purpose, and to accept an artifact, its form must be appropriate to its purpose. When engineers decide whether to accept a given form for an artifact, they approach the problem from two different perspectives, as Figure 2.22 shows. The first perspective is from the *operating environment* that'll use the artifact, and the second is from the *engineering environment* that'll produce it. This trinity of purpose, form, and environment, and how they relate to the acceptability of a design, is a central tenet of an engineering world view.

From the operator's perspective, the key question in evaluating whether the form of a design is acceptable is how well it serves its intended purpose in the operating environment. The *operating environment* of an artifact includes all characteristics of

Figure 2.21 René Magritte, *The Treason of Images* ("This is Not a Pipe"), 1928-29, copyright Los Angeles County Museum of Art.

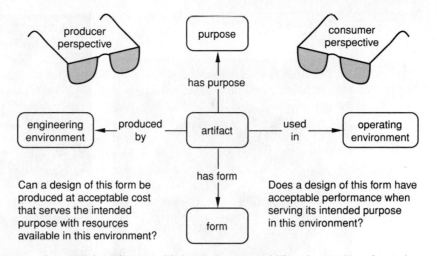

Figure 2.22 In an engineering "world view," the acceptability of an artifact depends upon its purpose, form, and environments that both produce and use it.

the surroundings in which the artifact is used. This includes naturally occurring and artificial phenomena such as:

- the effects of the physical environment, such as temperature, humidity, gravity, or radiation;
- socioeconomic conditions, including funding, building codes, environmental regulations, manufacturing and operating standards, and local mores and customs;
- the particular ways in which people interact with the artifact, including how they operate and maintain or repair the artifact;
- the ways other artifacts may interact with it, such as an automobile engine and its transmission.

From the producer's perspective, the acceptability of a design depends on whether the form can be produced at acceptable cost, given the resources available in the engineering environment. The *engineering environment* represents the complete organization that develops and produces an artifact together with all the resources it has at its disposal. This includes:

- the people involved in the design and manufacturing of the artifact;
- the tools and methodologies used in the design;
- available materials and technology;
- factories and production facilities.

When developing a design, engineers must look at its form from both perspectives, in order to ensure that it's acceptable in both the operating and engineering environments. To illustrate the relationship among purpose, form, environment, and acceptability, we use the example of a clock. As Simon states in *The Sciences of the Artificial* [Sim96], a simple description of the purpose of a clock is "to tell time." Over the years, people have designed many different forms of clocks, and clock designs acceptable in some environments may be unacceptable in others.

Figure 2.23   Is this a clock? A sundial will only serve the purpose of keeping time when the sun is out. This particular sundial, located on the Caltech campus in sunny, southern California, is surrounded by trees and is mostly in shade except for a few hours in the middle of the day.

The Sundial   As the first example, consider the form Figure 2.23 depicts. Is this a clock? In engineering terms, we restate the question as, "How effectively does this form serve the purpose of 'telling time' for a given environment?" The answer to this question depends on several factors, including how accurate the clock needs to be and where and when we will use it. A Simon notes, a sundial may work well as a clock most of the time in a sunny climate such as Phoenix, but it would work less often in Seattle, and wouldn't work anywhere at night [Sim96]. From the producer's perspective, a sundial can be made with the most primitive of tools, requires only a compass for calibration, and once set, can keep accurate time for centuries!

The Ship's Chronometer   In the eighteenth century, one of the greatest technical challenges was determining the longitude of a ship at sea. So acute was the problem of ships being lost that in 1714, the British government offered a prize of £20,000—equivalent to $6 million today—to the first person to demonstrate an effective scheme accurate to within one-half of a degree that would be tested on a ship sailing,

> ... over the ocean, from Great Britain to any such Port in the West Indies as those Commissioners Choose ... without losing their Longitude beyond the limits before mentioned ... tried and found Practicable and Useful at Sea.[1]

[1] http://www.rog.nmm.ac.uk/museum/harrison/longprob.html.

In theory, it's not difficult to calculate longitude; since 15 degrees of longitude equates to an hour of time difference, given the ship's local time and the local time at some reference point—say, Greenwich, England—the longitude is easy to determine. On-board, a navigator could determine the ship's local time from the position of the sun. The problem lies in accurately keeping track of the local time at the home port to within two minutes for the duration of a two-month journey at sea.

John Harrison, a carpenter and clockmaker from England, designed these clocks in the 1700s. Harrison began work on the design of a ship's chronometer for the prize in 1730. By the mid-1720s, Harrison had built England's most accurate pendulum clock—which kept time to within one second per month—but this design would not work on a boat because of the operating environment, specifically, the rocking motion and the changes in temperature and humidity. In 1765, 35 years after his first attempt, Harrison was awarded the first half of the prize for demonstrating a clock accurate to 39.2 seconds over a 47-day voyage between England and Jamaica. He claimed the remainder of the prize in 1773, at age 80, after the Board of Longitude was finally convinced that his results weren't a fluke, and that other clockmakers could build his design.

One of the most interesting aspects of Harrison's engineering efforts—aside from his sheer persistence—is the extent to which he devised innovations that not only made the design of a precision timepiece possible, but also advanced the design of many other kinds of machines. In essence, Harrison's contributions to research and development enriched the engineering environment for generations of engineers who succeeded him. One of his inventions, for example, was the caged roller race, which was a precursor to the modern caged ball bearing. Another invention, the bimetallic strip, compensated for changes in temperature. Harrison's bimetallic strip consisted of a strip of steel riveted to a strip of brass. Since brass expands more with changes in temperature than steel does, the strip will bend toward the steel side as the temperature rises and toward the brass side as it cools. Harrison used this effect to adjust spring tensions in his clock; today, the bimetallic strip is still used in thermostats to turn a furnace on or off to control temperature.

The Smartphone and Radio Clocks   Finally, we consider the form of a clock for a cell phone/personal digital assistant, as pictured in Figure 2.24. The operating environment for this clock is a portable electronic device that contains data processing and wireless communications hardware and software, and it's designed for a person who depends upon the device for communication, managing a schedule, and other functions while traveling between time zones.

Most cell phones today contain a radio clock that periodically synchronizes with a local time signal communicated over its cellular network. These signals, in turn, are based on clock signals broadcast from multiple global-positioning satellites (GPS). Each GPS satellite contains an extremely accurate atomic clock that references the frequency of oscillations between the nucleus and electrons of an atom, typically cesium-133. In fact, since 1967, the SI definition of a second has been 9,192,631,770 oscillations of the cesium-133 atom. Yet, despite the accuracy of a radio clock, a cell phone can't synchronize its time to a GPS atomic clock if it's out of range of its carrier's signal. In remote areas—say, out at sea or in a region outside a carrier's service range—a smartphone would need to rely on its internal clock, and its user would have to manually reset the time whenever he or she crosses a time zone.

Figure 2.24 World clock on a Palm Treo smartphone.

2.3.2 Requirements, Specifications, and the Forces That Shape a Design

During the design process, engineers will continually refine the form of an artifact until it's acceptable within the operating and engineering environments. One way of visualizing this process is to imagine "forces" in the environment that impel the engineer to change the shape of a design until it reaches an acceptable form. As pictured in Figure 2.25, forces in the operating environment drive the form of the design so that its actual *performance* acceptably serves its purpose, while forces in the engineering environment drive the form in a direction that keeps its development *cost* within available resources. Together, these forces "squeeze" the design into its final "shape."

For an engineer to produce an acceptable design, he or she must be able to determine how much and in which direction to change each aspect of the form in order to make it acceptable. This requires that the actual performance and performance goals, as well as the actual cost and cost goals, be expressed as quantities that can be objectively measured and compared. Figure 2.26 adds these ideas to our earlier

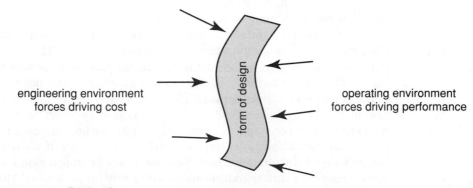

engineering environment
forces driving cost

form of design

operating environment
forces driving performance

Figure 2.25 Design forces.

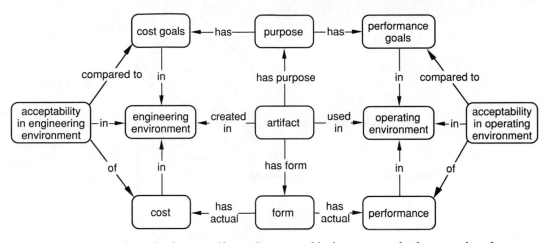

Figure 2.26 In order for an artifact to be acceptable, it must meet both cost and performance goals.

concept map. The values for the performance and cost goals form the basis of a kind of agreement or contract between the producer of an artifact and the consumer. To consumers, the goals are *specifications* that state how they can expect an artifact to perform in a particular environment, while for the producers, the goals are *requirements* that must be met. Clearly, both views of the goals need to be in sync for a product to succeed.

Constraints and Objectives   Engineers quantitatively express goals in two common ways—as constraints or as objectives. A *constraint* is a hard limit on a value, typically expressed as a mathematical inequality or equality. Some examples of constraints in the specification of a watch include:

- it's accurate to *within* ± 15 seconds per month
- it's water-resistant to a depth of *up to* 30 m
- its battery life is *at least* 2 years

An *objective* is a goal of minimizing or maximizing a value. The main difference between a constraint and an objective is that while a violation of a constraint makes the artifact unacceptable, an objective indicates a direction for a performance to make the artifact more or less desirable. We may also consider an objective a "soft constraint" that's desirable but not critical to meet. Some examples of objectives for the design of a watch include:

- its design should seek to *minimize* the manufacturing cost
- it should be *as thin as possible*

The Environment as a Mold for a Design   In *The Sciences of the Artificial*, Simon describes the environment as a "mold" for a design; that is, constraints within the environment determine its final shape [Sim96]. Just as a mold has two pieces—an

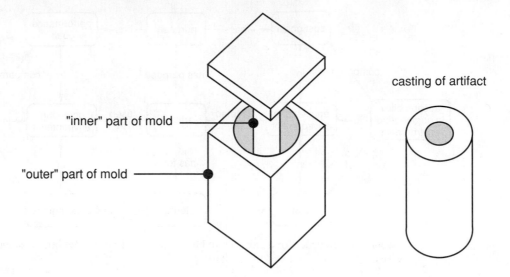

casting of artifact

Figure 2.27 The inner and outer parts of a mold.

"outer" piece and an "inner" piece that Figure 2.27 depicts—the design environment also has two parts—the operating environment and the engineering environment. By looking at this analogy closely, we form a mental image that gives us greater insight into the relationship among purpose, environment, and form.

The Operating Environment as the "Outer" Environment   Simon refers to the operating environment of an artifact as its "outer" environment because it contains forces on the design external to the engineering organization. Figure 2.28 illustrates a view of the operating environment as the outer part of a mold. Within the operating environment is a polygonal "hole," called the *acceptable region*, whose sides represent each of the individual constraints on the performance of the artifact. The artifact itself is drawn as an amorphous shape in which the actual performance is the outer surface of this shape. If the observed behavior fits entirely within the acceptability region—inside the constraints—then the artifact is acceptable in the operating

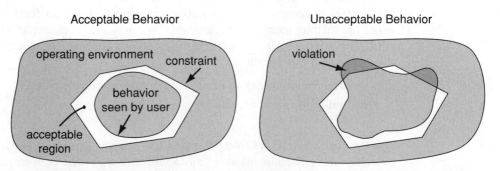

Figure 2.28 A diagram for visualizing the relationships among an artifact, its form, and its operating or "outer" environment.

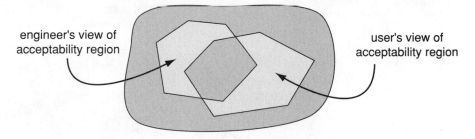

engineer's view of
acceptability region

user's view of
acceptability region

Figure 2.29   The effects of either misunderstanding the user's viewpoint on the purpose of an artifact or improper characterization of the operating environment.

environment. If it doesn't, then the artifact is unacceptable and the parts of the behavior that lie outside of the constraints signify violations.

From the diagram in Figure 2.28, we identify many possible forms for the artifact acceptable to the user; by contrast, a large number of forms won't be acceptable. If the engineers' goal is to produce an artifact acceptable to the user, it's not necessary to come up with a design that will fit the constraints exactly; rather, the design needs only to *satisfy* the constraints. In practice, when engineers design something, they usually don't attempt to seek the right design, just one "good enough" to satisfy the constraints.

Engineering projects fail when engineers either misunderstand the purpose of an artifact from the user's perspective, or overlook some aspect of the operating environment. In either case, the effect is the same—the engineers miscalculate the location of the acceptability region. Figure 2.29 illustrates this situation.

The Engineering Environment as the "Inner" Environment   The engineering environment consists of the set of all resources available to design and produce an artifact. The constraints imposed by the engineering environment limit the cost of the design, as measured in monetary, time, or other units. Simon refers to the engineering environment as the *inner environment* of an artifact—in other words, this environment holds the resources within the control of the engineering organization.

Figure 2.30 adds the engineering environment to our earlier drawing of the operating environment as the "inner" part of a mold. In the figure, the engineering environment—a polygon inside the acceptable region—further constrains its area. Each of the edges of the new polygon corresponds to a cost constraint, just as each of the sides of the outer polygon corresponds to a performance constraint. The acceptable region is now the space between the two environments. The artifact is drawn as a band in which the inner edge is its actual cost and the outer edge, as before, is its actual performance. In order to be acceptable in both the operating and engineering environments, the band must fit within the acceptability region without crossing over any of the constraints.

2.3.3   Design Hierarchies

The remainder of this section focuses on the "form" of an artifact, and how through a design process, it evolves to have a characteristic hierarchical, "boxes-in-boxes" shape. Specifically, we look at how we can break down a design problem into a

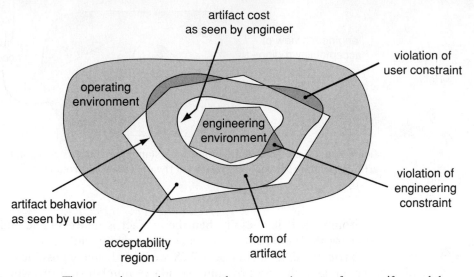

Figure 2.30 The operating environment as the *outer environment* for an artifact and the engineering environment as its *inner environment*. The artifact must conform to both to be "acceptable."

hierarchy of subordinate problem, and how engineers use both composition or parts hierarchies, and abstraction or classification hierarchies to synthesize a solution. For example, Figure 2.31 shows this problem:

> Design a system for a 125 lb person to lift a 250 lb weight from the floor and place it on a shelf 6 ft high in a closet 8 ft wide by 8 ft deep.

While you probably have a number of ideas already to solve this problem, we'll work through possible solutions *systematically*, introducing methods and terminology we can later apply to more complex problems.

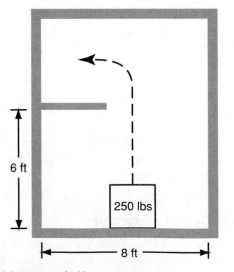

Figure 2.31 Lifting a weight onto a shelf.

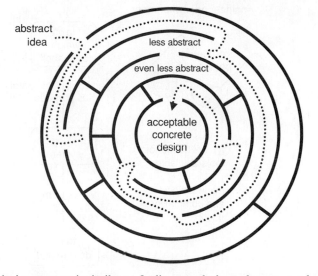

Figure 2.32 A design process is similar to finding a path through a maze, where advancing from outer to inner rings involves determining design decisions.

Decision Hierarchy Engineers rarely solve design problems all at once. Instead, they typically use multiple passes, refining both the formulation of the problem as well as its solution with each pass. As Figure 2.32 illustrates, we can compare a design process to finding a path through a maze. At the start, the designer knows only a basic set of requirements, and the goal is to find a fully realized design that satisfies those requirements. As the designer advances through each ring of the maze, he determines more of the unknowns and the design gradually transforms from an abstract concept to a concrete implementation. The first few passes, or the outermost rings in Figure 2.32, are sometimes called *conceptual design*, while the later passes or innermost rings are called *detailed design*.

Figure 2.33 shows a concept map of the initial design problem for a lifting system. The purpose of the system is "to lift weight onto shelf." The constraints of the operating environment include the dimensions of the closet, the height of the shelf, and the weights of the load and operator. The constraints and objectives in the engineering environment are to use common household tools and to minimize cost and complexity. The form of the design must satisfy these objectives and constraints in order to be acceptable.

As a designer searches for a path from the initial problem statement to a final, concrete design, he or she faces a series of decisions, where each decision pins down some aspect of the design and also limits choices later on. We can picture this process as finding a path through a set of branching roadways, as Figure 2.34 illustrates, where at each fork in the road, the designer decides which branch to use. Acceptable designs are at the ends of some roadways, but not others.

This hierarchy of choices is called a *decision hierarchy* or *decision tree*. It is a kind of classification hierarchy, where each node represents a class of related options or alternatives that satisfy a common goal. Figure 2.35 illustrates a decision tree that classifies some of the highest-level options. In this breakdown, we set the first major decision to be the choice of using either a powered or manual method, which are shown as two children of the root, connected with "is-a" relationships, as defined

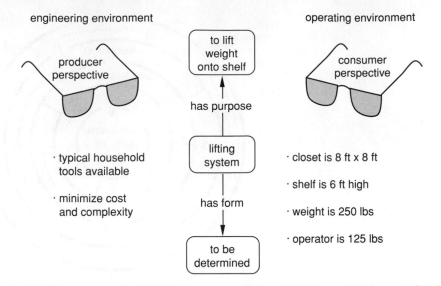

Figure 2.33 Concept map for the lifting system problem, showing goals and constraints in the engineering and operating environments.

in Section 2.2.3. Under each of these classes, there are further options, such as the choice of a forklift as a powered method or a lever or block-and-tackle as a manual method. Given the objectives from the engineering environment of keeping things simple, we rule out the powered approaches, but regard both of the manual methods as worthy of further consideration.

Structural Hierarchy Making a further decision between the two manual lifting methods—a lever or a block-and-tackle—requires more detail on their implementation in order to determine if either approach satisfies the constraints and objectives.

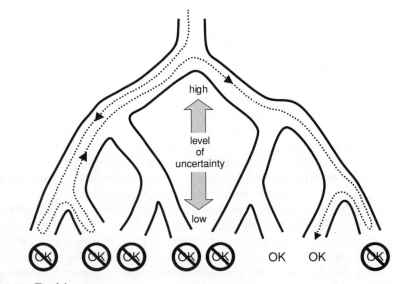

Figure 2.34 Decision tree.

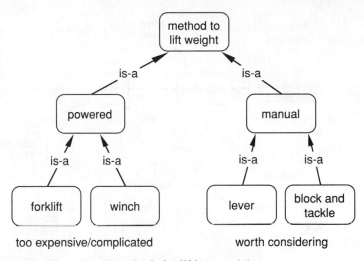

Figure 2.35   A class hierarchy of methods for lifting a weight.

One way to represent this additional detail is to break down the structures of the two simple machines, as Figure 2.36 shows. This *structural hierarchy* is a kind of composition hierarchy, and the higher-level nodes are connected to the lower-level nodes through "has-part" relationships. In the figure, a lever has a beam and a fulcrum, while a block-and-tackle has a pulley system and a rope. Each of these we can further break down to unveil more design questions. The "parts" in the second level of the structural breakdown are not physical components, but rather parts of the remaining information we need to specify remaining details of the design. These

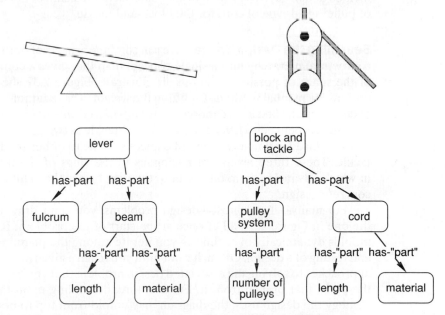

Figure 2.36   Structural breakdown of two manual lifting methods.

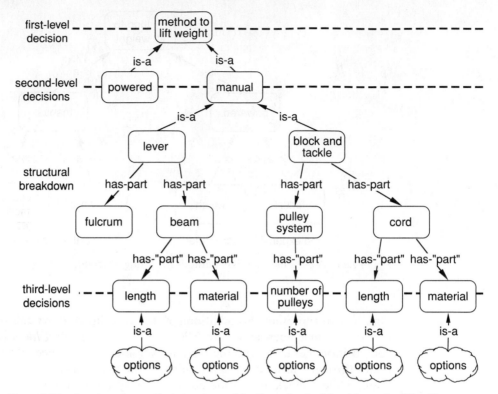

Figure 2.37   A concept map that uses a combination of a decision hierarchy ("is-a") relationships and a component hierarchy ("has-part" relationships) to illustrate the options for manual methods for lifting a weight.

include details such as the length and material of the beam of the lever, or the number of pulleys and type of cord for the block-and-tackle.

Searching the Design Space   We can combine aspects of both a decision tree and a structural hierarchy into a single concept map that gives a comprehensive picture of the space of possible solutions. To illustrate, Figure 2.37 shows a diagram with options for manual methods for lifting the weight. The diagram contains three levels of decisions: the first level chooses between powered and manual methods; the second level selects between a lever and a block-and-tackle as two different manual methods; and the third level involves design choices specific to either the lever or block-and-tackle. These three levels are analogous to the rings of the maze in Figure 2.32, in which advancing from one level to the next brings the solution closer to being a concrete design.

For many—if not most—design problems, you can't draw a diagram, such as the one in Figure 2.37, all at once at the start of the problem. Rather, the diagram unfolds dynamically over time as you iterate among the planning, implementation, and testing of a design. Often, it's not until you reach a given node and start exploring approaches to solving the associated design problem that you're ready even to define the paths that branch from it. Furthermore, depending upon the problem solving strategy the design uses, the diagram may unfold from top to bottom, from bottom to top, or from the middle outwards. For example, for the weight-lifting problem, a person who walks into the room and finds a wooden beam and a barrel might attempt a "bottom-up" solution using a lever first, whereas this person may pursue a different path if he or she started the design with a clean sheet.

In addition to the decision of moving upward or downward on the tree, there's also the decision of turning left or right. Complementary to the option of working in a top-down or bottom-up fashion, there's also the choice of exploring paths in a *depth-first* or *breadth-first* order. In a *depth-first* search, you'd follow a path from the root of the tree to a leaf, and then evaluate whether the design at the end of the road is acceptable. If it isn't then you'd retrace your steps and explore another path to a leaf—the approach the path in Figure 2.34 uses. In a breadth-first search, you'd move slightly down each path at a given fork in the road, and explore the full breadth of options at each level, and then decide which looks most promising. In other words, in a breadth-first search you spend extra time to observe each pathway, whereas the depth-first approach follows a path without interruption. Clearly, a breadth-first approach can save wasted time and effort, but sometimes it's impossible to determine the most promising path without proceeding slightly down toward the end.

In general, you may be unable to find an acceptable design by following a particular path because it's either physically or economically *infeasible* to meet requirements given certain design choices. In other words, starting from a given design concept, there's no combination of decisions that can be made down the road that can avoid violations of either performance or engineering constraints. Earlier, we decided that a powered approach wouldn't lead to an acceptable solution to the lifting problem because it was too expensive. Figure 2.38 illustrates the results of pursuing a manual approach using either a lever or block-and-tackle. Since the required lever arm is too long to fit in the room, the lever is infeasible. On the other hand, since the block-and-tackle is compact, there's a feasible solution down this path.

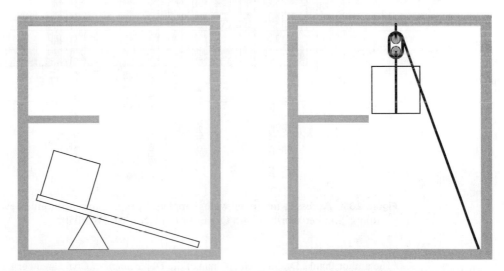

Figure 2.38 There is no feasible solution for using a lever to lift the weight to the required height in a small room, while there does exist a solution using a block and tackle.

EXAMPLE: WATER SUPPLY FOR RURAL COMMUNITIES IN DEVELOPING NATIONS

One of the most crucial engineering tasks worldwide is to provide clean drinking water. In much of the developed world, people generally take clean drinking water for granted. Often, these people are even unaware of the source of their tap water, let alone the municipal engineering infrastructure that delivers it to their homes and businesses. For much of the world population, however, access to fresh water is a daily struggle. According to the World Health Organization (WHO), today 2.6 billion people have no proper means of sanitation and 1.1 billion don't have access to improved drinking water [Wor05b]. In much of the developing world, the infant mortality rate is above ten percent, with the mortality rate by age five often much higher, and the majority of these deaths result from water-related diseases. The WHO estimates that 4,500 children die each day from a lack of water supply and sanitation services. Figure 2.39 shows access to improved water supplies by region and economic development grouping. Table 2.1 lists the countries with the lowest access to improved water supplies. Afghanistan has the lowest worldwide—only 13 percent of its population has any access to clean drinking water, and only two percent with a household connection. In North America, Haiti has the lowest access; only 71 percent of its population has access to clean drinking water and only 11 percent with a household connection.

In this section, we examine the issue of providing clean drinking water to rural areas in developing nations.[2] In addition to highlighting this important global problem

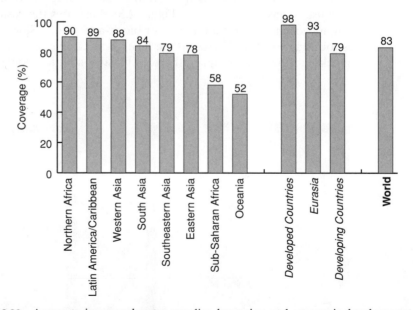

Figure 2.39 Access to improved water supplies, by region and economic development grouping. Source: World Health Organization/UNICEF [Wor05b].

[2]The author thanks Dr. Stephen Silliman of the Department of Civil Engineering and Geological Sciences at the University of Notre Dame, South Bend, Indiana for his assistance in developing this example. Dr. Silliman and his students have worked with rural communities in Haiti and Benin to develop adequate sources for drinking water.

TABLE 2.1 Countries with lowest access to improved drinking water supplies.

	Total (% coverage)	Household connection (% coverage)
Angola	50	5
Mali	48	10
Democratic Republic of the Congo	46	10
Niger	46	8
Madagascar	45	5
Equatorial Guinea	44	8
Lao People's Democratic Republic	43	8
Mozambique	42	11
Cambodia	34	6
Chad	34	5
Somalia	29	1
Ethiopia	22	4
Afghanistan	13	2
Haiti*	71	11

*Lowest in North America
Source: World Bank Water and Sanitation Program [Wor05a].

for its own sake, we demonstrate the application of the concepts we introduced in the first half of this chapter to characterize a multifaceted, real-world engineering problem. Specifically, we illustrate:

- the definition of acceptability in terms of both technical and socioeconomic constraints imposed by the operating environment and the engineering environment;
- breaking down a problem using hierarchies of components and alternatives;
- how an initial design problem unfolds into subproblems as we continue the breakdown;
- the use of concept maps as diagrams to support the analysis of a design problem.

In this section, we *don't* provide the kind of quantitative analysis we needed to decide among options or to assess whether a design is ultimately acceptable. In Chapter 4, we will return to the water system design problem, and show the use of *models* to explore engineering design tradeoffs and to make quantitative decisions. Finally, we stress that the analysis in this example is just *one* way of looking at the issue of providing safe drinking water and that there is no "right answer" to this complex problem. There are certainly many other alternatives, options, and related concepts in addition to those we discuss, and you may find other ways to organize the concepts we include that make more sense to you. Nonetheless, using the framework presented earlier in this chapter—specifically characterizing acceptability from the perspectives of the two environments, and using concept maps to help break the problem down into manageable pieces—greatly helped us obtain a better understanding of this complex problem. Thus, we encourage you to look at this example as a template to characterize other design problems.

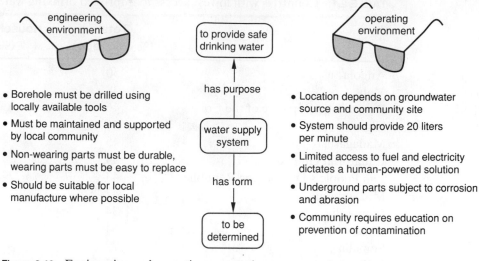

Figure 2.40 Engineering and operating perspectives on constraints.

2.4.1 The Top-Level Problem: Meeting Community Needs

Objectives and Constraints in the Operating and Engineering Environments

The first step in analyzing this problem is to define the characteristics of an "acceptable" solution—in other words, what are the engineer's and user's perspectives on this problem, and what are the goals we're trying to satisfy in each? Figure 2.40 illustrates a few key characteristics of acceptability from the engineering and operating perspectives. The sections below describe these perspectives in further detail.

The Operating Environment and Performance Goals Several sets of constrains shape the operating environment for a water supply system, including geological, technological, and social and economic constraints. The main geological constraint is the availability of a source of clean water. Figure 2.41 shows the relative abundance

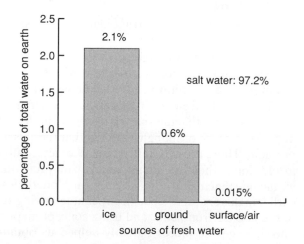

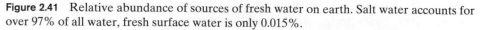

Figure 2.41 Relative abundance of sources of fresh water on earth. Salt water accounts for over 97% of all water, fresh surface water is only 0.015%.

of water from different sources on earth. Surface water is water found in ponds, streams, lakes, and rivers. While it's sometimes easily accessible, surface water is highly susceptible to contamination and still strenuous to collect, and so isn't considered an acceptable approach. Groundwater is more than an order of magnitude more plentiful than surface water and is typically cleaner, but requires wells for collection. Groundwater resources are still largely untapped in rural developing areas, and thus most of the work in providing drinking water in these areas focuses on access to groundwater.

Lifting groundwater from a well requires energy. For domestic use, the means for accessing water from a given well should be capable of delivering at least 20 liters of water per minute, and so one of the main technological constraints in the design of a water supply system is devising a power source able to lift the water from a well at an acceptable rate. In most rural developing communities there's neither reliable electricity nor a reliable fuel source for an engine, so a human-powered solution is usually the best choice. Furthermore, in such communities, women and children are largely responsible for gathering water, so any human-powered solution must take both the strength and size of the operator into consideration. Since the water is underground, the system has to function reliably in a highly abrasive and corrosive environment. Finally, the system must be designed so that contamination is not introduced in the process of removing the water.

Patterns of human activity also constrain the placement of a well. It should be located close enough to the population center to be convenient, but it's also necessary to control animal and human activity in its vicinity to avoid contamination. Similarly, the well must be either upstream or far downstream of latrines. In determining location, it's also important to consider local politics to ensure equal public access.

Education is also important in the design of the overall system and its operating environment. The community needs to understand the issues surrounding the water supply and public health, as well as the importance of preventing water contamination from its source to its use. The nature of the contamination can be difficult to understand, since the microbes that cause disease are imperceptible, and the community must learn to use the well even when surface water sources are available.

The Engineering Environment and Cost Goals   The most important consideration from the engineering perspective is not the initial development of a water supply system, but rather its maintenance. Even if a water supply project has been initiated from outside the community, experience has shown that in order for the project to have long-term success, ultimately, there must be local ownership. What this means is that it is important to regard the center of the engineering environment as located within the community, and that outside engineers, even if they conceive the project, should work as consultants to the local community. This concept of local ownership of the water supply system goes by the name *village-level operation and maintenance (VLOM)*. The key requirements of a VLOM groundwater pump system as described by the World Bank Water and Sanitation Program [Rey92] are:

- The borehole must be designed and constructed in a manner appropriate to the pump and the local conditions.
- The user community can perform routine maintenance, decide when to carry out repairs, determine who will do the work, and determine who will be responsible for paying for them.

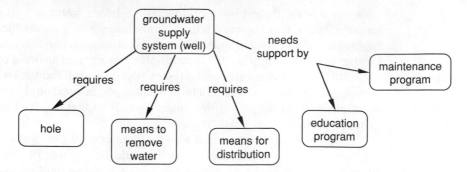

Figure 2.42 Top-level components of a water supply system for a rural community in a developing country.

- Non-wearing parts of the pump must be durable and reliable, and parts subject to wear must be easy to service and inexpensive.
- As far as possible, the pump must be suitable for manufacture using existing local industrial resources or facilities that can be readily established. Imported components are to be used only if they are critical to achieving other VLOM objectives.

Breakdown of a High-Level Design Concept Figure 2.42 illustrates a first-level breakdown of the components of a groundwater supply system. Two obvious components of the system are a hole that provides access to the water, and some means for removing water from the hole. The complete system, however, has additional components that must be considered. One of these is a means for distributing water removed from the well. This can be as simple as carrying the water from the well in buckets, but must ensure avoiding contamination thereafter. Many well projects fail when its community neglects the implementation of a maintenance program and an education program.

We can break down each of these concepts further to get a deeper understanding of both the components and alternatives in the design of the water supply system. Figure 2.43 adds a breakdown of concepts involved with both the hole and the means for removing water, along with comments on the various options as they relate to the objectives and constraints. A hole requires a means for digging or drilling it, which in turn leads to additional options. It also has dimensions of depth and diameter that depend on the level of the water table and how the hole is drilled.

Various means exist for removing water from the hole. One approach used since ancient times is lowering buckets or climbing down into a wide-diameter, hand-dug well to get water. One of the obvious problems with this approach is that the well becomes more difficult to dig the deeper the water source. Another problem, however, is that it is very susceptible to contamination, and thus we consider this alternative unacceptable. The preferred approach is to use a narrow borehole to reach the groundwater and then to remove it with a pump. Regardless of whether a bucket on the end of a rope or a pump is used, energy is required to lift the water from deep underground to the surface. Just as with the weight-lifting problem earlier in this chapter, this could be a manual method or one of several powered approaches. Because many rural communities in developing nations don't have adequate access to either a fuel supply or electricity, and because they are easy to maintain locally, manually operated methods often provide the best alternative.

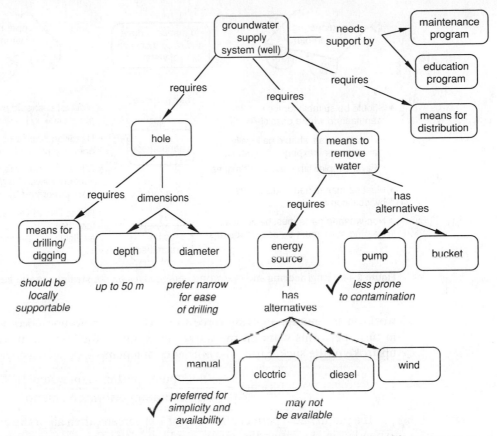

Figure 2.43 Further breakdown of concepts for a water supply system, with comments on options relative to objectives and constraints. A manually operated pump is an appropriate choice for further consideration.

2.4.2 A Lower-Level Problem: Design of a Handpump

Handpumps play a critical role in supplying water in many locations. In this section, we analyze the design of a handpump. As before, we begin by defining objectives and constraints in the operating and engineering environments. Next, we break down the structure of the design, examining alternatives and how these relate to the objectives and constraints.

Objectives and Constraints in the Operating and Engineering Environments
Fundamentally, a pump is a machine that is used to draw liquid into a pipe and raise that column of liquid to some height. There are a number of measures we could use to quantify the acceptability of a handpump, but three of the most important are the *efficiency*, the *reliability*, and the *cost*. Efficiency is a measure of how much water the pump produces relative to the energy expended in operating it. Availability is a measure of the percentage of time the pump functions properly and is available for use. Cost is the expense of producing the pump.

Efficiency In physics, work is defined as the energy expended in applying a force to move something over some distance. In operating a handpump, a person performs

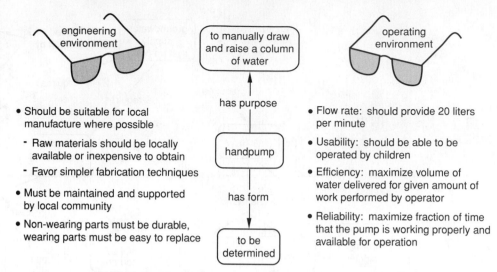

Figure 2.44 Engineering and operating perspectives on constraints on the handpump design.

work on the pump, applying force to move the handle up and down. The pump, in turn, performs work on the water below, lifting the column of water with each upstroke of the sucker rod. The efficiency of a pump is defined as:

$$\text{efficiency} = \frac{\text{work pump performs on water}}{\text{work operator performs on pump}}$$

If a pump had a perfect efficiency of 100 percent, then all of the energy a person expended in operating the pump would result in the pump raising an equivalent amount of water. In any real pump, however, some of the work performed by the operator is wasted. For example, with each upstroke, the operator has to lift not only the column of water, but also the sucker rod and the piston. Friction in the system, such as between the piston seal and the cylinder wall, also adds to the amount of work the operator must perform during each cycle. Finally, leaks in the system may result in a loss of water raised, even if the piston travels the full length of the cylinder.

Reliability Availability considers both how frequently a pump breaks down and the time it takes to perform a repair. To illustrate, Figure 2.45 shows the operating histories for two different pumps with the same availability. Although pump A breaks down less frequently than pump B, it takes longer to repair. Even though pump B

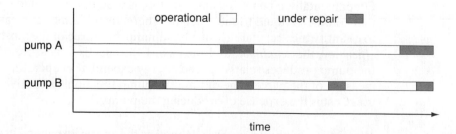

Figure 2.45 Operating histories for two pumps that have the same availability.

breaks down more often, it may still be the more desirable pump since the repair time—the time when there's no water service—is short.

In engineering, the average time between repairs is often called the *mean time before failure* or *MTBF*. While MTBF is a good indicator of the reliability of a system, it doesn't always provide enough information to determine if a design is *acceptable*, as the two pumps in Figure 2.45 illustrated.

Cost   The cost of a handpump includes the cost of the raw materials, the cost of tools and equipment, and the cost of skilled labor. When assessing the cost of candidate handpump designs, it's important to consider not only the initial cost of fabrication but also the continued cost of operation. As in our discussion of reliability, it's not inherently bad that parts fail occasionally if the repairs are inexpensive.

Breakdown of Handpump Design Concepts   Figure 2.46 shows a concept map for the design of a typical handpump with a breakdown of the basic structure and some of the principal alternatives. The following paragraphs detail these concepts.

High-Level Organization   We consider a pump as having four main components: a drop pipe, a cylinder, a sucker rod, and a handle assembly. The drop pipe is a long pipe that carries water between the underground source and the surface. The cylinder is the heart of the pump that forces water up the drop pipe, which we'll

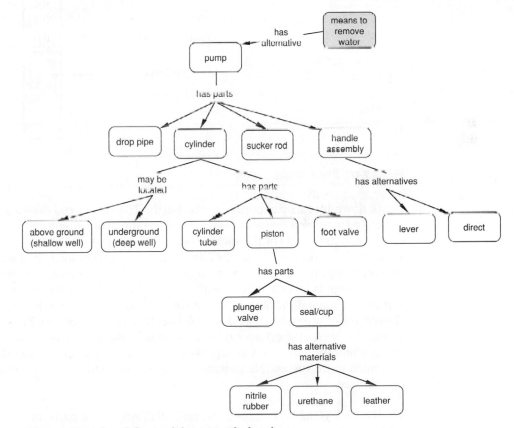

Figure 2.46   A breakdown of the parts of a handpump.

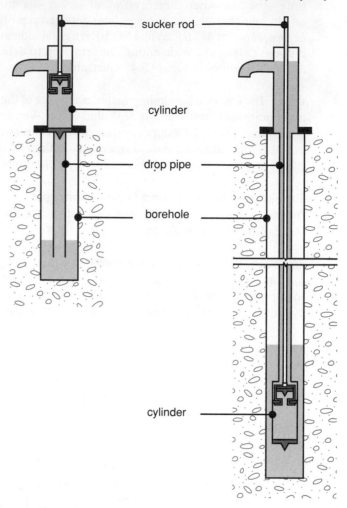

Figure 2.47   Pump configurations.

break down into its component parts shortly. The sucker rod drives a piston within the cylinder up and down to cause the "pumping action," and the handle assembly is attached to the end of the sucker rod. These components may be arranged in two main configurations, as Figure 2.47 illustrates. The first configuration is a *shallow-well suction pump*, in which the cylinder is located above ground. The main advantage of this design is its easy access to all moving parts, which simplifies installation and repair. The main disadvantage is that since this configuration relies on suction to lift the column of water, it can work only for depths of approximately 7 m or less [Wat99]. For deeper wells, the cylinder must be located at the bottom of the drop pipe. While this configuration can work at depths of 50 m or more, the underground moving parts are much more susceptible to wear and corrosion, and are more difficult to install and replace.

The Pump Cylinder   Figures 2.48 and 2.49 illustrate the parts and the operation of a pump cylinder, which are also included in the concept map in Figure 2.46. The

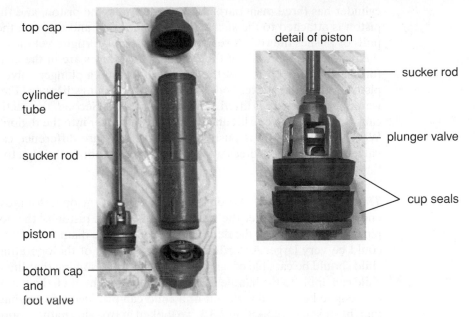

top cap

cylinder tube

sucker rod

piston

bottom cap and foot valve

detail of piston

sucker rod

plunger valve

cup seals

Figure 2.48   Components of a deep-well pump cylinder.

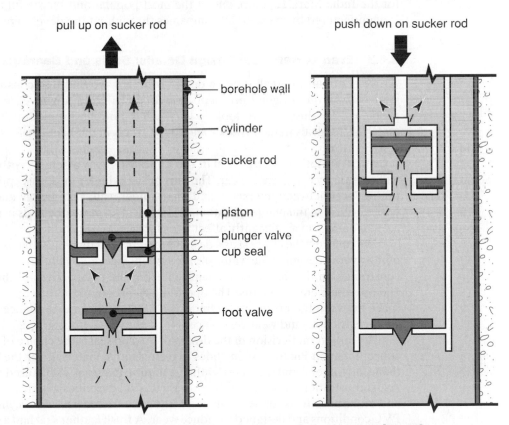

pull up on sucker rod

push down on sucker rod

borehole wall

cylinder

sucker rod

piston

plunger valve

cup seal

foot valve

Figure 2.49   Components and operation of a pump cylinder, based on [Wat99].

cylinder has three main parts: the cylinder tube, the piston, and the foot valve. The piston is attached to the sucker rod and slides up and down in the tube as its user pulls or pushes the rod. A seal or cup forms a watertight, yet movable seal between the piston and the walls of the cylinder. Two valves are in the cylinder assembly: a foot valve at the bottom of the cylinder tube and a plunger valve found within the piston. When the sucker rod is pulled up, two things happen. First, the column of water above the cup is lifted towards the surface. Second, the suction below the cup causes the foot valve to lift and open, drawing water into the region under the piston. When the sucker rod is pushed down, the pressure difference causes the plunger valve to open, which forces water into the region above the cup, to be lifted the next time the rod is pulled.

The Handle Assembly   When using a handpump, the operator must be able to apply enough force to counter the water pressure on the piston at the bottom of the drop pipe. Depending upon the depth of the well and the diameter of the piston, this force could be very large. According to the constraints of the operating environment, a child should be capable of operating the pump. Assuming that the maximum force a child can apply to the handle is his or her own weight, then the handle assembly must be designed so that the weight of a child can produce a force that is much heavier than he or she is. In Section 2.3.3, we looked at two alternative approaches to lifting a heavy weight, namely, a lever and a block-and-tackle and both approaches have been used in the design of pump handle assemblies. Figure 2.50 illustrates handle assembly for the India Mark II pump, one of the most popular and successful designs, which is manufactured by many small companies throughout the developing world.

2.4.3 Even Lower-Level Design Details: Seals and Bearings

As we continue to break down a design problem, we find that constraints and objectives from the original problem "trickle down" into low-level, detailed design problems. To illustrate, we look at two detailed design problems: the selection of materials for seals in the cylinder and for bearings in the handle assembly.

A Critical Detail: The Cup Seal in the Cylinder   Any assembly with moving parts will eventually experience wear. The part of the cylinder most susceptible to wear is the cup seal between the piston and the cylinder wall. This simple and inexpensive part is critical to the pump operation. It's a typical scenario one finds in many designs: a small detail can significantly affect the success of a project.

We can view the cup seal as a design problem in its own right that is a subproblem of the overall design of the water supply system. Figure 2.51 illustrates objectives and constraints on the design of the cup seal from the perspectives of the engineering and operating environments. The operating environment of the cup seal is the small space between the piston and cylinder wall and the key performance measures are leakage, friction, and wear resistance.

An important decision in the design of the cup seal is the choice of material. The concept map in Figure 2.46 includes two alternative materials for the cup seal: urethane, a synthetic material, and leather, a natural material, as pictured in Figure 2.52. As an illustration of tradeoffs, Reynolds [Rey92] describes a series of performance and endurance tests for seals made of a variety of materials in a cylinder made of PVC conditions and designed to induce wear. A fresh leather seal had a slightly better efficiency than a urethane seal, but the differences were insignificantly small. For the

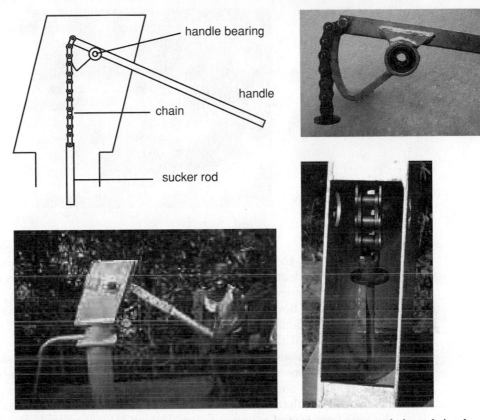

Figure 2.50   The India Mark II pump. The rugged design incorporates a chain and circular guide at the end of the lever arm that keeps the sucker rod vertically aligned. Clockwise, from top left: schematic view of design; closeup of chain and guide; connection to sucker rod, India Mark II pump in use in Benin, West Africa (photo copyright Steve Silliman).

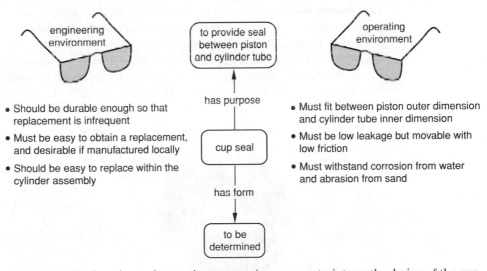

Figure 2.51   Engineering and operating perspectives on constraints on the design of the cup seal.

Figure 2.52 A leather cup seal.

endurance tests, the pumps were cycled until the output flow slowed to a trickle. Two samples of leather seals lasted for 620 and 807 hours, whereas two samples of urethane seals lasted for only 270 and 324 hours. Inspection of the pumps after the endurance tests, however, showed that the cylinder with the leather seal was heavily scratched and damaged, whereas the cylinder with the urethane seal remained in good condition, with only the seal itself worn.

The conclusions you may draw from this experiment and its design goals are complex. In terms of efficiency, the two seal designs are comparable. In terms of pump reliability and availability, the leather seals have a longer mean time before failure (MTBF) than the urethane seals, but since they lead to damage of the cylinder, the long-term cost of maintenance may be much higher. On the other hand, leather seals can be made from local materials, whereas the urethane seals may have to be imported. Thus one possible strategy would be to favor urethane seals, but to use leather seals "in a pinch" for short durations when urethane seals aren't available.

The Pump Handle Bearings   The design of the pump handle bearings also illustrates interesting tradeoffs among the efficiency, availability, and cost of a pump design. As Figure 2.50 showed earlier, the purpose of the handle bearing is to provide a low-friction collar that supports the handle about a pin, while allowing it to rotate freely. Figure 2.53 shows two possible designs for a handle bearing. The first is a ball bearing set. A ball bearing set consists of a set of steel balls set in a shallow circular

ball bearings   plastic top hat bearing

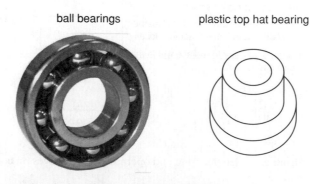

Figure 2.53 Alternative designs for pump handle bearing.

track called a race. When they work *properly*, ball bearings provide extremely low friction. Because they're made of steel, they're also durable. Fabricating a ball bearing set, however, requires the ability to shape metal to very precise tolerances, which in turn requires very specialized equipment and highly skilled labor. An alternative bearing design is a plastic "top hat" bearing, also shown in Figure 2.53. While this design would wear more quickly than steel ball bearings, it can be cheaply and easily fabricated with a mold. Thus the plastic top hat bearing could be easily manufactured in a developing nation, whereas the ball bearings would likely have to be imported, which has significant impact on cost.

PROBLEMS

1. **Short-Term Memory**
   Come up with an example, similar to "doing math in your head," where the limiting factor in solving a problem clearly seems to be the limited ability to keep information in your short-term memory.

2. **Why Won't My Car Start?**
   Write a set of 3–5 rules or productions that might help diagnose why someone's car won't start.

3. **Planning a Solution to a Problem**
   Think of some project that you've worked on in the past—either in school, at work, or for fun—where you were part of a team that spent at least a day or two in coming up with the solution to a problem.

   • Give an example of where you applied a divide-and-conquer strategy to solving the problem.
   • Give an example of working forwards (forward chaining), working backwards (backward chaining), and working from the middle in piecing together a solution for this problem.

4. **Using Graphs to Convey Information**
   Give a specific example or two of the use of a graph (nodes and edges) to convey information in some common, everyday application.

5. **Organization of a Parts Catalog**
   Go to the web site of some engineering parts supplier. Your instructor may help you pick one. Draw a hierarchical concept map with at least ten concepts that reflects how the supplier organizes its products. Is the organization primarily by form or by purpose?

6. **Construct a Concept Map: River, Beam, etc.**
   Organize the following words into a concept map and identify the domain:

river	paint	beams
cables	joints	foundations
bolts	steel rebar	catenary
concrete	roadbed	asphalt
forms	truss	toll

7. Construct a Concept Map: Sausage, Vegemite, etc.

Organize the following words into a concept map and identify the domain:

sausage	Vegemite	orange juice	congee	butter
beverages	coffee	yoghurt	cereal	sugar
Cap'n Crunch	fruit	oatmeal	over easy	grapefruit
toast	tea	jam	Wheaties	poached
scrambled	meat	banana	breakfast	bacon
bagel	beans	yeast	eggs	carbohydrates

8. Concept Map: U.S. Government

In the two columns below are listed concepts and relation words that characterize aspects of the U.S. government. Draw a concept map that shows the relationship between each of the concepts. Each relationship should be used at least once, and several may be used more than once.

Concepts	Relationships
Government	Has part
Legislative	Passes
Executive	Can veto
Judicial	Reviews
Laws	Appoints
Rulings	Approves

9. How Stuff Works

Go to the web site www.howstuffworks.com and find an article describing some engineering system of interest to you. Using the car example from Chapter 2 as a guide, construct a concept map with approximately 10 concepts in it. Describe the use of hierarchy in your final organization.

10. Product Specifications and Design Objectives

Look up online the specifications of some product of interest to you. Write down 3–5 elements of the specifications in the form of mathematical constraints. In addition, write down 3–5 objectives that are not listed as part of the specifications, but which you think would be important considerations in the design of the product.

11. Form, Purpose, and Environment

Similar to the example of clock used in Chapter 2, pick another type of artifact that can have a variety of forms. State what the purpose of the artifact is, and then give examples of how the environment—either engineering environment, operating environment, or both—affects the specific form. Some possibilities are (there are many more):

- automobiles
- yoghurt (different forms of packaging include fruit mixed in cups, fruit on the bottom of cups, tubes)
- video entertainment systems
- bridges

12. **Purchase of a Faulty Product**

    Discuss an example of what you consider to be a violation of a constraint in the operating environment for some product you bought. What was the constraint and what was the violation?

13. **An Ill-Conceived Product**

    Give an example of a product (good, process, or service) that you think "missed the mark" because the producer and the consumer had misaligned views on the acceptable region. Explain.

14. **Design Alternatives and Constraints**

    Pick a scenario similar to the problem of lifting a weight onto a shelf. Identify a set of constraints on the problem. Brainstorm a set of alternative approaches (3–5) and discuss whether or not each alternative is compatible with the constraints.

15. **Snow Removal System**

    A fact of life in many cold climates is snow removal during the winter. Snow must be removed from roads and driveways so that vehicles can pass. Heavy snows and especially ice should also be removed from roofs as it may cause damage.

    (a) Draw a diagram similar to Figure 2.22 describing the design considerations from the producer perspective and the consumer perspective for a system for removing snow from a driveway.

    (b) Draw a diagram similar to Figure 2.22 describing the design considerations from the producer perspective and the consumer perspective for a system for removing snow from a roof.

    (c) One way of removing snow is by melting it, using electric heating coils. Evaluate this as a possible solution for removing snow from a driveway and from a roof, according to the criteria that you defined above.

16. **The X Prize**

    Search the Ansari X Prize online and look up the basic rules of the competition. Analyze the winning entry built by Burt Rutan and Paul Allen and determine how the rules of the competition provided the constraints and design considerations that went into creating Space Ship One. In about 200–300 words, describe how the final design of the spacecraft was shaped by the engineering environment as well as the constraints imposed by the rules of the competition.

17. **Alternative Pump Designs**

    Come up with a variety of pump design concepts that could be operated by one or more children and could have better flow rates than a handpump.

18. **Gear Pumps**

    A piston pump is but one of several common pump designs. Another popular design is the *gear pump*.

    (a) Draw a schematic for an external gear pump and explain briefly how it works.

    (b) Draw a schematic for an internal gear pump and explain briefly how it works. What's another name for an internal gear pump?

    (c) List the advantages and disadvantages of piston pumps versus gear pumps. Describe an operating environment where a gear pump would be preferred over a piston pump and vice versa.

Learning and Problem Solving

LEARNING OBJECTIVES

- to explain some of the general differences between how experts and novices solve problems, to relate these concepts to the cognitive model in the previous chapter, and to describe some strategies for studying engineering that promote the development of expertise;
- to analyze problems in terms of the level of understanding that it requires, including application, analysis, and synthesis;
- to articulate a framework for solving engineering problems that includes the following steps: problem definition, background research, planning, implementation, checking results, generalizing, and documenting results, and to apply this framework to solving simple problems;
- to explain what a *heuristic* is and to give examples of common heuristics in solving engineering problems.

3.1   INTRODUCTION

Learning and problem solving are tightly intertwined. One might even say that the reason that we learn is to be able to solve problems. In school as well as the "real world," a person's level of understanding is measured by his or her problem-solving ability. The flip side of this is that you cannot learn new concepts—deeply and meaningfully—without putting in considerable effort using those concepts to solve problems. As we discussed in Chapter 2, we build up our knowledge by constructing a network of links between concepts in our long-term memories, and it takes solid intellectual activity—problem solving—to form these links.

The remainder of this chapter examines the issues of learning and problem solving and offers a practical approach, based on the ideas of knowledge representation from Chapter 2, to becoming more proficient at both. Section 3.2 first looks at the stages of the learning process and the nature of expertise. Section 3.3 then presents a system for assessing your level of understanding, called Bloom's Taxonomy, that's based on your ability to solve progressively more difficult problems. Section 3.4 presents a framework for studying, based on recommendations from the National Research Council, that will help you build a deep level of understanding. Subsequently, Section 3.5 presents a related framework with a general approach to solving the kinds of engineering problems that you're likely to encounter both in school and in a professional field, while Section 3.6 works through a detailed example using this methodology to calculate the amount of the greenhouse gas carbon dioxide that a typical automobile produces per year. Section 3.7 looks at solving larger problems, in this case a semester-long senior design project. Finally, Section 3.8 provides a listing of rules of thumb for solving problems, called *heuristics*, that suggest ways of getting "unstuck" on difficult problems.

3.2   EXPERTISE AND THE LEARNING PROCESS

Fundamentally, the practice of engineering is about bringing human ingenuity to bear in solving technical problems. The purpose of an engineering education is to enable students to develop as competent, and eventually expert engineers. What does it mean to be "expert" at something? The Oxford English Dictionary defines the adjective *expert* as "trained by experience or practice, skilled, skilful." We recognize experts in a field by their ability to come up with effective solutions to difficult problems quickly. Novices, by contrast, take much longer to solve problems than experts, and their solutions tend to be less insightful and complete.

As you begin your study of engineering, however, here's the bad news: there's no shortcut or easy path to becoming an expert, no matter how bright you are. Research has demonstrated that even among prodigies, it takes approximately ten years of hard work to become a world-class expert in a field. Mozart began composing music at age 4, but didn't produce a critically acclaimed masterpiece until he was a teenager. The same has been observed among chess grandmasters. The late Bobby Fischer, one of the most dominant chess champions, first learned to play chess at age 6, but did not reach grandmaster status until age 15. Ten years, of course, is more than twice as long as most students spend studying for a bachelor's degree in engineering, and in fact, by ten years after starting college, many professionals are on their *second* jobs. It's important to realize from the outset that your professional education will still be in its early stages when you complete a bachelor's degree, making it important to develop good habits and effective strategies for learning while you are still in school. This brings us to the good news: research has also shown that it's possible for people to learn strategies for organizing their thoughts that can make them dramatically more effective learners and problem solvers.

Research indicates that experts in a field are able to store and use approximately 50,000 pieces of information, including not only isolated facts, but also familiar patterns. A typical college graduate, who may be considered an "expert" at written and spoken English, has a vocabulary of between 50,000 and 100,000 words [DA94]. What distinguishes expert from novice behavior, however, is not the number of facts stored in long-term memory, but rather the ability to access and use them. Experts are far more adept than novices at locating information pertinent to solving a problem and then applying it. Problems that are complex to a novice may appear to an expert as simple as single-digit arithmetic is to us, and experts often come up with an answer so quickly that they can't even explain how they did it.

The process of learning a skill can be broken down into three stages, which we will call the *exposure* stage, the *association* stage, and the *automatic* stage. To illustrate what occurs during each of these stages, we'll consider the example of Billy, a child learning how to read. During the exposure stage, the learner inputs sensory information—sights, sounds, etc.—to short-term memory and makes the first few tentative links between the new information and existing concepts in long-term memory. When Billy is first learning to read, during the exposure stage he might be introduced to the shapes of individual letters and their sounds, and be encouraged to link these ideas to familiar objects, such as "d is for duck." His teacher may even draw the letter d' to look like a duck to help form the link. During this exposure stage, the ability to perform a skill is very low and quite laborious. Imagine Billy in this stage of learning trying to sound out a word: the child may know, tentatively, the sounds associated with each letter, but needs to recall these bits of information from long-term memory one letter at a time. Since it takes 1–2 seconds to recall a fact from long-term memory, the child has great difficulty linking the sounds together to form a recognizable word.

During the association stage, the learner improves his or her ability to access and use a particular concept by associating it with other concepts. Two things happen during this stage: first, new links are formed between concepts, making isolated facts part of larger chunks of information, and second, existing links become stronger as they are exercised. When learning to read, during the association stage, a child becomes more skilled at sounding words out and also begins to recognize entire words or parts of words, or anticipates a word from its context in a sentence—sometimes incorrectly. It is during the association stage that skills really begin to develop. The most important point about this part of the process, then, is that *making the associations demands substantial effort from the learner*. In order to build and strengthen associations, you need to practice applying the concept in a variety of contexts repeatedly, which requires dedication. Without an active effort, you won't make certain associations with certain of these skills.

As a learner continues to practice applying concepts during the association stage, her or his skills become more and more automatic. Individual facts are tightly bound up in higher-level patterns or chunks of knowledge. While during the exposure stage a learner is very aware of the process of locating each individual fact, during the automatic stage a learner performs a skill without consciously thinking about it, and often can't easily explain how or why he or she performed this skill. This is how nearly anyone reading this sentence reads; the words on the page seem to instantaneously take on meaning, and when we are deeply engrossed in reading, we don't even notice that we are looking at letters or even words.

3.3 WHAT DO YOU KNOW? LEVELS OF UNDERSTANDING

The goal of a learning process is to gain a deep and meaningful understanding of a body of knowledge. But what does it mean to have a "deep understanding" and how can we tell when we acquire it? These were among the questions that Benjamin Bloom, David Krathwohl, and a large committee of colleagues posed following the 1948 Convention of the American Psychological Association, in their research to classify the goals of the educational process. Their results were first published in 1956, and described a hierarchy of learning outcomes, ranked from the simplest to the most complex, known as *Bloom's Taxonomy*. Since the introduction of Bloom's Taxonomy, educators have widely used it as a tool to develop their teaching plans and evaluation strategies. In this section, we explain Bloom's Taxonomy to give you an "insider's advantage" on obtaining better results out of learning efforts, in particular:

- as a way to evaluate your own level of understanding of material,
- as a way to gain insight into the types of questions you may see on assignments and exams, why the questions are posed as they are, and what types of solutions professors are looking for,
- as a prelude to developing a strategy for studying that promotes good learning.

Bloom's Taxonomy, shown in Figure 3.1, defines six levels of understanding. They are ordered from the simplest behaviors that demonstrate only a shallow understanding, to more complex behaviors that demonstrate a deep understanding of a body of knowledge. Each successive level in the taxonomy builds on the previous ones, and a student should attain a given level of understanding before moving on to the next.

The following sections describe the levels in Bloom's Taxonomy, with examples of the kinds of tasks and questions that test knowledge at each level, and key words that help identify which level a test question might address. While Bloom's Taxonomy is a valuable tool for assessing levels of understanding, first keep in mind several important caveats:

- The levels in the taxonomy assess a person's understanding of the material in a particular domain—they do *not* assess a person's overall cognitive ability.

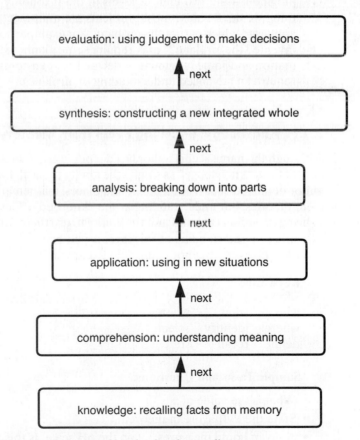

Figure 3.1   Bloom's taxonomy of levels of understanding.

Everybody has areas that they understand very well, and areas that they don't. While this may seem obvious, one of the dangers of using an assessment scale such as Bloom's Taxonomy is that some students may associate a low "score" with her or his own intelligence. This may lead to frustration and make learning more difficult. Be careful not to fall into this trap. On the other hand, if you're having trouble understanding something and it slows down your progress, be sure to put in the extra effort to improve, such as by asking questions.

- While "comprehension"—defined as the ability to understand the meaning of a concept—is low on the scale, don't forget that some concepts really are hard to grasp! In many situations, we need to be content with a *working understanding* of a concept that we can apply without fully understanding it. In his famous *Lectures on Physics*, Nobel Laureate Richard Feynman said about energy: "It is important to realize that in physics today, we have no knowledge of what energy is." Nonetheless, we're able to apply a partial understanding of energy concepts very effectively to many practical situations, even in beginning science and engineering courses. The notion of a working understanding applies as we move up to the other levels of the taxonomy as well.

- While the sample tasks and questions in the sections below provide examples that are focused on testing specific levels of understanding, in practice, solving any problem requires skills at all the levels. Planning strategies and exercising judgment—the two highest levels in the taxonomy—are part of almost everything we do. Because most problems are multifaceted, we must ask which skill applies to which part of the problem. For example, to answer the question "Who invented the airplane?" may require some planning to obtain background information and some judgment to decide between competing claims, but it doesn't require an advanced understanding of airplanes.

3.3.1 Knowledge: Recalling Facts from Memory

Although the name seems to imply a deeper level of understanding, Bloom's Taxonomy defines "knowledge" as simply the ability to recall facts. These facts may include minor details such as names, dates, and places, mathematical formulas, or even complete theories, but understanding at the "knowledge" level requires only the ability to bring these facts to mind, not the ability to use them for a higher-level purpose. An example of understanding at this level is you being able to recall Einstein's famous equation $E = mc^2$, but not being able to explain it.

Keywords:

who	what	when
where	identify	select
define	recall	state

Sample Tasks and Questions:

- What is the value of π
- State the Pythagorean theorem.
- A given liquid measures 3.2 on the pH scale. Is this liquid an acid or a base?
- What are the parts of a concept map?

3.3.2 Comprehension: Understanding Meaning

The next level, "comprehension," is understanding what something means. Comprehension constitutes explaining a concept or translating it from one form to another, such as rewriting a word problem as an equation.

Keywords:

state in your own words   summarize   explain
paraphrase   classify   what is the meaning of
interpret   give an example of   what part doesn't fit
which is the best explanation

Sample Tasks and Questions:

- In your own words, explain a given concept map.
- Draw a concept map that corresponds to a given paragraph description.
- A triangle has sides A, B, and C with lengths a, b, and c, respectively. Given that $a^2 - b^2 + c^2 = 0$, which side of the triangle is the hypotenuse?
- Write the expression $a^2 + b^2$ as a spreadsheet formula or Matlab statement.
- Give an engineering example that illustrates the expression, "an ounce of prevention is worth a pound of cure."
- Give an English description of the different regions of a graph of current versus voltage for a given electrical component.

3.3.3 Application: Using in New Situations

Application is defined as taking information that you have learned and using it in a new situation. In an "application" level of understanding, you recognize that a particular rule, law, or equation you learned in class can be used in solving given problems, whether on a homework assignment, in a lab, or in the workplace. Application, of course, presumes the lower levels of understanding; you can't apply a concept unless you can both recall and comprehend it!

Keywords:

construct   demonstrate   predict
explain how   what would happen if   show that
find, given assumptions

Sample Tasks and Questions:

- Given a right triangle with a hypotenuse of 10 and a width of 8, what is its height?
- What would happen if a wire were connected directly across a battery?
- How high would a projectile travel if launched vertically from earth with an initial velocity of 30 m/s? Assume negligible drag.

3.3.4 Analysis: Breaking Down into Parts

Bloom and his colleagues define the next level in the taxonomy, "analysis," as the ability to break a concept down into parts. Analysis requires the ability not only to comprehend individual concepts and to be able to apply them, but also to understand the relationships between concepts. An understanding at the analysis level also implies the ability to make inferences from a set of facts. The intuition that enabled us to decompose the concept of a car into a power train, electrical system, and passenger environment represented a transition from the comprehension and application levels to the analysis level of understanding. On the other hand, simply recalling the breakdown that we used for the car as a fact is a knowledge-level understanding, not analysis.

Keywords:

break down	what ideas apply	compare
contrast	what is the relationship between	what conclusions
decompose	what is the justification	what might cause

Sample Tasks and Questions:

- What is the rationale behind making the cross-section of an I-beam shaped like the letter "I"?
- If nothing happens when you turn the ignition key in a car, what might be the cause?
- The book *The Soul of a New Machine* by Tracy Kidder describes the development of a computer system at the Data General Corporation in 1979. Draw a concept map that illustrates the main parts of the development team and the relationships among them.
- Given a set of data indicating the amount that a sample of a material stretches when forces of different magnitudes are applied across its length, determine what the material might be.
- What are the main differences among a JPEG, GIF, and bitmap image file?

3.3.5 Synthesis: Constructing a New Integrated Whole

Synthesis is the ability to put parts together to form a new, integrated whole. There are many possibilities for this new "whole," such as a plan, report, proof, story, or design. Synthesis draws heavily on creativity and the ability to put together new structures. Drawing a concept map for a new design, as opposed to drawing a map that analyzes an existing design, requires an understanding at the synthesis level. Synthesis is tightly coupled with analysis, and a person must have a strong understanding of analysis if the object being synthesized is to serve its purpose effectively. In the synthesis process, a person will often come up with a new way to put something together to solve a problem—an "aha" moment—and then spend a great deal of time analyzing the proposed solution to determine if it is satisfactory before moving on to the next creative step.

While synthesis is a high-level skill, solving almost any problem requires some amount of synthesis, to the extent that we need to plan a strategy for coming up with the solution. For example, before solving a typical "application" problem, such as finding the length of the hypotenuse of a right triangle, you first need to develop a simple plan for attacking it, beginning with comprehending the problem and ending with clearly presenting the result.

Keywords:

create	compose	design
devise	develop	solve
formulate	prove	plan

Sample Tasks and Questions:

- Develop a strategy for studying for an exam.
- Design a lightweight truss structure that will span a distance of 10 m and will deflect by no more than 1 cm when a load of 100 kg is applied at its midpoint.
- Write an essay that either defends or disputes the use of genetic engineering in food production.
- Prove that the equation $x^n + y^n = z^n$ has no nonzero integer solutions for x, y, and z when $n > 2$.

3.3.6   Evaluation: Using Judgment to Make Decisions

Evaluation is making a judgment about the value of an idea. At first, it may not seem clear why evaluation is the highest level of understanding in the taxonomy—after all, isn't it more difficult to synthesize a new solution than it is to choose which of two alternatives is better? The answer to this question is that the act of choosing alone isn't evaluation, it's just the application of a rule. Answering the question "Which *costs more*, a house of straw or a house of bricks?" does not require evaluation in the sense of Bloom's Taxonomy. Instead, it involves the ability to analyze the cost of a house by applying certain formulas to the cost calculation, then applying the rule of which number is bigger. On the other hand, answering the question "Does it *make more sense* to build a house of straw or a house of bricks under a given set of circumstances?" does require evaluation. And depending upon the circumstances, either house may clearly be a better option, or the choice may be murky.

The main idea that separates simple comparison from evaluation is *judgment*, which involves managing *risk* and *uncertainty*. Unlike comparison, evaluation requires the ability to synthesize different possible scenarios, analyze their likely outcomes, and then pick one. Judgment decisions also have consequences, ranging from mild to severe. For example, if one of the key considerations in building a house is protection from predators, then the choice of a house of straw could be fatal. On the other hand, in a mild climate where other building materials are scarce, a house of straw may be the best possible alternative.

Keywords:

judge	optimize	what is the best
decide	appraise	critique
defend	evaluate	what is most appropriate

Sample Tasks and Questions:

- Which move should you make next in a game of chess?
- A startup company has created a new technology for developing computer chips that they claim has higher performance and is less expensive than existing technologies. Should your company use this new technology in an upcoming product? What are the risks involved?
- A friend has told you that he has a copy of tomorrow's exam. What should you do?

3.3.7 **Social and Societal Responsibilities of Decision Making**

The ability to make a sound evaluation is one of the most important requirements for technical leadership. In design teams that work well, each team member trusts that the other team members will make good decisions regarding their parts of the project. Conversely, if a team member doesn't have an evaluation-level understanding of his or her part of the project, it is extremely unlikely that the project will go smoothly.

For all but the most routine assignments, however, it's actually likely that some, if not all, members of a team will have weaknesses as decision makers. Even if a person is highly experienced in a task area, there is still almost always some degree of risk and uncertainty associated with each decision. In class projects in school, furthermore, all the team members generally have very little experience in the task at hand, and may have only a tenuous grasp of the material at even the synthesis or analysis level of understanding. Just because you may not be as prepared as you would ideally like before taking on a task, it does not necessarily mean that the task is inappropriate. Instead, it is important that you anticipate that there will be bumps in the road and you plan accordingly. The ability to make good decisions only comes with experience, and some of the most valuable learning experiences happen when things go wrong.

3.4 GETTING GOOD RESULTS FROM YOUR LEARNING EFFORTS

Armed with background on both the learning process and how to measure success, we now focus our attention on some tips on how to get the best results from the efforts that you put into learning. Following these guidelines should help you advance to deeper levels of understanding, and may also benefit you by improving your grades in school!

Our recommendations are based on the results of a multi-year study in the late 1990s and early 2000s conducted by the National Research Council (NRC) on how people learn and on best practices for teaching history, mathematics, and science, particularly at the K–12 levels [BBC00] [DB05]. Their work focused on three fundamental principles, quoted below:

1. Students come to the classroom with preconceptions about how the world works. If their initial understanding is not engaged, they may fail to grasp the new concepts and information, or they may learn them for the purposes of a test, but revert to their preconceptions outside the classroom.
2. To develop competence in an area of inquiry, students must (a) have a deep foundation of factual knowledge, (b) understand facts and ideas in the context of a conceptual framework, and (c) organize knowledge in ways that facilitate retrieval and application.
3. A "metacognitive" approach to instruction can help students learn to take control of their own learning by defining learning goals and monitoring their progress toward achieving them.

While these recommendations were directed at teachers, we believe that there are important messages for students in them as well. In the following sections, we consider what each of these guidelines suggest for you, the student.

3.4.1 Get Ready to Learn

The first recommendation in the NRC study focuses on engaging what you already know as you move on to learning new material. One of the main implications of this engagement is that you need to prepare before studying new topics.

Be Mentally and Physically Prepared   The first thing to prepare for learning is your attitude. Before you can engage in the learning process, you have to be in the right frame of mind. This isn't as easy as it may sound, given the busy lives and many distractions that most students face. Recall that during the first step in the learning process, the exposure step, you only have at most a few seconds to transfer what you see and hear from short-term memory to long-term memory, so it's critical to be focused and alert. It's also critical not to clog up your very limited short-term memory capacity with thoughts irrelevant to the task at hand.

It's also important to minimize any physical impediments to learning. Arrive at class on time and sit where you can easily follow the lecture. Bring whatever materials you need for class with you. Get a good night's sleep. You could come up with a list of suggestions along these lines, all of which are common sense, but also require planning and effort. If you can adjust your schedule to meet these basic conditions, it can tremendously impact your learning capacity, making it possible for you to juggle your classes with the rest of your activities.

Take Stock of What You Know   The first two steps in repairing a flat bicycle tire are finding the hole and then roughing up the surface around the hole so that the patch will stick better when you glue it on. An analogous approach applies to learning: to prepare for learning new material, you should first review what you already know that's relevant to the topic, have a sense of what you don't know, and "roughen up" some of the sites where you'll be attaching new knowledge by exercising related neural pathways.

Even a small amount of effort spent in reviewing old material and skimming new material can go a long way towards making a good connection between the two. Your mind has a natural tendency to wander among connected concepts once you stimulate it with a particular idea. One of the secrets of taking advantage of this effect is to allow enough time for it to take place. Try taking ten minutes before you go to bed, before you take a shower in the morning, or over lunch, to review your notes from your last class and skim through material for your next class and observe the results. You'll be amazed at the difference this practice can make in your understanding of the material with very little conscious—but plenty of subconscious—effort expended.

Acknowledge Misconceptions   In addition to taking stock of what you know, it's also important to acknowledge that there will be things that you *think* you understand, but don't—at least in the current context. For example, when told that the earth is round, children will sometimes picture it as a pancake shape, rather than as a sphere [VB89] [BBC00]. Such misconceptions can impede learning, and you need to be open to them.

3.4.2 Building a Good Structure for Knowledge

The second set of recommendations in the NRC study focus on the kind of structure that you should build to support your knowledge base. The first point in this area is that you have a "deep foundation of factual knowledge." The notion of a "deep foundation" itself has two implications: first, that you have access to *enough* factual information, and second, that you have a deep level of understanding of that information. The next points state that the material should be logically organized for easy retrieval and application. Figure 3.2 illustrates a recommended organization whose construction addresses all these considerations. The key features of this structure include:

- The new information is mapped hierarchically. This organizes the information into manageable pieces for easier retrieval and indicates which concepts are central to the topic and which are peripheral.
- The new concepts are linked to concepts in your prior knowledge. This extends existing pathways to provide access to new ideas, and vice versa, and serves to clarify and explain them.
- The *process* of organizing the information fosters a deep understanding at the higher levels of Bloom's Taxonomy.

In the remainder of this section, we describe an approach to building this structure, including ideas for coping with information overload, tips for organizing the new information into a hierarchy, and strategies for deepening your understanding. Earlier, we spoke of the importance of being mentally and physically prepared for learning, and the same advice applies here: specifically, you need to be prepared to do some hard work. Building a good knowledge structure and developing expertise in a field requires active learning—reading, writing, asking questions, making mistakes, and being willing to change. While some approaches to learning are more effective or efficient than others, there are no shortcuts, and a passive approach to learning won't lead to a deep level of understanding. On the other hand, if you put in the effort, you *will* see results.

Information Overload   Between June 2, 2004 and November 30, 2004, Utah software engineer Ken Jennings made history, earning $2,520,700 by winning a record

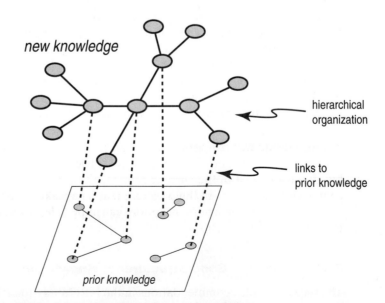

Figure 3.2   Desired shape for structuring new knowledge.

74 straight matches of the television game show *Jeopardy!* With the possible exception of Mr. Jennings, however, the NRC study indicates that most engineers and engineering students would be far better served by understanding a small number of concepts in depth, rather than a large number superficially.

This is an issue that frustrates teachers as well as students. There are a number of forces that push teachers towards trying to cover too many topics in a single course. One is preparing students with all the prerequisites for a variety of courses downstream. Another is standards that may mandate that certain topics be covered. The Accreditation Board for Engineering and Technology (ABET), responsible for accrediting engineering programs in the U.S., recognizes the downside of specifying detailed lists of topics. Instead, ABET puts more responsibility on individual programs—and in the end, teachers—to make wise decisions on what material to teach their students, and more importantly to implement ways of assessing that students have meaningfully learned this material. We hope that a similar logic will prevail at the K–12 levels.

Even if a professor pares the list of topics down to what she or he believes is a minimal set of core topics, and presents them in a logical structure, it may still seem like a torrent of information to you, the student, who doesn't—yet—have the same depth of understanding as the teacher. You should expect this while you're still on the bottom rungs of Bloom's Taxonomy with respect to this material, still struggling simply to recall facts and to understand their meaning. The first thing you need to do to cope with the flood of information is sort through it and determine what's centrally important, and what isn't. The next few paragraphs present techniques for doing so.

In a typical engineering course, information pours in from a variety of sources, including lectures, textbooks, labs, and tutorials. Your challenge is to extract the important information from these streams and assemble it in a structure like that in Figure 3.2. In general, we recommend taking an approach to making sense of the new material that is both active and dynamic. By *active*, we mean using techniques that get you physically involved in the learning process, creating some tangible artifacts that represent your understanding of the material. By *dynamic*, we mean using techniques that let you modify your representation as you progress through the material in multiple passes, gaining a deeper understanding with each pass. Historically, people have used many types of tools for recording ideas, including taking notes, highlighting, outlining, and drawing concept maps. Below, we describe a process for building a knowledge structure that uses a mix of these.

Notes on Notes Taking notes in class serves two purposes. First, it provides you with a written record that you can go back and review later, and more importantly, it gets you actively engaged in the material. Even if you are provided lecture notes by the teacher, *still* take notes on your own, either on the handouts or in a separate notebook.

Before reading a chapter or section from a textbook, briefly skim it to get a sense of its structure, which will help you determine which concepts are central and which are peripheral as you go on to read it in more detail. Also, look at its entry in the table of contents, which also conveniently gives an outline of the chapter. Finally, be sure to read the opening paragraphs carefully, which usually set the context for what lies ahead and often provide a synopsis of the material.

Highlighting is a useful technique for identifying key concepts, and we recommend highlighting—or listing concepts in a separate notebook if you don't want to mark up your book—as you read though a textbook or lecture notes. Focus on highlighting single words or short phrases that capture individual concepts, rather than entire paragraphs, which will be useful later when you build a concept map. While valuable, note that highlighting has its limitations. In terms of Bloom's Taxonomy, highlighting doesn't support an understanding beyond basic knowledge of facts and their meanings. In terms of knowledge structure, highlighting addresses identification of concepts but not linking them with relationships.

As you take notes, either in class or while reading a text, strive to include comments that amplify the basic material. Consider the different types of comments you might include in terms of both levels of understanding and knowledge structure. Notes that either clarify definitions or restate material in your own words reinforce comprehension. Notes that suggest connections to outside material are particularly valuable in achieving deeper levels of understanding, and in building a structure with a rich set of links to prior knowledge. Look for opportunities to include notes of the form, "this reminds me of . . . ," "look back at . . . ," "maybe this is why . . . ," or others that help build links between ideas.

Using Concept Maps   Concept maps are particularly useful tools for organizing ideas because they produce a visual representation that mirrors the logical organization you're trying to achieve. Some people have suggested that concept maps can replace other forms of note-taking, but we believe that they work best in conjunction with other techniques. While concept maps could be used as a form for taking notes in class, in practice this often doesn't work very well, because it is difficult to predict how the map will grow, which makes it difficult to fit on a page. We believe that the best time to draw a concept map is after you have made a first pass through the material and have a basic knowledge of facts and their meanings, but are not yet clear on how they fit together or relate to new situations. Thus, while traditional practices of highlighting and note-taking are both effective and convenient for moving through the first levels of understanding, we'd suggest that drawing concept maps is an ideal tool for moving to the higher levels of application, analysis, synthesis, and evaluation. The following steps outline an approach to building a concept map based on course materials that follows the steps described in Section 2.2.

1. *Preparation.* Decide on a title for the concept map, which may be expressed as a question. Gather your source materials, which may include your textbook, lecture notes, lab notebook, and assignments. Assemble your drawing tools, such as Post-It™ notes or a computer drawing program.

2. *List key concepts.* Try to list concepts from memory, and then go back through your notes and other sources and use highlighted terms to add to the list. Include concepts for each of the references to outside material that you may have written in your notes.

3. *Add initial relationships.* Draw an initial set of relationships between concepts, first without looking back at your source materials, and then adding to them with your notes and text open. At this stage, don't worry too much about the shape of the map you're forming.

4. *Refine the map into a hierarchy.* Identify concepts and relationships that help organize the map into a hierarchy of chunks. In particular, look for opportunities to apply the principles of decomposition and classification as discussed in Section 2.2.3. You may find that you need to add concepts to organize chunks, much as we added the concepts of a "power train" in the car example in Figure 2.14. If you're having trouble labeling the relationships, it probably indicates

that you don't understand them as well as you should; go back to your notes or your textbook and reread the relevant sections. You may find that either renaming or subdividing concepts will help.

5. *Add cross-links.* Look for links between different parts of the map to help reinforce your understanding of the material, much as we found links between the power train and electrical system of a car in Section 2.2 and Figure 2.15. In particular, if you haven't done so already, try to link in key concepts from examples, lab, or projects, using relationships such as "illustrates," "demonstrates," or "is an application of." Many engineering and science professors have received student feedback that laboratory exercises "didn't have anything to do with the course." Sometimes labs aren't as well-conceived as they could be, but in general, the issue is a lack of understanding of the key concepts in the lab and their relationship to other concepts in the course. Labs and design projects provide some of the best opportunities for linking new material to outside or prior knowledge, and often produce some of the most "memorable" pathways for accessing new knowledge. Finally, add a few concepts and relationships that link the new ideas into a broader context. For example, think about the last topic or couple of topics that you covered in the course and form links to the current material. Another idea is to go back and look at the course syllabus and identify cross-links between concepts in this topic area and others that you've already covered.

3.4.3 Metacognition: Monitoring Your Own Understanding

The focus of the third NRC recommendation is "metacognition"—an awareness of your own level of understanding. As discussed in the introduction to this chapter, metacognition is a critical ingredient for your development as an independent learner. There are two aspects to metacognition. The first is a basic understanding of how your "cognitive machinery" works, so that you can use it effectively—the major theme of this entire chapter. The second aspect is carrying on an internal dialog where you continuously monitor your own level of understanding of material and generate new questions that lead to enhanced understanding. In this section, we look at several approaches to assessing your understanding.

Concept Maps and Levels of Understanding   Two of the advantages of drawing a concept map are that it engages you in an internal dialog and fosters a deeper understanding of material. In the last section, we described a process of drawing a concept map from course materials as a study aid, the steps of which address understanding at specific levels. When you work on recalling concepts and relationships from memory, before going back and looking at your source materials, you are testing yourself at the knowledge level of understanding in Bloom's Taxonomy. The act of drawing a concept map itself is restating information in another form, which exercises a comprehension-level understanding of the material. Linking concepts through relationships labeled "is an application of" or "is an illustration of" draws on an application level of understanding, while arranging a concept map into hierarchically organized chunks or breaking a single concept down into component pieces demands an analysis level of understanding. In short, drawing a concept map not only provides you with a convenient diagram of course material you can go back and study, it also engages you in a process that methodically tests your level of understanding and provokes the kinds of questions that enable you to improve it.

Reciprocal Teaching Discussing ideas with a friend is of course an excellent way to monitor and improve your understanding. Not only do you get direct feedback from another person—as well as his or her added perspective—but discussion also provides a means for *practicing* a dialog that you may later conduct with yourself! The NRC study suggests a particular framework for discussion called *reciprocal teaching* [PB84]. In this approach, you play the role of a teacher, explaining material to another person, who plays the role of a student and asks questions. You then reverse roles. The "lecture" topic could be a synopsis of a class, a textbook chapter, a lab, a homework problem, or any of a wide variety of other subjects. The "student's" questions may focus on points of clarification, but most importantly she or he should ask you to explain why you did something the way that you did. In answering these questions out loud, you may discover alternative, simpler, or better ways of doing something. The student should also ask questions that probe understanding at the different levels of Bloom's Taxonomy, using the sample questions in Section 3.3 as a guide. After engaging in reciprocal teaching sessions with others, try the approach on yourself; you may be surprised at the insight it provides.

You can also combine drawing concept maps with reciprocal teaching. Have you and a friend or small groups of friends independently draw concept maps, and then explain them to each other and as a group try to merge them. This is easiest if you've drawn your maps using PostIt notes. You'll likely come away from this exercise with a shared understanding of the material very different from what you had individually.

Classifying Homework and Test Problems Another "metacognitive" activity that you might try is classifying homework and test problems in terms of both their content and the level of understanding that they address. Which concepts does a given problem test? Does it emphasize particular relationships among concepts? Referring to the sample questions and keywords in Section 3.3, try to determine the level of understanding that a question tests.

When you're stuck on a problem in a homework or an exam, this type of analysis may help you get moving. Researchers at the University of Cape Town, South Africa, for example, cite the multiple-choice exam question in Figure 3.3 that was highly unpopular with students [CDM96]. Student reaction to the question was that "it was impossible to remember the statistics for all the countries that were discussed in class and given to them in handouts," indicating that students viewed this as a "knowledge" question. They were surprised when told that such detailed knowledge was in fact not expected of them, but that they were to examine the table and perform a ranking on the basis of analysis and application of concepts related to the classification of countries into economic categories. The correct answer, by the way, is (3).

3.5 A FRAMEWORK FOR PROBLEM SOLVING

Engineers solve problems. Though right now you may be most interested in the types of problems you'll see in school, you'll learn that many of the academic problems you solve as part of particular courses prepare you for problems you'll encounter in the "real world." Some of these real-world problems may not look like your homework, exams or projects, but they contain many of the same elements. Keep in mind that engineers also need to solve many kinds of non-technical problems in their daily activities, including financial problems, personnel problems, communication

Look at the following table and indicate which countries' statistics are being reported in rows A, B and C.

Country	GNP per capita, 1991 (US $)	Growth rate of GNP per capita, 1980-91	Pop. growth rate, 1980-91	Structures of total employment 1980-85 (percentages)		
				Agricult.	Indust.	Service
A	500	2.5%	1.5%	51	20	29
B	1570	5.8%	1.6%	74	8	8
S. Africa	2560	0.7%	2.5%	17	36	36
C	25110	1.7%	0.3%	6	32	32

(1) A is South Korea; B is Kenya; C is Canada.
(2) A is Sri Lanka; B is Germany; C is Thailand.
(3) A is Sri Lanka; B is Thailand; C is Sweden.
(4) A is Namibia; B is Portugal; C is Botswana.

Figure 3.3 Multiple-choice question reported in [CDM96] that tests understanding at the application and analysis levels, rather than just the knowledge level, to the students' surprise.

problems, and so on. Interestingly enough, some of the concepts you'll learn in solving the technical problems can be applied to these other non-technical areas with success.

One of the most important skills that engineers need to develop is being able to find solutions to problems that are ill-specified, and that may have many possible acceptable solutions. Many of the problems typically encountered in academic settings are crafted to have a unique "right" answer. The instructor has been careful to state the problem in a fashion that clearly identifies the "knowns" and "unknowns," and further, the number of equations that describe system behavior in the model is the same as the number of unknowns. Such problems without any ambiguities are sometimes called *closed* problems. Most real engineering problems, on the other hand, are *open-ended* and have many possible solutions. Open-ended problems are not as well defined or as well posed as the closed academic problems mentioned above. Early in the solution process, there are very few knowns and very many unknowns; most significantly, there are more unknowns than there is information to pin down their values. In order to arrive at a single solution, engineers need to use strategies for closing the problem down. This typically involves making assumptions for some of the unknowns, using engineering estimates or simply educated guesses.

In this section, we present a framework for solving many of the kinds of problems you're likely to encounter, both as a student and in your professional field. While every problem has its own unique characteristics, there is nonetheless a standard approach to solving engineering problems that you need to learn, practice, and then use regularly. The specific framework we present is based on one developed by Wankat and Oreovicz of the School of Chemical Engineering at Purdue University [WO93], which in turn is based on a strategy developed by Woods and colleagues at McMaster University in Ontario, Canada [WWH+75] [WCHW79]. Both of these were motivated by seminal work in teaching problem-solving skills developed by Stanford University mathematician Georg Pólya in the 1940s and published in his book *How to Solve It* [POL45]. The strategy has eight steps—seven working steps and one motivational step—listed below:

0. I can.
1. Define.
2. Explore.
3. Plan.
4. Implement.
5. Check.
6. Generalize.
7. Present the results.

In adopting this approach, you'll not only become a more proficient problem solver, but you'll also learn to present your results in a logical manner easily understood by other engineers—including your professors! The general approach applies to a broad range of problems, from short homework problems to multi-year projects. Some of these problems will be done using simply pencil and paper and will involve only a single technical concept, with results presented on a single sheet of paper. Others may involve many different technical concepts, tools, and methods. Many problems are solved—or their results presented—with the assistance of a computer.

While using this framework will help you stay on track, keep in mind that problem solving doesn't necessarily entail an orderly sequence of steps. There are often misstarts and dead ends encountered and the elements presented below are not necessarily addressed in the orderly sequence listed. One might have to work through them a first time, revisiting and revising certain issues in some of them as a solution is developed. In Section 3.8, we describe a set of rules of thumb called heuristics that will help you get moving when you are stuck.

3.5.1   Problem Solving Step 0: I Can

As discussed in Section 3.4, the first step in learning is being prepared, and this includes having a positive attitude. While some problems may seem impenetrable at first, with perseverance and some help if you need it, you will be able to make progress. Try to view each problem as a challenge, and don't give up too easily.

3.5.2   Problem Solving Step 1: Define

Before you begin trying to solve a problem, you need to fully understand it. First, restate the problem in your own words, calling out what's known and unknown, and draw a diagram or diagrams that help you visualize the problem. Each of these points are further discussed below.

Identify the "Knowns"   Every problem statement provides some amount of information that helps define the problem—what's "known" about the situation. Sometimes this information is stated explicitly, while other times it's implied. A careful reading of the problem—if it's a homework or exam problem—will often divulge information not initially obvious. You may find listing all the information given in the problem statement to be helpful to identifying what's known, but realize that not all of this may actually be applicable to solving the problem.

Identify the "Unknowns"   Every problem has at least one unknown that constitutes the answer to the problem. Since most engineering problems require a series of steps, however, there are other less obvious unknowns along the path to a solution. First you must determine "A," then "A" and other information are used to determine "B" and so on. In such cases, you will typically add to the list of unknowns as your solution strategy unfolds.

State in Simpler Terms   Restate the problem in terms of the knowns and unknowns identified in the previous step.

Develop a Diagram, Schematic, or Visual Representation for the Problem
Some of the most useful tools available to you are simple sketches, illustrations and drawings that pictorially represent aspects of the problem at hand. In some cases, such as a schematic for an electrical circuit, it would be extremely difficult to understand the problem without them. In other cases, such as a concept map, they help you visualize the structure of the problem and its solution. These pictures and other representations will also be most useful in communicating what you have done to others who will look at your solution. You can use these to define terms or symbols you may use in your solution and in many cases they can help you to establish important relationships between elements of the problem.

3.5.3   Problem Solving Step 2: Explore

This step in the problem-solving process can be viewed as a pre-planning step, where you pause to think about what the problem is actually asking you to solve, what additional information you might need, and general strategies that might be applicable.

Does the Problem Makes Sense?   Once you've defined your interpretation of the problem, look back over it and decide if it makes sense. Does the known information provided seem to be consistent with the unknowns that you're asked to find? Are there any "red herrings"—information provided that may lead you down a false path? If the problem doesn't make sense, go back and review it, looking for something that you may have missed. This is the time to ask if the problem even needs to be solved; in complex problems that are broken down into many pieces, for example, you may be able to reuse a result from an earlier solution rather than solving the problem again from scratch. For example, it's usually more cost-effective to buy a wheel from a parts supplier than to re-invent one.

Assumptions   In their field of work, many engineers rely on the use of *assumptions*. Often, the statement of the problem isn't complete or some information hasn't been provided with the detail you might desire. Even though some information may be "missing," it is still necessary for you to be able to proceed with the solution of the problem. The fact that not all the information will always be available adds an element of uncertainty in the problem-solving process. Learning to deal with this uncertainty is important and requires engineers to carefully consider the results of their problem-solving efforts, as mentioned below.

Assumptions may be related to the physical properties of certain materials or the influence of certain physical phenomena, such as friction. Knowing when to make assumptions is a skill you learn with experience, and the only way you can develop that experience is by solving problems. So the more problems you solve, the more experience you gain and the more effective you will become. Knowing when and how to make assumptions requires a sound understanding of the relevant physics and mathematics, so as you gain a greater insight into a particular discipline, you will become more skillful at making assumptions. In practice, few engineering problems have unique, exact solutions. This contradicts what many students have grown to expect from their problem-solving experiences in high school—so prepare yourself for this uncertainty.

Note that it's always important for you to explicitly state your assumptions when you document your problems.

What are the Key Concepts and Possible Approaches?   Once you have established the overall objectives, you need to begin to formulate the solution by attempting to classify or characterize the problem. In the case of the kinds of problems you encounter in a specific course, you usually have a pretty good idea of the topics you are currently studying. Developing the ability to look at a problem and then say "that problem deals with X" is a skill that you are working to develop and will only be able to do with practice.

As the problems become more involved—and in particular as you begin working on design problems—you will find that understanding the concepts and selecting the approach becomes more difficult as a given problem will most likely involve multiple concepts. Does it involve a conservation law? Is it conservation of energy, momentum, mass, species, etc.? Again, your ability to answer these questions will improve with experience, and you should be prepared to learn from each problem you solve.

What Level of Understanding is Tested?   What level of understanding does the problem address according to Bloom's Taxonomy? The examples in Section 3.3 may help you decide.

3.5.4   Problem Solving Step 3: Plan

Planning is determining the sequence of steps that you will take in solving a problem *before* committing resources to its implementation. Novices—and even experienced engineers—often try to jump past the planning step. Hoping to finish an assignment sooner, students may start plugging numbers into seemingly relevant equations, typing statements into a computer, or connecting wires, typically with disastrous results. Rushing to implement without planning almost never saves time and almost always produces sloppy results that are notoriously difficult to debug. Yet while it averts trouble later on, planning is also the most taxing step in the problem-solving process, drawing on your ability to analyze the problem, to synthesize a process for solving it, and to evaluate whether or not this process is likely to work. Because of uncertainty, plans must often be changed as you encounter unforeseen roadblocks, and you may have to start and stop repeatedly as you follow a few false paths. The disorder of the planning process usually isn't apparent, however, when you see the final form of a worked solution. Sometimes when you look at an example problem in a textbook or lecture and see the steps taken, you might ask yourself, "I see how they did that, but why those steps, and why in that order?" Well, the actual solution may *not* have followed that order, but with the hindsight provided by working through to the solution, the author felt this was the best way to present the concept, sparing you the details of failed attempts. As you gain more experience in problem solving, you too will develop the ability to work through more complex problems, with the confidence that occasionally reaching dead-ends is a part of the process.

As discussed in Section 2.1.3, problem solving typically involves chaining together a sequence of steps between what is known and what is unknown, or between some initial state and a goal. One very effective tool for developing a plan is to draw a concept map that identifies the initial state and the goal, and then finding a path through intermediate steps that connects them. Drawing a concept map to help planning serves several purposes, including:

1. It gets your ideas down on paper as you work with them; remember that you can only keep a few concepts in short-term memory at a time.
2. It stimulates the pathways in your memory, which may help you find the concepts and methods that you need to solve the problem.
3. It assesses your understanding of the problem; if you are having difficulty drawing a concept map, you probably don't understand the problem well enough yet to be able to solve it.
4. It provides a nice form of documentation for keeping track of the status of your solution and for presenting your plan to others.

Developing a plan this way is similar to the way that we found a flight itinerary between two cities, or found a solution to joining two strings hanging from the ceiling in Section 2.1.3. In general, the map should have a hierarchical shape, with known information and assumed values at the leaves, the main goal at the root, and subgoals in between, as shown in Figure 3.4.

Now that we've identified the general shape of a plan, the next issue is how to determine its content. While the plan as a whole for solving a particular problem may be a new and unique creation, the elements of the plan typically are not. In Section 2.1.3, we saw how small chunks of "how to" or procedural knowledge, each solving a simple problem, could be chained together to solve complex problems. Given

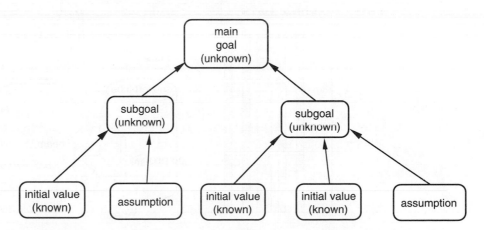

Figure 3.4   General shape of a plan for solving a problem.

that we have vast stores of procedural and declarative knowledge in our long-term memories, however, the question becomes, how do you find the bits of information that you need?

The basic approach is the same that we recommended in Section 3.4 for learning new material, namely, to run through an internal dialog with the goal of eliciting the information. We can frame this dialog around a set of "rules of thumb" or *heuristics* that described general problem-solving techniques. In Section 3.8, we provide a list of heuristics with examples of their application. These include suggestions such as:

- divide and conquer
- work forward
- work backward
- try a simpler problem
- take a break

Generally speaking, these heuristics were discovered by experienced problem solvers who simply noted that they "worked," and added them to the list for that reason. Upon deeper inspection, however, you will see that many may be explained in terms of aspects of the cognitive model described in Section 2.1, and address ways of getting better performance out of your cognitive machinery. Others address your understanding of the problem itself, and finding ways to transform it so you can more readily relate it to patterns you already know.

Example 3.1 illustrates the development of a plan for painting a water tower.

Example 3.1	**Painting a Water Tower**

Painting a Water Tower

Develop a plan for determining how many gallons of paint would be needed to apply a single coat to the water tower shown in Figure 3.5. The water tower consists of a tank on top of a tower. The tank is a sealed cylinder. The tower consists of eight

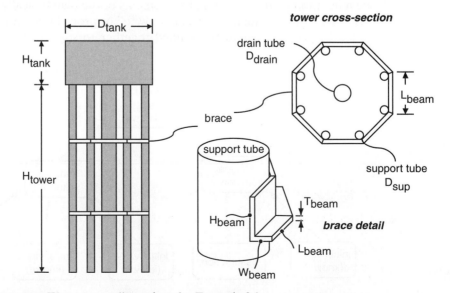

Figure 3.5   Water tower dimensions for Example 3.1.

support tubes arranged symmetrically around a wider drain tube. Two octagonal braces made from L-beams (that is, beams shaped like the letter "L") surround the set of support tubes at levels one-third and two-thirds up the height of the tower.

Solution The initial state in this problem is that we know the dimensions of the water tower. The goal state is to know the quantity of paint required to cover the tower. In developing the plan, we'll mainly use two heuristics: "working backwards" from the goal toward the initial state, and "divide and conquer" to break the tower into component parts.

Figure 3.6 illustrates the development of the plan in three steps. The first step simply states that finding the amount of paint depends on the surface area of the water tower and the coverage of the paint, which is the number of gallons required per unit area. The second step of the plan continues to work backwards, breaking the tower down into component pieces, and determining that the paint coverage should be obtained by looking it up from a supplier. The final step adds the operations for calculating the areas of the parts of the tank from the given dimensions. It also identifies the geometric shapes of the components, and notes the formulas needed to calculate the areas.

3.5.5 Problem Solving Step 4: Implement

Often this step is what a student considers as "solving the problem," but for us, this is the *fourth* step in the process, which begins *after* planning. This is the implementation step, where you solve the equations, write the computer program, wire up a circuit, or assemble a structure. While planning makes implementation go much more smoothly, as discussed earlier, you should still expect occasionally to encounter problems during implementation that force you to go back and change your plans. In solving complex problems, there is often a fair amount of iteration between the planning and implementation steps   the trick is to do a good enough job in planning so that these iterations are not too costly in terms of time or money.

Even with advance planning, you should still count on making mistakes as you implement your solution. For this reason, it's extremely important to work methodically during implementation. This will not only minimize the likelihood of making mistakes, but will also make it easier to find them when you do. Taking small steps and documenting your work as you proceed are very important. When working on numeric problems, solve equations symbolically before substituting numbers.

3.5.6 Problem Solving Step 5: Check

Once you've produced an answer to a problem, you still need to assure yourself and others that it's acceptable. *You* are the first person who must check your answer. For homework problems where there are answers in the back of the book, checking your results is easy; in practice, however, things are rarely that simple, and you'll need to learn to rely on a variety of techniques to determine whether or not an answer is reasonable before you go on to use or submit it.

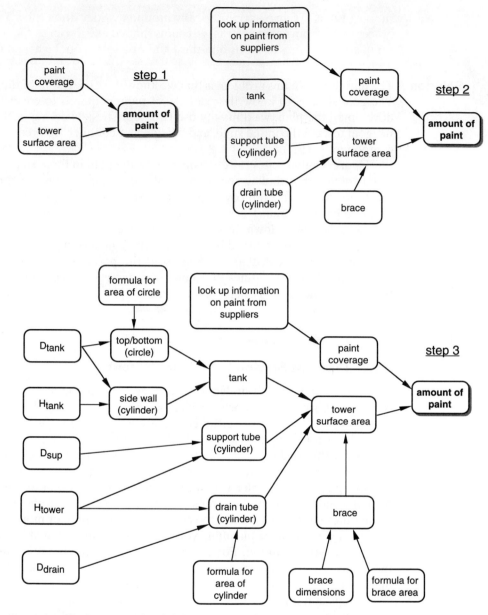

Figure 3.6 Refinement of a plan in three steps for determining the amount of paint for Example 3.1.

Sanity Checks The first step you should take in checking an answer is to see if it even makes sense, with what is commonly called a *sanity check*. For example, if you're solving for the time that it'll take for a projectile launched from the ground and the result turns out to be negative, something has probably gone wrong. Sometimes the problem is a typographical error or an improper entry into a calculator. Sometimes it's an algebraic mistake. Other times, it may be that you used a principle or assumed conditions that are not applicable to the current situation. In any case, you will need to retrace your work, perhaps all the way back to the "define" step, to identify and correct the problem. Careful, detailed documentation of your work is especially important here.

Test Cases   Another approach to checking an answer is using *test cases*. A test case is a known solution to a particular instance of a general problem. For example, suppose that you've been asked to develop a set of equations that describe the position and velocity of a projectile after it has been launched with a given initial velocity in a given direction. One very simple case to test with your equations is that at time zero, the distance traveled should also be zero. Another test case would be that if you assume that there are no forces other than gravity acting on the projectile after launch, then its final velocity when it reaches the ground should be the same as its initial velocity. Test cases are especially important when producing tools such as computer programs. For example, if you have written a computer program to simulate the flow of current in an arbitrary electrical circuit, you might test that program by simulating a circuit that is simple enough to analyze by hand. Software companies typically maintain large suites of test programs, called *regression tests*, which they run before a new version of a program is released, to ensure that the new and old versions produce the same results from parts of the program unaffected by the changes.

3.5.7   Problem Solving Step 6: Generalize

Getting the answer to a given homework problem may not be nearly as important as developing an understanding of the approach and solution methodology, and their impact on solving other problems. The last step in the problem-solving process, to generalize, is critical to learning from the experience.

During the generalization step, you should consider points relative to both the problem and the process of solving. Some questions you might ask yourself include:

- What specific facts have you learned about the content?
- Could this problem have been solved more efficiently? Were there steps taken that, in retrospect, were unnecessary? For example, was some effect so small that it could have been ignored?
- Could you have applied something you learned in solving this problem to another problem that you'd seen in the past?
- Were there any problems or bugs that you encountered that you should remember, in case you run into them again?

3.5.8   Problem Solving Step 7: Present the Results

Communication and presentation play an extremely important role in engineering practice. Central to this is the fact that engineering is both a *social* and *technical* profession. Engineers must be able to communicate with colleagues in appropriate and precise technical language, but must also be able to discuss their recommendations with a lay audience, including government agencies and clients who make substantial investments based on these recommendations. Engineers produce many kinds of documents—technical memos, reports, schematics, proposals, editorials—each with a different style and emphasis. Cutting across these, however, are several important maxims to keep in mind: show your work, give good directions, and be neat.

Show Your Work. Ultimately, decisions about engineering matters are made by people. Before someone will accept your recommendations, especially in critical situations, she or he will want to know your rationale. When firms bid on a construction project, the potential client considers not only the cost and the schedule presented in the bid, but also looks very closely at where those numbers came from to determine if they are realistic. At any point in a design process, unclear communications could lead to two kinds of bad decisions, both potentially disastrous: good ideas being rejected, and bad ideas being wrongly accepted.

Every homework assignment and project report is an opportunity to improve and refine your communication skills, and accordingly, how clearly you show your work will be an important part of the assessment. Below are some suggestions that should help you present your work:

- Don't just write down numbers and equations; provide enough supporting comments so you—later on—or the person who reviews your work, understands your problem-solving process. Be concise in your statements—no need for complete sentences.
- You must use units with all numbers. If you refer to a parameter that's a measure of length and write down "23.4," it will mean nothing. You *must* write it as 23.4 m (or 23.4 in, 23.4 mi, 23.4 km, etc.). Write units next to all numbers; you will likely run into trouble if you try to "do the units in your head." Don't simply assume because all the units are in SI or another system that you can ignore them, because others reviewing your work might not understand your results. Appropriately using units will prove to be a major asset in checking your work. It's possible that an engineer may find more mistakes by "checking units" than any other single available technique.
- Label the sections of your problem and clearly identify any assumptions you are using. Help the person following your work with a roadmap!

Give Good Directions. The product of a design process is a plan for implementation. If you implement a product in the way its design was intended, the description of the plan must be clear and unambiguous. This includes both the language with which the plan is described, as well as accompanying diagrams. There are many, many situations in which engineers are called upon to give directions: to technicians carrying out experiments, to shop foremen in factories, to construction firms, and to computers. A computer program is simply a set of directions, written in a specialized language, that direct the operation of a machine. For better or for worse, unlike people, a computer will always do *exactly* what you tell it to, without trying to determine what you really meant. While this sometimes seems harsh, one of the advantages of using computer-based tools in solving problems is that used properly, they can provide clear and unambiguous documentation of your work, since the computer will only interpret the program in one way.

Be Neat. In any engineering document or presentation, neatness counts. Not only is a sloppy document difficult to read, it also conveys the impression—right or wrong—that the underlying ideas are disorganized, as well. For handwritten documents, work on your penmanship and print if necessary. For word-processed documents, take care that included figures are clear, with legible fonts.

Some more complex problems require that results be presented in tables, plots or other visual forms. All graphs should be labeled and contain appropriate legends and units on the axes.

3.6 HOW MUCH CO_2 DOES A TYPICAL PASSENGER CAR PRODUCE?

Greenhouse gas emissions are a serious problem that threatens our environment by depleting the ozone layer, which in turn leads to global warming. Automobile emission of carbon dioxide, CO_2, accounts for a major portion of this problem. In this problem, we explore how much a single car contributes to the problem. Specifically, we want to estimate the mass of CO_2 that a typical passenger automobile produces per year. The following sections work through a solution to this problem.[1]

3.6.1 Define

In this step, we need to determine what's known and unknown in our problem, and define the goal of the solution.

Knowns Very little is known initially from the specification of the problem, other than that the car and driver scenario we should analyze is "typical." The assumptions we'll need to make to define the problem further should be consistent with this.

Unknowns The only unknown specified directly is the mass of CO_2, which corresponds to the main goal of the problem. As we plan our strategy for solving the problem, we'll add other intermediate unknowns corresponding to subgoals to the list.

State in Simpler Terms What is the mass of CO_2 produced as a result of burning the volume of gasoline that a typical car consumes in a year?

Develop a Diagram, Schematic, or Visual Representation for the Problem Figure 3.7 shows an initial concept map for the problem, which we'll later refine into a detailed plan to solve it.

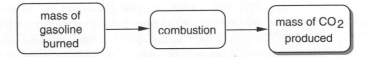

Figure 3.7 Initial diagram for the problem of determining the mass of CO_2 produced by a typical car in one year.

---

[1]Thanks to Ed Maginn of the Department of Chemical Engineering at the University of Notre Dame for suggesting this problem.

3.6.2   Explore

The explore step is the "pre-planning" step in which we double-check that we understand the problem, determine assumptions that must be made, and identify key concepts.

Does the Problem Make Sense?   Although little is specified, the problem does make sense. We know that cars do produce CO_2 as a result of burning gasoline; we just haven't determined yet what the relationship is between the mass of each of these substances.

Assumptions   The plan for solving this problem will require a number of assumptions, estimates, and approximations. Below is a list of these, and ways that we might determine their values.

- *What is the mass of gasoline burned?* We can determine this by making assumptions on miles driven per year by a typical driver and the fuel efficiency in miles per gallon for a typical car.
  - We'll assume that the typical car travels 50 mi per day.[2] This equates to 18,250 mi or 29,370 km per year.
  - We'll assume that the fuel economy of a typical passenger car is 25 mi per gallon, or 10.6 km/L.
- *What is gasoline and what does it mean to "burn gasoline?"* We obtained the answer to this question by searching the Internet, on the Newton BBS site—a K-12 educational resource provided by the U.S. Department of Energy and Argonne National Laboratory [U.S05]. Gasoline is a mix of hydrocarbons—molecules composed of hydrogen and carbon. For this analysis, we approximate gasoline as composed completely of isooctane, C_8H_{18}. When isooctane burns completely, it combines with oxygen to produce carbon dioxide and water according to the following chemical equation:

$$C_8H_{18} + 12.5O_2 \rightarrow 8CO_2 + 9H_2O \qquad (3.1)$$

If the isooctane doesn't burn completely, it'll produce other compounds, such as carbon monoxide, CO. For our analysis, we'll assume complete burning, which would slightly overestimate the amount of carbon dioxide.
- We will also likely need some physical constants such as the density of gasoline (to calculate mass from volume), as well as information about the atomic mass of the elements involved in the burning of gasoline.

What are the Key Concepts and Possible Approaches?   The core technical concept in the problem is the one illustrated by the simple diagram drawn earlier in Figure 3.7, which is the chemistry of burning gasoline. Although we haven't figured this out in detail yet, it'll involve being able to balance chemical equations. Other concepts are mass, volume, density, and unit conversions.

What Level of Understanding is Tested?   The problem will test the core technical concepts listed above at the application level. Developing a plan for a solution is one of the more challenging aspects of this problem. Although the planning doesn't

---

[2]This is close to the figures published by the U.S. Department of Transportation [HrR04].

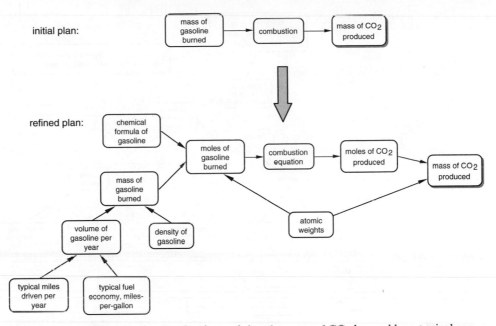

Figure 3.8 Formulation of a plan for determining the mass of CO_2 burned by a typical car per year.

require a deep knowledge of advanced technical concepts, plan formation requires analysis and synthesis. Another difficult aspect of this problem is that it requires a number of assumptions. Even though the assumptions are largely related to simple concepts, such as the average number of miles a person drives per year, this still requires an evaluation level of understanding.

3.6.3 Plan

We developed the plan for solving this problem by refining and extending the concept map of our initial diagram in Figure 3.7. The result, shown in Figure 3.8, was a two-step transformation:

1. First, we did the "easy part" of adding the steps to calculate the mass of gasoline, working backward (backward chaining) from that concept to primary inputs.
2. Next, we refined the steps in the combustion of gasoline to include the steps in calculating the masses of compounds in a simple chemical reaction. We did this by looking back at a few high-school-level chemistry examples using these concepts.

3.6.4 Do It

The first step toward the solution is to make the estimates, approximations, and assumptions we identified earlier. We found that the most convenient way to document these was to add them to the concept map of the plan, as shown in Figure 3.9. The result is a complete plan that includes all the information needed to formulate equations and substitute values.

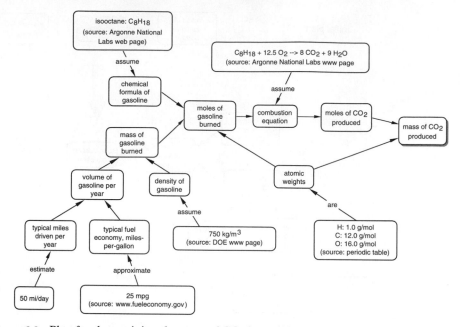

Figure 3.9 Plan for determining the mass of CO_2 burned by a typical car per year, annotated with assumptions, estimates, and physical constants.

The next step is to set up the equations and solve them symbolically. First, we need to define the symbols we used in the equations, which are listed below:

Symbol	Description	Value
distance	typical distance driven per year	29370 km
fuel economy	typical gasoline milage	10.6 km/L
ρ_{gas}	density of gasoline	0.75 kg/L
AW_H	atomic weight of H	1 g/mol
AW_C	atomic weight of C	12 g/mol
AW_O	atomic weight of O	16 g/mol
AW_{gas}	atomic weight of gasoline	unknown
AW_{CO2}	atomic weight of CO_2	unknown
$mass_{gas}$	mass of gasoline	unknown
$mass_{CO2}$	mass of CO_2	unknown
mol_{gas}	moles of gasoline	unknown
mol_{CO2}	moles of CO_2	unknown

Now, the equations. First, we'll calculate the mass of gasoline burned per year. The volume of gas burned per year is equal to the distance driven divided by the fuel economy:

$$V = \frac{distance}{fuel\ economy}$$
$$= 29370\,km \times \frac{1}{10.6}\frac{L}{km}$$
$$= 2771\,L$$

The mass of gasoline is the product of this volume and the density of gasoline:

$$\text{mass}_{\text{gas}} = \rho_{\text{gas}} \times V$$
$$= 0.75 \frac{\text{kg}}{\text{L}} \times 2771 \text{ L}$$
$$= 2078 \text{ kg}$$

Next, we'll convert the mass of gasoline consumed to moles. For this, we'll first need to find the atomic weight of gasoline, C_8H_{18}:

$$\text{AW}_{\text{gas}} = 8 \times \text{AW}_C + 18 \times \text{AW}_H$$
$$= 8 \times 12 \frac{\text{g}}{\text{mol}} + 18 \times 1 \frac{\text{g}}{\text{mol}}$$
$$= 114 \frac{\text{g}}{\text{mol}}$$

$$\text{mol}_{\text{gas}} = \frac{\text{mass}_{\text{gas}}}{\text{AW}_{\text{gas}}}$$
$$= \frac{2078 \times 1000 \text{ g}}{114 \text{ g/mol}}$$
$$= 18228 \text{ mol}$$

Now is the key step in the analysis, where we determine the number of moles of CO_2 produced. According to the reaction for gasoline burning (3.1), each molecule of gasoline burned produces 8 molecules of CO_2. Thus,

$$\text{mol}_{\text{CO2}} = 8 \times \text{mol}_{\text{gas}}$$
$$= 8 \times 18228$$
$$= 145824 \text{ mol}$$

Finally, given the number of moles of CO_2 produced, and its atomic weight, we determine the mass:

$$\text{AW}_{\text{CO2}} = \text{AW}_C + 2 \times \text{AW}_O$$
$$= 12 \frac{\text{g}}{\text{mol}} + 2 \times 16 \frac{\text{g}}{\text{mol}}$$
$$= 44 \frac{\text{g}}{\text{mol}}$$

$$\text{mass}_{\text{CO2}} = \text{mol}_{\text{CO2}} \times \text{AW}_{\text{CO2}}$$
$$= 145824 \text{ mol} \times 44 \frac{\text{g}}{\text{mol}}$$
$$= 6416 \text{ kg}$$

3.6.5   Check

One striking feature of the result is that the mass of CO_2 produced, 6,416 kg, is greater than the mass of gasoline consumed, 2,078 kg. Does this make sense? Checking back on the calculations, we see that 8 moles of CO_2 are produced for every mole of gasoline burned, and that the atomic weight of CO_2 is a little less than one-third of the atomic weight of gasoline. So yes, this answer does make sense.

3.6.6   Generalize

In solving this problem, we made a number of assumptions regarding what's "typical," such as the fuel economy of a car in miles per gallon (mpg), and the distance driven

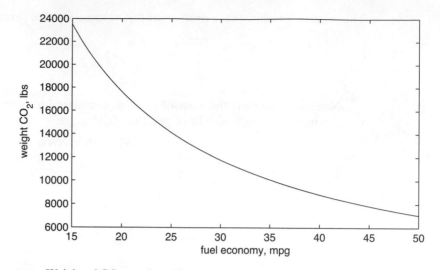

Figure 3.10   Weight of CO_2 produced by a car per year versus vehicle fuel economy, assuming that a car travels a distance of 18250 mi/yr.

per year. One question we might ask is how the mass or weight of CO_2 produced would change if our assumptions changed. One way of addressing this is to vary the values of the assumptions over a range and plot the results. This type of analysis is commonly called a *parameter sweep* and provides different insights into the results of a problem from simply finding a single-point solution. Figure 3.10 shows the results of sweeping the fuel economy over the range of 15 mpg, common in a large sport utility vehicle, through 50 mpg, common in a hybrid. Over this range, the mass of CO_2 varies from over 22000 lbs down to approximately 7000 lbs, still assuming that a car is driven a distance of 18250 miles per year.

3.6.7 Present the Results

In the previous sections, we detailed our thought process in solving the problem; you might consider this as our notes or rough draft. This isn't written to the same standards as a homework solution you might turn in. In a homework solution, you need to provide sufficient detail that the reader/grader can follow your analysis, but generally don't need the extra discussion of how you arrived at this analysis. Figures 3.11 and 3.12 show an example format for a homework solution that's informative, concise, and neat. It presents the solution with the following sections:

Given: What is given in the problem?

Find: What is the unknown you need to find?

Diagram: Provide a diagram. In this example, we use the first-version concept map; in other problems, this may be a force or circuit diagram, etc.

Plan: A synopsis of the plan.

Analysis: The "do it" part of the solution. This should clearly follow the steps in the plan, with brief comments added as necessary to clarify your work.

Comments: This is where you can comment on whether your answer seems reasonable according to your "check" step, or perhaps provide insight that comes from the "generalize" step.

Given   A "typical" passenger car travelling a "typical" distance per year

Find   Mass of CO_2 produced per year

Diagram

Plan

1. Estimate the mass of gasoline burned per year by a typical car
2. Calculate the number of moles of gas burned from the mass
3. Using the chemical reaction for burning gasoline, determine the number of moles of CO_2 produced
4. Calculate the mass of CO_2 from moles

Assumptions

annual distance = 50 mi/day x 365 days

= 18250 mi = 29370 km

fuel economy = 25 mi/gal = 10.6 km/l

Gasoline is isooctane C_8H_{18} and burns according to the reaction equation

$$C_8H_{18} + 12.5\, O_2 \rightarrow 8\, CO_2 + 9\, H_2O$$

Physical Constants

Atomic weights (AW), g/mol

$AW_H = 1$ $AW_C = 12$ $AW_O = 16$

density of gasoline = 0.75 kg/L

Analysis

1. Estimate mass of gasoline burned per year

$$\text{volume of gas per year} = \frac{\text{distance per year}}{\text{fuel economy}}$$

$$= 29370 \text{ km} \times \frac{1}{10.6} \frac{L}{km}$$

$$= 2771 \text{ L}$$

mass = density x volume

$$= 0.75 \frac{kg}{L} \times 2771 \text{ L}$$

$$\boxed{= 2078 \text{ kg}}$$

Figure 3.11   Example homework format.

2. Convert mass of gasoline to moles

First, we need the atomic weight of gasoline, C_8H_{18}

$$AW_{gas} = 8 \times AW_C + 18 \times AW_H$$
$$= 8 \times 12 + 18 \times 1$$
$$= 114 \text{ g/mol}$$

$$\text{moles of gas} = \frac{\text{mass}}{AW_{gas}}$$

$$= \frac{2078 \times 1000 \text{ g}}{114 \text{ g/mol}} = \boxed{18228 \text{ mol}}$$

3. Calculate moles of CO_2 produced

From the reaction equation for burning gas,
8 moles of CO_2 are produced for every mole of gas

$$\text{moles of } CO_2 = 8 \times \text{moles of gas}$$
$$= 8 \times 18228 = \boxed{145824 \text{ mol}}$$

4. Calculate mass of CO_2 produced from moles

First, we need the atomic weight of CO_2

$$AW_{CO_2} = 1 \times AW_C + 2 \times AW_O$$
$$= 1 \times 12 + 2 \times 16$$
$$= 44 \text{ g/mol}$$

$$\text{mass of } CO_2 = \text{moles of } CO_2 \times AW_{CO_2}$$
$$= 145824 \text{ mol} \times 44 \text{ g/mol}$$
$$\boxed{= 6416 \text{ kg}}$$

Comments

The estimated mass of CO_2 is 6416 kg (14145 lbs). This is a little
more than 3 times the mass of gasoline burned. This is because
the carbon molecules from the gasoline combine with oxygen from the
air, resulting in the greater mass of the gas produced.

Figure 3.12 Example homework format (continued).

3.7 PLANNING LARGER PROJECTS

The steps we introduced in Section 3.5 apply to solving engineering problems of
all different sizes, from paper-and-pencil homework problems you do yourself to
complex team projects that require weeks, months, or years to complete. As the
complexity of a project grows, so does the importance of the planning step. In particular, for any project with a tight budget and a strict deadline, the difference between
success and failure is often how effectively a team distributes the work and man-
ages its time. Project management is a complex topic, and many books have been

Figure 3.13 The SolderBaat team ponders their design on their project homepage.

written on methods for keeping projects on track. In this section, we look at some of the issues and offer suggestions by examining a senior design project by a team of mechanical engineering students at Notre Dame. Also, we introduce a very common tool, called a *Gantt chart*, for developing and presenting project schedules.

3.7.1 SolderBaat—A Circuit Board Assembly and Test System

AME40463 is the capstone Senior Design course in the Department of Aerospace and Mechanical Engineering at Notre Dame [BR06].[3] Students work in teams of 4–6 engineers, with each team managed as if it were a small business. Each design group is responsible for the development of a product concept, preparation of appropriate documentation, and fabrication of a prototype that should demonstrate the feasibility of the product's design. The work must be completed on a tight schedule: the course is only one semester long, and students meet for five hours in class and at least seven hours outside of class each week. Each year, the class is given several project areas from which to choose. In 2006, one of the options was a circuit-board assembly, solder, and test system. The goal of the project was to produce an automated system that could take an electronic component, insert it into a predetermined set of holes in a circuit board, solder it in place, and then test that a proper electrical connection has been made. One team of five students—Nick Frohmader, Brad Shervheim, Jeff Lammermeier, Mike Lavery, and Dave Rowinski—developed a design called the *SolderBaat*. Figure 3.13 shows their project home page.

Before describing the *process* by which the students developed SolderBaat— the main focus of this section—we'll first give an overview of the result. Figure 3.14

[3]The author thanks Professor Stephen Batill of the Department of Aerospace and Mechanical Engineering at Notre Dame, who taught the course in 2006, for providing this example.

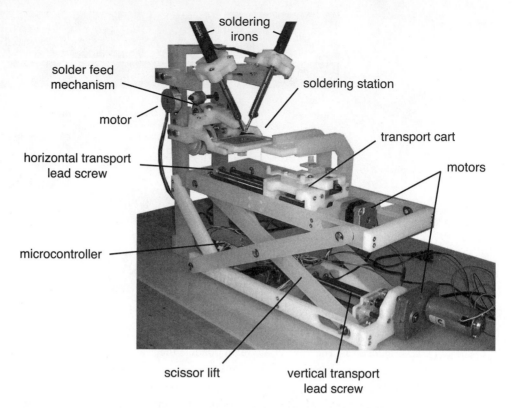

solder feed mechanism

soldering irons

soldering station

motor

transport cart

horizontal transport lead screw

motors

microcontroller

scissor lift

vertical transport lead screw

Figure 3.14 The SolderBaat.

shows a photograph of the completed SolderBaat prototype, while Figure 3.15 shows a functional breakdown. The design has two main subsystems, the **mechanical system** and the **electrical/control system**, which share motors, sensors, and the soldering irons. These two subsystems, together with *their* main subsystems, are described below:

- The **mechanical system** forms the structure of the SolderBaat.

  - The **transport cart** collects a component and holds it for positioning, while the horizontal and vertical **transport mechanisms** move the cart into place. The horizontal transport mechanism uses a **lead screw** to slide the cart along a pair of **guide rails**. The vertical transport mechanism uses a **scissor lift** to raise the cart as a second lead screw pulls the arms at the base of the lift together. **Motors** drive each of the lead screws and **touch sensors** determine when the transport cart is in the proper position for soldering.

  - The **soldering station** solders the positioned electrical component to the printed circuit board. It lifts the circuit board with the component to two fixed **soldering irons** that heat the connections at either end of the component. A **feed mechanism** uses motorized rollers to draw solder from spools to the tips of the irons. A lamp on the circuit board glows when a proper electrical connection is made.

- The **electrical/control system** uses a **microcontroller** running a **software** program to sequence the operation of the SolderBaat. It checks the status of a variety of **sensors**, such as positioning sensors in the transport mechanisms, heat sensors

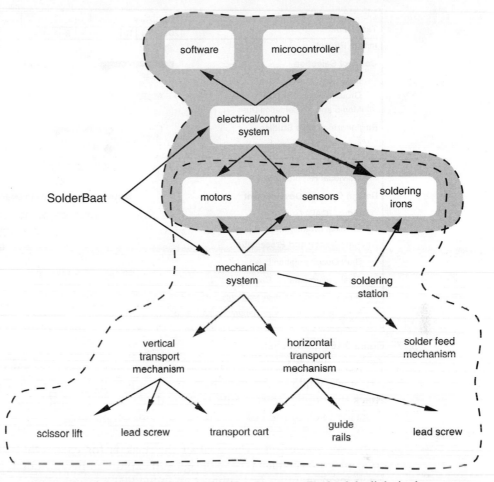

Figure 3.15   Breakdown of SolderBaat and its subsystems. Each of the links in the concept map represents "has-part" relationships

on the **soldering irons**, and a light sensor on the soldering station, to determine when to turn the **motors** and soldering irons on or off.

3.7.2   Task Scheduling

In order to complete their project within the semester, students in the capstone design class must carefully manage their time. Professors Steve Batill and John Renaud provide a detailed course manual that outlines certain activities and milestones that all project teams must meet to stay on track. This is designed to guide students through the process of brainstorming ideas and evaluating their feasibility before they invest their time and money in building the prototype. The professors present the general project plan using a common schedule diagram called a *Gantt chart*, together with a set of task descriptions. A Gantt chart is a set of timelines for each task stacked together, so that we can see the beginning and end dates for all the tasks in the project at once. Figure 3.16 shows a Gantt chart for the senior design project. The leftmost column in the chart lists the tasks in outline form with subtasks indented. To the right are the timelines for each task. Solid gray bars indicate the durations of

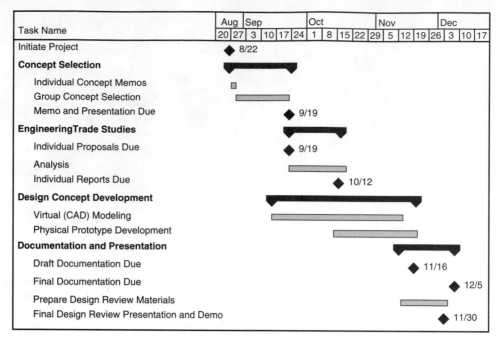

Task Name	Aug	Sep					Oct					Nov				Dec		
	20	27	3	10	17	24	1	8	15	22	29	5	12	19	26	3	10	17
Initiate Project	◆ 8/22																	
Concept Selection																		
Individual Concept Memos		▪																
Group Concept Selection																		
Memo and Presentation Due					◆ 9/19													
EngineeringTrade Studies																		
Individual Proposals Due					◆ 9/19													
Analysis																		
Individual Reports Due										◆ 10/12								
Design Concept Development																		
Virtual (CAD) Modeling																		
Physical Prototype Development																		
Documentation and Presentation																		
Draft Documentation Due														◆ 11/16				
Final Documentation Due																◆ 12/5		
Prepare Design Review Materials																		
Final Design Review Presentation and Demo															◆ 11/30			

Figure 3.16 Gantt chart.

a subtask, while heavy black bars with points indicate the durations of task groups. Black diamonds indicate milestones and due dates. The following paragraphs briefly describe each of the tasks in the general project plan.

Concept Selection The goal of this task is for each group to pick a design concept—a basic idea for the design—to explore in further detail for the semester. Each team member first submits an individual concept memo with a hand-drawn sketch of her or his idea. Using these as a starting point for discussion, the groups then spend several weeks selecting a concept to pursue further. The group as a whole then submits a memo and gives a presentation to the class on their concept. Figure 3.17 shows two of the early concept sketches for SolderBaat.

Engineering Trade Studies Engineering trade studies are experiments that provide quantitative justification for decisions made during the design process. For this task, each student on the team chooses some aspect of the design to analyze. Before they begin their analyses, the students first submit a proposal; this forces the students to think about the aspects of the design that merit further study, and also provides them with feedback from the instructors. The SolderBaat team performed the following trade studies:

- **Lift Mechanism Analysis:** Nick's trade study addressed the issue of designing the lift mechanism. The purpose was twofold: to determine dimensions necessary for the correct lift elevations, and to determine a suitable motor to drive the lift. Figure 3.18 illustrates some of the options he considered. The results of his analysis directly determined the lift design he used in the final prototype.
- **Gripper Mechanism Analysis:** Jeff's trade study addressed the issue of designing the gripper mechanism. Using both experimental data from gripper strength

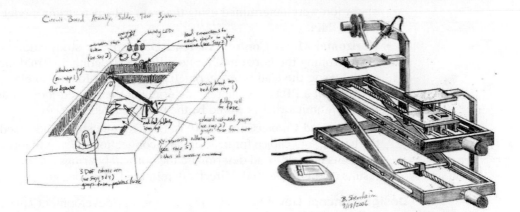

Figure 3.17 Early concept sketches for the SolderBaat design. The sketch on the left envisioned using a robotic arm to pick a component from a tray and place it on the printed circuit board. The sketch on the right uses transport mechanisms similar to the final design.

tests and a theoretical model of the gripping action, Jeff determined a quantitative justification for the design of the gripper and the solenoid requirements.

- **Soldering System Analysis:** Mike's trade study addressed the issue of designing the soldering system. The choice for the soldering iron power, the type of solder, the length of the feed solder, and the time for the soldering

Option	Key Features	Key Challenges	Image
Scissor lift driven directly	-greatest mechanical advantage (scissor) -simple design	-developing enough torque to hold the scissor in place	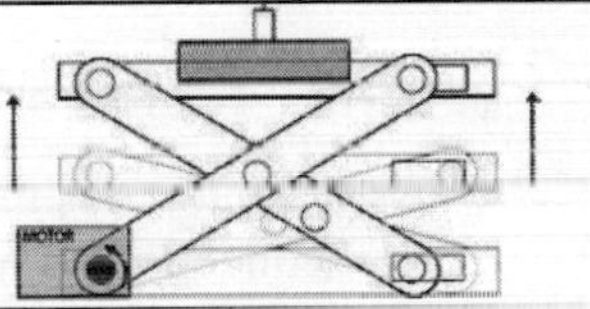
Scissor lift driven by lead screw	-greatest mechanical advantage (scissor) -greater power transfer with lead screw	-creating a driving mechanism	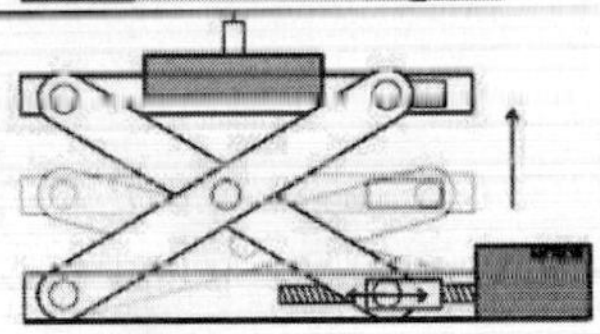
Cam	-completely customizable motion	-complex design	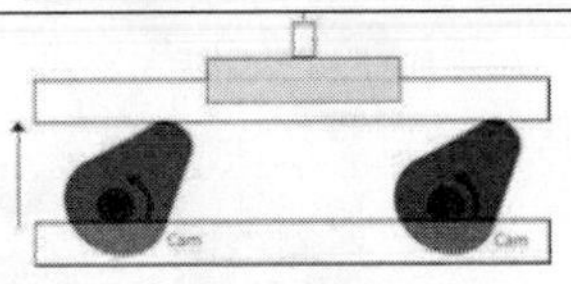
Vertical lead screw	-linear motion -similar control to horizontal motion	-difficult to place -too much weight needed to be raised	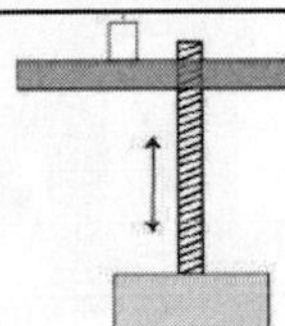

Figure 3.18 Options for the design of the vertical transport mechanism analyzed in a trade study.

operation were determined using theoretical heat-transfer models and experimental data.

- **Horizontal Motor Control Analysis:** Dave's trade study addressed the issue of controlling the horizontal motion. By constructing a virtual motion control model using the lead screw geometry, stepper motor parameters, and the application of a PID controller, he determined that closed-loop feedback control based on limit switches would be the most feasible.
- **Bearings and Connections Analysis:** Brad's trade study addressed the issue of the best design choice for bearings and connections. Using theoretical machine elements models, Brad determined the optimal bearings and pin joints, and the results of this trade study directly determined these design decisions.

Design Concept Development   Design concept development constitutes "fleshing out" the design concept. This task begins before the trade studies and continues through the development of the physical prototype. A major part of this task is the development of a virtual model for the system, using the computer-aided design and manufacturing (CAD/CAM) package called Pro/ENGINEER [Par] pictured in Figure 3.19. Using this tool, the students enter engineering drawings of their designs into the computer, that they can then simulate to produce working animations of their systems in operation. The CAD/CAM tools also provide input directly to equipment that automatically cuts, drills, and shapes the components for their physical prototypes.

Documentation and Presentation   The final task is to prepare the documentation for the project and to present a final design review and demonstration to a panel of outside experts from the industry. Figure 3.20 shows sample pages from the SolderBaat user's manual.

3.7.3 Teamwork and Results

As a senior design project, the SolderBaat was successful in two ways: as a design concept and as a learning experience. While the Notre Dame Aerospace and

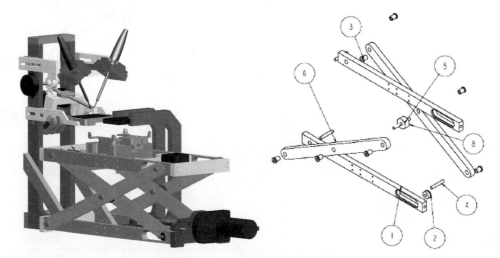

Figure 3.19   Left: Virtual model of the SolderBaat in Pro/E. Right: Exploded view of the scissor lift, also from Pro/E.

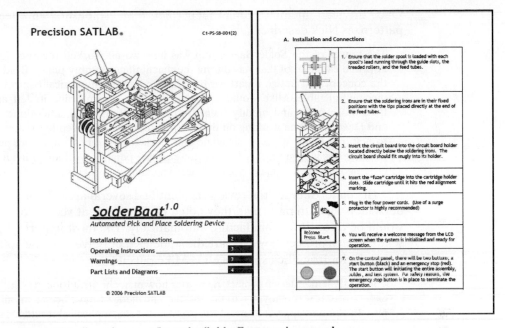

Figure 3.20 Sample pages from the SolderBaat user's manual.

Mechanical Engineering senior design course emulates an industrial design project, it differs in important ways. First, the project doesn't incorporate the key initial step of market research and defining requirements, which the instructor provided. Also, many, but not all, industrial projects begin with the selection of an experienced team leader, who then builds a team with the right mix of skills. For each member of the SolderBaat team, this was his first experience with a project of this complexity, and they had to learn how to work together along the way. Still, they met their milestones, stayed within budget, produced excellent documentation, and received favorable reviews from a visiting panel of industry experts. Dave describes this example of teamwork:

> Looking back on the development of SolderBaat, I notice two key elements that contributed to our success. The first is an early development of objectives and division of duties. Although these decisions did not consume most of the project time, they drove it the entire way. Halfway through the project, some groups were still modifying their basic objectives so that their design required drastic changes. By establishing objectives early on, our design evolved piece-by-piece as new complexities were discovered, but the general design remained the same. The second is that though this was certainly not an easy project and there were many setbacks, on the whole this was a fun experience. Though no one on our team knew each other beforehand, we formed a great group dynamic. There were times when one group member wanted to do something differently, but cooperated in the interest of keeping everyone happy. When decisions really made a difference, these other ideas wouldn't be abandoned without some technical analysis, but for smaller problems, cooperation was key.

The diverse mix of skills on the team was also an asset. Brad and Mike were the best machinists and Dave was the best programmer. Jeff handled much of the assembly and Nick served as the group leader, handling most of the administrative tasks.

Once the team members found their roles, they established an effective working pattern, as Nick describes:

> The power of the SolderBaat group was how we all worked together for one goal. A prime example of that was during the machining/assembly phase. Brad would take the concept and design a part for the CNC [computer numerical control, automated machining] table; Mike would take that drawing and machine it; Jeff and I would take those parts and modify them by hand (drilling holes, smoothing edges, etc); and Dave would be working on the controls to move the assembly. Everyone worked to help the team. We set a schedule and stuck to it the best we could, pushing back deadlines as a group rather than because of one person. Working individually for the group goal helped us make SolderBaat from concept to reality.

The panel of industrial reviewers identified several areas for improving the SolderBaat design. In particular, they noted that the current version of the SolderBaat would be difficult to manufacture automatically. The most important products of a senior design project, however, are the students themselves and the SolderBaat team learned a lot from the experience. As Mike puts it:

> The AME senior design project, no matter how many hours it took, was still a very cool experience. It was enjoyable to take all the knowledge I have been compiling over the past 3 years and apply it to a real-life problem. Seeing a physical result, rather than just numbers on paper, was very rewarding.

3.8   HEURISTICS

Mathematician Georg Pólya recalled that as a student, he

> listened to lectures, read books, tried to take in the solutions and facts presented, but there was a question that disturbed him again and again: Yes, the solution seems to work, it appears to be correct, but how is it possible to invent such a solution?' [Pol45]

Years later, as a professor at Stanford, his desire to try to explain the motives and procedures for problem solving to his students led him to write his book *How to Solve It*. Pólya frames his method as a dialog between a teacher and a student, which he explains as follows:

> The way from understanding the problem to conceiving a plan may be long and tortuous. In fact, the main achievement in the solution of a problem is to conceive the idea of a plan. This idea may emerge gradually. Or, after apparently unsuccessful trials and a period of hesitation, it may occur suddenly, in a flash as a "bright idea." The best that the teacher can do for the student is to procure for him, by unobtrusive help, a bright idea. The questions and suggestions we are going to discuss tend to provoke such an idea [Pol45].

The questions and suggestions that Pólya presents are basic "rules of thumb" for problem solving called *heuristics*. In this section, we provide a list of some of the most commonly used heuristics, which is based on the work of Pólya and others, including Rubinstein [Rub75], Wankat and Oreovicz [WO93], and Lumsdaine and Lumsdaine [LL95].

As we discussed in Section 3.4, one of the most important skills for you to develop is to learn how to become your own teacher. When stuck on a problem, try running through an internal dialog similar to that between Pólya and his student, using this list of heuristics as a guideline to help generate that spark of a "bright idea."

3.8.1 Write It Down

As demonstrated by the example of multiplying a 4-digit by 2-digit number in Section 2.1, the limited capacity of your short-term memory makes it very difficult to work problems with more than a few concepts in your head. Don't rely on your memory—write things down. Just seeing parts of a problem explicitly on paper can help you get past roadblocks.

3.8.2 Restate in Simpler Terms

Restating a problem in simpler terms improves your comprehension and eliminates extraneous information. As suggested in [Rub75]:

- As you restate a problem, try to package related information in chunks.
- If a word problem can be rewritten in symbolic or mathematical form, do so.

3.8.3 Draw a Picture

You should have already drawn *a* picture or diagram supporting this problem in the "problem definition" step, but as your plan progresses, additional diagrams are helpful. This suggestion is really just a variant of restating the problem in simpler terms. Example 3.2 illustrates the value of a picture.

Example 3.2 **Draw a Picture**

Given the following description of a tic-tac-toe board, determine what X's next move should be.

The first X is placed in the center of the board. The first O is placed one square to the left and one square up from the first X. The second X is placed in the square that is two squares to the right of first O. The second O is placed immediately below the first O.

Solution It's difficult to solve this problem without drawing a picture of the board. With a picture it's easy.

3.8.4 Do You Know a Related Problem?

This is probably the most widely used heuristic: looking for a problem that is in some sense similar to the problem at hand. When stumped on a homework problem, one of the most profitable places to look is at examples in the textbook or those that were worked on in lecture, since many routine homework problems are designed to exercise your comprehension and ability to apply the same concepts that these examples illustrate.

Solving a problem out of context is clearly more difficult, as is finding a related problem. As Pólya notes, there are many ways that problems could have common points between them, and the challenge is finding a problem that's related in a way relevant to obtaining a solution. Pólya's suggestion is to look at the unknown, and to try to think of a familiar problem having the same or similar unknown. While very useful, you can't apply this heuristic blindly, as illustrated in Example 3.3.

Example 3.3	**Related Problems**

Related Problems

Which two of the following three problems have the most in common?

1. A right triangle has legs of lengths 3 m and 4 m. What is the length of its perimeter?

2. A rectangle has sides of lengths 3 m and 4 m. What is the length of its perimeter?

3. A person is rowing a boat across a river. If the person is rowing at a speed of 3 mph perpendicular to the bank and the river is flowing at a rate of 4 mph, what is the speed of the boat in the direction of travel?

Solution   Although problems 1 and 2 both involve calculating the perimeters of polygons and problem 3 involves calculating the speed of a boat, the solutions to problems 1 and 3 both require using the Pythagorean theorem to calculate the hypotenuse of a right triangle. Having the experience of solving problem 2 would not help you solve problems 1 or 3, but having the experience of solving either of these would help you solve the other. Therefore, problems 1 and 3 have the most in common.

3.8.5 Work Backwards/Forwards

Working backwards, also known as backward-chaining, is starting with the goal and working backwards to primary inputs. This is usually the best approach to consider first for most engineering problems, because it makes you focus on goals and ways to refine them to subgoals. One example of a backwards or top-down approach was the development of a plan for determining the amount of paint needed to cover a water tower in Example 3.1. Working forward, also called forward-chaining, is first to look at what you have available, and then think of ways to use it to move closer to the goal. Example 3.4 illustrates both the forward- and backward-chaining approaches to solving the same problem.

Example 3.4   **How Many Games in a Tournament?**

The NCAA college basketball tournament starts with 64 teams. If teams are eliminated after a single loss, how many total games are played in determining the winner?

Solution   The working-forward approach to solving this problem is to start with the 64 teams and systematically reduce them to 1 winner, adding up the number of games along the way, beginning with 32 games in the first round, 16 in the second round, and so on until you get to 1 game in the finals.

Before you add this up and get a result, consider working backwards from the goal. At the end of the tournament, there is 1 winner and 63 losers. Every game played produced 1 loser, so there must have been 63 games. Now go back and complete the forward-chaining approach and see that you get the same answer.

3.8.6 Work Top-Down/Bottom-Up

The terms "top-down" and "bottom-up" refer to the direction in which you work through a problem that has a hierarchical organization when drawn as a tree diagram with the root at the top and the leaves on the bottom. A top-down approach involves taking the initial problem and splitting it into subproblems (much the same as divide and conquer), while a bottom-up approach is to take existing pieces and combine them.

One problem that illustrates the difference between a bottom-up and top-down approach is meal planning, as illustrated in Figure 3.21. Planning a holiday meal for a large number of guests is typically solved in a top-down fashion: you start with the goal of serving a meal, then break this down into courses, such as appetizers, main meal, and dessert, and then break the courses down into specific dishes and ingredients. By comparison, the next few meals *after* the holiday meal are typically planned in a bottom-up fashion, looking for ways to use up the leftovers!

Alternating between working forward and backward and meeting somewhere in the middle can also be a fruitful approach that's actually frequently practiced.

3.8.7 Divide and Conquer

Breaking a complex problem into pieces that can be solved independently is an essential engineering problem solving skill. It's also an advanced skill, because it requires understanding the central concepts of the problem at an analysis level. Divide-and-conquer works when a problem can be decomposed into *independent* pieces—that is, pieces whose solution does not have an effect on the solution of another piece. Breaking the water tower down into components that could be painted separately in Example 3.1 was one illustration of this. Another example of divide and conquer is factoring a polynomial in order to find its roots, as illustrated in Example 3.5.

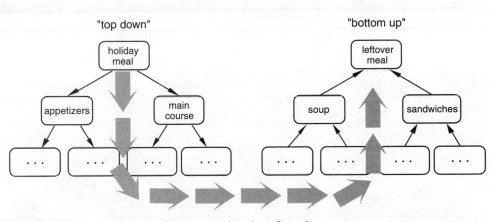

Figure 3.21 Top-down versus bottom-up planning of meals.

Example 3.5

Divide and Conquer: Finding the Roots of a Polynomial
Find the values of x that solve the following equation:

$$x^2 - 7x + 12 = 0$$

Solution

Using a divide and conquer approach, we can factor the second-order polynomial into two first-order polynomials, and then find the values of x for which either of these equal zero. Because the factors are independent, we can find the roots separately.

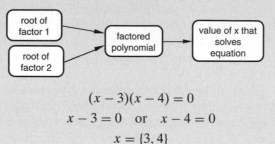

$$(x - 3)(x - 4) = 0$$
$$x - 3 = 0 \quad \text{or} \quad x - 4 = 0$$
$$x = \{3, 4\}$$

It's important to note that not all problems can be easily factored into parts that can be solved independently. The design of an airplane is one example of this. You can't design the fuselage (body) independently from the wings, because the entire structure needs to have an aerodynamic form so that it can fly. In fact, any aspect of the design that affects the weight of the plane, including how many passengers and how much fuel it can carry, has an impact on the aerodynamics. To solve problems such as this generally requires techniques that consider a number of variables simultaneously. Example 3.6 illustrates this in determining voltages across components in a simple electrical circuit.

Example 3.6

Voltages in an Electrical Circuit
Determine the voltage drops v_1 and v_2 across the two resistors R_1 and R_2 in the circuit below. Consider the following laws in determining your solution.

- *Kirchhoff's Voltage Law:* The sum of the voltage drops (or negative voltage gains) in a loop equals zero.
- *Kirchhoff's Current Law:* The sum of the currents flowing into a node is zero.
- *Ohm's Law:* The current through a resistor equals the voltage across it divided by the resistance.

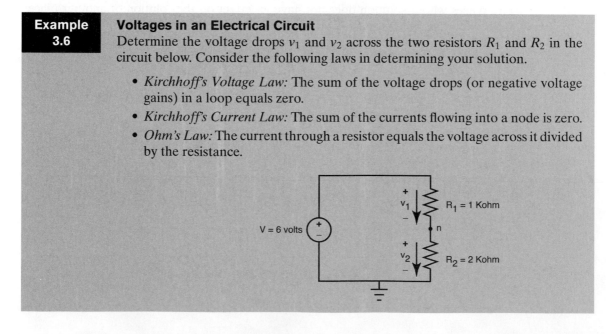

Solution   Perhaps thinking of other problems in which you were able to break a system down into separate components, such as the water tower problem in Example 3.1, you may be tempted to try to solve for v_1 in terms of R_1 and v_2 in terms of R_2 independently. The problem with this thinking, however, is that v_1 and v_2 each depend on *both* R_1 and R_2, as shown below:

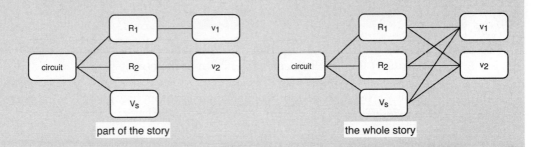

part of the story   the whole story

This is because the voltage that drops across the resistors depends on the current flowing through the circuit, which flows through both resistors and therefore depends on both. Systems with mutual dependencies such as this circuit typically lead to a system of several equations in several unknowns that must be solved *simultaneously*. In this case, we could formulate two equations using Kirchhoff's voltage and current laws in terms of the unknown voltages v_1 and v_2. First, according to Kirchhoff's voltage law, the sum of the voltage drops across the resistors minus the voltage gain across the power supply equals zero:

$$v_1 + v_2 - V = 0$$

Using Kirchhoff's Current Law and applying Ohm's Law to determine the current through a resistor, the current flowing into node n through R_1 minus the current flowing out of n through R_2 equals zero:

$$\frac{v_1}{R_1} - \frac{v_2}{R_2} = 0$$

Substituting values, we get the two equations:

$$v_1 + v_2 = 6$$

$$v_1 - \frac{1}{2}v_2 = 0$$

Solving, we get $v_1 = 2$ volts and $v_2 = 4$ volts.

3.8.8 Check for Unnecessary Constraints

You can make a problem more difficult than it actually is by assuming constraints exist that aren't really there. This heuristic could also be called "think out of the box." Example 3.7 illustrates a classic example of a problem that requires this heuristic, as suggested in [NS72] and [Rub75].

Example 3.7

Thinking Out of the Box
Without lifting your pencil from the paper, draw four straight lines that pass through each of the nine dots arranged in a box as shown below:

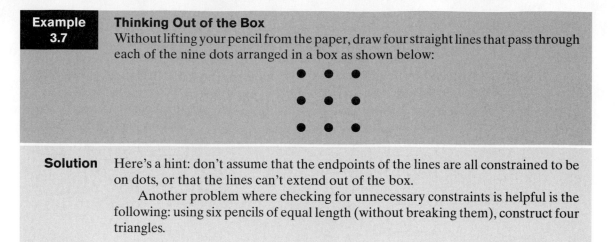

Solution   Here's a hint: don't assume that the endpoints of the lines are all constrained to be on dots, or that the lines can't extend out of the box.

Another problem where checking for unnecessary constraints is helpful is the following: using six pencils of equal length (without breaking them), construct four triangles.

3.8.9   Discuss

Discussing a problem with another person can often help you get moving when you're stuck. Oftentimes, just describing the problem to someone else will give you new insights. People who are skilled at asking the right questions, such as a teaching assistant or professor, can be particularly helpful. Friends can also help each other by brainstorming aloud, or even running through a set of questions based on this list of heuristics. When discussing a problem, be sure to *listen* carefully for ideas—this often takes a conscious effort when you're preoccupied with the problem.

3.8.10   Try Solving a Scaled-Down Version of the Problem

Sometimes composing and solving a scaled-down version of the problem that is easy to solve will help you see a pattern that can be applied to the larger problem. As an example, let's reconsider the problem of determining the total number of games played in a tournament, shown earlier in Example 3.4.

Example 3.8

Number of Games in a Tournament, Solved Another Way
Try solving the problem of determining the total number of games played in a tournament, presented earlier in Example 3.4, by first solving simple cases with only a few teams entered.

Solution   The sketch below solves the problem for the simple cases of 2 through 6 teams, which are small enough to be solved by sketching the tournament brackets in their entirety and simply counting the games.

   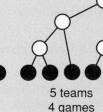 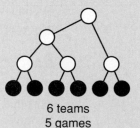

2 teams   3 teams   4 teams   5 teams   6 teams
1 game   2 games   3 games   4 games   5 games

We can see a pattern emerging. Although this is not a formal proof, we might infer that the number of games is one less than the number of teams. Asking ourselves why this might be true, we may arrive at the "solving backwards" approach shown in Example 3.4.

3.8.11   Try Solving a Simpler but Related Problem

Solving a simpler problem that retains some of the concepts of the original problem, but discards others may provide the insights that you need to solve the more complex problem. One example of this, from Pólya [Pol45], is shown in Example 3.9. In this example, solving a two-dimensional problem that is related to the original three-dimensional problem helps establish the key concept needed for the solution.

Example 3.9	**Finding the Diagonal of a Box**

Given a rectangular box with dimensions A, B, and C, what is the length of the body diagonal?

Solution   In *How to Solve It* [Pol45], the author describes an imaginary discussion between a teacher and a student who is stuck on this problem. The teacher asks the student if he can't come up with a simpler problem that reminds him of the original problem. With some prodding, the student comes up with the problem of finding the length of the diagonal of a rectangle, which he knows can be solved using the Pythagorean theorem. Then the teacher asks the student if he can find a way to apply this result to the solution of the original problem. Now the student sees a path to a solution, illustrated below. Do you see it?

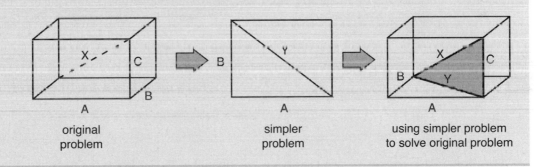

original   simpler   using simpler problem
problem   problem   to solve original problem

3.8.12   Use Models

A model is an approximation to the real system, such that when actions are performed on the model, it will respond in a manner that allows us to estimate the behavior of the original system. Models can range in accuracy from crude to extremely accurate, and there's generally a trade-off between the accuracy of the model and the effort and cost required to develop it. Models can take on many forms, including physical models, equations, or computer simulations.

Even very simple models can make solving a problem easier, as Example 3.10 shows.

Example 3.10

Hobbits and Orcs

Three Hobbits and three Orcs are on one side of a river. They all have to get across, using a canoe that can hold only two of these creatures at a time. At no point can more Orcs than Hobbits be left on one side of the river, as the Orcs would kill the Hobbits. Develop a plan for transporting all six creatures across the river that keeps the Hobbits alive.

Solution

Using simple physical models such as different coins for the Hobbits and Orcs is much easier than trying to solve this problem in your head, and more convenient than using pencil and paper. This problem is also often used to illustrate the point that sometimes you need to take a step that seems to move you further *away* from the goal in order to solve the problem.

Note that the Hobbit/Orc version of this problem is a more "culturally sensitive" version presented in [And80] of a classic artificial intelligence problem originally called "missionaries and cannibals" [NS63].

3.8.13 Guess and Check

Ironically, for some problems that are very difficult to solve directly, it can be fairly easy to test if a candidate solution is correct. In these situations, guessing an answer and checking if it is acceptable may be your best approach.

The key to using this technique is to guess wisely, and when possible to use feedback from wrong guesses to come up with better guesses. To see the potential power of a guess-and-check strategy, consider the game "Twenty Questions." In this game, one player thinks of something—anything—and the other player can ask up to twenty yes-or-no questions to try to guess what player one is thinking of. Despite the vast number of possibilities, a skilled guesser will usually come up with the correct solution.

Example 3.11

Guess and Check to Find a Root of a Polynomial

Find a value for x, accurate to one decimal place, that solves the following equation:

$$f(x) = x^3 - 7x^2 + 20x - 100 = 0$$

Solution

In Example 3.5, we factored a polynomial to find its roots. In this case, however, factoring isn't so easy. Instead, we'll use a guess and test approach. First of all, for extremely large negative values of x, the left-hand side of the equation will be negative, and for extremely large positive values it will be positive, so it must equal zero somewhere in between. Our strategy will be to keep guessing values of x, with each guess narrowing the region where the left-hand side crosses from negative to positive. First, we'll guess $x = 0$, which yields $f(0) = -100$. Next, we guess $x = 10$, which yields $f(10) = 400$. Thus $f(x)$ must equal 0 somewhere between $x = 0$ and $x = 10$. We try the middle of the range, $x = 5$, which yields $f(x) = -50$. Now we know that there's a root somewhere between $x = 5$ and $x = 10$. We continue making guesses in this fashion, dividing the range in half each time, as shown in the table below:

Guess	x	$f(x) = x^3 - 7x^2 + 20x - 100$
1	0	−100
2	10	400
3	5	−50
4	7.5	78.1
5	6.25	−4.3
6	6.4	3.4
7	6.3	1.8

Thus $x = 6.3$ is an accurate solution to the equation to within one decimal place. Note that there are a number of ways that we could have been "smarter" with our guesses. For example, we could have plotted the function at $x = 0$, $x = 5$, and $x = 10$ and then interpolated to see where the plot crosses the x-axis. This would have saved us several guesses.

3.8.14 Use an Analogy

An analogy is a relationship between two situations, where concepts in one situation have a direct correspondence or *mapping* to concepts in the other. Analogies are powerful tools for learning and problem solving, particularly when you are familiar with concepts in one area and see a way that they can be used to understand concepts in another area with which you are less familiar. As an example, water flow is often used as an analogy for explaining concepts in electrical circuits, as shown in Figure 3.22.

Engineers have long used analogies from the physical and biological world as inspiration for solving problems and for creating new inventions. In 1948, for example, Swiss inventor George de Mestral had the inspiration for the design of a two-sided fastener after seeing his dog covered in burrs during a nature hike. After returning

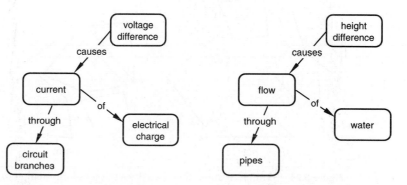

Figure 3.22 Water flowing through pipes is an analogy for electrical current.

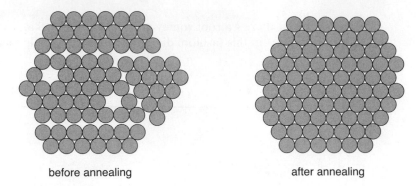

before annealing   after annealing

Figure 3.23 In physical annealing, a material sample is heated to a high temperature to allow the atoms to move about. As the sample slowly cools, the atoms find stable arrangements, thus repairing defects.

home, he noted the tiny hooks on the ends of the burrs under his microscope, and in collaboration with a weaver and textile manufacturer in France, developed a fastener using tiny hooks and loops. He gave his invention a name that was a combination of "velour" and "crochet"—VELCRO™.

More recently, scientists and engineers have looked to natural processes as analogies for solving combinatorial optimization problems, such as how to efficiently pack components of an automobile in the small space under the hood, or how to route billions of wires between devices in an integrated circuit. A popular technique for solving such problems is called *simulated annealing* [MRR⁺] [KGV83]. In materials science, annealing is a process of heating materials such as glass, crystals, or metals and then cooling them at a controlled rate in order to make them more workable or less brittle. Figure 3.23 illustrates a material sample before and after annealing. Before annealing, the atomic structure of the sample has various defects—sections of the structure are "dislocated." These defects are instabilities in the material that cause internal stresses, weakening it. When the sample is heated to a high temperature, the atoms move around freely, and find more stable positions that relax the stresses. As the sample cools, the defects are repaired.

In simulated annealing, a computer program uses the analogy of physical annealing to find stable, "stress-free" configurations of virtual objects, as shown in Figure 3.24. A simulated annealing program uses random numbers and probabilities to guide

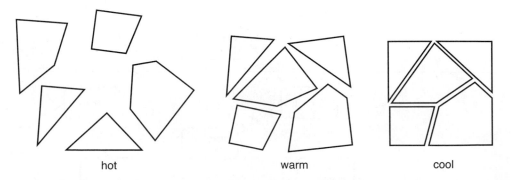

hot   warm   cool

Figure 3.24 Simulated annealing uses the analogy of physical annealing to find stable configurations among virtual objects.

the process. At high simulated temperatures, there is a high degree of randomness. As the simulated temperature cools, the program is less likely to accept "stressful" configurations, and the system eventually settles into a stable arrangement.

Other analogies from nature have also been used to solve combinatorial optimization problems. *Genetic algorithms* [Hol75] [Hol92], for example, simulate the processes of genetic mutation and natural selection to generate candidate solutions to a problem, and select the fittest for "survival."

3.8.15   Change Your Perspective

Looking at a problem from a different perspective can lead to an unexpected solution. Consider, for example, the problem of landing an unmanned exploration vehicle safely on Mars. Conventional thinking would interpret the problem as how to land the vehicle slowly using a parachute. Because of the thin atmosphere on Mars, a parachute would need to have 38 times the area of a parachute designed to work in Earth's atmosphere to achieve the same ability to slow a descent. This would add greatly to both the weight and volume of a landing capsule.

For the Mars Pathfinder mission, which landed on Mars on July 4, 1997, NASA engineers changed their perspective on the problem from how to land the vehicle gently to how to cushion the vehicle if it doesn't. The result was the development of the airbag scheme pictured in Figures 3.25 and 3.26. A few minutes after entering the Mars atmosphere, a small parachute opened to slow the descent of the lander to to 65 m/s (145 mph). At an altitude of approximately 300 m the giant airbags inflated and at 100 m, solid rocket motors fired to bring the descent to a near stop. From there, the vehicle crash-landed and bounced until it came to rest on the surface of Mars.

Figure 3.25   A change in perspective on the Mars landing problem, from how to land gently to how to cushion the blow—NASA engineers perform an inflation test of the Mars Pathfinder airbag system at the Jet Propulsion Laboratory (JPL), Pasadena, California. Photo courtesy of NASA.

Figure 3.26   The Pathfinder lander and Sojourner rover safe on Mars, with the airbags now deflated. Photo courtesy of NASA.

The example problem of tying two strings hanging from the ceiling described in Section 2.1.3 and illustrated in Figure 2.5 illustrates a classic solution in which a change in perspective is useful. In this example, changing the question from asking how you can get to both strings to asking how you can make a string come to you, held the key to solving problem. Only 40 percent of the subjects in Maier's experiment were able to complete the task within 10 minutes [Mai31]. One obstacle of finding a solution was what psychologists call functional fixedness: seeing the pliers as something other than just a tool for grabbing.

3.8.16   Look at the Big Picture

Sometimes in solving problems—especially evaluation problems—we are stuck on small details that upon reflection may not seem critical to the overall solution. An example of this phenomenon is making an expensive purchase, such as a car or a house. The fact that you don't like the wallpaper in the bedroom shouldn't overly influence your evaluation of a house that you're considering, since this can be easily changed. If you are stuck on an evaluation problem, try stepping back and looking at the big picture to get a better focus on the problem.

3.8.17   Do the Easy Parts First

Problems with many parts often seem intimidating because of their sheer size, even if many of the parts are easy. Doing the easy parts first can make the remainder of the problem seem manageable, and also builds confidence and momentum.

3.8.18   Plug in Numbers

Although in general it is best to solve numerical problems symbolically first and then substitute numbers later, Wankat [WO93] suggests that plugging in numbers early can make an abstract problem more concrete, and thus help stimulate a solution.

3.8.19 Keep Track of Progress

For complex problems with many parts, don't rely on your memory to keep track of which parts of the problem have been solved and which remain. A simple "to do" list is one way to check your status. If you've drawn a concept map of your plan, it provides an excellent means for recording your progress towards the goal.

3.8.20 Change the Representation

A *representation* of a problem is a logical structure that relates the parts of the problem in a manner that supports its solution. Having a good representation is critical to solving a problem and for many types of problems, representations have been devised that are almost always used in solving them. Being able to represent the parts of a problem in tabular form is particularly useful. Gaussian elimination, for example, is a technique for solving systems of equations that involves arranging the coefficients of the equations in a table, and manipulating rows and columns using a simple set of rules to get the table into a triangular form. This is illustrated in Example 3.12.

Example 3.12	**Representing and Solving Equations Using a Table**

Solve the following system of equations:

$$2x + 4y = 1$$
$$x + 5y = 2$$

Solution Using Gaussian elimination, we can write the equations in what's called an augmented matrix form, where each equation has a row and the coefficients are arranged in columns, with the "right-hand side" of the equations in a column separated by a bar. We can then manipulate the table to form an equivalent system of equations using the following rules:

1. you can multiply a row by a constant
2. you can add one row to another

Using these rules to transform the table into a triangular shape as shown below produces a system that is easy to solve.

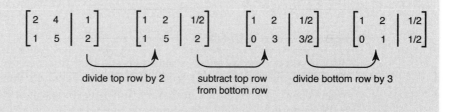

divide top row by 2   subtract top row from bottom row   divide bottom row by 3

$$x + 2y = 1/2$$
$$0 + y = 1/2$$

By inspection, $y = 1/2$. Substituting this into the top row, we obtain $x = -1/2$.

Another example that also uses a table and has been used frequently to illustrate the value of a good representation of a problem is a game called *Number Scrabble* [NS72]. This is shown in Example 3.13.

Example 3.13	**Number Scrabble** In this game, two players share a set of nine tiles, numbered 1–9, face up. Each player takes turns selecting a tile. The first player to collect a set of tiles that adds up to 15 wins.

Solution   Number scrabble can be a challenging game to play, until you realize that it can be represented as a tic-tac-toe game! The trick is to arrange the numbers 1–9 in a 3×3 table called a magic square, so that all rows, columns, and diagonals add up to 15, as illustrated below.

2	7	6
9	5	1
4	3	8

Once you have the magic square, since most people are "experts" at tic-tac-toe, the game becomes easy to play.

3.8.21   Replan

Sometimes when you can't get traction in moving forward toward a solution, the cause of the problem is that your plan is not good.

3.8.22   Pay Attention to Hunches

If you have a hunch about a path to a solution, don't be afraid to try and pursue it.

3.8.23   Take a Break

When you are really stuck, sometimes the best thing to do is to just take a break. Working past the point of frustration and exhaustion does occasionally produce results and is sometimes necessary, but oftentimes it doesn't help. Tenacity is a very important quality for engineers, but you also need to recognize your physical limitations. This author has on many occasions worked late into the night trying in vain to solve a problem, such as debugging a computer program, only to come up with the answer quickly the next morning after a night's sleep. Stepping away from a problem for a few hours also lets your mind explore new ideas in the background.

PROBLEMS

1. **Concept Map: Expertise and the Learning Process**

   Draw a concept map of Section 3.2, "Expertise and the Learning Process."

   - Start by either listing or highlighting approximately 10–12 key concepts from the section.
   - Add relationships to the concept map and organize the map roughly into a hierarchy.
   - Extend the map to include links to approximately five additional concepts from Section 2.1, "What We Think About How We Think."

2. **Concept Map: Engineering Disciplines**

   As part of a team of 3–4 students, develop a concept map that encompasses 3–4 of the disciplines described in Section 1.4, "Engineering Disciplines and Majors." Each student should individually develop a concept map based on one discipline, and then the group should meet and combine these into a single map. The final map should include connections between disciplines wherever they can be added meaningfully.

3. **Studying with Music**

   Do you ever study with music in the background? Do you find it helpful or distracting? What impact do you suspect music would have in terms of the cognitive model presented in the chapter?

4. **Levels of Understanding: Making Copies**

   Determine the levels of understanding needed in each of the steps in the following scenario and give reasons for your assessment.

   (a) Mike gets a job working as a copyboy. Today is his first day, and within the first hour, the copier runs out of paper. Mike refills the copier.

   (b) Mike needs to make copies of 10 articles for one of his business's employees.

   (c) The paper jams in the middle of making the copies. Mike needs to fix the copier and finish making the copies. Luckily, the screen on the copier gives explicit directions for removing the paper jam.

   (d) Mike is having a very trying day. When he tries to remove the paper jam, he accidentally changes some settings on the copier and the copies are not the same as the original ones. Mike rectifies this situation.

   (e) A message shows up on the copier panel telling Mike that the toner is low. He must fix this problem as well.

   (f) Finally, Mike's day ends. He is upset with the issues with the copier, and he is contemplating getting a new job that he enjoys more. In the process of finding a job, he does a lot of research and eventually finds the perfect job: as a go-kart tester.

5. **Levels of Understanding: Making Up Your Own Questions**

   Make up example questions that test understanding at each of the levels of Bloom's taxonomy:

   (a) based on material that you've covered thus far in this course,

   (b) based on material that you've covered in other courses that you are currently taking or have taken recently.

For each example, give a brief justification for why that question illustrates the stated level of understanding.

6. Hmm . . .
What level of understanding does this problem test?

7. Practice with the Problem-Solving Framework
The following problems come from a standard pre-algebra textbook. We'll use them to practice applying the problem-solving framework introduced and illustrated in Sections 3.5 and 3.6. Strive for clarity in writing up your solutions, patterned after the example in Figures 3.11 and 3.12. Pay special attention to giving a succinct description of the plan, even if it seems obvious to you. In your comments at the end of the solution, note any heuristics that you may have used.

- In one long-distance phone call, Amy talked to her parents for twice as long as her brother talked. Her sister talked for 12 minutes longer than Amy. If Amy's phone call was 62 minutes long, how long did each person talk on the phone?
- Roberto needs to draw a line that is 5 inches long, but he does not have a ruler. He does have some sheets of notebook paper that are each 1.5 inches wide and 11 inches long. Describe how Roberto can use the notebook paper to measure 6 inches.
- As the hands of a clock move from 6:00 AM to 6:00 PM, how many times do the hands form a right angle?
- Alice, Nathan, and Marie play in the school band. One plays the drum, one plays the saxophone, and one plays the flute. Alice is a senior. Alice and the saxophone player practice together after school. Nathan and the flute player are sophomores.

8. Assumptions and Approximations in Solving Energy Problems
Solve each of the following problems, making an assumption or approximation as necessary. Use the framework presented in Section 3.5 and illustrated with the example in Section 3.6 as a guide. If you need to look up any information in making your assumptions, be sure to cite your sources.

(a) Estimate the cost of the electricity bill in your household last month. Find out what it actually was. How close was your estimate? Note that power consumption for electrical appliances and equipment is typically measured in watts (W), which has units of energy per unit time, while electricity is typical sold by utility companies in units of kilowatt-hours.

(b) How much ethanol could a corn field of area 1 square kilometer produce? How much gasoline would it take to produce the corn, from preparing the fields through harvest, and to deliver it to an ethanol processing plant?

(c) How far would one have to walk to burn the calories contained in your favorite candy bar?

9. Storing Text on a DVD
One way of representing text in a computer is to use what is called the *ASCII code*, where each letter requires one *byte* of data. (A byte is a string of 8 bits of data.) Approximately how many DVDs would it take to store the text of all of the books in your school's library in ASCII format? Pattern your solution after

the example in Section 3.6 and be sure to cite all sources that you may have used in determining your answer.

10. **Problem Solving Strategies and Heuristics**
Solve each of the following problems, which come from a pre-algebra textbook [EDNK92]. State which heuristics you used in determining your solution, and give a brief justification of why the heuristic applies.

(a) There are 10 hockey players on the ice at the end of a game. Each player shakes hands with each of the other players. What is the total number of handshakes?

(b) Connect all points in the figure below by drawing exactly four line segments without lifting your pencil off the paper. Do not draw through any one point more than once.

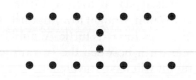

11. **Proofs Come in Handy: The Mutilated Chessboard**
The Mutilated Chessboard is a classic problem devised by Martin Gardner, who wrote the "Mathematical Games" column of *Scientific American* magazine from 1956 to 1981. In this exercise, we'll use this example to probe some of the strengths and limitations of physical demonstrations in solving problems. The figure below illustrates both a standard chessboard and a "mutilated chessboard." The standard chessboard has 64 squares arranged in 8 rows and 8 columns, while the mutilated chessboard has two diagonally opposite corners removed, for a total of 62 squares.

standard chessboard   mutilated chessboard

(a) First, suppose that you have a collection of dominoes that are each exactly the same size as two squares on a chessboard. Is it possible to cover all of the squares on a standard, non-mutilated board with exactly 32 dominoes? How do you know?

(b) Now, is it possible to cover all of the squares in a mutilated chessboard with exactly 31 dominoes? How do you know? If it turns out that it's not possible, how can you prove this? *Try to solve this problem without looking up the answer, even though it's readily available online.*

(c) Explain how this problem illustrates both strengths and limitations of physical demonstrations in solving problems.

12. Guess and Check to Determine an Average

The Guess and Check method (Section 3.8.13) can be useful when you are faced with limited "computing" resources, or are restricted in the operations that you may use. To illustrate, we'll find the average of a small group of numbers, but without using division. The approach is as follows: you first guess an average; you then multiply the average by the length of the list of numbers, and see how far off this product is from the sum of the numbers. That will then lead you to modify your initial guess, and you proceed from there to converge to the correct average.

Consider the numbers 16, 28, 12, 17, 23, 21. Find their average using Guess and Check. Again, you may add, subtract, and multiply, but may not use division. Show all your steps—how many guesses did it take?

13. Rabbit and Turtles

Two turtles, A and B, start 300 m apart and move towards each other. Turtle A moves at 80 m/h, train B at 70 m/h. A rabbit starts hopping from turtle A, moving at 137 m/h, straight towards turtle B. When it reaches turtle B, it reverses direction and runs towards turtle A, until it reaches it. The rabbit continues running back and forth between the two turtles until they meet. Find the distance that the rabbit will have traveled.

Note how this problem can be solved either with great difficulty or very easily, depending on whether you look at it from a distance perspective or a from a time perspective.

14. Does Multiplication Equal Addition?

You want to test your friend by making the claim that adding two numbers is the same thing as multiplying them. He of course doesn't give any credence to your claim, so you tell him that $2 + 2$ and 2×2 give the same result. He still doesn't take you seriously in spite of that one very well known example. So you tell him that $3 + 1.5$ and 3×1.5 also produce the same result. And the same is true for $5 + 1.25$ and 5×1.25. You tell him that you can actually come up with an infinite number of similar examples to make your point. All in good jest of course, since your initial claim is ludicrous. Come up with a formula that produces pairs of numbers to support your initial claim. Explain why even when you can support a claim with an infinite number of examples, it does not make the claim true. Come up with another example of an obviously false claim that can nonetheless be made to appear true with an infinite number of examples.

15. What Do These Problems Have in Common?

Which two of the following three problems have the most in common? Explain.

(a) Find the acceleration produced on a given mass by a given force.
(b) Find the current flowing through a given resistor when a given voltage is applied across it.
(c) Find the volume of a container that houses a gas at a given pressure and given temperature.

16. Graphical Insights on Quadratic Equations

In Section 3.8.20, you saw how a change of representation can help you see a problem in a better light. Graphical representations are often used for that very purpose. A common example is with quadratic equations, where plotting the corresponding parabola curve can help you visualize whether or not there are any real roots, and where they are.

For each of the following three quadratic equations, plot the curve on a Cartesian coordinate system. See how the existence and number of roots (two, one, or none) corresponds with the parabola intersecting the x axis, being tangent to it, or not intersecting it. And note of course how the x-intercepts correspond to the roots.

(a) $x^2 - 5x + 6 = 0$

(b) $x^2 - 2x + 1 = 0$

(c) $x^2 - 3x + 4 = 0$

17. Substitution of Variables to Change a Problem into Something Familiar

In mathematics, a change of variables is often used to transform a new problem into a familiar one. Consider the following equation:

$$x^3 - 3x\sqrt{x} + 2 = 0$$

You are asked to find its roots, that is, the values of x that make it true. Solve the problem by rewriting the equation in terms of a variable y that is a function of x so as to lead you to a known problem, in this case a quadratic equation. From that solution, determine the roots of the original equation.

18. Combining Common Sense with Algebra

When solving problems, a combination of common sense and pragmatism can be helpful. That often occurs if some information seems to be missing, or for instance if you have fewer equations than unknowns.

Consider a problem where you have 100 beads. Some are red, some white, some blue, and they weigh, respectively, 20 grams, 6 grams, and 1 gram. The total weight of all 100 beads is 200 grams. Find how many beads there are of each color. Solve the problem using math and common-sense logic, not trial and error.

19. A Plan for Applying to College

Draw a concept map that illustrates the plan that you followed for applying to college, from the beginning of the spring semester of your junior year of high school through the first day of college classes. Draw a Gantt chart that shows a schedule for each of these tasks. Write up your solution to this problem in a manner such that you could use it to help explain the process to a high school student who is just beginning the college application process.

20. Gantt Chart

Engineering organizations and societies such as the Institute of Electrical and Electronics Engineers (IEEE) and American Society of Mechanical Engineers (ASME) sponsor a wide variety of professional conferences where researchers present their latest work in the field. The typical conference is a 3–4 day event, divided into sessions that may contain either papers, panel discussions, or tutorials. Papers are submitted by authors months in advance, and are evaluated by several reviewers before they can be accepted for presentation. In addition to presentations at the conference, the papers are also published in a book or CD that is disseminated at the conference. Conferences also typically provide some meals for attendees, as well as a banquet with a guest speaker.

Given this description of a typical conference, make a list of 5–10 tasks that would be part of organizing one. Draw a Gantt chart that schedules these tasks.

PART II

MODEL-BASED DESIGN

Laws of Nature and Theoretical Models

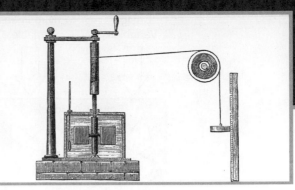

LEARNING OBJECTIVES

- to explain what a "theory" is, and the relationship between theory and experimentation;
- to discuss the evolution of basic theories of motion, conservation of energy, and conservation of mass;
- to discuss the interplay between practice and theory in the development of the piston engine, and to use simple theoretical models to analyze aspects of its behavior;
- to use theoretical models in solving a design problem, such as finding sizings for the lever handle of a handpump.

4.1 ENGINEERING MODELS

Some engineering students have been fortunate enough to participate in pre-engineering programs such as the First LEGO™ League robotics design competition or American Society of Civil Engineering bridge-building contests. In addition to fostering creative problem-solving skills, such projects also introduce students to the important notion that seemingly good ideas don't always work out in practice. Often in such programs, students have ample opportunity to test and modify their designs before they formally evaluate them. If the design doesn't work, then like a sculptor working with clay, the designer adds something here or removes something there until the design is acceptable.

This cut-and-try methodology is also sometimes used in industry, particularly in circumstances where the design is simple, or where the risk or cost of failure is low. In many situations, however, there is no second chance in the event of failure.

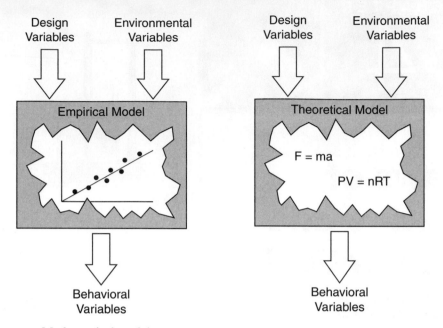

Figure 4.1 Mathematical model

For engineering systems such as buildings, bridges, or airplanes—to name just a few—failure to meet specifications could mean a loss of life. For others—such as an integrated circuit chip—the cost of fabrication is so high that a company may not be able to afford a second chance. In these situations, it's critical for the engineering team to be highly confident that a design will be acceptable *before* it's built. To achieve this, engineers use *models* to predict the behavior of their designs. A model is an approximation to a real system, such that when actions are performed on the model, it will respond in a manner similar to the real system. Models can have many different forms, ranging from physical prototypes such as a crash-test dummy to complex computer simulations. In this part of the book, we introduce some of the basic forms of mathematical models used widely in engineering analysis and design.

We can think of a mathematical model as a kind of virtual system, as Figure 4.1 illustrates, whose input is a set of variables that represent either aspects of the design or aspects of the environment, and whose output is a set of variables that represent the behavior of the system. Inside is a set of mathematical relationships that describe the operation of the system.

Broadly, we can classify mathematical models into two categories, according to the nature of these relationships. *Theoretical models*, the focus of Chapter 4, describe how a system will behave in terms of physical laws and relationships, using fundamental theories from physics, chemistry, or other sciences. *Empirical models*, on the other hand, discussed in Chapter 5, are based on observation or experience, often without regard to understanding of how a system works. Consider for example, two different ways of modeling a golf shot. In order to design a high-performance golf club, as Figure 4.2 illustrates, engineers at an equipment manufacturer would use theoretical models based on laws of physics including Newton's Laws of Motion, Hooke's Law, conservation of momentum, conservation of energy, and laws of aerodynamics to model what happens when a club strikes a ball. A typical golfer, on the other hand,

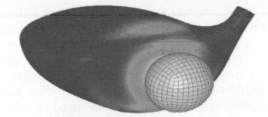

Figure 4.2 Engineers at Karsten Manufacturing, maker of PING golf clubs, used theoretical models running on a Cray supercomputer to analyze the performance and structural integrity of a new driver design [Mor07].

would use a simple empirical model to select a club for a shot, perhaps based on the average distance of each determined by hitting a bucket of balls at a driving range.

Sometimes, an engineering model may be as simple as a single equation. For example, we could construct a reasonable model for a spring scale—such as that pictured in Figure 4.3—from Hooke's Law, which states that the extension of the spring is proportional to the force suspended from it, or $F = kx$. Oftentimes, however, it takes a collection of many equations to describe the relationships between the parts of a system, all subject to multiple laws of nature. Consider, for example, modeling how a bridge deforms when a heavy truck passes over it. A common way of modeling a structure such as a bridge is as a network of springs, as shown in Figure 4.3. Each spring in the network still obeys Hooke's Law. But the force across each spring depends upon the displacements of the joints at either end, which in turn depends upon the forces. Further, Hooke's Law alone isn't enough to properly model the deformation; we also need Newton's Laws of Motion to determine the balance of forces that keep the bridge in place, as well as laws for material properties that let us represent steel beams as springs. We find the same type of situation in modeling the behavior of an electrical circuit, where we need to consider the currents through all elements and the voltages at all points in the circuit all together. Chapters 6 and 7 provide two detailed studies of modeling interrelationships in systems, the one for a type of lightweight structure known as a *truss* and the other for *digital electronic circuits*. Mathematically, the analysis of both kinds of systems lead to a set of interdependent equations that must be solved simultaneously.

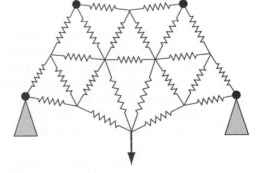

Figure 4.3 Springs.

Predicting how far a golf ball will fly or how long our known oil reserves will last requires being able to model how a system changes over time. The final chapter in this part of the book, Chapter 8, looks at ways of modeling change. We use a technique developed by the 18th-century mathematician Leonhard Euler that models a changing system as a sum or accumulation of small changes over time. Using Euler's method, we show how we can model changing systems using a spreadsheet that complements traditional methods based on calculus.

4.2   EVOLUTION OF THEORY

Consider the following three problems:

- A 10 kg block is held at a height of 5 m. What is its potential energy?
- A 1 liter cylinder contains air at 100° C and at atmospheric pressure. What would be the pressure inside the tank if the cylinder were cooled to 0° C?
- A steel spring stretches by 2 cm when a 20 N weight is suspended from it. What would be the total amount of stretch if an additional 30 N were suspended from the spring?

These are typical of the kinds of "word problems" that often students see in high-school science classes. To solve these, students learn to look for the right formula, and then input values to obtain a result. Students who have taken a high-school physics course, for example, would likely recognize that solving the first problem requires the formula for gravitational potential energy, which states that potential energy of a mass m held at some height h is equal to the product of its weight and height, or

$$E = mgh$$

where g is the acceleration due to gravity. Students who have taken a high-school chemistry course would probably realize that they can solve the second problem using the formula for the Ideal Gas Law, which states that for a gas in a container, the pressure inside the container times its volume is proportional to the number of molecules of the gas times its temperature, or

$$PV = nRT$$

Finally, the third problem requires Hooke's Law, which states that the stretch of a spring is proportional to the force applied to it, or

$$F = kx$$

The difference between solving a "word problem" such as the ones above, and building and using a theoretical model is subtle but important, and illustrates the process of maturing from the kind of thinking expected of high-school students to that required of engineers. Taking a step back, we can view each question above as asking how a man-made object—a block or a cylinder or a spring—would behave under conditions in nature. In answering *this* question, an engineer is not just plugging numbers into a formula; rather, he or she draws upon what we know about how nature works to make a prediction.

Much of what we know about how nature works has been conveniently packaged into what we call "Laws of Nature," such as the Law of Conservation of Energy, the Ideal Gas Law, or Hooke's Law. In high school, we often learn these laws as facts, as truths about how the universe operates. In reality, however, these laws are *theories* that scientists have constructed and described mathematically, and that they continue to test. In his *Lectures on Physics*, Richard Feynman describes the origin of these laws as follows:

> Each piece or part of the whole of nature is always merely an approximation to the complete truth, or the complete truth so far as we know it. In fact, everything we know is only some kind of *approximation*, because *we know that we do not know all the laws* as yet. Therefore, things must be learned only to be unlearned again or, more likely, to be corrected.
>
> The principle of science, the definition, almost, is the following: *The test of all knowledge is experiment.* Experiment is the *sole judge* of scientific "truth." But what is the source of knowledge? Where do the laws that are to be tested come from? Experiment, itself, helps to produce these laws, in the sense that it provides us with hints. But also needed is *imagination* to create from these hints the great generalizations—to guess at the wonderful, simple, but very strange patterns beneath them all, and then to experiment to check again whether we have made the right guess [Fey94].

It comes as a surprise to many to learn that the typical Law of Nature did not emerge all at once to one person in a "eureka" moment; in many cases, it took centuries of scientists collecting, examining, and discarding hints before the simple statements that we learn today took shape. In between, as Feynman stated, it took imagination or inspiration to fit the pieces of theory together. While we might initially think of the three laws presented above, the Ideal Gas Law, The Law of Conservation of Energy, and Hooke's Law, as three completely separate ideas, in reality, the history of their development is tightly intertwined, involving many of the same people, places, and problems. Further, the inspiration for each advance came from across the spectrum of human motivations, ranging from a desire to be the first with a new idea or to disprove an old one, to a desire to develop a successful product or business or a desire to better understand the world in the context of one's religious faith.

In the remainder of this chapter, we illustrate the evolution of a set of theoretical models by tracing some of the history behind the Ideal Gas Law, the Law of Conservation of Energy, Hooke's Law, and the Conservation of Mass. We show how early ideas behind them led to the development of the steam engine, and how improvements to the steam engine in turn led to new and better models. Rather than presenting an inventory of laws as facts, we instead focus on how for most of the history of their development, the models proposed by some very smart people—such as Aristotle, Galileo, Descartes, and Hooke—were either *wrong* or at least inaccurate when applied in an unforeseen context. Further, there's no reason to doubt that many of the scientific theories in circulation today will undergo similar revisions. That's not to say that engineers shouldn't trust the laws of gravity when building a bridge or theories of genetics when designing new medical treatments. The reason for this is that these theories have been tested over and over again, and each time come up with successful predictions, proving them to be very good approximations, if not facts. On the other hand, engineers need to be skeptical about basing a critical decision on a theory that they don't understand, and in fact, have a professional responsibility to raise questions.

Figure 4.4 shows a timeline of some of the philosophers, scientists, and engineers that we mention, which is a nearly unbroken chain from the mid-1500s until today. We begin by looking at some of the early models for matter and motion Aristotle proposed in 335 BC that had dominated until they were challenged by Galileo in the

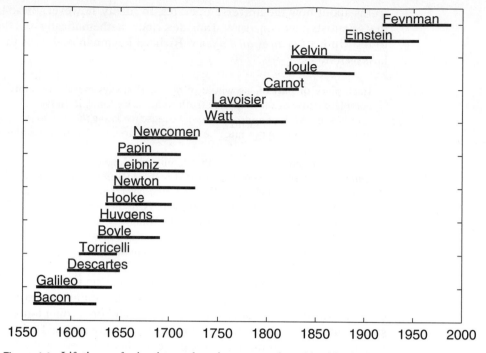

Figure 4.4   Lifetimes of scientists and engineers mentioned in this chapter.

early 1600s. Then, we focus on a group of scientists and mathematicians from across Europe during the 1600s who worked on a series of overlapping problems that laid the foundations for each of these laws. Many of these men, including Robert Boyle, Robert Hooke, Christiaan Huygens, Isaac Newton, Gottfried Leibniz, and Denis Papin, were connected through the Royal Society of London, and found themselves enmeshed in the same kinds of collaborations as well as conflicts that we find in almost any science and engineering enterprise today. Next, we describe how this group directly contributed to the design of the first modern steam engines and look at some of the improvements to the design by Thomas Newcomen and James Watt over the 1700s. From there, we discuss how a desire to understand the limitations of the steam engine led men including Sadi Carnot, James Joule, and William Thomson (Lord Kelvin) to develop new theories in the 1800s that produced the field of study of heat and energy known as thermodynamics. Finally, we jump back to 17th-century London and trace how the quest to turn lead into gold led to modern chemistry and the law of conservation of mass, and how the experiments of early alchemists addressed some of the same questions that we do today as we consider alternative fuels for the modern internal combustion engine.

4.3   MODELS OF MOTION

4.3.1   Aristotle's Physics

Today, scientific knowledge and experimentation go hand in hand, but just a few generations before the technology that produced the steam engine, this linkage was still a new idea. Until the late Renaissance, the ideas of the ancient Greek philosophers, and in particular Aristotle, guided the scholar's perspective on the natural world. Like his teacher Plato, Aristotle believed that the greatest human faculty was reason, and that the path to knowledge was through contemplation.[1] In Plato's Academy, the most important subjects of study were logic and mathematics. When Aristotle founded his school called the *Lyceum* in 335 BC, he added to the curriculum the study of the natural world, including physics and mechanics and biology, and in doing so created the first organized program in the sciences.

For Aristotle, the study of nature was a search for *causes*. In his view, every object, either living or inanimate, was perfectly crafted to serve a particular purpose. According to Aristotle, everything in nature was composed of four basic elements: earth, water, air, and fire. Motion was also a central idea in Aristotle's philosophy, and for him, it meant the process of something seeking its natural state in an orderly world. This very broad concept applied equally to the process of a person aging from a child to an adult, as it did to the process of a stone dropped from the hand and falling to the ground. Aristotle made a distinction between natural motion, or things moving of their own accord, and violent motion, or things being pushed by an external force. The natural motion of objects was tied to the natural order of the elements of which they were made. Thus, earth moves downwards most strongly, and water flows downwards too, but not so strongly, since a stone will fall through water. In contrast, air rises in water and fire rises most strongly of all, since it shoots upward through air. Since wood drops through air but floats in water, Aristotle would conclude that it contains both earth and air.

Aristotle was the first to think quantitatively about the *speed* of movements, and asserted several laws of motion:

1. Heavier things fall faster, with a speed proportional to their weight;
2. The speed of a falling object is inversely proportional to the density of the medium through which it falls;
3. For violent motion, the speed of a moving object is in direct proportion to the applied force.

These rules have an elegance and simplicity that would appeal to both the senses and to reason in Aristotle's world: after all, a leaf falls slower than a stone, a stone falls faster in air than in water, and the speed of a cart depends upon how hard you push it. Today, of course, we understand that each of these laws is incorrect. Nonetheless, Aristotle's work was of such a quality that it was widely accepted and even became part of the official orthodoxy of the Christian church, which posed problems for scientists such as Galileo who questioned it nearly two thousand years later.

4.3.2 Galileo and the Scientific Method

During the Renaissance, the Aristotelian world view that dominated for centuries faced its first serious challenges. The English philosopher and statesman Francis Bacon (1561–1626) wrote a series of books and essays that argued forcefully to trust experimental evidence over logic and reason alone, and thus established the philosophical basis for the *scientific method* as we know it today. Around the same time in

[1] The author thanks Dr. Michael Fowler, Beams Professor of Physics, University of Virginia, for use of materials from his course "Physics 109: Galileo and Einstein" in writing this section on Aristotle's physics.

Italy, Galileo Galilei (1564–1642) embarked on a course of study that would prove much of Aristotle's physics wrong. While Galileo is best known today for his discoveries in astronomy and physics—as well as for his conflicts with the Church—his greatest contribution to modern science was perhaps the pattern he established for conducting careful experiments and making detailed observations, for giving precise descriptions of his methods and data, and for constructing convincing arguments based on these results. From his writings, we find that Galileo had little patience for "scientific" statements that couldn't be supported by data, and seemed to relish the opportunity to confront and overwhelm his intellectual opponents. Under threat of torture, the Church forbade Galileo from advocating the Copernican theory that the earth revolves around the sun, and in order to publish his results in astronomy, he agreed to present all viewpoints. In his *Dialogue Concerning the Two Chief Systems of the World: Ptolemaic and Copernican*, Galileo couched his arguments in the form of a discussion among three fictional characters, "Salviati," who represented Galileo's viewpoint, "Sagredo," a learned colleague, and "Simplicio," who represented the Church's view and whose arguments were easily dismissed by the other two. The Church was understandably furious and banned the book. Undaunted, Galileo employed the same device again in his *Discourses on Two New Sciences*, in which he criticized Aristotle's physics of motion [Gal14]:

SALVIATI: I greatly doubt that Aristotle ever tested by experiment whether it be true that two stones, one weighing ten times as much as the other, if allowed to fall, at the same instant, from a height of, say, 100 cubits, would so differ in speed that when the heavier had reached the ground, the other would not have fallen more than 10 cubits.

SIMPLICIO: His language would seem to indicate that he had tried the experiment, because he says: We see the heavier; now the word see shows he had made the experiment.

SAGREDO: But I, Simplicio, who have made the test, can assure you that a cannon ball weighing one or two hundred pounds, or even more, will not reach the ground by as much as a span ahead of a musket ball weighing only half a pound, provided both are dropped from a height of 200 cubits.

Legend has it that Galileo tested his theory by dropping two balls from the Leaning Tower of Pisa. There's no record that he actually did that, but in *Two Sciences*, he describes a much more sophisticated set of experiments using pendulums and inclined planes to study the motion of falling objects. Making careful timing measurements with a water clock, Galileo showed that the speed of a falling ball *increased* at a constant rate, whereas Aristotle had stated that the speed remained constant. Galileo also had the insight that a ball falling from a given height would acquire sufficient *momentum* to carry it back up to the same height, regardless of the path it follows, as long as there isn't any friction. He demonstrated this with an "interrupted pendulum" as shown in Figure 4.5. Further, he calculated that the square of the speed at the bottom of the pendulum was always proportional to the height from which it was released, or

$$h = \text{constant} \times v^2 \qquad (4.1)$$

Example 4.1 illustrates the kinds of calculations Galileo would have made in his experiments.

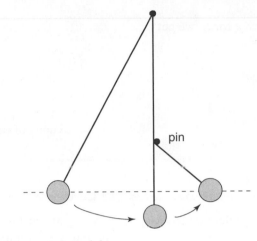

Figure 4.5 Galileo's interrupted pendulum.

Example 4.1	**Speed of a Ball Down a Ramp**

Galileo used a water clock that drained water at a steady rate from a tank into a cup to accurately measure time in his experiments. Suppose that Galileo let a ball roll down a ramp and then opened the spigot on the water clock as soon as it reached the bottom and closed it once the ball rolled an additional two cubits. If there were 4 units of water in the cup after this experiment, how much water should be in the cup if he repeated the experiment with a ramp twice as high?

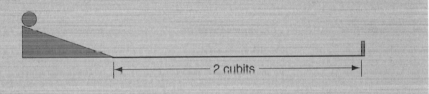

Solution

Given: Ramp R_1 of height h_1 and roll time $t_1 = 4$ units. Ramp R_2 of height $h_2 = 2h_1$ and roll time t_2.

Find: t_2 in equivalent units of water.

Plan: Use Equation (4.1), to specify the proportionalities between the heights of the ramps and the roll times for the two experiments and solve for t_2.

Analysis: According to Equation (4.1), if v_1 and v_2 are the velocities of the ball at the bottom of the ramp for each of the two experiments, then

$$\frac{h_1}{v_1^2} = \frac{h_2}{v_2^2}$$

Since velocity is inversely proportional to the roll time,

$$h_1 t_1^2 = h_2 t_2^2$$

Solving for t_2, we get

$$t_2 = t_1 \sqrt{\frac{h_1}{h_2}}$$

$$= 4\sqrt{\frac{1}{2}}$$

$$= 2.8 \text{ units of water}$$

4.3.3 René Descartes and Conservation of Motion

For Galileo, discovering that the square of the speed of a falling object was proportional to the distance of the fall was an important result, because it disproved Aristotle's theory. Galileo, however, didn't go on to answer the question of *why* this might be true, or even try to calculate the constant of proportionality between h and v^2. He used the term 'momentum' to explain how the speed of the ball of a pendulum always carries it to the same height, but he didn't give the term a precise definition and sometimes used it in contradictory ways. Today, we would recognize Equation (4.1) as a consequence of the equivalence of gravitational potential energy—mgh—and kinetic energy—$\frac{1}{2}mv^2$—but it would take more than a century before scientists understood this result in those terms. In short, Galileo didn't think of his observations in terms of something being *conserved*.

A different kind of thinker, René Descartes, took the next step beyond experimental observation toward posing a theory of conservation. Born in France in 1596, Descartes lived and worked during the same time as Galileo. Both men profoundly affected the future of science, but their contributions and methods greatly differed. Whereas Galileo opened the door to using experimentation as a way of testing ideas, Descartes brought new life to using reason and mathematics in the manner of Aristotle for thinking about how nature *might* be organized. Using today's terminology, we would call Descartes a *theoretician* and Galileo an *experimentalist*, and the two approaches are complementary parts of the scientific discovery process.

While Galileo fought with the Church, Descartes drew inspiration from his religious views. In his *Principia Philosophiae* of 1644, Descartes asserted a law of conservation of motion that was based upon his religious faith:

> *Descartes' Law of Conservation of Motion:* That God is the primary cause of motion; and that He always maintains an equal quantity of it in the universe [Des91].

Descartes imagined a universe completely filled with particles of earth, air, and fire, bouncing around in clusters as objects in a sea of a fourth kind of celestial particle called *aether*. When two objects or clusters of particles collided, one could impart a quantity of motion to the other, but the total amount of motion remained the same. Descartes defined the measure of a quantity of motion as follows: "when one part of matter moves twice as fast as another twice as large, there is as much motion in the smaller as in the larger [Des91]." Thus, we could write Descartes' definition of the quantity of motion in an object as

$$\text{quantity of motion} = \text{size} \times \text{speed}$$

where to Descartes, size referred to volume, not weight or mass. After his conservation law, Descartes introduced three more laws of nature that further describe his theory of how things move [Bla66]:

1. Each thing, insofar as in it lies, always perseveres in the same state, and when once moved, always continues to move.

2. Every motion in itself is in a straight line, and therefore things which are moved circularly always tend to recede from the center of the circle which they describe.

3. If a moving body A collides with moving body B and if A has less force to continue in a straight line than B has to resist it, A will deflect in the opposite direction and, retaining its own motion, will lose only the direction of its motion. If, however, A has a greater force than B does, then body A will move with body B and give it as much of its motion as it loses.

The first two laws refer to the motion of individual objects. The first law describes the principle of *inertia*, which means "resistance to change," and jibes with our everyday experience that the motion of an object won't change unless something interacts with it. Descartes illustrates the second law with the example of a stone thrown from a rotating sling that flies off in a direction tangent to the circular motion, which also matches our expectations. Upon close inspection, however, experience tells us that Descartes' third law is wrong. Descartes himself gives several examples to illustrate his third law that not only contradict what we would observe in practice, but also seem to contradict his own conservation law and definition of motion! In one such example, Descartes considers a moving object B that collides with a larger object C which is initially at rest. We can picture these objects as two metal balls suspended from strings as pendulums, as illustrated in Figure 4.6. In this example, Descartes states that B would bounce off C with its original speed but in the opposite direction, and that C would remain at rest. According to his third law and the definition of motion, however, since B has more motion than C (which has none because it's at rest), B should continue along the same line at a lower speed and C should move along with it at the same speed. But what would actually happen, however, if we

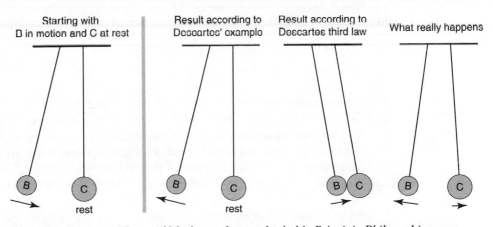

Figure 4.6   Descartes' Laws of Motion and examples in his *Principia Philosophiae* were wrong about collisions.

performed the experiment ourselves, is that B would bounce back at a reduced speed and C would move forward at a speed slower than B.

Even though Descartes' model of motion didn't agree with experimental evidence, his methods and ideas profoundly influenced the future of scientific thought. The next generation of scientists studied his work carefully and by the end of the 1600s, would develop models that accurately predicted the speeds and directions of objects after a collision. While doing so, they would develop an entire theoretical model of motion that could not only predict how long it would take for a falling apple to hit the ground, but that could also explain why the moon revolves around the earth once every 28 days. Further, they would lay the foundation for two separate conservation laws: conservation of momentum and conservation of energy.

4.3.4 The Royal Society

During the mid-1640s, a reading group began meeting in London to discuss the works of Descartes, Galileo, and Bacon, and in general "for the Promoting of Physico-Mathematicall Experimentall Learning" [The]. In their time, these men were known as "natural philosophers"—the term "scientist" wasn't coined until the 1800s—and it was common for them to work not only across different branches of science, but also across the arts, philosophy, and theology. One of the group's organizers, Robert Boyle, was both an alchemist who opened a laboratory in Oxford, and a deeply religious man who wrote extensively on his religious views. Boyle's assistant, Robert Hooke, was a skilled painter, musician, and craftsman who had a hand in developing inventions as diverse as the vacuum chamber and the wind-up clock, and whose 1664 best-seller *Micrographia* contained detailed drawings of insects, fossils, and other specimens that he observed through his improved microscope. As the group's membership grew, their meetings became more formal and in 1660, they obtained a charter from King Charles II to form the "Royal Society of London for Improving Natural Knowledge" [The]. During its early years, the Royal Society elected many of Europe's leading scientists as members, including the Dutch physicist and astronomer Christiaan Huygens, the German mathematician and philosopher Gottfried Leibniz, and the Lucasian Professor of Mathematics at Cambridge who would become one of the most influential scientists of all time, Isaac Newton. Today, the Royal Society remains one of the world's most prestigious scientific organizations, whose membership includes the current holder of the Lucasian Chair, Stephen Hawking.

As its motto, the Society selected the Latin phrase "*Nullius in Verba*," which translates roughly as "don't trust anyone's word on anything." This proclaimed the Society's dedication to the importance of experimental evidence over book learning alone, but also foretold some of the controversy and distrust that arose amongst its early members. Within the first few decades, Hooke would accuse Huygens of stealing his ideas for a spring and accuse Newton of stealing his ideas in optics. Newton, in turn, removed Hooke's portrait from the gallery when he became the Society's president, and fought with Leibniz over credit for the invention of calculus.[2] Yet despite these conflicts, The Royal Society flourished as a hub for the exchange of ideas from across

---

[2]It was Leibniz who devised the notation that we use for calculus today, the long 'S' symbol for integration $\int$ from the Latin *summa* and d/dx for differentiation.

Europe, and the basic laws of nature that today bear the names of its early members, including Boyle's Law for gases, Hooke's Law for springs, and Newton's Laws of motion, grew from these collaborations.

The Society met regularly to discuss a range of topics, and one of the early issues that they considered was challenges to Descartes' theory of conservation of motion. They were intrigued by Descartes' basic idea that *something* related to motion was conserved in nature, but realized that the details of his model weren't correct. It is important to understand the difficulties scientists faced when they tried to model the dynamics of colliding objects during the 1660s; they had neither the consistent definitions of physical concepts nor the mathematical tools that we learn in high school today.

On the physics side, there still wasn't a clear understanding of the difference between mass and weight. Further, there was confusion over the definition of a "force," where some used this term or its Latin translation "*vis*" in the same sense that we do today, while others used it to mean the concept that we would call "energy" or the concept that we would call "momentum." Our common definitions of mass and force come from Newton's *Philosophiae Naturalis Principia Mathematica*, which wasn't published until 1687, although he worked out most of the concepts between 1666 and 1668, when Cambridge was closed as a precaution against the Plague. Newton defined the *mass* of an object as "the quantity of matter ... arising from its density and bulk conjointly" [New29]. He defined a *force* as something that causes a body to accelerate uniformly, where the acceleration is proportional to the force and inversely proportional to the mass of the body. Today, we express this relationship as $F = ma$, although Newton himself never wrote it down this way, instead describing his laws in text as was common at the time. Newton also realized that gravity was a force, and that the weight of a body was the force of gravity upon it. Today, we express this as $W = mg$, where g is the acceleration due to gravity.

As for mathematics, arithmetic algebra, in which letters represent numeric values in equations, was still a new idea that was just catching on, and it wasn't until the 1800s that mathematicians began to solve algebraic equations for unknown variables using new-found laws such as the commutative, associative, and distributive properties. Instead, during the late 1600s, mathematicians would convert algebraic expressions to geometric figures, and then analyze the figures using the same laws of geometry that Euclid formulated around 300 BC. Descartes first introduced this technique of "graphing" algebraic expressions, in which an expression such as x^2 might be drawn as a line segment of a certain height, in 1637, in the appendix *Geometry* to his *Discourse on Method*—the same book where he famously wrote "I think, therefore I am" [Des37].

4.3.5 Huygens' Improvements to Descartes' Model

Christiaan Huygens first figured out a correct solution to the problem of collisions between hard objects using an approach very different from Descartes' [Goe75]. From his correspondence, it's clear that he had a solution by 1655 when he was only 23, but he didn't present his result to the Royal Society until 1668, and his full study *De Motu Corporum ex Percussione* [Huy88], or *Of the Motion of Bodies in Collision*, wasn't published until after he died. Huygens based the first part of his solution on the assumption that the perception of motion is relative to the observer. Galileo first stated this *principle of relativity* in his *Dialog* in 1632, and Einstein made use of it again nearly three hundred years later in 1905, with the added twist that the speed of light always appears to be the same to any observer. Huygens imagined two objects colliding on a boat, with one person observing the collision from the boat and another observing it from the shore. While both objects might appear to be moving to the left or right to the person on the boat, depending on the boat's motion, one could appear to be at rest to the person on the shore. Huygens reasoned that if a body collided with another body that remained at rest before and after the collision—such as a wall—it would bounce off at the same speed in the opposite direction. Thus, according to the principle of relativity, the relative speeds at which the two bodies separate after the collision would be the same as the relative speeds at which they approached each other before the collision. Algebraically, we can express this condition as

$$\text{Principle of relative velocity:} \quad v_{1_i} - v_{2_i} = v_{1_f} - v_{2_f} \qquad (4.2)$$

where v_{1_i} and v_{1_f} are the initial (before collision) and final (after collision) velocities of the first body, and v_{2_i} and v_{2_f} are the initial and final velocities of the second body. In his original formulation, Huygens considered only the *speed* of the bodies, which has a positive value, whereas we're using *velocity*, which can be positive or negative. We'll adopt the common convention that a body moving to the left has a negative velocity and that a body moving to right has a positive velocity.

In Equation (4.2) the velocities before the collision v_{1_i} and v_{2_i} are known and the velocities after the collision v_{1_f} and v_{2_f} are unknown. Thus we have only one equation but two unknowns, and so the principle of relative velocity alone is not sufficient to determine a unique solution to the collision problem. Huygens, therefore, needed to devise another "law" that applied to the situation. He rejected Descartes' Law of Conservation of Motion and instead considered Galileo's work on pendulums. An individual pendulum will always swing back to the same height, but Huygens wondered if a similar relationship was possible regarding the heights of two pendulums before and after a collision. The answer to this question resulted from a discovery Galileo's assistant, Evangelista Torricelli, made. In it, Torricelli determined that for any machine consisting of a set of interconnected weights, their *center of gravity* can't ascend on its own. As Figure 4.7 illustrates, the center of gravity is essentially the balance point between the weights. If we assume that two weights W_1 and W_2 are connected by a lightweight, rigid bar to form a lever, then the center of gravity is that point along the bar where

$$\frac{W_1}{W_2} = \frac{d_2}{d_1} \qquad (4.3)$$

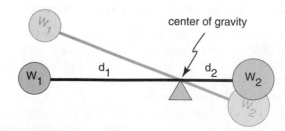

Figure 4.7   Center of gravity.

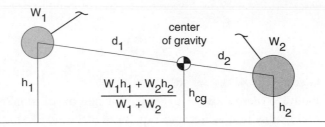

Figure 4.8 Calculation of the height of the center of gravity between two objects.

such that the lever is balanced. Regardless of whether the bar is horizontal or tipped up or down, the lever will still be balanced at the same point. On the other hand, raising (or lowering) the center of gravity would require applying some external force to the lever as a whole.

Huygens determined that Torricelli's principle would also apply to a system of colliding pendulums, such that if two pendulums released from two different heights collided at the bottom of their swing, when they return to the top of their swing, their center of gravity would be at the same height as it was before they were released, even though the individual pendulums may be at different heights. Using Equation (4.3) together with the geometry of similar triangles, it may be shown that if two weights W_1 and W_2 are at heights h_1 and h_2, as illustrated in Figure 4.8, then the height of their center of gravity h_{cg} will be

$$h_{cg} = \frac{W_1 h_1 + W_2 h_2}{W_1 + W_2} \tag{4.4}$$

According to Torricelli's principle, h_{cg} and hence the quantity $W_1 h_1 + W_2 h_2$ must have the same value before and after the collision. Further, as Galileo determined, the height of a pendulum swing is proportional to the square of its velocity at the bottom. Since v_{1_i} and v_{2_i} are determined by the heights before the collision and v_{1_f} and v_{2_f} determine the height after the collision, and since weight is proportional to mass, the following statement must be true:

Huygens' conservation of mv^2: $m_1 v_{1_i}^2 + m_2 v_{2_i}^2 = m_1 v_{1_f}^2 + m_2 v_{2_f}^2 \tag{4.5}$

Now, between the principle of relative velocities from Equation (4.2) and Huygens' principle of the conservation of mv^2 from Equation (4.5) we have two equations in two unknowns and can solve for the velocities v_{1_f} and v_{2_f} after the collision. Example 4.2 illustrates Huygens' method by checking Descartes' results.

Example 4.2	**Collisions According to Descartes and Huygens**

Collisions According to Descartes and Huygens

In his *Principles of Philosophy*, to illustrate his third law of motion, Descartes gives the example of a collision between two objects of equal size, one moving to the right at 4 degrees of speed and the other moving to the left at 6 degrees of speed, and states that after the collision they will both move to the left at 5 degrees of speed. According to Huygens' principles of relative velocity and conservation of mv^2, is this correct?

Solution

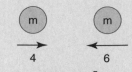

Given: $v_{1_i} = 4 \quad v_{2_i} = -6 \quad v_{1_f} = v_{2_f} = -5$

Find: Do these satisfy Huygens' principles of collision?

Plan: Substitute values into Equations (4.2) and (4.5).

Analysis: According to the principle of relative velocity,

$$v_{1_i} - v_{2_i} = v_{2_f} - v_{1_f}$$
$$4 + 6 = -5 + 5$$
$$10 = 0$$

According to the principle of conservation of mv^2,

$$m_1 v_{1_i}^2 + m_2 v_{2_i}^2 = m_1 v_{1_f}^2 + m_2 v_{2_f}^2$$
$$m(4)^2 + m(-6)^2 = m(-5)^2 + m(-5)^2$$
$$52m = 50m$$

Both statements are false, and thus both of Huygens' principles are violated by Descartes' example, even though they satisfy Descartes' Law of Conservation of Motion.

Huygens thus determined two laws governing collisions without using Descartes' Law of Conservation of Motion, which he realized was incorrect. It turned out, however, that Descartes' law could be fixed with a small change. By taking size to mean mass and replacing speed with velocity, which could either be positive or negative, we get a new quantity called *momentum* that *is* conserved:

$$\text{Conservation of momentum:} \quad mv_{1_i} + mv_{2_i} = mv_{1_f} + mv_{2_f}$$

Including conservation of momentum, we now have three laws we can use to solve for the two unknown velocities after a collision, and it's sufficient to pick any two. Example 4.3 illustrates a solution using the principles of conservation of momentum and relative velocity.

Example 4.3	**Determining Velocities After a Collision** Using the principles of conservation of momentum and relative velocity, determine the velocities after the collision for the bodies in Example 4.2.

Solution

Given: $m_1 = m_2 = m \quad v_{1_i} = 4 \quad v_{2_i} = -6$

Find: v_{1_f} and v_{2_f}

Plan: Solve as two linear equations in two unknowns

Analysis: We use the principle of relative velocity to obtain our first equation in the two unknowns.

$$v_{2_f} - v_{1_f} = v_{1_i} - v_{2_i}$$
$$= 10$$

We use conservation of momentum to obtain our second equation in the two unknowns.

$$mv_{1_i} + mv_{2_i} = mv_{1_f} + mv_{2_f}$$
$$v_{1_f} + v_{2_f} = v_{1_i} + v_{2_i}$$
$$= -2$$

When we solve the two equations in two unknowns, then $v_{1_f} = -6$ and $v_{2_f} = 4$. Thus, the velocities of the two bodies are swapped after the collision.

4.3.6   Newton's Laws of Motion

Huygens was aware that momentum was conserved, but he chose not to consider this in his paper. John Wallis presented the first paper on collisions using conservation of momentum to the Royal Society in 1668. It was Newton, however, who fully developed the concept of momentum and used it in his Laws of Motion, paraphrased from the first English translation of his *Principia* [New29]:

1. Every body at rest remains at rest, and every body in motion remains in motion, unless it's compelled to change that state by forces impressed upon it.
2. The change in motion (momentum) of a body is always proportional to the force impressed upon it.
3. For every action, there is an equal and opposite reaction, or the mutual forces of two bodies upon each other are always equal in magnitude and pointed in opposite directions.

While Descartes included a special case in his laws of motion to define rules for collisions, Newton's Laws are general enough to handle collisions without specifically mentioning them. Further, Newton's Laws cover forces and motion in *any* direction, whereas Descartes, Huygens, and Wallis only considered collisions between objects moving towards each other in a horizontal line. In order to model a collision with Newton's Laws, we need to use all three of them. First, according to Newton's Third Law, for as long as two bodies are in contact during a collision, each exerts an equal and opposite force on the other, $F_{A\text{-on-}B}$ and $F_{B\text{-on-}A}$, as shown in Figure 4.9. In reality, the magnitude of these forces changes over time as the two bodies squash together imperceptibly and then spring apart, but we can model them as an average constant force applied over some collision time, t_c.

Next, according to Newton's First Law, these forces will cause a change in the motion of each body. Finally, according to Newton's Second Law, the change in momentum for each of the bodies is proportional to the force acting on it. Thus, for body A,

$$F_{B\text{-on-}A} = \frac{\text{momentum after collision} - \text{momentum before collision}}{\text{collision time}}$$
$$= \frac{m_A v_{A_f} - m_A v_{A_i}}{t_c}$$

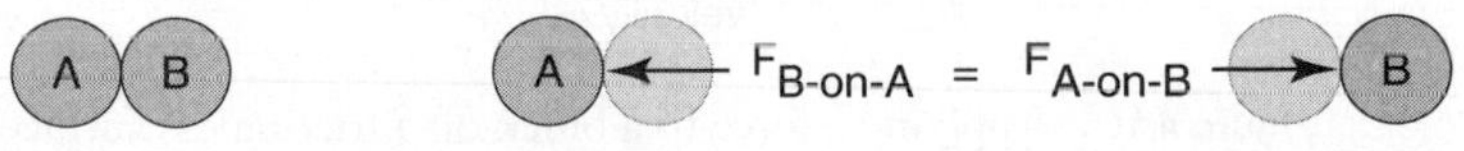

Figure 4.9   Newton's Third Law.

Similarly, for body B,

$$F_{\text{A-on-B}} = \frac{m_B v_{B_f} - m_B v_{B_i}}{t_c}$$

But since $F_{\text{B-on-A}} = -F_{\text{A-on-B}}$ according to the Third Law, we get

$$\frac{m_A v_{A_f} - m_A v_{A_i}}{t_c} = -\frac{m_B v_{B_f} - m_B v_{B_i}}{t_c}$$

$$m_A v_{A_i} + m_B v_{B_i} = m_A v_{A_f} + m_B v_{B_f}$$

which is the Law of Conservation of Momentum.

In deriving the conservation of momentum during a collision from Newton's laws, we considered the application of equal and opposite force over the *time* that the two bodies were in contact. To derive the conservation of mv^2, we consider the application of the forces over the *distance* that the bodies are in contact, that is, over the very short distance where the two bodies squash and spring off against each other. Before looking at a collision, however, let's first model the situation of a force pushing a single body over a given distance, such as the block sliding across a frictionless surface shown in Figure 4.10. Suppose that the body has a mass m and an initial velocity v_i, and then we apply a force F to it until it moves by some distance x and reaches a final velocity v_f at time t. From Newton's Second Law, the force is equal to the rate of change of momentum, or

$$F = \frac{\text{change in momentum}}{\text{change in time}}$$

$$= \frac{m(v_f - v_i)}{t}$$

Multiplying both sides by the distance x, we get

$$F \cdot x = \frac{m(v_f - v_i)}{t}x \qquad (4.6)$$

Next, we express x in terms of velocity and time. If the block moves at a constant velocity, the distance is simply the product of velocity and time. Since it accelerates uniformly, however, the distance is the product of its *average* velocity and the time. Thus,

$$x = \frac{(v_f + v_i)}{2}t$$

Substituting this result for x into Equation (4.6), we get

$$F \cdot x = \frac{m(v_f - v_i)(v_f + v_i)}{2}$$

$$= \frac{1}{2}mv_f^2 - \frac{1}{2}mv_i^2 \qquad (4.7)$$

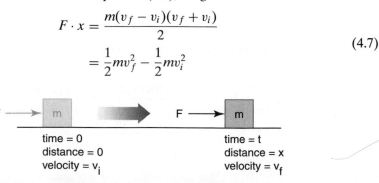

time = 0
distance = 0
velocity = v_i

time = t
distance = x
velocity = v_f

Figure 4.10   Applying a force to a block on a frictionless surface.

With our 21st-century eyes, we recognize Equation (4.7) as an important statement about the equivalence between mechanical work, $F \cdot x$, and kinetic energy, $\frac{1}{2}mv^2$. Newton, however, wouldn't have seen it this way; to him, it would simply be a theorem we derived from his laws of nature, so we'll continue with the model without calling further attention to it—for now.

Given the result in Equation (4.7), we return to the case of a collision between two bodies, A and B, as pictured earlier in Figure 4.9. Again, according to Newton's Third Law, the force that A exerts on B over the distance x is equal and opposite to the force that B exerts on A, so

$$F_{\text{B-on-A}} \cdot x = -F_{\text{A-on-B}} \cdot x$$

Applying Equation (4.7) and rearranging terms, we get

$$\frac{1}{2}m_A v_{A_f}^2 - \frac{1}{2}m_A v_{A_i}^2 = -\left(\frac{1}{2}m_B v_{B_f}^2 - \frac{1}{2}m_B v_{B_i}^2 \right)$$
$$m_A v_{A_i}^2 + m_B v_{B_i}^2 = m_A v_{A_f}^2 + m_B v_{B_f}^2$$

which is the law of conservation of mv^2. As before, given the two equations for conservation of momentum mv and conservation of mv^2, we have enough information to solve for the two unknown velocities after the collision.

4.3.7 Leibniz and the "Living Force," Work and Energy

Huygens and Newton both realized that the quantity mv^2 was conserved in a collision, but neither placed any particular significance on this; unlike the quantity mv, which was called momentum, they didn't even bother to give mv^2 a name. To the German philosopher and mathematician Gottfried Leibniz, however, mv^2 held special meaning. Leibniz argued that mv^2 should be the true measure of motion conserved in the universe, and not size times speed as Descartes had suggested. In 1686, Leibniz published a paper with the inflammatory title,

> *Brief Demonstration of a Notable Error of Descartes and Others Concerning a Natural Law, According to Which God is Said Always to Conserve the Same Quantity of Motion; A Law Which They Also Misuse in Mechanics* [Ilt71].

Leibniz's paper really didn't demonstrate anything new, as it was published roughly twenty years after Huygens and Newton had found the error in Descartes' Law of Conservation of Motion and had also determined that both mv and mv^2 were conserved in collisions. Like Descartes' work, Leibniz's argument was more of a philosophical statement about a way of understanding the organization of the universe. Leibniz gave the quantity mv^2 the Latin name "*vis viva*" or the "living force," and in particular noted how for a falling object, *vis viva* is converted between height and motion. For much of the 18th century, Leibniz's theory of the conservation of *vis viva* was a controversial topic. This was partly due to other disputes that arose between Leibniz and members of the Royal Society, especially the dispute between Newton and him over credit for the invention of calculus.

While, as we showed, the conservation of *vis viva* could be derived from Newton's laws of motion, treating it separately became more important during the 19th century, when scientists and engineers placed greater emphasis on understanding the theory behind the machinery that powered the Industrial Revolution. In 1807,

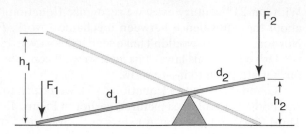

Figure 4.11   A lever.

the English scientist Thomas Young[3] coined the term "*energy*" to mean the same thing as Leibniz's *vis viva* [You07], and roughly 20 years later, the French physicist Gaspard Coriolis[4] coined the term "*travail*" or "*work*" in English for the process of moving a force over a distance. These are the same terms we still use today, and thus Equation (4.7) would state that the quantity of work $F \cdot x$ is equivalent to the (kinetic) energy $\frac{1}{2}mv^2$.

One of the curious properties of work is that the amount of work performed in moving a small force over a long distance can be the same as the work performed in moving a larger force over a shorter distance. To illustrate this principle, consider the example of using a lever to lift a heavy object, as shown in Figure 4.11. If just enough force F_1 is applied to the left end of the lever to lift a weight F_2 at the right end of the lever, then

$$\frac{F_1}{F_2} = \frac{d_2}{d_1}$$

By similar triangles,

$$\frac{d_2}{d_1} = \frac{h_2}{h_1},$$

and so

$$F_1 \cdot h_1 = F_2 \cdot h_2$$

Thus the amount of work done in pushing the left end of the lever down is equivalent to the amount of work done in lifting the weight at the right end. At either end the amount of work *must* be the same, or else a lever—run in one direction or the other— would create energy, which doesn't make sense. In other words, a lever converts work in the form of one kind of motion to an equivalent amount of work in the form of a different kind of motion. Any other simple machine operates similarly, such as a block-and-tackle or a gear train. This idea grew extremely important during the 1700s and 1800s as engineers invented mechanisms for using a simple up-and-down piston engine to power factories.

---

[3]Young is perhaps best known by engineers today as the name behind *Young's modulus*, which describes how much a given material will stretch or compress under pressure. Beyond this, Young also extended Huygens' theories of optics, and made the first translations of Egyptian hieroglyphics from the Rosetta Stone.

[4]Coriolis is best known for his work on the motion of rotating bodies, and today we use the term *Coriolis effect* to describe why hurricanes or cyclones rotate counterclockwise in the northern hemisphere and clockwise in the southern hemisphere.

4.4 | MODELING THE "SPRING OF AIR"

4.4.1 The Horror of the Vacuum

To Aristotle, a vacuum—a region completely devoid of matter—was impossible, since logically, everything had to be made of *something*. Further, according to Aristotle's laws of motion, the speed of an object falling through a medium with zero density would be infinite, a result he considered absurd. The notion of a vacuum also bothered the Church, since according to doctrine, a "void" shouldn't exist on earth following Creation. This concept was known as *horror vacui*, literally the "horror of the vacuum" or "nature abhors a vacuum," and people believed that if ever a vacuum was created, something would immediately rush in to fill it. *Horror vacui*, for example, explained how a siphon or straw worked; when the air was sucked out of a tube, the "horror" of the vacuum caused water to fill the void.

In 1630, a nobleman and amateur scientist from Genoa named Giovan Battista Baliani wrote a letter to Galileo asking why the siphon he designed couldn't carry water over a hill 21 meters high [Ins07] [Mid63]. Through experiment, Galileo determined that the maximum height that a suction pump and siphon could lift a column of water was only about 11 meters. He argued that this was because there was indeed a vacuum inside the tube, and that the vacuum exerted a force on the water, but like a rope, the tube of water could withstand only so much force before breaking. Galileo sent his conclusions to Baliani, who wasn't convinced by this explanation. Baliani, in fact, had his own theory, which he wrote back to Galileo. He imagined that the surface of the earth was at the bottom of a sea of air, and just as a diver under water would find himself pressed in all directions by the force of water above him, so must everything on the ground be pressed by the weight of air in the atmosphere. He figured when air was sucked out of the end of siphon above the water, the remaining air somehow became lighter, so that the heavier air below the water could push it upwards, as illustrated in Figure 4.12. What bothered Baliani, however, was that he

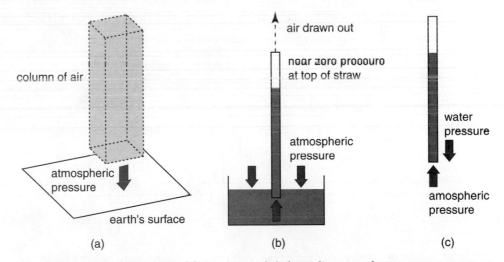

Figure 4.12 How a straw works. (a) A column of air from the atmosphere exerts pressure on earth's surface. (b) As air is sucked out of the top of the straw, the pressure decreases, and the greater atmospheric pressure is able to force a column of liquid up the straw. (c) Liquid will remain in place in a straw as long as the pressure difference between the "void" at the top and the air pressure below balances the weight of the column of liquid.

figured that the pressure at the bottom of the sea of air should be great enough to raise the water more than 11 meters.

Both Galileo and Baliani's theories, while not quite right according to what we know today, moved us closer to our understanding of how a suction pump works. Baliani's "sea of air" model comes very close to our current view of air pressure, but he overestimated how great that pressure should be. On the other hand, the space at the top of the siphon didn't just contain "lighter" air; it really was nearly empty—a vacuum—as Galileo suggested. Galileo's assistant, Evangelista Torricelli, studied the correspondence between his mentor and Baliani, and reasoned that if Baliani was correct in his sea-of-air theory, then a siphon would be able to raise a liquid heavier than water to an even lower height [Mid63]. In 1643, he took a tall glass tube with one end sealed and filled it with mercury, inverted it and set the open end in a bowl of mercury. Sure enough, some of the mercury drained out, until it reached a fixed level inside the tube with an empty space above, which was the first sustained vacuum. The height of the column of mercury was indeed much lower than 11 meters, as Example 4.4 explores. Torricelli observed that the mercury level changed daily with the weather, which he realized was related to changes in air pressure, and thus had produced the first working barometer. A few years later, in 1648, the French scientist Blaise Pascal convinced his brother-in-law to carry a barometer up a mountain; he saw the mercury level dropped as he ascended, confirming the "sea-of-air" theory of the atmosphere [Gal81]. Today, the SI unit of pressure is the Pascal (Pa), where $1 \, \text{Pa} = 1 \, \text{N/m}^2$.

Example 4.4	**Torricelli's Barometer** Galileo reported that the maximum height a siphon could support a column of water is approximately 11 m. Assuming that this is correct, what would be the expected height of a column of mercury in Torricelli's barometer?

Solution   In solving this problem, we'll assume we are comparing two different "barometers" that measure the same air pressure: the first filled with water (H_2O) and the second with mercury (Hg). For generality, we'll assume that the cross-sectional areas for the tubes of each barometer may be different (but we'll see that this doesn't matter). Given this, we'll restate the problem as follows:

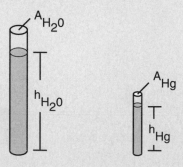

Given: Air pressure P, a water barometer with cross-sectional area A_{H_2O} and column height $h_{H_2O} = 11 \, \text{m}$, and a mercury barometer with cross-sectional area A_{Hg} and column height h_{Hg}.

Find: h_{Hg}

Plan: For either barometer, the air pressure is equal to the weight of the fluid column divided by its cross-sectional area.

1. Express air pressure in terms of the dimensions of both columns of fluid and the density of the fluid.
2. Set the air pressure measurement according to each barometer to be equal, and then solve for h_{Hg}.

Analysis: Given the densities

$$\rho_{H_2O} = 1000 \; \frac{kg}{m^3} \qquad \rho_{Hg} = 13579 \; \frac{kg}{m^3}$$

for the water barometer,

$$
\begin{aligned}
P_{H_2O} &= \frac{weight}{A_{H_2O}} \\
&= \frac{\rho_{H_2O} \times volume \times g}{A_{H_2O}} \\
&= \frac{\rho_{H_2O} \times A_{H_2O} \times h_{H_2O} \times g}{A_{H_2O}} \\
&= \rho_{H_2O} \times g \times h_{H_2O}
\end{aligned}
$$

Similarly, for the mercury barometer,

$$P_{Hg} = \rho_{Hg} \times g \times h_{Hg}$$

Setting $P_{H_2O} = P_{Hg}$ and solving for h_{Hg}, we get

$$
\begin{aligned}
h_{Hg} &= \frac{\rho_{H_2O}}{\rho_{Hg}} \times h_{H_2O} \\
&= \frac{1000}{13579} \times 11 \, m \\
&= 0.81 \, m
\end{aligned}
$$

This is considerably less than the 11 m for water.

4.4.2 Boyle's Law

By the mid-1600s, scientists across Europe had begun to investigate the properties of a vacuum. One of these men was Robert Boyle of Oxford, who was one of the founders of the Royal Society. Boyle was born in Ireland in 1627 and like Descartes, his religious beliefs framed his work. In his autobiography, Boyle tells of waking one night to a thunderstorm so powerful that he "began to imagine ... the Day of Judgement's being come" [MA06]. He vowed to devote himself to a religious life, and over the course of his career, he wrote extensively on the relationship between God's power and man's perception of it; for Boyle, careful and unbiased scientific experimentation would lead to better understanding of God's power [Hun]. Given this outlook, the existence of a vacuum was just the type of physics problem that Boyle would find interesting.

Much of Boyle's greatest work grew from his long and successful collaboration with his student, Robert Hooke. Physics students today best know Hooke as the

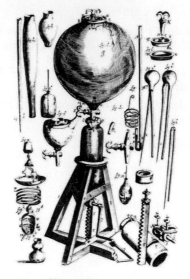

Figure 4.13 Hooke's air pump, from [Boy60].

name behind "Hooke's Law," which states that the stretch of a spring is proportional to the force across it. In fact, Hooke's research into springs began with his work with Boyle, investigating the relationship between the pressure and volume of air in a pump. As a child, Hooke was a quick learner who displayed an early talent as an artist, musician, and craftsman. In fact, he built himself an assortment of clever mechanical toys. He studied at Oxford, where he met Boyle and the other men who would found the Royal Society. A few years after graduating from Oxford, in 1658, Hooke came to work for Boyle as a laboratory assistant and applied his mechanical skills to building instruments.[5]

In one of his first projects with Boyle, Hooke devised a glass vessel with an air pump as an experimental vacuum chamber. Pictured in Figure 4.13, the chamber was large enough to contain objects ranging from burning candles to small animals to observe how they reacted as the air was pumped out. Hooke himself once squeezed inside a large bell jar as the air was removed in order to experience the effects of a vacuum firsthand. One of the more ingenious features of Hooke's design was a brass knob, liberally coated with "sallad oyl," which could be turned to control experiments inside the chamber while still maintaining a vacuum.[6] Using this apparatus, Boyle and Hooke studied the relationship between pressure and volume. In another series of experiments using a column of mercury to compress a volume of air trapped in a glass tube—to be discussed in detail in Section 5.2.1—Boyle and Hooke discovered that at constant temperature, the pressure and volume of air inside a closed container are inversely proportional. Today, we would express their result algebraically:

$$PV = \text{constant, at constant temperature}$$

Boyle described the results of their experiments in his 1660 manuscript, *New Experiments Physico-Mechanical, Touching the Spring of the Air, and Its Effects* [Boy60]. Among their results, for example, Boyle and Hooke found that the air

---

[5]A well-referenced biography of Hooke with discussion of his experiments appears in [Cha96].
[6]from [Cha96]

inside a lamb's bladder could expand to 152 times its original size inside the vacuum chamber. Hooke and Boyle were aware that air expanded with temperature, but it would be more than a century before Jacques Charles would conduct the first systematic experiments in 1787 to determine that the change in volume was proportional to the change in temperature at fixed pressure. Joseph Louis Gay-Lussac confirmed in 1802 that the change in pressure was also proportional to the change in temperature at fixed volume, which gave the combined result

$$PV = kT,$$

where k is a constant. Finally, in 1834, Benoît Paul Émile Clapeyron combined this result with Avogadro's hypothesis that equal volumes of gases at the same temperature and pressure contain equal numbers of molecules to arrive at the Ideal Gas Law,

$$PV = nRT,$$

where

- P is pressure in Pascals (Pa),
- V is volume in m^3,
- n is the amount of gas in moles (mol),
- T is the temperature in degrees K,
- R is the gas constant, $8.31\ m^3 \cdot Pa \cdot K^{-1} \cdot mol^{-1}$.

By burning different substances inside Hooke's vacuum chamber, Boyle would eventually make many important discoveries that laid the foundation for modern chemistry, as we'll discuss further in Section 4.7.

4.4.3 Hooke's Law

In an addendum of a paper in 1676, Hooke included an anagram, "ceiiinosssttuv," which was listed as a discovery that he planned to write about in the future [Cha96]. This was a fairly common trick at the time, so that a scientist could stake a claim to an idea, yet still continue to work on it in secret. Hooke decoded the anagram in another paper two years later as "*ut tensio, sic vis*," or "as the extension, so the force." In this 1678 treatise, *Lectures de potentia restitutiva, or Of Spring, Explaining the Power of Springing Bodies* [Hoo78], Hooke wrote,

... in every springing body ... the force or power thereof to restore itself to its natural position is always proportionate to the Distance or space it is removed therefrom...

Today, we would express this as

$$F = kx,$$

in which

- F = the force on the spring
- x = the extension
- k = the spring constant

In *The Power of Springing Bodies*, Hooke describes several types of springs that he states obey his law, including a "coyl or helix" of wire, a long straight wire, and "a body of Air, whether it be for the rarefaction or for the compression thereof." Today, we know that Hooke's Law is a good approximation for the first two cases, but how well does it apply to a "body of Air?" Example 4.5 examines this case.

Example 4.5

A Pneumatic Spring

An ideal pneumatic spring consists of a cylinder fitted with a frictionless, yet airtight piston. The cylinder is 10 cm long and its cross-sectional area is 1 cm², and when the piston is fully retracted, the air inside is at atmospheric pressure (approximately 100 kPa). Determine a mathematical relationship for the force on the piston versus the distance it travels in the cylinder. Does the pneumatic spring obey Hooke's Law?

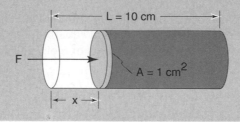

Solution

Given: Cylinder length $L = 0.1$ m, cross-sectional area $A = 0.0001$ m², $P = 100$ kPa when displacement x equals 0.

Find: Force F in terms of x. Is F proportional to x?

Plan: See if you can write Boyle's Law in the form $F = kx$.

1. Express the pressure P inside the cylinder in terms of the force F and area A.
2. Express the volume V inside the cylinder in terms of the length of the tube L, the displacement x, and the area A.
3. Use Boyle's Law to express the relationship between pressure and volume in terms of F and x.
4. Check if F is proportional to x. If it is, a plot of F versus x should be a straight line.

Analysis: If the piston isn't moving, then the force F over the area of the piston must equal the pressure P inside. Thus,

$$P = \frac{F}{A}$$

Note that this assumes that F includes not only the applied force, but also the force that's due to air pressure outside of the cylinder. The volume inside the cylinder is

$$V = A(L - x)$$

According to Boyle's Law,

$$PV = c$$

$$\frac{F}{A}[A(L - x)] = c$$

Solving for F in terms of x, we get

$$F = \frac{c}{L - x}$$

Thus, we see that F is *not* proportional to x; it is inversely proportional to $(L - x)$! We can plot F versus x if we solve for c and substitute in values. We can get c by applying Boyle's law when $x = 0$

$$
\begin{aligned}
c &= PV \\
&= P \times AL \\
&= 100 \, \frac{\text{N}}{\text{m}^2} \times (0.0001 \, \text{m}^2)(0.1 \, \text{m}) \\
&= 0.001 \, \text{N-m}
\end{aligned}
$$

Thus,

$$
F = \frac{0.001}{0.1 - x} \, \text{N}
$$

which is plotted in Figure 4.14. This clearly is not a straight line, and in fact, the force becomes infinite as the piston is pushed to the far end of the cylinder. Hooke's Law, then, doesn't apply to a "body of Air," even though Hooke claimed that it should.

There's no obvious explanation for why Hooke wrote that his spring law should hold for a body of air, when in fact it doesn't. He clearly understood that pressure and volume are inversely proportional, since he helped discover Boyle's Law. Moyer provides an interesting discussion of this contradiction in [Moy77]; one possibility is that Hooke was simply lazy with his mathematics and didn't care about the distinction between two quantities being proportional or inversely proportional, something that Newton would never have tolerated!

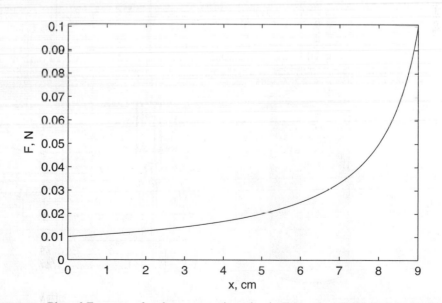

Figure 4.14 Plot of F versus x for the pneumatic spring in Example 4.5. While it is a spring, F is not proportional to x.

THE BIRTH OF THE PISTON ENGINE

It wasn't long after Boyle's and Hooke's experiments that people developed the idea of using heated gases to power an engine. Denis Papin was a French scientist who first became interested in the topic while working with Leibniz and Huygens in Paris. He visited the Royal Society in 1675, and for the next several years worked closely with Boyle and Hooke. During this period, he invented a pressure cooker, and while seeing the steam release valve open and close, had the idea of using steam and vacuum pressure to move a piston up and down in a cylinder. Huygens had earlier proposed using gunpowder to drive the piston, but when this proved impractical, Papin modified Huygens' design to use steam. In 1690, Papin published a paper for the design of the rudimentary piston steam engine shown in Figure 4.15(a). Papin's design consisted of a cylinder with a movable, yet close-fitting piston inside, with a small amount of water at the bottom of the cylinder. The cylinder was then heated with a fire below and as the water boiled, the expanding steam drove the piston upward. When the piston reached the top, the fire was removed and as the steam cooled, a vacuum was created inside the cylinder and air pressure pushed the piston back down.

4.5.1 Newcomen's Engine

There were several problems, however, with Papin's design. First, it was slow; Papin calculated that heating the water to boiling would take a minute. Second,

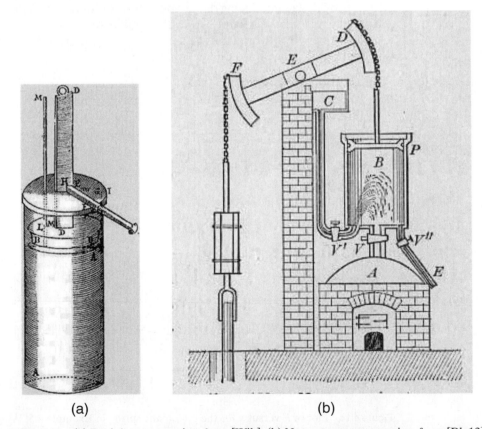

(a)   (b)

Figure 4.15   (a) Papin's steam engine, from [Wik]. (b) Newcomen steam engine, from [Bla13].

Papin didn't have access to the metalworking technology to build an engine large enough to do useful work with close enough tolerances between the piston and the cylinder to maintain a vacuum. Thomas Newcomen, an English blacksmith, solved both problems; in 1712, he devised an engine for pumping water from mine shafts. Figure 4.15(b) illustrates Newcomen's engine, attached to the sucker rod of a pump via a rocker arm. A weight attached to the sucker rod pulls it down, while the engine raises the rod by pulling down on its end of the rocker arm. At the center of the engine is cylinder B containing a piston P. As the piston rises from the weight of the sucker rod, steam flows into the cylinder from boiler A through valve V. At this point, the pressure inside the cylinder is the same as the atmosphere outside the cylinder. When the piston reaches the top of the cylinder, valve V closes and valve V' opens, letting cold water spray into the cylinder from tank C. The change in temperature in the cylinder causes the pressure in the cylinder to drop, and air pressure pushes the piston down, raising the pump sucker rod. When the piston gets near the bottom of the cylinder, valve V opens briefly to eject the now warm water inside, and then the process repeats.

The Ideal Gas Law governs the relationship among the temperature, pressure, and volume inside the cylinder below the piston. Example 4.6 illustrates how a change in temperature can produce a sufficient force to move a weight.

Example 4.6	**Using a Piston Engine to Lift a Weight**

Suppose that a piston engine similar to the Newcomen engine pictured in Figure 4.13(b)—except that it uses air rather than steam as the working gas—must lift a weight of 2000 N (450 lbs) attached to the end of the rocker arm.[7] The diameter of the cylinder is 0.5 m and its length is 2 m. If the gas inside the cylinder is heated to 200° C at atmospheric pressure, to what temperature must it be cooled before the piston starts to move?

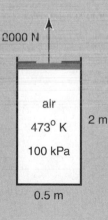

Solution The piston will begin to move once the difference between the atmospheric pressure outside the cylinder and the pressure inside the cylinder produces a downward force on the piston greater than 2000 N. Given this, we can restate the problem.

Given: Cylinder length $L = 2\,\text{m}$, diameter $d = 0.5\,\text{m}$, initial temperature $T_1 = 473°\,\text{K}$, initial pressure $P_1 = P_{\text{atm}} = 100\,\text{Pa}$, applied force of 2000 N.

Find: Temperature T_2 at pressure P_2, so that the net force on the piston is zero (i.e., it begins to move).

Plan: We can solve this problem in two steps:

1. First, determine what pressure is needed inside the tank so that the net force on the piston is zero.
2. Then, use the Ideal Gas Law to determine the temperature that will produce this pressure.

Analysis: The figure below shows the forces acting on the piston.

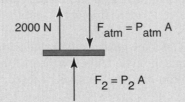

If the net force on the piston is zero, then

$$2000 - F_{\text{atm}} + F_2 = 0$$
$$2000 - P_{\text{atm}}A + P_2A = 0$$

When we solve the problem, we get

$$P_2 = P_{\text{atm}} - \frac{2000\,\text{N}}{A}$$
$$= 100\,\text{kPa} - \frac{2000\,\text{N}}{\pi \times (0.25)^2 \text{m}^2}$$
$$= 89.8\,\text{kPa}$$

For a fixed cylinder volume (because the piston hasn't moved yet), according to the Ideal Gas Law:

$$\frac{P_1}{T_1} = \frac{P_2}{T_2}$$
$$T_2 = \frac{P_2}{P_1}T_1$$
$$= \frac{89.8\,\text{kPa}}{100\,\text{kPa}} \times 473°\,\text{K}$$
$$= 425°\,\text{K}$$

Thus, the air in the cylinder would need to be cooled to 425° K or 152° C in order for the engine to begin to lift a 2000 N weight.

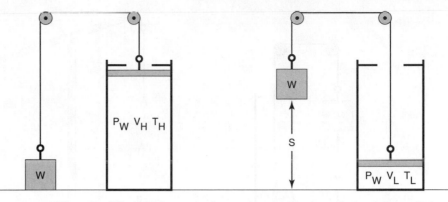

Figure 4.16   When the difference between the pressure inside the cylinder drops sufficiently below the atmospheric pressure outside the cylinder, the piston is able to raise a weight.

As Example 4.6 shows, once the temperature inside the cylinder of a Newcomen-style piston engine drops low enough, the difference between the atmospheric pressure outside the engine and the gas pressure inside it is large enough that the piston starts to fall, raising a weight as it does. Once the piston begins to fall, Newton's Laws of Motion and the Ideal Gas Law together determine what happens next. If we don't change the temperature, the piston will remain where it is. In fact, as we saw in Example 4.5, the gas inside the cylinder will act as a spring—not one that obeys Hooke's Law, but still a spring—that holds the piston in place. If we pressed down on the piston, the cylinder volume would decrease and the pressure inside would increase according to $PV = $ constant, so that if we let go of the piston, the increased pressure would cause it to spring back up to its original position. In order to continue to pull the piston down, we need to continue to decrease the temperature. After we've cooled the gas as much as we can, the piston will come to a stop somewhere lower in the cylinder. The pressure in the cylinder will be the same as it was when the piston began to move—just enough to hold the weight in place—but the volume will be lower, again as determined by the Ideal Gas Law at the lower temperature, which Figure 4.16 illustrates.

- $W = $ the weight
- $P_W = $ the pressure that balances the weight
- T_H and $V_H = $ the high temperature and corresponding volume
- T_L and $V_L = $ the low temperature and corresponding volume
- $s = $ the distance the piston has traveled

By heating and then cooling the gas, the engine has moved the weight W by a distance s. Further, by attaching the engine to an appropriate mechanism such as a lever as illustrated in Figure 4.17, it could be used to perform equivalent work, applying a different force over a different distance.

4.5.2   James Watt's Improvements to Newcomen's Design

Each cycle of Newcomen's steam engine required changing the temperature inside the cylinder. This repeated heating and cooling required burning a lot of coal, and

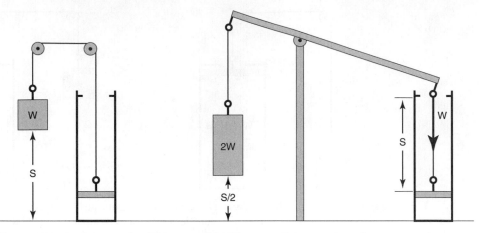

Figure 4.17 A piston engine lifting a weight W over a distance s does the same work as an engine lifting a weight $2W$ over a distance $s/2$.

the quest to make Newcomen's design more efficient reflects the same basic issues that engineers face today in designing engines with improved fuel economy. An instrument maker from Glasgow, Scotland named James Watt made the first major improvement to the steam engine in the mid-1760s, about 50 years after Newcomen's first design. While experimenting with a Newcomen engine, Watt determined that as much as 80 percent of the burned fuel went into reheating the cylinder after it was cooled with water, when all that really needed to be heated and cooled to move the piston was the gas inside it. Watt devised a way to cool the steam in a separate chamber outside of the cylinder called a *condenser*, which allowed the cylinder itself to remain hot, greatly reducing fuel consumption. Watt built the first prototype of his engine in 1765, and it took a decade before he and business partner and foundry owner Matthew Boulton found a way to manufacture reliable steam engines profitably. Boulton and Watt booked their first orders for engines in 1775, and by the end of the century, their firm had sold between 300–400 of them to a broad range of industries, including mining, water works, iron works, and textile mills.

Watt needed a way to quantify the output of his engines so that he could price them accordingly. He decided to rate his engines according to their *power* output, which is defined as the quantity of work per unit time, or

$$\text{power} = \frac{\text{work}}{\text{time}}$$
$$= \frac{\text{force} \times \text{distance}}{\text{time}}$$

At the time, many factories in Britain used teams of horses to power their equipment, and in a fine example of engineering approximation, Watt decided to base his unit of measure on the power output of a horse. Consulting with the owner of a paper mill, Watt estimated that a workhorse could walk approximately 2.5 times per minute around a circular path with a diameter of 24 feet, while pulling a weight of 180 lbs.

Rounding the value of pi down to 3, Watt calculated the power output of one horse as

$$\text{power output of 1 horse} = \frac{\text{force} \times \text{distance}}{\text{time}}$$

$$= \frac{(180 \, \text{lbs}) \times (2.5 \times \pi \times \text{diameter})}{1 \, \text{min}}$$

$$= \frac{(180 \, \text{lbs}) \times (2.5 \times 3 \times 24 \, \text{ft})}{1 \, \text{min}}$$

$$= 32,400 \, \text{foot-pounds per minute}$$

Rounding up to 33,000 foot-pounds per minute, Watt defined this unit of power measurement as 1 "horse," which today we call 1 *horsepower*.[8] In his honor, the MKS unit of power was named the Watt, where

$$1 \, \text{W} = 1 \frac{\text{N} \cdot \text{m}}{\text{s}}$$

4.6 THE SCIENCE OF THERMODYNAMICS

4.6.1 Sadi Carnot and the Limits of Engine Efficiency

Even as the steam engine developed into a commercial success in the early 1800s, scientists still understood very little of the principles of work and energy that today we learn in high-school physics classes. Watt and other engineers experimented with various approaches to increase the horsepower and reduce the fuel consumption of steam engines, yet without better theoretical models of how an engine worked, they still didn't know what limits—if any—there were to these values. In 1824, five years after James Watt's death, a 28-year-old French physicist named Sadi Carnot wrote a short book called *Reflections on the Motive Power of Heat* that would dramatically change this.

Sadi Carnot was the son of Lazare Carnot, an engineer, mathematican, and military leader. Because both of his parents were artistic, they named their son for the 12th-century Persian poet and moralist, Sheikh Saadi of Shiraz. Lazare Carnot served as "Organizer of Victory" in the French Revolution and was appointed by Napoleon first as his Minister of War and later as his Minister of the Interior. When the British army and her allies, commanded by the Duke of Wellington, defeated Napoleon at Waterloo in 1815, the elder Carnot was exiled to Germany, never to return to France. Sadi, who had recently graduated from the École Polytechnique in Paris, was impressed by the contribution of the steam engine to Britain's economic and military might, and despaired that France had fallen so far behind in this technology. Further, as an educated French scientist, Carnot was especially perturbed that the steam engine was invented by a few British engineers with little formal training, and that there was still no solid theory to explain how it worked.

---

[8] As discussed at http://www.sizes.com/units/horsepower_british.htm, there are a number of versions of the story of how Watt came to define his unit of power, the horse. This version, based on entries in his "Blotting and Calculation Book, 1782 & 1783," is most likely correct.

Reflections on the Motive Power of Heat is a beautifully written book, where in accessible language, Carnot explains some of the basic principles that serve as the foundation of the field of *thermodynamics*—the study of heat and energy. Carnot opens the book with this introduction, translated into English in 1890 by Robert H. Thurston, a founder and first president of the American Society of Mechanical Engineers (ASME) [Car90]:

> Everyone knows that heat can produce motion. That it possesses vast motive-power no one can doubt, in these days when the steam-engine is everywhere so well known. To heat also are due the vast movements which take place on the earth. It causes the agitations of the atmosphere, the ascension of clouds, the fall of rain and of meteors, the currents of water which channel the surface of the globe, and of which man has thus far employed but a small portion. Even earthquakes and volcanic eruptions are the result of heat. From this immense reservoir we may draw the moving force necessary for our purposes....
>
> Already the steam-engine works our mines, impels our ships, excavates our ports and our rivers, forges iron, fashions wood, grinds grain, spins and weaves our cloths, transports the heaviest burdens, etc. ... Notwithstanding the work of all kinds done by steam-engines, notwithstanding the satisfactory condition to which they have been brought to-day, their theory is very little understood, and the attempts to improve them are still directed almost by chance. ... The question has often been raised whether the motive power of heat is unbounded, whether the possible improvements in steam-engines have an assignable limit, a limit which the nature of things will not allow to be passed by any means whatever; ... We propose now to submit these questions to a deliberate examination.

Carnot later introduces the abstract idea of a *heat engine* that might produce motion by heating and cooling anything that expands and contracts with temperature, from a gas such as steam or air, to even a liquid or solid. To heat and cool this material, he assumed that his engine contained two bodies at two different, constant temperatures, which he called the furnace and the refrigerator. He famously argued that heat must always flow from bodies at higher temperatures to bodies at lower temperatures—say from the furnace to the gas and then from the gas to refrigerator—because to do other could produce perpetual motion, which can't happen. He showed that what he called the "motive power" of a heat engine—which today we call work—depends on the difference between its maximum heating and minimum cooling temperatures. Further, he reasoned that the *efficiency* of an engine, the percentage of the energy stored in heat that it converts to motion, is limited by the cooling temperature. In other words, it's *impossible* to convert into motion all of the heat that burning fuel generates.

Carnot's work remained largely unknown until after his death in 1832, but later scientists visited it, including Clapeyron—who also formulated the Ideal Gas Law—Rudolf Clausius, and Lord Kelvin, and it ultimately became a cornerstone of the science of thermodynamics. Today, we express the *Carnot efficiency* of an engine with the simple formula

$$\text{Carnot efficiency} = \frac{T_H - T_C}{T_H} \tag{4.8}$$

where T_H and T_C are the hot and cold temperatures, in degrees Kelvin, of the engine. Example 4.7 illustrates an application of this law.

Example 4.7	**Efficiency of a Steam Engine**
	Suppose that a steam engine operates with steam at atmospheric pressure and is cooled by ice water. What is the best possible efficiency that it could achieve?

Solution

Given: A hot temperature T_H of the boiling point of water and a cold temperature T_C of the freezing point of water.

Find: The Carnot efficiency

Plan: Substitute numbers into the formula.

Analysis: Using the formula for Carnot efficiency η, where $T_H = 373°$ K and $T_H = 273°$ K

$$\eta = \frac{T_H - T_C}{T_H}$$

$$= \frac{373 - 273}{373}$$

$$= 0.268$$

Thus the maximum possible efficiency is 26.8%. In a real engine, the efficiency would typically be much lower, because not all of the heat that goes into producing the steam is converted into useful work by the engine.

4.6.2 James Joule: From Building a Better Brewery to a Theory of Heat and Energy

Carnot successfully advanced the theoretical understanding of heat and energy. Yet, parts of his theory didn't quite agree. In particular, Carnot subscribed to an earlier theory that heat was a basic substance—called *caloric*—that was always conserved; in other words, heat could be moved around, but never created or destroyed. Ultimately, it was the work of another British engineer with little formal schooling that disproved the caloric theory.

James Joule was working as the manager of his father's brewery in Manchester, England; here, he began to consider replacing its steam engines with the newly invented electric motor. Born on Christmas Eve in 1818, Joule was sickly as a child and home-schooled until he was 14, when his father sent him and his brother for tutoring in science and mathematics with the eminent chemist John Dalton. Joule developed an interest in electricity, and continued to experiment with it while he worked at the brewery. He published his first paper on improvements to electromagnets for motors when he was 20, and over the next five years, wrote a series of papers on the results of his experiments in electricity, work, and heat. These papers reveal the extraordinary care and skill that Joule brought to controlling his experiments and making measurements. In an 1840 paper on the heating effects of an electrical current flowing through a wire, Joule describes his temperature measurement technique:

> The thermometer which I used has its scale graduated on the glass stem. The divisions were wide, and accurate. In taking temperatures with it, I stir the liquid gently with a feather; and then, suspending the thermometer by the top, ... I bring my eye to a level with the top of the mercury. In this way a little practice has enabled me to estimate temperature to the tenth part of Fahrenheit's degree with certainty [Jou63b].

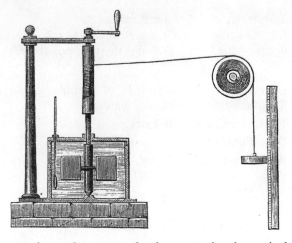

Figure 4.18 Joule's experimental apparatus for demonstrating the equivalence of work and heat, from [har69].

It was well known that electrical current flowing through a wire heated it. The caloric theory explains this phenomenon; it says that the heat must have been transferred to the wire from elsewhere in the system. By 1843, however, by carefully monitoring temperatures throughout his apparatus, Joule conclusively showed that heat was *generated* by the electrical current and wasn't transferred from somewhere else. Contrary to the beliefs of the leading academic scientists in London, Cambridge, Glasgow, and Paris, the 25-year-old brewer from Manchester asserted that caloric—heat—wasn't conserved. Not surprisingly, the academic community at first didn't seriously recognize Joule's work.

Joule suggested that heat was not a substance, but rather a state of vibration that could be induced by mechanical means, either directly through friction, or indirectly, as by cranking a generator to produce an electrical current in a wire. In his most famous experiment, in 1845, Joule demonstrated the equivalence of kinetic energy and heat [Jou63c] [Jou63d]. In this experiment, Joule devised a paddle-wheel in a can of water and used a system of weights, string, and pulleys to turn it, as illustrated in Figure 4.18. He attached a 4 lb weight to the string and let it descend 12 yards, repeating this 16 times before taking the temperature of the water using a very accurate thermometer. Normalizing his results, he found that "when the temperature of a pound of water is increased by one degree of Fahrenheit's scale, an amount of *vis viva* is communicated to it equal to that acquired by a weight of 890 pounds after falling from the altitude of one foot" [Jou63c].

Joule persisted with his experiments, and within a few years produced enough evidence to convince the leading expert on heat and temperature, William Thomson, Lord Kelvin, of his results. Like Descartes before them, Joule and Lord Kelvin shared a religious view of the conservation of energy. In a lecture at St. Ann's Church in Manchester in 1847, Joule noted that many scientists believed *vis viva* could be destroyed, such as when a falling object hits the ground. To Joule, this was an affront: "we might reason *a priori*," he stated, "that such absolute destruction of the living force cannot possibly take place, because it is manifestly absurd to suppose that the powers with which God has endowed matter can be destroyed any more than that they can be created by man's agency" [Jou63a]. Once Lord Kelvin accepted Joule's theory of a mechanical equivalence for heat, the rest of the community did as well. After building on the work of Joule and Carnot over the next 50 years, scientists stated and restated their findings as two fundamental principles of heat, work, and energy; these evolved into the First and Second Laws of Thermodynamics.

First Law of Thermodynamics   The *First Law of Thermodynamics* is conservation of energy, the idea that the amount of energy in the universe is constant, and that it can neither be created nor destroyed. Beginning with the equivalence between mechanical work and heat, the law expanded to cover a wider jurisdiction as scientists realized that energy could appear in other forms. In an 1852 paper, Lord Kelvin commented on Carnot's proposition of a "waste" of mechanical energy when heat transferred from a warm to a cold body:

> As it is most certain that Creative Power alone can either call into existence or annihilate mechanical energy, the 'waste' referred to cannot be annihilation, but must be some transformation of energy.

He then listed some of the different "stores" of energy, including "static" forms of energy such as a "quantity of weights at a height, ready to descend and do work when wanted, an electrified body, [or] a quantity of fuel," as well as "dynamical" forms of energy such as "masses of matter in motion, a volume of space through which undulations of light or radiant heat are passing, a body having thermal motions among its particles (that is, not infinitely cold)" [TLK52].

Second Law of Thermodynamics   By using a falling weight to spin a paddlewheel, Joule demonstrated that mechanical work could be converted to heat. If Joule's paddlewheel then began spinning on its own, raising the weight and leaving the water at its original temperature, this would not be a violation of the First Law of Thermodynamics. We know from experience, however, that this will never happen because of the *Second Law of Thermodynamics*. The Second Law has been stated and misstated many times over the years—there's even a web site that lists over one hundred versions. One of the simplest is Rudolf Clausius' original statement in 1850:

> Heat cannot of itself transfer from a colder to a hotter body [Cla79].

To understand how this statement of the Second Law explains why Joule's apparatus can't spontaneously run in reverse, we can think of it in terms of an idealized heat engine as defined by Carnot. Recall that Carnot's heat engine uses two heat reservoirs, called the furnace and the refrigerator, to expand and condense a working fluid to produce motion, and that the efficiency of the engine is proportional to the difference between their temperatures. In Joule's apparatus, the water was in an insulated container, so if we consider the heated water inside it to be a furnace, then there is no corresponding refrigerator to which the heat can flow. Alternatively, even if we think of the water on either side of the paddlewheel as two separate heat reservoirs, their temperatures are the same, so the efficiency of an engine constructed from them would be zero.

That's not to say that we couldn't design an engine that used the heated water to lift the weight. If the water started at the same temperature as the environment, after the weight falls, it would be at a higher temperature than its surroundings, and so the water could act as a furnace and the air outside could act as a refrigerator to heat and cool a gas inside of a piston engine. The problem, however, is that since the temperature of the furnace would only be slightly higher than the temperature of the air outside, the efficiency of this engine would be very low, and only a small fraction of the energy stored in the form of heat in the furnace could convert back to work. As a result, the engine could only lift the weight a very short distance. Further, if we repeated the process, the efficiency of the engine a second time would be even worse, since the temperature of the water would be lower than it was after the first cycle, and after many repetitions, the temperature of the water would be the same as the temperature of the air outside; thus, the apparatus could no longer function as an engine.

4.7   CONSERVATION OF MASS

4.7.1   Robert Boyle and *The Sceptical Chymist*

When Robert Boyle conducted his experiments on the relationship between the pressure and volume of air in a container, he had no notion of air being a mixture of gases of different elements—nitrogen, oxygen, and carbon dioxide—let alone any concept of an ideal gas. During the late 1600s, European scientists still largely viewed matter as composed of the four "classical elements," Earth, Water, Air, and Fire, an idea dating back to ancient Greece. Like many of his scientific contemporaries, Boyle was an alchemist who believed that different metals such as lead, silver, and gold could transform into each other, and so he designed many of his experiments with the purpose of achieving this goal. With more than three centuries of hindsight, today we often think of the alchemists as backwards or superstitious, but some curious observations motivated their quest, which, given their understanding of matter, made their hypotheses seem plausible.

One particularly well-known phenomenon before Boyle's time is that when certain metals are heated in a crucible, they would reduce to a type of ash called *calx*. The calx of lead in particular was a yellowish powder that curiously weighed more than the original lead sample. Further, if calx was burned with wheat, it would transform back to the original metal. Boyle had little use for theory unsupported by experimentation, and in 1661, he wrote *The Sceptical Chymist* [Boy61], which ridiculed the methods of alchemists who relied on theory alone, and pleaded for them to use experimental methods to determine the composition of materials. Using the vacuum chamber he developed with Hooke, Boyle conducted detailed experiments on how different materials burned with and without air. He observed that lead wouldn't burn once air was removed from the chamber, but that a mixture of either sulphur or charcoal with saltpeter—the components of gunpowder—would. He deduced that something in saltpeter, known today as potassium nitrate, KNO_3, is also found in air, enabling carbon or sulphur to burn. He also determined that this substance required for burning is also necessary for respiration, since the small birds and animals he placed in his vacuum chamber couldn't survive after the air was withdrawn. Boyle didn't know what this substance was, but he suggested that there may be more elements than the classical ones of Earth, Water, Air, and Fire. His own view was that matter was composed of "corpuscles" of some number of basic elements, and that different materials had different forms depending upon the shapes, arrangements, and motion of these corpuscles. His experiments also led him to conclude that there was some kind of "fire particle" that attached to metals when they burned, which explained the weight increase.

Two of Boyle's contemporaries in Germany, Johann Becher and his student Georg Stahl, also advanced a theory of a fire particle, which they called *phlogiston*. For the next century or so, the phlogiston theory of combustion enjoyed a level of credibility on a par with Newton's theory of gravity. Over the course of the 1700s, scientists realized that air was not a single gas, but a mixture of different gases with different properties. Joseph Priestley, a teacher and clergyman from Leeds, England made some of his first discoveries quite literally over a beer. Living next door to a brewery, Priestely determined that the gas produced by fermenting beer could extinguish a flame, and also found that he could infuse this gas into water to create a new beverage by placing a container of water over a vat of beer. Priestly called this gas—which we now know as carbon dioxide—"fixed air." In 1774, Priestly discovered another gas given off by heating a sample of the calx of mercury with light focused through a magnifying glass. He named his discovery "dephlogisticated air" and saw that it would cause a candle to burn more brightly and also could sustain a mouse under glass. A few years later, the French chemist Antoine Lavoisier mistakenly attributed the source of the acidity to Priestley's gas, and he renamed it *oxygen*, Greek for "becoming sharp."

4.7.2 Antoine Lavoisier

Like Boyle and Hooke before him, Lavoisier was a diligent and devoted experimenter. He carefully weighed both the reactants and products of combustion before and after burning them. From this, Lavoisier observed that the weight of oxygen consumed when burning a metal corresponded exactly with the increased weight of the calx. Repeating these experiments with other materials, Lavoisier found that after careful measurement, the weights of the reactants and the products were always the same. With these results, he disproved the phlogiston theory, and replaced it with a law of conservation of matter or mass. During the first year of the French Revolution, 1789, Lavoisier published his *Traité Elémentaire de Chimie* or *Elementary Treatise on Chemistry*, today regarded as the first modern chemistry text [Lav89]. In addition to presenting the Law of Conservation of Mass, the book defined an *element* as a basic substance that can't be reduced further, and included a list of more than thirty elements, most we still today regard as elements and some as compounds, while others—heat and light—are now viewed altogether differently.

While Lavoisier's greatest contributions were through science, he earned his living as a tax collector, a fateful choice at the time of the French Revolution. Labeled a traitor, he was arrested in 1794 and beheaded on the guillotine the same day. In his eulogy for Lavoisier, the French mathematician Joseph Louis Lagrange stated, "It took only a moment to cut off that head, but may take more than a hundred years to produce another like it."[9]

---

[9]J. B. Delambre, "Éloge de Lagrange," *Memoires de l'Institut*, 1812, p. XIV.

4.8   ANALYSIS EXAMPLE: THE INTERNAL COMBUSTION ENGINE

By the late 1800s, advances in thermodynamics, chemistry, and manufacturing technology led to the design of lighter, more efficient engines that, in turn, gave rise to the automobile. Rather than using an external boiler to produce steam, these new *internal combustion* engines burned liquid fuels inside the cylinder itself to produce the heat to expand the air and move the piston. In this section, we'll explain the operation of the most common design for an internal combustion engine, called a *four-stroke engine*, and use theoretical models that we developed earlier in the chapter to analyze its performance.

4.8.1   Operation of a Four-Stroke Engine

Italian engineers Eugenio Barsanti and Felice Matteucci in 1854 produced the first working internal combustion engine. In 1876, the German engineers Nikolaus Otto, Gottlieb Daimler and Wilhelm Maybach developed a four-stroke engine, where the piston travels up and down twice each time that fuel burns in the cylinder, with the combination of momentum and the power from other cylinders carrying the crankshaft through two full revolutions. A few years later, Karl Benz used a similar design in the first production of automobiles, and today, the four-stroke Otto cycle remains the basis for most automobile engine designs. Figure 4.19 illustrates the steps in the Otto cycle. The first stroke is the *intake* stroke, where a mix of fuel and cool air enters the cylinder through an intake valve as the piston moves down the cylinder. The second stroke is the *compression* stroke, where the piston compresses the air-fuel mixture into a small space at the top of the cylinder. When the piston reaches *top dead-center*, the spark plug fires, igniting the fuel. This causes an instant increase in pressure, driving the piston downward during the *expansion* or power stroke. After the piston reaches *bottom dead-center*, the exhaust valve opens and the exhaust stroke begins as the rising piston clears the cylinder of burnt fuel and hot

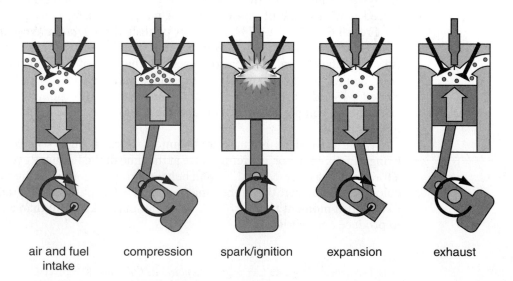

air and fuel   compression   spark/ignition   expansion   exhaust
intake

Figure 4.19   The four-stroke Otto cycle.

gases, leaving the cylinder ready to begin another cycle with an intake stroke. Of the four strokes in the cycle—intake, compression, expansion, exhaust—the expansion stroke is the only one where the burning fuel delivers power to the engine, and the goal of the other three is to maximize the efficiency of the power stroke.

4.8.2   Efficiency of the Intake Stroke and Air/Fuel Ratio

During the intake stroke, either a mechanical *carburetor* or an electronic *fuel injector* sprays a mist of fuel and air into the cylinder, and one of the most important factors in engine performance is the ratio of fuel to air in this mixture. Initially, one may naively assume the more the fuel the greater the power, but a closer look reveals why this is not so. First, fuel requires oxygen to burn, and unless there's enough oxygen in the mix, then the fuel will not burn completely. Second, air itself moves the piston; fuel merely provides the energy source required to heat and expand it. Ideally, for a given volume of the air and fuel mixture injected into the cylinder, there should be just enough fuel that the combustion process completely consumes both the fuel and the available oxygen in the air. The ratio of the mass of air to the mass of fuel required to meet this condition is known as the *stoichiometric ratio* for the fuel. Example 4.8 illustrates the calculation of the stoichiometric ratio for isooctane—the main ingredient in gasoline.

Example 4.8	**Stoichiometric Ratio of Isooctane**

Determine the stoichiometric ratio of isooctane, C_8H_{18}, assuming that air is 23.2 percent oxygen by weight.

Solution Solving this problem involves calculating the mass balance for the combustion of isooctane. We'll assume that isooctane burns according to the reaction equation

$$C_8H_{18} + 12.5O_2 \rightarrow 8CO_2 + 9H_2O$$

The atomic weights of the elements in this reaction are as follows:

element:	H	C	O
atomic weight (g/mol):	1	12	16

Given this information, we can now solve the problem.

Given: Combustion equation for isooctane and the percentage of oxygen in air by mass

Find: Ratio of the mass of air to mass of isooctane for complete combustion

Plan: Our plan has 4 steps

1. Calculate weight of 1 mol of isooctane
2. Calculate the weight of oxygen required to burn one mol of isooctane
3. Calculate the weight of air that contains the required amount of oxygen
4. Divide the weight of the air by the atomic weight of isooctane

Analysis: First, from the provided atomic weights, the mass of 1 mol of isooctane is

$$\text{mass of isooctane} = 8(\text{mass of C}) + 18(\text{mass of H})$$
$$= 8(12) + 18(1)$$
$$= 114\,\text{g}$$

According to the reaction equation, 12.5 mols of O_2 are required to burn 1 mol of isooctane, thus

$$\text{mass of burned oxygen} = (12.5)(2)(\text{mass of O})$$
$$= (12.5)(2)(16)$$
$$= 400\,\text{g}$$

Since air is 23.2 percent oxygen by mass,

$$\text{mass of air} = \text{mass of burned oxygen}/0.232$$
$$= 1724.1\,\text{g}$$

Finally, the stoichiometric air-to-fuel ratio is

$$\text{stoichiometric ratio} = \frac{\text{mass of air}}{\text{mass of isooctane}}$$
$$= \frac{1724.1}{114}$$
$$= 15.12$$

Thus the stoichiometric ratio for isooctane is 15.1. We note that gasoline is actually a mixture of chemicals, and its stoichiometric ratio is usually quoted as approximately 14.7.

4.8.3 Efficiency of the Compression Stroke and the Compression Ratio

Of the four strokes in the Otto cycles, the compression stroke has probably the least obvious, but also most significant, impact on the overall efficiency of an engine. The chief metric of the compression stroke is called the *compression ratio*, r_c, as illustrated in Figure 4.20. The compression ratio is the ratio of maximum to minimum volume inside the cylinder, or

$$r_c = \frac{\text{cylinder volume at bottom dead center}}{\text{cylinder volume at top dead center}}$$

Qualitatively, we can begin to see the significance of the compression ratio by comparing the operation of an internal combustion engine to Carnot's ideal heat engine. First, imagine an ideal piston heat engine, as pictured in Figure 4.21. In order to raise the piston, we slowly add an amount of heat energy Q_H to the gas inside the cylinder, such that the gas pressure inside the cylinder is always maintained at the same value P, which raises its temperature to T_H as the volume increases to V_H. In

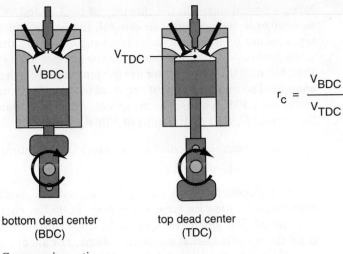

Figure 4.20 Compression ratio.

order to lower the piston, we remove an amount of heat energy Q_C, which reduces the temperature back to T_C as the volume decreases to V_C.

The efficiency η of this ideal engine is equal to the ratio of the work that we obtain from the engine divided by the energy consumed, or

$$\eta = \frac{W}{Q_H}$$

According to the Law of Conservation of Energy, the amount of work done by the engine is equal to the difference between the heat added and the heat removed, or

$$W = Q_H - Q_C$$

so,

$$\eta = \frac{Q_H - Q_C}{Q_H} \qquad (4.9)$$

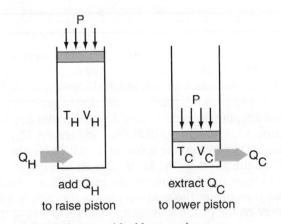

Figure 4.21 A piston and cylinder as an ideal heat engine.

Next, we want to relate the quantity of heat added to or removed from the gas to its change in temperature. In general, the amount of heat required to change the temperature of an object varies from substance to substance. For example, it takes 4.186 Joules—or 1 calorie—to raise the temperature of 1 gram of liquid water by 1° K, but only 0.140 Joules to raise the temperature of a gram of mercury by the same amount. The quantity of heat required to raise the temperature of a fixed mass of a material by 1° K is known as the *specific heat* of the material, usually represented by the symbol C, with metric units of J/(g°K). Thus,

$$\text{heat} = \text{specific heat} \times \text{mass} \times \text{temperature change}$$
$$Q = Cm\Delta T \tag{4.10}$$

The specific heat of a material varies with its condition or state; as an extreme example, a melting ice cube will remain at 0°C as energy is added until all the ice has melted. For gases, the specific heat at constant pressure has a different value from the specific heat at constant volume. For air at constant pressure, the specific heat C_p ranges from 1.0 to 1.1 J/(g °K) over the temperature range 0–500° C. If we approximate C_p for air as constant, then by combining Equations (4.9) and (4.10) the efficiency may be written as

$$
\begin{aligned}
\eta &= \frac{C_p T_H - C_p T_C}{C_p T_H} \\
&= \frac{T_H - T_C}{T_H} \\
&= 1 - \frac{T_C}{T_H}
\end{aligned}
\tag{4.11}
$$

Note that this is equivalent to the Carnot efficiency—the difference between the high and low temperatures reached inside the cylinder limits the efficiency. We can take this analysis one step further and relate it back to the compression ratio r_c by applying the Ideal Gas Law. Because the pressure of the gas is constant,

$$\frac{T_C}{T_H} = \frac{V_C}{V_H} = \frac{1}{r_c}$$

Substituting this result back into Equation (4.11), we obtain the engine efficiency in terms of the compression ratio:

$$\eta = 1 - \frac{1}{r_c} \tag{4.12}$$

According to Equation (4.12), for an ideal heat engine, the greater the compression ratio r_c, the more efficient the engine. But how does the efficiency of the ideal engine compare with that of a real four-stroke engine based on the Otto cycle? In the ideal engine, we burned the fuel slowly, gradually raising the temperature of the gas and keeping the pressure inside the cylinder constant, while in a real engine, the fuel burns very rapidly. As illustrated in Figure 4.22, in the ideal engine, the pressure and hence the force on the piston stay constant as the piston moves through the cylinder. Thus, the gas does the same amount of work on the piston over every incremental step from the beginning to the end of the stroke. For the realistic engine, however, the pressure in the cylinder shoots up rapidly when the fuel ignites, and thus the force on the piston is greater at the beginning of the stroke than it is at the end. The implication of this is that the engine does more work at the beginning of the stroke

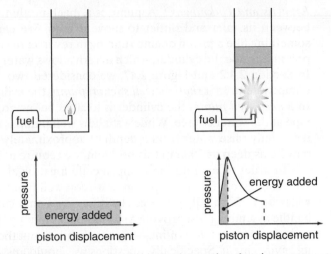

Figure 4.22   The idealized engine adds heat slowly, moving the piston at constant pressure, while a realistic engine adds heat rapidly, causing a spike in pressure as the fuel is ignited.

than it does at the end. Regardless of whether the fuel is burned slowly or rapidly, however, the engine does the same total amount of work for a given heat input. The efficiency still increases with the compression ratio, although the formula is slightly more complicated than Equation (4.12) and depends on the specific heat of the gas in the cylinder.

4.9   DESIGN EXAMPLE: THE HANDPUMP

In this section, we revisit the design of a handpump that we began in Chapter 2 as another example of a problem that involves the concepts of pistons, cylinders, forces, pressure, work, and efficiency. As we noted in Chapter 2, children have much of the responsibility for gathering water in many communities, so a handpump must be designed so that a child can operate it. Thus, we state the design problem as:

> Design a lever handle assembly for a deep-well pump that could be operated by a child. Assume that the water source may be as deep as 50 m below the surface, and that the pump cylinder has a diameter of 50 mm.

This example is intentionally somewhat verbose, in order to fully illustrate the problem-solving framework presented in Chapter 3. We begin by better defining the problem, then come up with a plan for solving it, and finally execute the plan.

4.9.1   Problem Definition and Plan of Attack

The first three steps in our problem-solving framework are to *define* the problem, *explore* possible approaches, and *plan* a path for solving it.

Define   The first step in solving this problem is to define it more clearly. In general, this means identifying the known and unknown quantities in the problem, restating the problem in precise terms, and drawing some simple diagrams.

Knowns and Unknowns A pump is a machine that uses a difference in pressure between its inlet and outlet to move a fluid. We can think of a piston pump as something like a piston engine running in reverse: raising the piston causes a drop in pressure inside the cylinder, which in turn allows water to be drawn into the cylinder. In Section 2.4.2 and Figure 2.47, we considered two different configurations for a pump system. In a *shallow well suction pump*, the cylinder is located above ground. In a *deep well pump*, the cylinder is located below ground at the bottom of a drop pipe at the water source. While a shallow well pump is easier to install and maintain, it can only raise water from a depth of approximately 10 m. Since the water supply may be as deep as 50 m in this problem, we require a deep well pump.

In order to operate the pump, a child must be able to provide enough force to raise the piston. Since the piston of a deep well pump is located below a column of water up to 50 m tall, the forces bearing down on it can be quite large, and thus the handle assembly must provide a substantial amount of leverage. What is unknown—and what we must determine—are key details about the design of the handpump and its environment. Specifically, questions we should answer include:

- How much force does it take to raise the piston?
- How much force could a child apply to a pump handle and how far could a child move it?
- How should a lever be sized to enable a child to lift this weight?
- How much water would be delivered with each pump stroke?

Draw a Diagram and Define Variables Figure 4.23 illustrates key variables in the handpump design problem, which are also listed in Table 4.1. The pump handle is shown as a lever with a pivot as the fulcrum. L_1 and L_2 are design variables representing the lengths of the lever arm on the handle and load sides of the pivot, respectively. F is the force a child could apply to the end of the handle and d_1 is the distance he or she could move it up and down. We consider both of these to be environmental variables, since the capabilities of the operator are part of the environment. The behavioral variable d_2 is the distance that the load end of the lever moves when the handle end is moved over distance d_1.

W is the total weight that the lever must lift, which consists of the force required to counteract the pressure on the piston as well as the pump hardware below ground. W is a behavioral variable, because its value depends upon the values of other variables. The first of these is the depth of the well, h, which is an environmental variable because it is determined by the location of the water source. The weight W also depends on the radius of the piston, which is the same as the radius of the cylinder, $r_{cylinder}$. These are design variables because we choose their values.

Restate the Problem in Clearer Terms Given the definition of these variables, we can restate the design problem as follows:

> Determine pump handle lengths L_1 and L_2 for a deep-well handpump, assuming a maximum depth h for the well of 50 m and a cylinder radius $r_{cylinder}$ of 25 mm. Assume that a child operator can apply a force F to the handle over a distance d_1. As part of the analysis, estimate the volume of water dispensed in each pump cycle.

Explore The "explore" step is the pre-planning step where we check that the problem makes sense, note any assumptions that must be made, identify key concepts

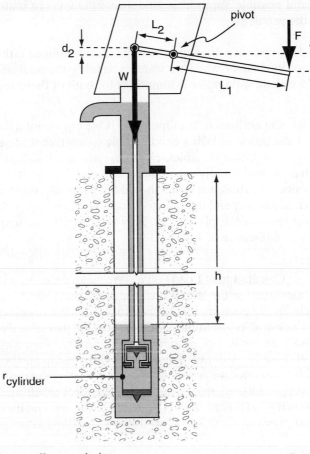

Figure 4.23 Deep well pump design.

TABLE 4.1 Variables in the handpump design problem.

Name	Description	Type
W	weight at the load end of the lever	behavioral
F	force that a child can to apply to the handle end of the lever	environmental
L_1	distance from the handle end of lever to pivot (fulcrum)	design
L_2	distance from the load end of lever to pivot	design
d_1	distance a child can move the load end of the lever	environmental
d_2	distance the load end of lever moves	behavioral
h	distance from ground surface to water source (well depth)	environmental
$r_{cylinder}$	radius of pump cylinder	environmental

and possible approaches, and assess the level of understanding required in solving the problem.

Assumptions In general, we make assumptions either when parts of the problem are unclear or when we want to simplify the analysis. In solving the pump design problem, we'll make assumptions for both of these reasons.

Capabilities of the Operator   One important question to consider in the design of the pump is, Will a child be able to operate it? Specifically, we've identified two environmental variables, F and d_1, which are the force a child could apply to the handle and the distance that a child could move it. It's unclear, however, what the values for these variables should be. Naturally, these capabilities vary from child to child and depend upon factors such as age, height, and weight. We'll assume a child operator can apply at least 50 N (between 11–12 lbs) of force to the pump handle over a distance of 1 m.

Calculation of Load Force   We make a second set of assumptions to simplify the calculation of the load W that the pump must lift. As mentioned above, this force includes the pressure on the piston as well as the weight of the pump hardware. Clearly, it's desirable to minimize the weight of the pump hardware, so that most of the operator's energy is applied to raising water, rather than wasted on moving the machinery. Without more detailed knowledge of the pump hardware—which we couldn't possibly have since we're still in the beginning stages of design—it's difficult to estimate its weight. Instead, we'll assume that the force resulting from pressure on the piston is much greater than the weight of the hardware, and base our estimate of W on the water pressure alone. Later, we'll check how these assumptions affect the overall design.

What are the Key Concepts and Possible Approaches?   Stepping back, the main concept in solving this problem is to construct models for a system and use them to inform a decision. The models themselves will involve a variety of physical concepts, listed below.

- We'll need to apply principles of **mass, volume, density, and pressure** in order to calculate the forces on the piston.
- We'll need to use the **principles of a lever** in order to determine the sizings for the handle.
- Determining the volume of water dispensed per cycle depends upon the **efficiency** of the pump, which in turn involves the concept of **work**.

What Level of Understanding is Tested?   As a design problem, the pump problem draws on abilities at all levels of understanding. Even though the individual concepts involved in this problem are not very complex, this problem is nonetheless difficult because it involves pulling many concepts together and requires an advanced level of understanding. Specifically, key skills required for solving this include:

- **application** of the physical principles listed above,
- **analysis** of the pump system in breaking it down to component parts,
- **synthesis** of models and an experimentation plan for using them,
- **judgment** in interpreting results from the models and making design decisions.

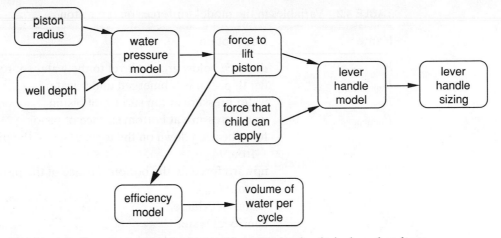

Figure 4.24 Concept map for the first version of a plan for designing a handpump.

Plan Having explored the main features of the handpump design problem, we're now ready to develop a detailed plan for solving it.

Figure 4.24 illustrates a first version of a plan for using models to design the pump. First, we'll construct a model for the forces on the piston based on the depth of the well and the radius of the piston. Next, we'll build a model for the lever, and from the weight of the water column and the force a child could apply to the lever handle, we'll calculate lever sizings. Finally, we'll develop an efficiency model, and using that with the water weight, we'll determine the volume of water delivered per cycle.

4.9.2 Modeling Forces on the Piston

Our first task is to build a model for the force W required to raise the piston in terms of the well depth and piston or cylinder radius. Figure 4.25 illustrates the forces acting

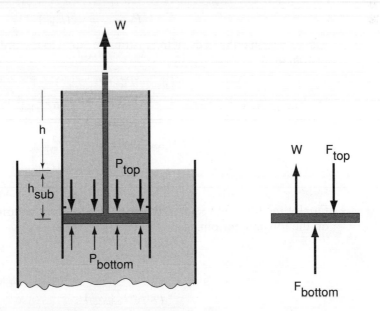

Figure 4.25 Forces on the piston of a deep well pump.

TABLE 4.2 Variables in the model for forces on the piston.

Name	Description
h	distance below ground level to the water source
h_{sub}	depth piston is submerged under water
P_{top}	water pressure at top surface of piston
P_{bottom}	water pressure at bottom surface of piston
F_{top}	force bearing down on the top surface of the piston (resulting from P_{top})
F_{bottom}	upward force on the bottom surface of the piston (resulting from P_{bottom})
$r_{cylinder}$	radius of cylinder
ρ	density of water
g	gravitational acceleration

on the cylinder and Table 4.2 defines the variables in the model. As pictured, the cylinder of a deep well pump is completely submerged into the water at the bottom of the well, and the drop pipe that connects the cylinder to the surface is filled with water.

As we saw in Section 4.4.1, the formula for the pressure on an object immersed in a fluid is

$$\text{pressure} = (\text{density of fluid}) \times g \times (\text{depth})$$

Thus, the pressure at the top surface of the piston is

$$P_{top} = \rho g (h + h_{sub})$$

while the pressure on the bottom surface of the piston is

$$P_{bottom} = \rho g (h_{sub})$$

The forces on either side of the piston are equal to the pressure times the area of the piston, or

$$F_{top} = P_{top}(\pi r_{cylinder}^2)$$
$$= \rho g (h + h_{sub})(\pi r_{cylinder}^2)$$

$$F_{bottom} = P_{bottom}(\pi r_{cylinder}^2)$$
$$= \rho g (h_{sub})(\pi r_{cylinder}^2)$$

Finally, the force W required to raise the piston is the difference between the forces on the top and bottom:

$$W = F_{top} - F_{bottom}$$
$$= \rho g [(h + h_{sub}) - h_{sub}](\pi r_{cylinder}^2)$$
$$= \rho g h (\pi r_{cylinder}^2)$$

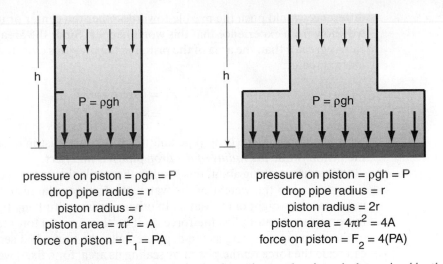

Figure 4.26 The pressure of the water in the drop pipe on the piston is determined by the length of the drop pipe and is independent of its radius. The pressure is uniform across the surface of the piston. The force on the piston is proportional to the pressure and the piston area.

Using this model, we can now calculate the force W required to raise the piston, assuming a cylinder diameter of 50 mm and a maximum well depth of 50 m.

$$W = \rho g h (\pi r_{\text{cylinder}}^2)$$

$$= 1000 \, \frac{\text{kg}}{\text{m}^3} \times 9.81 \, \frac{\text{m}}{\text{s}^2} \times 50 \, \text{m} \times 3.14 \times (0.025 \, \text{m})^2$$

$$= 963 \, \text{N}$$

The force required to raise the piston is thus 963 N or approximately 217 lbs. We'll round this up to 1000 N for the design of the lever handle.

We'll make one final but important observation before leaving the model for the force on the piston. In Figure 4.25, we pictured the drop pipe as having the same diameter as the pump cylinder, whereas in Figure 4.23, the drop pipe was narrower than the cylinder. What effect does the diameter of the drop pipe have on the force on the piston? To answer this question, consider the two cases in Figure 4.26. In the case on the left, a drop pipe of radius r is connected to a cylinder of same radius. As we know, the pressure P increases the deeper we descend into the drop pipe, with $P = \rho g h$. Just as we calculated before, the total force on the piston is the pressure times the area or

$$F_1 = (\rho g h)(\pi r^2)$$

Now, consider the case on the right, where the radius of the drop pipe is r but the radius of the cylinder, and hence the piston, is $2r$. As we'd expect, the pressure on the piston directly below the drop pipe is $P = \rho g h$; after all, the pressure is the result of the weight of the column of water above the piston. But the pressure is also uniformly the same value of P over the *entire surface* of the piston, even the part that is not under the drop pipe! If it weren't, then if we dropped a small object such as a marble into the pipe and let it sink to the surface of the piston, then the pressure difference would push the marble towards either the center or toward the wall, and we know from experience that this won't happen. Since the area of the piston in this case is greater than the area of the piston in the previous case, the force on the piston is also greater:

$$F_2 = (\rho g h)(4\pi r^2)$$
$$= 4F_1$$

Thus, the depth of the drop pipe and the area of the piston determine the force on the piston, while *the radius of the drop pipe has no effect.*

The curious thing about this result is that while the force on the piston in the first case is equal to the weight of the water, the force on the piston in the second case is 4 times the weight of the water! In other words, multiplying the area of the piston by some factor multiplies the force on it by the same factor, regardless of the area of the drop pipe, as long as the depth remains the same. If it seems strange that we can scale the force on the piston by scaling its area, for a fixed weight of water in the drop pipe, consider that this is no stranger than scaling the force at the end of a lever by moving the fulcrum or scaling the force at the end of a rope by adding a pulley to a block and tackle. In fact, the principle of scaling a force by changing the area of a piston is the central idea behind *hydraulic machines,* a topic that we will explore further in the problems at the end of the chapter.

4.9.3   Modeling the Handle Lever Arm

Now that we have estimated the weight of the water column, we turn to the design of the handle lever arm for the pump. As before, we'll first define the variables and then write equations for the model.

Defining the Variables   Figure 4.27 shows a schematic of the lever handle that illustrates the variable in the problem, which Table 4.3 further describes. At this point, there are no "known" values, but we assume that the weight of the load could be as high as 1000 N. The force a child can apply to the handle and the distance he or she can move it are environmental variables whose values are unknown at this time, and we will vary these values after we build the model to explore tradeoffs in the design. The design variables are the lengths of the lever arm on either side of the pivot or fulcrum, L_1 and L_2. We define the *lever ratio*, R, to be the ratio of the two lengths

$$R = \frac{L_1}{L_2} \tag{4.13}$$

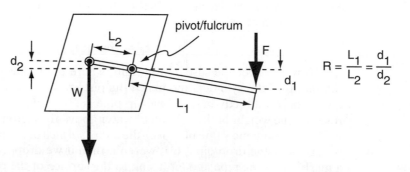

Figure 4.27   Variables in the handle lever arm model.

TABLE 4.3 Variables in the model for the lever handle.

Variable	Description	Value
W	load weight	up to 1000 N (behavioral variable)
F	force that a child applies to the handle end	at least 50 N (environmental variable)
R	lever ratio	unknown (design variable)
L_1	distance from the handle end of lever to pivot (fulcrum)	unknown (design variable)
L_2	distance from the load end of lever to pivot	unknown (design variable)
d_1	distance that a child can move the load end of the lever	at least 1 m (environmental variable)
d_2	distance that the load end of lever moves	unknown (behavioral variable)

Constructing the Model   From Equation (4.3), the lever handle will be balanced when

$$\text{lever balance condition: } FL_1 = WL_2 \tag{4.14}$$

From this relationship, and given the definition of the lever ratio in Equation (4.13), we can express the model for the weight W the handle can lift as a function of the lever ratio R and the applied force F:

$$W = W(R, F) = RF$$
$$R = \frac{W}{F} \tag{4.15}$$

A second aspect of the pump handle model is the relationship between the distance the handle end is moved d_1 and the distance the weight is moved d_2. Referring to Figure 4.27, from similar triangles, we identify

$$R = \frac{L_1}{L_2} = \frac{d_1}{d_2}$$

Thus,

$$d_2 = \frac{d_1}{R} \tag{4.16}$$

Selecting Handle Sizings   Given models for the water column weight and the pump handle lever arm, we can now begin to select an acceptable design. In order to do so, we need to determine values for some of the variables that are unknown.

We begin by selecting the load weight. From the water pressure model for a deep well pump, we find that 1000 N is a reasonable minimum load, since this ensures that the pump could raise water from a well 50 m deep with a cylinder radius of 25 mm.

Next, we consider the force a child might apply to the pump handle, which we've assumed could be as little as 50 N. This gives us two constraints on the handle design:

$$\text{weight constraint: } = W \geq 1000 \, \text{N}$$
$$\text{applied force constraint: } = F \geq 50 \, \text{N.}$$

Note that both constraints are expressed as "greater than or equal to" relationships. This means that we would accept designs that could lift more than 1000 N, as well as support operators who could apply more than 50 N of force to the handle, but minimally, at least these values must be met. Thus, from Equation (4.15),

$$R = \frac{W}{F}$$
$$= \frac{1000}{50}$$
$$= 20$$

Given the selected value for the lever ratio $R = 20$, the final step is to select values for the lengths of the lever arm on either side of the fulcrum, L_1 and L_2. At this point, the only constraint on the problem is that L_1/L_2 must equal R. To narrow the problem down, we either need to make some assumptions or set additional constraints. Two considerations in choosing the dimensions are:

- we'd like to maximize the amount of water raised with each pump stroke;
- the design of the handle assembly should be as compact as possible while still delivering good performance.

For now, we'll assume that the sucker rod of the pump is attached to the load end of the lever arm by some mechanism such that the water column is raised by distance d_2 whenever the handle end of the lever is depressed a distance d_1. While it may seem that by increasing the lever arm lengths L_1 and L_2 that we could increase d_2, this isn't necessarily the case. From Equation (4.16), we saw that d_2 depends only on d_1 and R. In fact, as Figure 4.28 illustrates, unless the lever arm is shorter than the height of the fulcrum, then d_1 and d_2 are limited by the fulcrum height rather than the length of the lever. Thus, with the goal of keeping the design as compact as possible, it doesn't make sense to make L_1 much longer than the fulcrum height. If we assume that the maximum distance d_1 over which a child could move the pump handle up and down

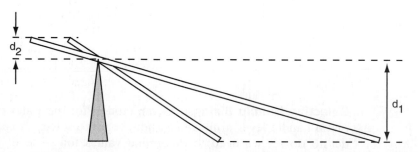

Figure 4.28 Diagram showing that the distance d_2 is limited by the height of the fulcrum and is independent of the length of the lever arm. The lever ratio R is identical for both the long and short lever arms pictured.

TABLE 4.4 Final values for variables in the pump handle design problem.

Variable	Description	Value
W	load weight	up to 1000 N
F	force that a child applies to the handle end	at least 50 N
R	lever ratio	20
L_1	distance from the handle end of lever to pivot (fulcrum)	1.5 m
L_2	distance from the load end of lever to pivot	7.5 cm
d_1	distance that a child can move the load end of the lever	1 m
d_2	distance that the load end of lever moves	5 cm

is 1 m, then we'll set the length of the pump handle L_1 slightly longer than that at 1.5 m, and thus,

$$L_2 = \frac{L_1}{R} = 7.5 \, \text{cm}$$

$$d_2 = \frac{d_1}{R} = 5 \, \text{cm}$$

Table 4.4 summarizes the values for handle variables.

4.9.4 Modeling Pump Efficiency

Our final task is to determine a model for the efficiency of a handpump. As we defined it earlier, pump efficiency is a measure of how much water the pump produces relative to the energy expended in operating it. In operating a handpump, a person performs work on the pump: he or she applies force to move its handle up and down. The pump, in turn, performs work on the water below, raising water with each upstroke of the sucker rod. We can thus more precisely define handpump efficiency as the ratio of the work that the pump actually performs in raising water versus the work a person performs in operating the pump, or

$$\text{efficiency} = \frac{\text{work pump performs on water}}{\text{work operator performs on pump}}$$

If a pump had a perfect efficiency of 100 percent, then all of the energy a person expended in operating the pump would result in the pump raising an equivalent amount of water. In any real pump, however, some of the operator's work is wasted. For example, with each downstroke of the hand, the operator has to lift not only water, but also the sucker rod and the piston itself. Friction in the system, such as between the piston seal and the cylinder wall, also adds to the amount of work the operator must perform during each cycle. Finally, leaks in the system may result in a loss of water raised, even if the piston travels the full length of the cylinder.

In order to quantify the efficiency of a pump, we need to be able to measure both the work the pump performs in raising water as well as the work the operator

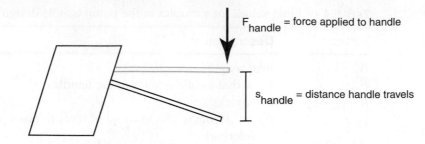

Figure 4.29   The work an operator performs on the pump during each cycle is $F_{\text{handle}} \cdot s_{\text{handle}}$.

performs on the pump. Recall that if a constant force of magnitude F is applied over some distance s, then the amount of work performed is

$$\text{work} = F \cdot s$$

Let's first consider the work the operator performs on the pump. In order to raise the sucker rod, the operator pushes down on the handle with some minimum required force and the handle, in turn, travels some distance, as shown in Figure 4.29. When the handle reaches the bottom, the operator lets the weight of the water on top of the piston return the handle to its starting position, doing no additional work herself. Thus, the work the operator performs in each cycle is

$$\text{work operator performs on pump} = F_{\text{handle}} \cdot s_{\text{handle}} \tag{4.17}$$

Now, let's consider the work the pump does in raising water. During each cycle, the pump adds the volume of water in the cylinder to the drop pipe. In doing so, the piston moves against the force of the water pressure over the length of the cylinder. Thus, the work that the pump does is

$$\text{work pump performs on water} = W \cdot s_{\text{cylinder}} \tag{4.18}$$

where W is the force of the water pressure on the piston and s_{cylinder} is the length of the cylinder. Suppose that a well has a depth h and the radius of the cylinder is r. From Equation (??), we saw that force on the piston is

$$W = \rho g(\pi r^2 h) \tag{4.19}$$

One way to determine the distance s_{cylinder} is to measure the amount of water dispensed in each cycle. As Figure 4.30 illustrates, the volume of water dispensed is the same as the volume drawn up into the drop pipe when the piston is raised. This space at the bottom of the pipe is a small cylinder with volume $V_{\text{dispensed}}$:

$$V_{\text{dispensed}} = \pi r^2 s_{\text{cylinder}}.$$

Thus,

$$s_{\text{column}} = \frac{V_{\text{dispensed}}}{\pi r^2} \tag{4.20}$$

Substituting Equations (4.19) and (4.20) into Equation (4.18), we get

$$\begin{aligned} \text{work pump performs on water} &= \rho g V_{\text{dispensed}} h \\ &= W_{\text{dispensed}} \cdot h \end{aligned} \tag{4.21}$$

which is simply the weight of the water dispensed times the depth of the well. Note that this result is independent of the diameter of the cylinder. Example 4.9 illustrates

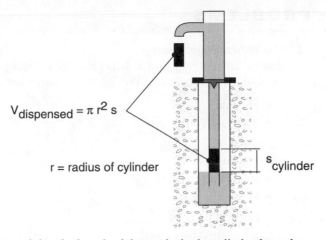

Figure 4.30 Determining the length of the stroke in the cylinder from the amount of water dispensed.

the use of the model to determine the efficiency for a given set of pump variable values.

Example 4.9	**Efficiency of a Handpump**

Efficiency of a Handpump

A handpump is used to draw water from a well 50 m deep. In pushing the lever handle of the pump down, a child applies 60 N (13.5 lbs) force through a distance of 1 m. If 100 mL of water is dispensed with each stroke, what is the efficiency of the pump?

Solution The efficiency of the pump is defined as

$$\text{efficiency} = \frac{\text{work pump performs on water}}{\text{work operator performs on pump}}$$

From Equation (4.17)

$$\text{work operator performs on pump} = F_{\text{handle}} \cdot s_{\text{handle}}$$
$$= 60\,\text{N} \cdot 1\,\text{m}$$
$$= 60\,\text{Joules}$$

From Equation (4.21)

$$\text{work pump performs on water} = \rho g V_{\text{dispensed}} h$$

$$= \left(1\frac{\text{kg}}{\text{L}}\right)\left(9.81\frac{\text{m}}{\text{s}^2}\right)(0.1\,\text{L})(50\,\text{m})$$

$$= 49\,\text{Joules}$$

Thus the efficiency is

$$\text{efficiency} = \frac{49}{60} = 0.82$$

This would be a rather good value for the efficiency of a real pump.

PROBLEMS

1. **WWW Scavenger Hunt**
   Draw a concept map that links (at least) the concepts listed below. Use web-based tools such as Google and Wikipedia to search for clues.

   - *vis viva*
   - Thomas Young
   - calculus
   - pyramids
   - stress and strain
   - Robert Hooke
   - mission to Titan

2. **Sagredo, Salviati, and Simplicio Discuss Hybrid Cars**
   Using Galileo's method of making a case by presenting several viewpoints, write a short discussion among three people who are debating the merits of hybrid cars. One of the protagonists is a scientist taking the pro view, the second is a scientist taking the con view, and the third is a non-scientist.

3. **Galileo's Interrupted Pendulum**
   Explain what would happen if the pendulum bob were released from a height higher than the pin in Galileo's interrupted pendulum setup.

4. **Alternative Solution to Collision Problem**
   Consider a collision between two bodies of equal mass, where one body is initially moving to the right at 4 m/s and the other is initially moving to the right at 6 m/s. In Example 4.3, we determined the velocities after the collision using the principles of conservation of momentum and relative velocity. For this exercise, determine the final velocities using the principles of conservation of momentum and conservation of energy instead. Compare your results with the original solution.

5. **Perfectly Inelastic Collisions**
   Earlier, we considered the case of a perfectly *elastic* collision, where the two colliding bodies bounce off each other, satisfying the principles of conservation of momentum, conservation of kinetic energy, and preservation of relative velocity. In this problem, we consider the case of a perfectly inelastic collision, where the two bodies stick together after they collide. We can think of such a collision as beginning with two bodies having masses m_1 and m_2 and initial velocities v_1 and v_2 and finishing with a single body with mass $m_1 + m_2$ and final velocity v_f.

   (a) If we know the masses and initial velocities before the collision, how many equations or laws do we need to determine the velocity after the collision? Compare this with the case of a perfectly elastic collision.

   (b) A mass of 6 kg moving at 3 m/s to the right has a perfectly inelastic collision with a mass of 4 kg moving at 5 m/s to the right. The combined mass after the collision ends up moving at a speed of 0.2 m/s to the left.

      i. Does conservation of momentum hold for this collision? If not, explain any difference before and after the collision.

ii. Does conservation of kinetic energy hold for this collision? If not, explain any difference before and after the collision.

6. An Elastic Collision

A 1500 kg body moving at 40 km/hr runs into a stationary 1800 kg body. Assuming the collision to be elastic, find the velocities of the two bodies after the collision.

(a) solve the problem using the principle of relative velocity and the conservation of momentum.

(b) solve the problem using conservation of momentum and conservation of kinetic energy, but without using the principle of relative velocity.

7. Playing Darts in an Elevator

You are playing darts in an elevator that is rising at a constant speed. Should you still aim at your target the way you would normally do, or should you aim higher, or lower? Explain.

8. Billiard Cue Balls

Billiard (pool) tables that do not require coins to be operated usually have all fifteen balls of the same size, so that they can all roll down to the same location when pocketed. On the other hand, in coin-operated billiard tables the white ball (cue ball) is slightly larger, so that it is the only ball that is separated and rolls down to the player, while the other balls roll down a different path under the table.

Describe what effect a slightly larger cue ball will have on the game and on a player's strategy and analysis of collision angles, as compared to a regular-sized cue ball (a billiard ball is 2 1/4″ in diameter and weighs around 5.5 oz; an oversized cue ball is 2 3/8″ in diameter).

9. Tricks in Urban Rail Station Design

One of the "tricks" used in the design of urban railway systems is to locate the stations at the top of a slight rise, to help the trains decelerate as they enter the station. Assuming that a car weighs 32,000 kg and has a maximum speed of 90 km/h, estimate the effect that elevating the track in the station by 2 m would have in slowing down a train with four cars.

London Dockland Light Rail Station (left) and Bangkok Sky Train (right).

Photographs courtesy of David Shallcross, University of Melbourne.

10. Disc Brake Stopping Time and Distance

A disc brake system consists of a metal disc connected to a wheel along with a caliper that closes on the disc to slow the rotation of the wheel by friction. Assuming that the brake calipers apply exactly the same pressure to the discs in both case, which would take a longer *time* and which would take a longer *distance* to come to a complete stop?

- a truck weighing 1800 kg traveling at 90 km/h
- the same truck traveling at 60 km/h carrying a load of 2250 kg

11. Inflating a Football

A football's air pressure is measured to be 11 psi, well below the recommended game-playing pressure of 13 psi. Air is quickly pumped into the ball until it is at game-play condition. If the ball was originally at an ambient temperature of 70 degrees Fahrenheit, and neglecting the heating effect on the needle, what is the immediate temperature increase inside the football?

12. Pressure in a Balloon

A balloon is filled with an ideal gas at room temperature at a pressure of 200 kPa.

(a) What would be the volume of the gas if the pressure remained constant and the temperature increased by 20 percent?

(b) What would be the temperature inside the balloon if the volume remained constant but the pressure were increased by 15 percent?

(c) What would be the pressure inside the balloon if the temperature remained constant but the volume was decreased by 10 percent?

13. Supporting a Weight with a Piston

A 50 kg object is sitting on top of a piston that is supported by a cushion of air in a cylinder. The radius of the piston is 1 m.

(a) In order to support this weight, how much force is needed?

(b) What does the air pressure inside of the cylinder have to be in order to produce enough force to support the weight?

(c) If you added 5 kg more to the mass on the piston, assuming that the temperature remains constant, what would be the change in height of the piston?

14. Units of Measure

Find the units of each quantity below as a combination of the units given. Note that some of the units may need to be squared or cubed to obtain the correct result.

(a) momentum (m, kg, s)

(b) momentum (N, s)

(c) force (m, kg, s)

(d) pressure (m, kg, s)

(e) pressure (N, m)

(f) momentum (pa, m, s)

15. More Units of Measurement

Find the units of each quantity below as a combination of the units given. Note that some of the units may need to be squared.

(a) energy (m, kg, s)

(b) energy (N, m)

(c) work (m, kg, s)

(d) work (N, m, s)

16. **Stored Potential Energy**

Calculate the potential energy stored in the following systems.

(a) A column of water 5 meters high (assume density of water $= 1000 \text{ kg/m}^3$).

(b) A spring stretched 3 cm with a spring constant k of 10 kg/s^2.

(c) 500 grams of air at room temperature and pressure in a closed vessel is heated by 30 K. Assume heat capacity of air $C_p = 1.012 \text{ J/(g K)}$ and density of air at room temperature and pressure $= 1.165 \text{ kg/m}^3$.

(d) 500 grams of air at room temperature and pressure is pressurized to 30 kPa above ambient pressure (100 kPa).

17. **Hydroelectric Power Plant**

The outlet of a dam for a hydroelectric power plant is located 50 m below the surface of the reservoir behind the dam.

(a) Calculate the potential energy per unit volume of water stored behind the dam. Assume that the density of water is 1000 kg/m^3.

(b) Calculate the velocity of the water exiting at the dam's outlet.

(c) Calculate the power transmitted by the water if it passes through a turbine at a rate of $2 \text{ m}^3/\text{s}$.

18. **Ethanol Air-to-Fuel Ratio**

Ethanol combusts according to the following reaction:

$$C_2H_5OH + 3O_2 \rightarrow 2CO_2 + 3H_2O$$

(a) Determine the stoichiometric air-to-fuel ratio for ethanol.

(b) Ethanol releases approximately 27 MJ (megajoules) of energy for every kilogram burned, whereas gasoline releases approximately 45 MJ/kg. If a car has a fuel efficiency of 12 km/l (28 mi/gal) for gasoline, estimate its fuel efficiency for ethanol.

(c) Suppose that during the intake stroke, a certain engine could draw in an equal volume of an air-fuel mixture of either gasoline (isooctane) or ethanol at their stoichiometric ratios. Which fuel would produce more work during the expansion or power stroke?

19. **Slow Versus Fast Combustion in a Piston Engine**

Consider two scenarios for burning fuel in a piston engine, the first where a given quantity of heat is added slowly at constant pressure, and the second where the heat is added very quickly as in an explosion. Assuming an ideal engine where there is no friction, answer the following questions. (Note: even though this problem requires only a few basic concepts from this chapter, based on trying it out on a few faculty colleagues, this is a difficult problem!)

(a) Sketch a rough plot of the position of the piston in the cylinder versus time for both cases.

(b) Will the piston reach the same maximum height in the cylinder in both cases? Explain why or why not.

(c) Will the pressure in the cylinder be the same in both cases when the piston reaches its maximum height? Explain why or why not.

20. Pump Forces

Which of the following scenarios requires the greatest force on the sucker rod to raise water from the well? Explain.

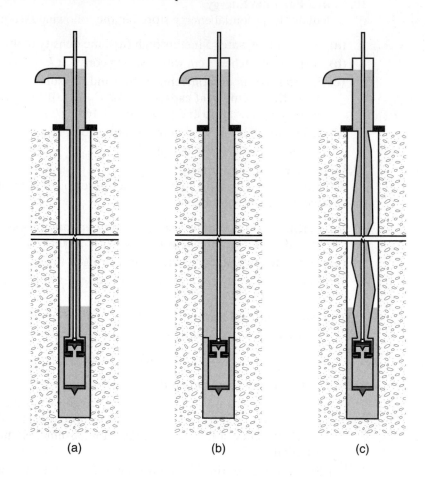

(a)   (b)   (c)

21. Water Pressure

A column filled with water is 10 m tall, with both its top and bottom open to the atmosphere.

(a) Calculate the pressure at the base of the column of water.

(b) Calculate the velocity at which the water exits the column. Hint: the kinetic energy of a "blob" of water leaving the column must be the same as the loss of potential energy of a blob of water of the same size from the top of the column.

22. Hydraulic Machines

The system illustrated below consists of two piston-and-cylinder assemblies filled with oil, connected by a narrow pipe. The piston on the left has a diameter of 2 cm and the one on the right has a diameter of 5 cm.

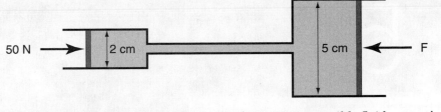

(a) Assuming that the oil inside the system is an *incompressible fluid*—meaning that its volume doesn't change when pressure is applied—if the smaller piston is pushed to the right by 1 cm, by how much will the larger piston move? Be sure to cite any "laws of nature" that you use in finding your solution.

(b) Now suppose that in order to push the smaller piston to the right, you needed to apply a force of 50 N. According to the principles of work and energy, what must the opposing force F be?

(c) Assuming the forces on the pistons from the previous problem, what is the pressure on each of the pistons? From this, what can you deduce about the pressure distribution throughout the oil?

(d) Explain what this problem has to do with the brake system in an automobile.

(e) Describe another engineering application that takes advantage of the principles illustrated by this problem. Draw a concept map of the system and provide a paragraph or two that explains the operation of the system in terms of the principles in this problem.

Data Analysis and Empirical Models

LEARNING OBJECTIVES

- to use basic mathematical and graphical techniques to determine how well experimental data fits a theory;
- to construct and use an empirical model that fits an equation to experimental data;
- to use elementary concepts from probability and statistics to analyze uncertainty in experimental data;
- to explain what a trade study is, and to show how "maps" of a design space can be used to find acceptable designs.

5.1 INTRODUCTION

As we saw in Chapter 4, even great minds like Aristotle and Descartes sometimes have been wrong. Today, we can say with confidence that Aristotle's and Descartes' laws of motion were incorrect. The reason that we can is because experimental evidence proves them so. In short, coming up with a theory is just part of the process of modeling; the rest of it lies in collecting and analyzing the data and drawing conclusions from it. Even for a very good theory, however, data collected in the real world often doesn't line up exactly with the values that the theory predicts, and there is an art to data analysis. This chapter introduces some of the mathematical and graphical tools that scientists and engineers use to analyze data when building and using models.

We begin the chapter by considering two theories three hundred years apart—Boyle's Law for gasses and Moore's Law for integrated circuit manufacturing—and use these to demonstrate ways of testing how well a theory fits the data. Next, we look at building and using empirical models, which involves finding equations that fit experimental data, even if there is no formal theory that describes it. Following this, we introduce techniques for quantifying the uncertainty in experimental data using statistics and probability. Finally, we present a graphical method for visualizing tradeoffs between design options.

5.2   THEORY AND DATA

5.2.1   Validating Boyle's Law

Robert Boyle and Robert Hooke had a theory that the pressure of a gas was inversely proportional to its volume at constant temperature. In order to convince their fellow scientists that the idea had merit, they needed to run an experiment and publish the results. In this section, we first look at the experiment Boyle describes in his paper *New Experiments Physico-Mechanicall, Touching the Spring of the Air, and its Effects*, published in 1690 [Boy60] [Wes05]. Then we look at a more modern way of analyzing the same data, using graphical techniques that hadn't yet been invented when Boyle published his results.

To test their theory, Boyle designed an experiment using a U-shaped glass tube, as pictured in Figure 5.1. The long leg of the tube was nearly 8 feet long and open-ended, while the short leg was approximately 1 foot long with its end sealed. Inside each leg of the tube, Boyle had placed strips of paper marked with 1/4 inch divisions. Working in a stairwell, Boyle poured enough mercury into the open end of the long leg of the tube until the bend was just full, trapping a volume of air in the sealed short leg. He then continued to slowly add mercury, so that the pressure from the weight of mercury in the longer leg compressed the trapped air. Each time the mercury level reached another 1/4 inch marking in the short leg, Boyle noted the corresponding level in the long leg, which he estimated to 1/4 of a division or 1/16 of an inch. He summarized his measurements and calculations in a table, which is excerpted in Table 10.3.8.

Column A of Boyle's table was the height of the column of trapped air in the short leg of the tube. Since the tube had a uniform diameter, this indicated the volume of air, which we'll denote by V. Column B was the height of the column of mercury in the long leg of the tube. This corresponds to the pressure exerted by the weight of the mercury, which we'll call P_{Hg}. The pressure exerted on the trapped air is equal to the sum of the pressure from the weight of the mercury together with the external

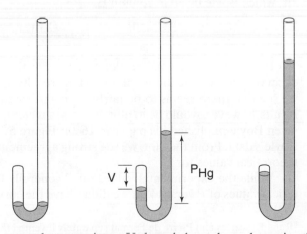

Figure 5.1   Boyle poured mercury into a U-shaped glass tube to determine the relationship between the pressure and volume of a gas.

TABLE 5.1 Data from Boyle's experiment, from [Boy60] and [Wes05].

A	B	C	D	E
			P, experiment	
V (tube-inches)	P_{Hg} (inches Hg)	P_{atm} (inches Hg)	$P_{Hg} + P_{atm}$ (inches Hg)	P, theory (inches Hg)
12	0	$29\frac{1}{8}$	$29\frac{2}{16}$	$29\frac{2}{16}$
10	$6\frac{3}{16}$	$29\frac{1}{8}$	$35\frac{5}{16}$	35
8	$15\frac{1}{16}$	$29\frac{1}{8}$	$44\frac{3}{16}$	$43\frac{11}{16}$
6	$29\frac{11}{16}$	$29\frac{1}{8}$	$58\frac{3}{16}$	$58\frac{2}{8}$
5	$41\frac{9}{16}$	$29\frac{1}{8}$	$70\frac{11}{16}$	70
4	$58\frac{2}{16}$	$29\frac{1}{8}$	$87\frac{14}{16}$	$87\frac{3}{8}$
3	$88\frac{7}{16}$	$29\frac{1}{8}$	$117\frac{9}{16}$	$116\frac{4}{8}$

air pressure of the atmosphere, P_{atm}. Using a barometer, Boyle determined that the atmospheric pressure at the time of his experiment was the equivalent of a column of 29 1/8 inches of mercury, which he listed as column C in his table. Thus the total pressure P, which he listed in column D, in units of inches of mercury, is the sum of columns B and C, or

$$P = h_{Hg} + P_{atm}$$

$$= h_{Hg} + 29\frac{1}{8}$$

If V_i and P_i are the values in the ith rows of columns A and D of the table, then according to the theory that PV is constant,

$$P_1 V_1 = P_i V_i$$

Solving this for P_i, we get the expected theoretical pressure for a given volume of air, which Boyle listed in column E.

$$P_i = \frac{P_1 V_1}{V_i}$$

Boyle then asked the reader to compare the values in columns D and E to observe how well the theory predicted the results of his experiment. From the table, we see that there seems to be fairly good agreement. A better way to visualize the results, however, would be to plot the data, a technique that hadn't yet been invented when Boyle published his paper in 1660.[1] Figure 5.2 shows a plot of P versus V for Boyle's data. From the plot, we see strong agreement between the experimental and theoretical values for P.

While the experimental values for P generally fall along the curve of the theoretical values of P versus V, if we didn't have the curve of theoretical values, it would

---

[1] René Descartes and Pierre de Fermat separately invented the system that we now call "Cartesian coordinates" in 1637. One of the earliest known x-y data plots was a graph of altitude versus barometric pressure by Edmond Halley in 1686.

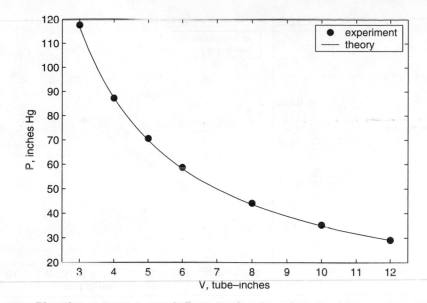

Figure 5.2   Plot of pressure on trapped air versus air volume from Boyle's experimental data.

be very difficult to tell at a glance whether or not the plot of experimental data had the right "shape." What should the expected shape be? According to the theory,

$$PV = k,$$

where k is an unknown constant. If we think of the data plot as a graph of a function of the form $y = f(x)$, then the shape of the experimental data in Figure 5.2 should follow the shape of a plot of

$$P = f(V)$$
$$= \frac{k}{V}$$

The function $f(x) = k/x$ isn't a shape we easily recognize. On the other hand, we can easily recognize a straight line. Thus, if we could somehow transform the experimental data into a form where it plotted along a straight line if the theory were correct, then we wouldn't even need a separate plot of the theoretical data to see if the experimental data validates the theory.

In general, a plot of a function will be a straight line if it can be written in the form

$$f(x) = mx + b$$

where m is the slope and b is the y-intercept. In the case of Boyle's law, transforming the data is easy: we can simply plot $1/P$ versus V instead of P versus V. In doing so, we get

$$f(V) = \frac{1}{P}$$
$$= \frac{1}{k}V$$

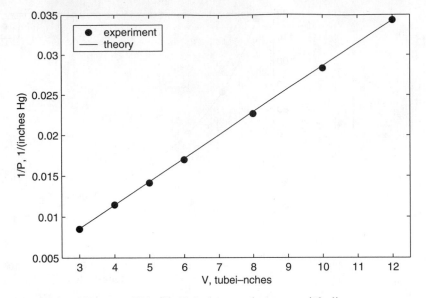

Figure 5.3 Plotting $1/P$ versus V for Boyle's data produces a straight line.

where the slope of the line passing through the data is $1/k$. Figure 5.3 shows the resulting plot. We can calculate the slope of the line, and its intercept—which should theoretically be zero—given two points on the line. Taking the two end points of the line as points (x_1, y_1) and (x_2, y_2), the slope is

$$m = \frac{y_2 - y_1}{x_2 - x_1}$$
$$= \frac{.034 - .009}{12 - 3}$$
$$= 2.87 \times 10^{-3}$$

Since the slope of the line is $1/k$,

$$k = \frac{1}{m}$$
$$= 349$$

We can then determine the intercept of the line given the slope and one point. Using the left endpoint (x_1, y_1), we obtain

$$b = y_1 - mx_1$$
$$= .009 - (.00287)(3)$$
$$= .00039$$

which is very close to zero.

5.2.2 Exponential Change, Log Plots, and Moore's Law

Boyle's experiment of compressing air in a U-tube provided a simple example of "linearizing" experimental data to see if it validates a theory. The theory predicted that pressure should be inversely proportional to volume. We plotted the reciprocal

TABLE 5.2 Transistor counts for Intel microprocessors.

Year	Product Name	Transistors	$\log_2$(transistors)
1971	4004	2,300	11.2
1972		2,500	11.3
1974		4,500	12.1
1978	8086	29,000	14.8
1982		134,000	17.0
1985		275,000	18.1
1989		1,200,000	20.2
1993	Pentium	3,100,000	21.6
1997		7,500,000	22.8
1999		9,500,000	23.2
2000		42,000,000	25.3
2001	Itanium	25,000,000	24.6
2003		220,000,000	27.7
2004		592,000,000	29.1

of pressure versus volume in order to transform this theory into a function in the form of a straight line. With similar ease, we can transform other common theoretical relationships, such as $f(x) = x^2$ or $f(x) = 1/x^2$, into linear functions. A particularly common form of theoretical relationship in science and engineering is the case where one quantity changes *exponentially* with respect to another. To illustrate, we'll consider the theory proposed in 1965 by Gordon Moore, one of Intel's founders, that the number of primitive switches called *transistors* would double approximately every two years for leading-edge integrated circuits. Table 5.2 shows the number of transistors on a series of Intel microprocessors between 1971 and 2004, and Figure 5.4 plots

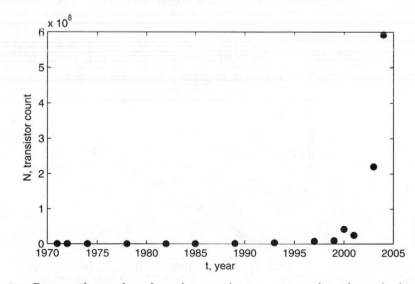

Figure 5.4 Because the number of transistors varies over many orders of magnitude, a plot of transistor count versus time is difficult to read.

this data. As the plot shows, the data is highly nonlinear. Further, because the transistor count varies from just over two thousand to more than half a billion, the early data points all lie very close to the x-axis. As a result, it's practically impossible to determine visually whether Moore's theory—known as *Moore's Law*—proves true.

To test Moore's theory, we'll assume that the transistor count does double at regular intervals, but that it's unknown whether this occurs every two years. Mathematically, we can express this theory as

$$N(t) = N_0 2^{t/k}, \tag{5.1}$$

where

- N = the number of transistors
- N_0 = the number of transistors in the starting product
- t = the time in years
- k = the interval between each doubling in transistor count.

This function is clearly not linear, since the independent variable t is in the exponent, and thus we say that $N(t)$ grows exponentially with t.

To transform Equation (10.1) into a linear equation, we find the logarithm of both sides:

$$\begin{aligned} \log_2 N(t) &= \log_2(N_0 2^{t/k}) \\ &= \log_2 N_0 + \log_2(2^{t/k}) \\ &= \log_2 N_0 + \frac{1}{k}t \end{aligned} \tag{5.2}$$

Equation (5.2) *is* a linear equation, with a slope of $1/k$ and intercept of $\log_2 N_0$. If we plot the logarithm of the transistor count versus time, the data should lie along a straight line if Moore's theory is really true, and further, we can determine the doubling interval k from the slope of the line. The last column in Table 5.2 shows the values for $\log_2 N$ and Figure 5.5 plots the results. After we find the logarithm of the transistor count, the data points do in fact lie approximately along a straight line.

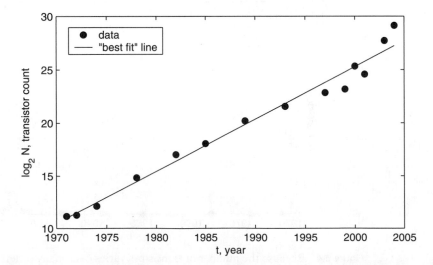

Figure 5.5 Plot of $\log_2$ of transistor count versus time for Intel microprocessors is approximately a straight line.

The line in Figure 5.5 is a "best fit" to the data; we determine this by approximating the line with a ruler. Another way to do this is through a mathematical technique known as *linear regression*, which calculates the coefficients of the line that minimizes the error between the data points and the line. For our purposes, a simple approximation of the best fit line is adequate.

The best fit line in Figure 5.5 passes through the data points at 1971 and 2000. From these two point, we calculate the slope of the line as

$$m = \frac{y_2 - y_1}{x_2 - x_1}$$

$$= \frac{25.3 - 11.2}{2000 - 1971}$$

$$= 0.49$$

Since the slope of the line equals $1/k$, the doubling interval is

$$k = \frac{1}{m}$$

$$= \frac{1}{0.49}$$

$$= 2.04 \text{ years}$$

which is very close to what Gordon Moore predicted in 1965. Since Moore was the chairman of Intel during much of this time, however, many would say that Moore's Law was a self-fulfilling prophecy!

While plotting the logarithm of transistor count enabled us to determine that Intel's microprocessor line satisfied Moore's prediction, in Figure 5.5 we can't read the number of transistors directly from the graph. In order to plot the logarithm of the data and read the actual data values, we use a semi-logarithmic or semilog plot, as shown in Figure 5.6. In a semilog plot, the spacing of values along the y-axis is transformed logarithmically, so that if the y-values have an exponential dependence on the x-values, the plot will be a straight line.

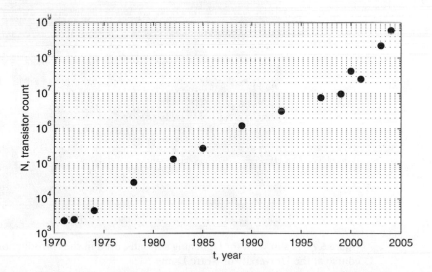

Figure 5.6   Semilog plot of transistor count in Intel microprocessors versus time.

5.3 EMPIRICAL MODELS

5.3.1 Introduction

In many situations, it's either difficult or impractical to predict how a system will behave according to basic scientific theories. In these cases, engineers often use *empirical models* based on experimental data. Sometimes, engineers will design and run a specific experiment with the goal of collecting data for building an empirical model. Other times, the data comes from the results of previous designs related to future designs. The aerospace industry, for example, has decades of experimental data on how different wing shapes perform under different conditions, and uses this information to develop new designs.

Unlike theoretical models, empirical models don't attempt to explain *how* or *why* a system behaves as it does, yet still can accurately predict *what* it will do under given conditions. To illustrate these principles we'll examine an empirical model for using a large slingshot for launching a softball at a target downrange. Figure 5.7 shows a picture of the slingshot. It consists of a frame with a horizontal base and vertical supports. Attached to these supports is a spring made of rubber tubing with a pouch that holds the ball during launch. Throughout the example, we'll assume that the tubing is attached at a fixed height on the vertical supports, and that the only adjustable setting is the horizontal pullback, which sets the point along the base to which the tubing is stretched before the ball is launched. We'll also ignore the effects of wind in building our model. Thus our model will have only two variables, as listed below:

Variable Name	Description	Type
X	distance spring pulled back along base	design
D	expected distance that softball will travel	behavioral

5.3.2 Running an Experiment

After defining variables, the next step in building an empirical model is to design and run an experiment to determine the relationship between the inputs and outputs of

Figure 5.7 A slingshot for launching softballs, used in the Introduction to Engineering course at the University of Notre Dame.

TABLE 5.3 Results of 6 trials for launching a softball from the slingshot with different pullback settings.

Trial	X	Distance
1	0.25	1
2	0.50	4
3	0.75	10
4	1.00	18
5	1.25	27
6	1.50	36

the model. The basic idea is to change the values of the input variables in a controlled fashion and then to observe the corresponding changes in the outputs. Each run of the experiment with a given combination of input values in known as a *trial*. When a model has many input variables, it may require very many trials to determine the effects of each on the output, and learning how to properly decide which trials to run is an important topic in advanced courses on the design of experiments. Our slingshot example, however, has only one input variable, so we'll simply sweep this over a range of values.

Through physical intuition, we expect that the distance the softball travels will increase smoothly as we increase the pullback distance X, but the shape of this function isn't immediately obvious. In fact, it's unlikely to be a straight line, because of the trigonometry involved in the relationship between X and the launch angle. In addition, the graphs of sines and cosines certainly aren't straight lines. Also, physical effects may influence the relationship between how much the spring is stretched, the launch velocity of the ball, and how the ball travels through the air. If we knew or strongly suspected that the relationship between X and D was a straight line, we could run test launches at two points and then connect them to get a model. In this case we don't, so we'll run more trial launches. Each trial has a cost, so there is also a motivation to keep the number of trials small. Assuming that our budget allows us to run six trials, we'll use them to evenly cover the full range of the launcher, with X ranging from 0.25 to 1.50 m in 0.25 m increments. Table 5.3 shows the results of the trials.

Figure 5.8 illustrates a plot of the data from Table 5.3, with the design variable X on the horizontal axis and behavioral variable D on the vertical axis. Note that the points of D versus X do *not* follow a single straight line. For short horizontal pullbacks, the slope of the curve is less than it is for long pullbacks. As the pullback increases, however, the curve approaches a line.

5.3.3 Interpolation and Fitting a Line to the Data

The data in Table 5.3 and Figure 5.8 span a range of values for the pullback X from 0.25 m to 1.50 m and corresponding launch distances D from 1 m to 36 m. Because we plotted only 6 points, however, there are sizeable gaps between data points. To estimate the distance D for a pullback X that lies in one of these gaps, we'd need to *interpolate* values between two points.

Graphical Method Although the data points don't lie along a straight line, we might still *approximate* the curve between two data points with a line, and the closer

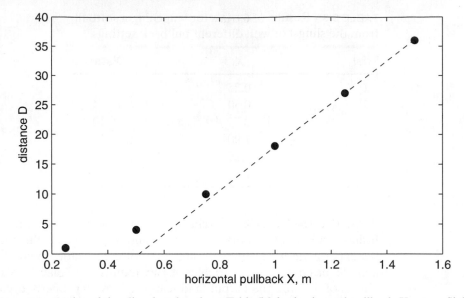

Figure 5.8 A plot of the slingshot data from Table 5.3 for horizontal pullback X versus flight distance D. Note that the data points do *not* lie along a straight line.

the data points, the better the approximation. Figure 5.9 shows a plot of the launch data with adjacent points connected by line segments. Such a plot is sometimes called a *piecewise-linear* model, since it's composed of pieces that are lines, even though the model as a whole isn't a line. Given a piecewise-linear plot, it's easy to interpolate a value between two points, as illustrated by Figure 5.9. If we want to estimate the distance a ball would travel for a launcher setting of $X = 0.6$ m, we simply read the corresponding value for D from the point on the line segment, which in this case is 6.4 m. Conversely, we could use the reverse analysis to find the X setting to launch a ball at a target 6.4 m downrange.

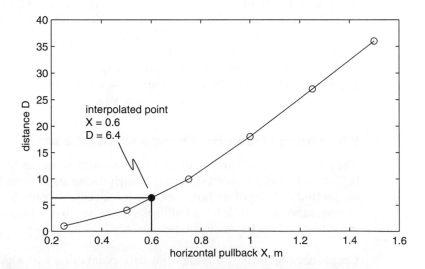

Figure 5.9 Using a piecewise linear plot to interpolate between two data points.

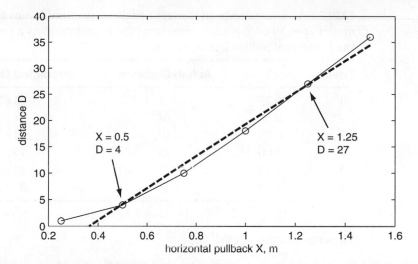

Figure 5.10   A line passing through the points that are one in from the extremes in the data set yields a fairly good fit to the data.

Numerical Method   Plotting experimental data, as in Figure 5.9, is extremely valuable because we can *see* how the behavior of a system changes as the design or environmental variables change. Graphical models do, however, have their limitations. One of these is that a person needs to physically read a graph to evaluate the model. Also, interpolation with a graph, while conceptual simple, can be tedious and time-consuming. What we'd prefer in many cases is to be able to plug values for the input variables into a formula and obtain values for the behavioral variables. We can do this by finding a function $D = f(X)$ that's a best fit to the data, much as we did for Boyle's Law and Moore's Law, except that this time we don't have a theory with which to start. Depending upon the shape of the experimental data, we might choose one of many different kinds of functions for f, such as a polynomial. In this case, as Figure 5.10 shows, a line passing through the second and fifth trial points from Table 5.3 provides a fairly good fit to the data, except for very small values of X. Thus, we will use these two points, whose (X, D) coordinates are (0.5, 4) and (1.25, 27), to find the equation of the line that will serve as our numerical model of the slingshot.

Using these two points, as we did before, we can solve for the equation of the line in slope-intercept form

$$y = mx + b$$

where

$$m = \frac{D_2 - D_1}{X_2 - X_1}$$

$$= \frac{27 - 4}{1.25 - 0.5}$$

$$= 30.67$$

$$b = D_1 - mX_1$$

$$= 4 - m(0.5)$$

$$= -11.33$$

TABLE 5.4 Comparison of launcher trials versus predictions from the numerical model $D = 30.67X - 11.33$. Note that the model predicts a negative distance when the horizontal pullback X is 0.25 m.

Trial	X	Actual Distance	Predicted Distance	Error
1	0.25	1	−3.67	−4.67
2	0.50	4	4.00	0.00
3	0.75	10	11.67	1.67
4	1.00	18	19.33	1.33
5	1.25	27	27.00	0.00
6	1.50	36	34.67	−1.33

Thus the equation of the line is

$$D = 30.67X - 11.33$$

Table 5.4 compares predictions from the numerical model with the actual launcher values for each of the six trial points. First observe that the error for points 2 and 5 is zero; this is expected since the line was *defined* to pass through these points. The magnitude of the error for points 3, 4, and 6 is less than 2. Most interesting, however, is the prediction for point 1. It has a large error, as expected, but the model predicts that the distance would be *negative*! This clearly demonstrates one of the major weaknesses of numerical, empirical models in general; they only deal with numbers and don't consider what is or isn't physically possible. For this reason, it's important to pay close attention to any result that comes from a numerical model. Using a graphical model in combination with the numerical model helps us understand this anomaly. From a strictly numerical viewpoint, getting a negative result at point 1 is the price we pay for fitting a straight line to data that's inherently non-linear.

5.4 USING STATISTICS TO QUANTIFY UNCERTAINTY

The model from the last section predicts that if the spring on the slingshot were pulled back 1 m, then the softball would land a distance of 18 m downrange. But how accurate and repeatable is this prediction? To examine this question, suppose that we ran a second experiment using the slingshot the day after building the initial model, consisting of 20 trials with a pullback of 1 m, with the results shown in Table 5.5. Not only do the trials not all land at 18 m, but they span a range of distances between 14.2 and 19.3 m. This means that there's a degree of *uncertainty* in the slingshot model. Uncertainty is common to many engineering problems, and a critical part

TABLE 5.5 Results of 20 trial launches with slingshot spring pulled back 1 m

trials 1–5:	17.5	19.0	16.4	19.3	16.6
trials 6–10:	16.0	17.4	16.7	18.1	17.5
trials 11–15:	15.1	14.2	17.4	15.7	17.8
trials 16–20:	19.3	18.5	15.7	17.9	17.0

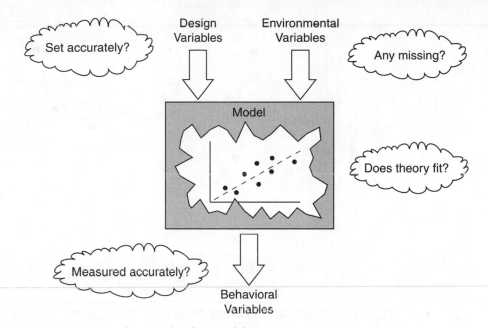

Figure 5.11 Sources of uncertainty in a model.

of engineering is being able to manage it. This section introduces some of the basic principles of quantifying uncertainty using *statistics* and discusses some of the ways engineers use statistics to justify decisions.

5.4.1 Sources of Uncertainty

As Figure 5.11 illustrates, every part of the process of building and using a model is subject to uncertainty. Beginning with the initial experiment, once source of uncertainty is *measurement error*, either in setting the values of the input variables, or in determining the values of the output variables. In the case of the slingshot, for example, in any given trial either the pullback of the spring or marking and measuring where the softball lands could be slightly off as a result of simple human error. Even if all of the measurements are completely accurate, a model may still predict erroneous results if the theory or *basis function* of the model doesn't consistently fit the data. If we used a single straight line as the basis for the slingshot model, there will clearly be cases where the model predicts inaccurate values, since the flight distance isn't strictly linear with respect to the pullback. Finally, even if we have made reliable measurements and have found a basis function that fits the data perfectly, a model might still make bad predictions if it fails to include a significant input variable. Suppose, for example, that the wind conditions on the day that we ran our initial experiment were very different from those on the day that we ran our second experiment. Since the model doesn't take wind velocity into consideration, we might reasonably expect its accuracy to vary with the weather.

Any and all of the sources of uncertainty we describe above can affect the accuracy of any given model. By analyzing the experimental results, we can find clues that indicate which of these may be involved, and then use this knowledge to reduce the uncertainty, or at least to factor it into our predictions. As a first step, we display the

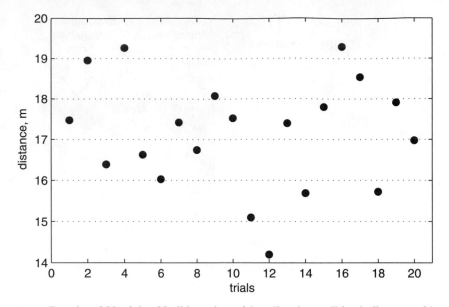

Figure 5.12 Results of 20 trial softball launches with a slingshot pull-back distance of 1 m.

data in a *scatter plot*, as Figure 5.12 shows, that simply graphs the launch distance for each of the 20 trials in the order in which they were performed. Inspecting the plot, we see that the trials appear to be randomly distributed over a range of distances, but the bulk of the distribution seems to be short of the predicted distance of 18 m.

5.4.2 Mean and Standard Deviation: Systematic and Random Error

While a scatter plot gives a good overall impression of the data, statistics provide a numerical summary that help us relate the results to our model. A *statistic* is simply defined as a number calculated from a sample of data values, such as a minimum, maximum, or average value. There are several ways to define the average of a sample of data, but the most common is the *mean*—the sum of the data values divided by the number of samples. More formally, if there are n samples of data and $x_i, i = 1 \ldots n$, are the values of the samples, then the mean $\bar{x}$ is defined as

$$\text{mean}: \quad \bar{x} = \frac{1}{n}\sum_{i=1}^{n}x_i \tag{5.3}$$

For the data in Table 5.5 and Figure 5.12, the mean value for the launch distance, d is

$$\bar{d} = \frac{17.5 + 19.0 + 16.4 + \cdots + 17.0}{20}$$
$$= 17.16\,\text{m}$$

Thus, the mean distance from the second slingshot experiment is in fact closer to 17 m than it is to the predicted distance of 18 m. This indicates that it's quite possible that there's some source of uncertainty that consistently causes the model to predict a longer distance, in combination with other sources that account for additional variations. We call this type of consistent error a *systematic error* or a *bias* in the data.

Brainstorming a bit, we can come up with a list of possible candidates that might be the cause of this systematic error:

- Maybe there was a stronger tailwind at the time of the first experiment or a stronger headwind at the time of the second.
- Maybe there was a difference in how the launch distance measurements were made, such as maybe the tape measure wasn't pulled taut in the first experiment.
- Maybe there was a difference in launch setup, such as setting the pullback distance according to the position of the back of the softball versus the front of the softball.
- Maybe the chosen "best fit" line inherently overpredicts the distance for a pull-back of 1 m.

Let's look more closely at one of these possibilities, that there was a difference in the launch setup. If there were a difference of a softball width in pullback distance, how much would this affect the flight distance? To answer this question, we determine the *sensitivity* of launch distance with respect to pullback distance from the numerical model of the launcher. In determining the equation of the best fit line through the data, we calculate the slope of the line as

$$m = \frac{\text{change in flight distance}}{\text{change in pullback}} = 30.67$$

According to this result, it would only take a change in pullback distance of 1/30 m—just a few centimeters—to change the launch distance by more than 1 m. Thus, differences in launch setup is a plausible explanation of the systematic error, and it's certainly worth checking further. Alternatively, a user might reasonably choose to compensate for the shorter launch distances by pulling back the spring a few centimeters more than the model suggests.

As we saw above, a bias in the mean or average value of experimental data may indicate a systematic error that can be corrected to reduce the uncertainty of a model. Another important statistic is a measure of how much the experimental data varies about the mean. Figure 5.13 illustrates both the mean and error about the mean in the scatter plot. If d_i is the launch distance of the ith trial and $\bar{d}$ is the mean distance, then the error of the ith trial about the mean is

$$e_i = d_i - \bar{d}$$

Just as we calculated the average value of the launch distance, we can also calculate an average for the error about the mean. If we calculated the average as the mean of the errors, however, the positive and negative errors would cancel each other out, and—as can be easily shown—the mean value of the e_i's would be zero. Instead, to measure the average *magnitude* of the error, we commonly take the mean of the *squares of the errors* and then take the square root of this value. This quantity is called the *standard deviation* of the data.[2] More formally, for a set of n sample data

---

[2]Note that statisticians sometimes use an alternate definition of the standard deviation that divides the sum of the square errors by $n-1$ rather than by n. For very large n, the values for both are close, but when using a software package or calculator, be aware of which definition is built-in.

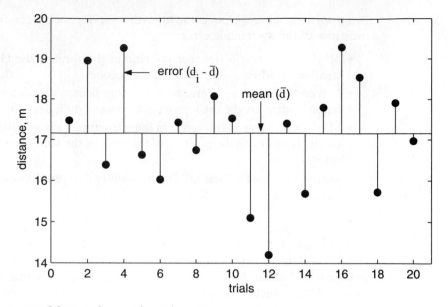

Figure 5.13 Mean and error about the mean

values x_i, $i = 1 \ldots n$, the standard deviation σ is:

$$\text{standard deviation:} \quad \sigma = \frac{1}{n}\sqrt{\sum_{i=1}^{n}(x_i - \bar{x})^2} \tag{5.4}$$

For the second launch experiment, the standard deviation is

$$\sigma = \frac{\sqrt{(17.5 - 17.16)^2 + (19.0 - 17.16)^2 + \cdots + (17.0 - 17.16)^2}}{20}$$
$$= 1.35 \,\text{m}$$

5.4.3 Estimating Probability

Taken together, the mean and the standard deviation provide a compact way for describing the results of the launch experiment. We would need more information, however, in order to answer the question:

> How likely is it that a softball launched with a pullback of 1 m will land within 1 m of 18 m?

Specifically, to answer this question, we need to be able to estimate the *probability* that a softball will land within a given range. Blaise Pascal and Pierre de Fermat first discussed a mathematical theory of probability in a series of letters in 1654. Two years later, Christiaan Huygens wrote the first paper on probability theory, *Van Rekeningh in Spelen van Geluck* (On Calculations in Games of Chance) which was translated and published in Latin as part of a textbook by his former mathematics professor at Leyden University. Huygens' manuscript was translated into English in 1692 by John Arbuthnot, a British physician, mathematician, and satirist who was a close friend of Jonathan Swift, author of *Gulliver's Travels*.[3] In the preface to his book, entitled, *Of the Laws of Chance, or a Method of the Calculations of Hazards of Game, Plainly demonstrated, And applied to Games at present most in Use, Which may be easily be extended to the most intricate Cases of Chance imaginable* [Arb92], Arbuthnot introduces—in his typically sardonic (and often ribald) prose— the connection between probability and uncertainty:

> You will find here a very plain and easie Method of the Calculation of the Hazards of Game, which a man may understand, without knowing the Quadratures of Curves, the Doctrin of Series's, or the Laws of Centripetation of Bodies, or the Periods of the Satellites of Jupiter; yea, without so much as the Elements of Euclid. There is nothing required for the comprehending the whole, but common Sense and practical Arithmetick; saving a few Touches of Algebra, ... where the Reader ... may make use of a strong implicit Faith.
>
> Every man's Success in any Affair is proportional to his Conduct & Fortune. Fortune (in the Sense of most People) signifies an Event which depends on Chance, agreeing with my Wish; and Misfortune signifies such an Event contrary to my Wish: an Event depending on Chance, signifies such an one, whose immediate Causes I don't know, and consequently can neither fortel nor produce it ... It is impossible for a Dye, with such a determin'd force and direction, not to fall on such a determin'd side, only I don't know the force and direction which makes it fall on such a determin'd side, and therefore I call that Chance, which is nothing but want of Art; that only which is left to me, is to wager where there are the greatest number of Chances, and consequently the greatest probability to gain;

Although we use the term "probability" commonly in everyday conversation, surprisingly, mathematicians still disagree over its meaning. The most common definition, based on the work of Pascal, Fermat, and Huygens, is the *relative frequency* of possible outcomes. In this view, every trial of an experiment is like a game with two possible outcomes: win or lose. Thus, if a person bets that a coin toss will come up heads, or that a roll of a dice will fall on six, or that a launch of a softball from the slingshot will land within 1 m of a target, he or she either will or will not be successful. According to the relative frequency definition, the probability of an outcome is the percentage of times that this outcome would occur in a very large number of trials. Thus, if we say that the probability of a coin coming up heads for a given toss is 50 percent, this means that in a very large number of tosses, we'd expect half to be heads.

It's important to realize that for any real experiment that has a fixed number of trials, the best we can do is estimate a probability. For example, if we tossed a coin three times, heads would come up in 0, 1, 2, or 3 of the trials. Thus from this experiment, our estimate for the probability of heads would be 0, 1/3, 2/3, or 1—but not 1/2. The accuracy of the estimation of probability increases with the number of trials. If we tossed a coin 1 million times, we wouldn't expect it to come up heads exactly 500 thousand times, but we would expect the ratio of the number of heads to the number of tosses to be very close to 0.5. With this understanding, we state the

---

[3] Arbuthnot was also appointed guardian of "Peter the Wild Boy," who was found in the woods near Hamelen, Germany in 1725 subsisting on acorns, berries, and tree bark, and then delivered to King George I of England. Swift and Arbuthnot wrote a satire on the uproar that surrounded Wild Peter entitled, "The Most Wonderful Wonder that ever appeared to the Wonder of the British Nation."

estimated experimental probability of an event as

$$\text{estimated experimental probability} = \frac{\text{number of successful trials}}{\text{total number of trials}}$$

Example 5.1 illustrates the estimation of probabilities of slingshot launches falling within given distance ranges.

Example 5.1	**Estimating Probabilities for the Slingshot**

Given the experimental data in Table 5.5 for launching a softball with 1 m pullback from the slingshot estimate the probability of a launch (a) landing less than 1 m away from a target at 18 m, (b) landing short of this range, (c) landing beyond this range.

Solution

Given: The experimental results of 20 trial launches

Find: Estimated probabilities for launches in the ranges (a) $d \leq 17$, (b) $17 < d < 19$, (c) $d \geq 19$.

Plan: Count the number of launches that landed in each of the three ranges and divide by the number of trials, which is 20.

Analysis: The number of trials in each of the ranges and the corresponding probabilities are as follows:

$$d \leq 17: \quad 9 \text{ trials} \quad \Rightarrow P = \frac{9}{20} = 0.45$$

$$17 < d < 19: \quad 8 \text{ trials} \quad \Rightarrow P = \frac{8}{20} = 0.40$$

$$d \geq 19: \quad 3 \text{ trials} \quad \Rightarrow P = \frac{3}{20} = 0.15$$

Note that the sum of the three probabilities is 100 percent. This is to be expected, since the three ranges span all possible distances, without overlapping.

In his *Games of Chance*, Huygens considered in particular the question of the expected gain or loss from a bet. He states that if there are p chances of winning (or losing) a sum of money a and q chances of winning (or losing) a sum of money b, all chances having equal weight, then the expected payoff from the bet is

$$\text{expected gain} = \frac{pa + qb}{p + q} \tag{5.5}$$

Huygens illustrates this with the following numerical example: if he has 3 chances to win 13 and 2 chances to win 8, then his expected gain is

$$\frac{(3 \times 13) + (2 \times 8)}{3 + 2} = 11$$

Interestingly, Huygens wrote *Games of Chance* at the same time that he was working on the problem of modeling collisions between objects. Further, as historians have noted, the development of the argument in *Games of Chance* is strikingly similar to that in *Motion of Bodies in Collision* [Hol84]. In particular, the formula for expected gain, Equation (5.5) has the exact same form as the formula for center of gravity, Equation (4.4). Thus, it is quite likely that Huygens used an analogy of balancing weights on either side of a lever beam as he worked through his theories of probability. In Example 5.2, we consider a hypothetical bet on the slingshot, and calculate who gets the short end of the stick.

Example 5.2	**A Fair Bet?** Suppose someone offers you a bet that he will pay you \$1.25 if you can launch a softball that will land less than 1 m from a target at 18 m, and otherwise, you must pay him \$1.00. Should you take the bet?

Solution   We can solve this problem using Huygens' formula for expected gain together with the results of the second launch experiment to estimate the chances of winning or losing. Out of the 20 trials in the experiment, 8 landed less than 1 m from a target at 18 m and 12 landed outside of this range. With this estimation, the solution is as follows:

Given: 8 chances of winning \$1.25 and 12 chances of losing \$1.00

Find: the expected gain

Plan: Substitute values into Huygens' formula, Equation (5.5)

Analysis: For this example,

$$p = 8 \qquad a = 1.25 \qquad q = 12 \qquad b = -1.00$$

$$\text{expected gain} = \frac{pa + qb}{p + q}$$

$$= \frac{(8 \times 1.25) + (12 \times -1.00)}{8 + 12}$$

$$= -0.10$$

According to this analysis, you would expect to lose 10 cents for every time that you play the game, so this would not be a good bet. On the other hand, from our earlier analysis, we determined that there is likely a systematic error that is causing launches to land about 1 m short of what the model predicts. If we pulled the spring back a few extra centimeters with each launch, we could imagine shifting the distances of each of the trials from Table 5.5 out by approximately 1 m. In this case, there would be 10 trials within the winning range and 10 trials outside of it. This changes the expected gain to

$$\frac{(10 \times 1.25) + (10 \times -1.00)}{10 + 10} = 0.125$$

which says that you could expect to win 12.5 cents for every game played, which is a pretty good bet. Ultimately, the choice is yours!

5.4.4 Frequency of Results and Histograms

In the solution of Example 5.1, we essentially sorted the results of the slingshot experiment in Table 5.5 into three "bins" according to distance—launches less than 17 m, between 17 m and 19 m, and greater than 19 m—and then counted the number of items in each bin. We can get a more detailed picture of the distribution of

TABLE 5.6 Counts or frequencies of launch distances sorted into bins.

bid ID i	range	count N(i)	probability P(i)
14	$14 \leq d < 15$	1	0.05
15	$15 \leq d < 16$	3	0.15
16	$16 \leq d < 17$	4	0.20
17	$17 \leq d < 18$	7	0.35
18	$18 \leq d < 19$	2	0.10
19	$19 \leq d < 20$	3	0.15
Total	$14 \leq d < 20$	20	1.00

launch distances by sorting them into finer bins. Table 5.6 shows the counts and corresponding probabilities of sorting the results of the 20 launches from the slingshot experiment into bins with ranges 1 m wide, spanning the range of distances from 14 m to 20 m.

In Table 5.6, the variable i represents the bin ID and the function $N(i)$ represents the count or frequency of trials in bin i. The function $P(i)$ represents the probability that a given trial is in bin i, which is equal to $N(i)$ divided by the number of trials, in this case 20. Since the histogram accounts for all of the trials in the experiment, in the count or frequency version of the histogram the sum of the bin counts must equal the number of trials or

$$\sum_i N(i) = \text{number of trials}$$

Similarly, the sum of the probabilities must equal 1, or

$$\sum_i P(i) = 1$$

Plotting the values in Table 5.6 help us to quickly visualize characteristics of the distribution of the data. Figure 5.14 shows plots of the bin counts and associated probabilities as a series of vertical bars. A plot of this form is known as a *histogram* from the Greek word "istos," meaning the mast of a ship. Since the area covered by a histogram is the sum of the individual bars, the total area of the frequency histogram equals the number of trials and the area of the probability histogram equals 1. Thus, the probability that a data value lies within a given range is equal to the fraction of the total area of the histogram determined by that range. Figure 5.15 shows two examples of this. In the case on the left, the probability that the distance is less than 17 corresponds to the area of the first three bars, which represent 40 percent of the area of the histogram. The median of the data, the point at which there is an equal probability of being either above or below it, lies somewhere in the middle of the fourth bar.

5.4.5 The Theory of the Bell Curve

We see that the shape of the histogram for the launch data approximates the familiar *bell curve* shown in Figure 5.16, more formally known as a *normal* or *Gaussian distribution*. Just as Boyle's Law describes a theory for how the pressure of a gas varies

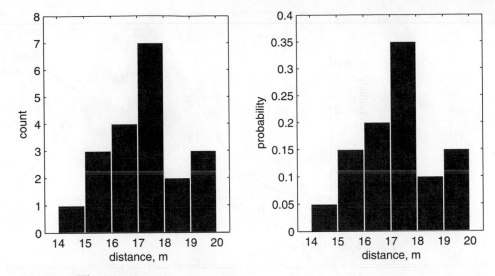

Figure 5.14 Histogram

with volume, or Moore's Law describes a theory for how the number of transistors on a chip will change over time, the normal distribution is a theory of what the probabilities of values falling in a given range would be for certain kinds of experiments. The French mathematician Abraham de Moivre (1667–1754) first described this theory in 1734 and later published it in the second edition of his book the *Doctrine of Chances*—which was popular among gamblers—in 1738 [Hal90]. The theory was later studied and refined by the German mathematician and scientist Karl Friedrich Gauss (1777–1855), for whom the distribution is named.

The basic idea behind de Moivre's theory was to consider an experiment where we toss a fair coin a very large number of times—say 1 million—and let the variable X represent the number of times that it comes up heads. The normal distribution or bell curve is then a plot of the probability that X will have a certain value. The greatest probability is that the coin will come up heads around half of the time, or in our example around 500,000 thousand times. This is the mean value of the distribution,

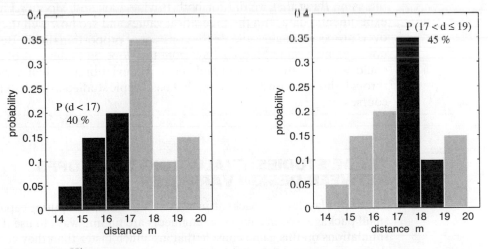

Figure 5.15 Histogram

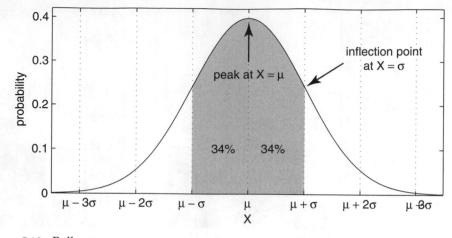

Figure 5.16 Bell curve

μ. The probability decreases as X gets either smaller or larger than μ, gradually at first, and then falling off more sharply. Thus, there is still a fairly large probability that the number of heads would be in the range around 490,000 or 510,000, but an extremely low probability that it would be close to 0 or 1 million. de Moivre calculated these probabilities to produce the bell curve, which has some interesting properties as illustrated in Figure 5.16. First, the curve is symmetric about the mean value μ, which is also the peak of the curve. Next, the inflection point where the shape of the curve changes from being concave down to concave up is one standard deviation σ to either side of the mean. Finally, regardless of the value of the mean or the standard deviation, the fraction of the area under the curve between the mean and one standard deviation is approximately 34 percent; or in other words, the probability that the value of X lies between $\mu - \sigma$ and $\mu + \sigma$ is approximately 68 percent.

If an experiment behaves according to the theory of a normal distribution, then we can take advantage of the properties of the bell curve to calculate probabilities. Looking at the data from the launch experiment, it appears to have a similar shape, but in order to determine how well it fits, we would need to do essentially the same thing that we did for both Boyle's Law and Moore's Law: compare the experimental data with the theoretical values and see how well they line up. Moreover, just as we could estimate the constant of proportionality in Boyle's Law or the doubling interval in Moore's Law from the slope and intercept of a best fit line, so could we estimate the mean and standard deviation of a bell curve. We won't go through this exercise here, but this is a basic topic studied in probability and statistics courses.

5.5 TRADE STUDIES: EVALUATING TRADEOFFS BETWEEN DESIGN VARIABLES

The Sony Playstation 3 packs an incredible amount of computing capability into a very small package—so much so that engineers are finding ways to use it to run scientific simulations on this game console that are much faster than they could run them on their conventional PCs. Behind this is the Cell Broadband Engine, a chip developed

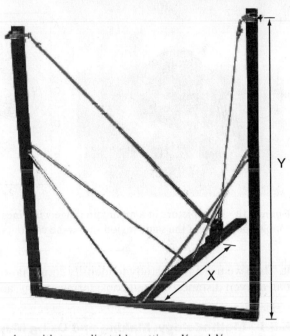

Figure 5.17   A slingshot with two adjustable settings X and Y.

in a collaboration between IBM, Sony, and Toshiba. According to Peter Hofstee, the lead architect of the Cell chip, one of the greatest challenges in developing the Cell was managing the tradeoffs between design variables and adapting its specifications to ensure that the final product would be acceptable to the engineers developing the Playstation and other products that would use the Cell. For example, increasing the number of transistors or the clock speed would enable more calculations per second, but it would also increase the amount of power consumed, causing the Cell to run hotter, which in turn would require the Playstation to have a more sophisticated way of cooling it.

The challenge faced by the Cell design team in making tradeoffs is common to many engineering projects, and in this section, we introduce a methodology called a *trade study* that provides a way for visualizing the impact of changes in design variables, so that designers can locate an acceptable combination. To illustrate the process, we'll use the example of a more complex version of the slingshot problem, where we can adjust both the pullback distance X along the base of the launcher, as well as the height Y at which the spring is attached to the uprights, as shown in Figure 5.17. Because both X and Y are adjustable, there are many possible combinations of values for these variable for launching a ball at a target a given distance away, for example using either a high, steep trajectory or a low, flat trajectory. To narrow down the options, we add an additional constraint of finding a combination of X and Y that minimizes the launch energy. In a slingshot, the launch energy is stored in the stretched spring and is transferred to the ball when the spring is released. Thus, the less the spring is stretched, the lower the launch energy.

The remainder of this section is organized as follows. First, we give an overview of the methodology for performing a trade study. Next, using the framework from Chapter 3, we clearly define the slingshot problem and develop a plan for attacking

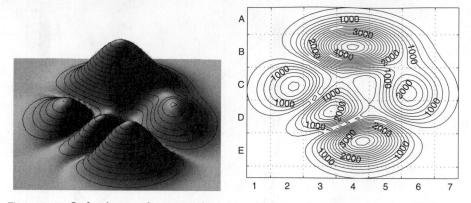

Figure 5.18 Left, picture of a mountain range with lines of constant elevation. Right, a topographic map of the same region shows the contours lines of constant elevation.

it. Then we use a trade study to identify combinations of slingshot setting to launch a ball a given distance without considering energy, and final add the energy constraint.

5.5.1 Methodology: Making and Using Maps

Just as civil engineers use maps to help decide where to locate a building or a road, we also use maps in trade studies to help find an acceptable design. Instead of longitude and latitude, however, the axes in these maps represent variables in models, where points on the map represent the coordinates of possible designs. Engineers often call such a "region," whose dimensions are variables of a design problem, a *design space*.

Before we consider maps of a design space, let's first take a look at an example of a geographical map, and some of the information that it conveys. Figure 5.18 on the left illustrates a small mountain range. On the mountains, we've drawn a set of *contour lines* or rings at constant elevations. The map of the terrain on the right in Figure 5.18 is known as a *topographic map*, and is like a top view of the terrain showing the contour lines. In the map, contours are drawn at increments of 250 feet in elevation. From the contours in the topographic map, we can identify a number of features in the terrain such as the highest point (in quadrant B4) or the location of all points higher than 3000 feet (around the peaks in B4 and E4). We can also determine the steepness of the terrain by noting the spacing between contours, where the closer the spacing, the steeper the terrain. For example the steepest route up the peak in C2 is from the southeast, where the contours are packed closest together, while the gentlest ascent is approaching from the west.

The first step in mapping a geographical area is to survey the terrain and measure elevations at intervals of longitude and latitude. When constructing maps of a design space, the first step is to gather performance data for different combinations of design variables. For empirical models, this data collection step often involves running experiments on either the actual system or a prototype. When constructing maps for theoretical models, the data collection step involves running experiments on the models. These experiments are sometimes called *parameter sweeps*, because they involve sweeping parameters or variables in the models over ranges of values. Table 5.7 shows a parameter sweep of the model for the stretch of the spring in a slingshot, $S = \sqrt{X^2 + Y^2}$. In the table, the rows correspond to values of Y varying

TABLE 5.7 Table with results of a parameter sweep for the stretch of a spring versus the X and Y displacements.

$\sqrt{X^2 + Y^2}$	$X = 0$	$X = 0.50$	$X = 1.00$	$X = 1.50$	$X = 2.00$
$Y = 0.00$	0	0.5000	1.0000	1.5000	2.0000
$Y = 0.50$	0.5000	0.7071	1.1180	1.5811	2.0616
$Y = 1.00$	1.0000	1.1180	1.4142	1.8028	2.2361
$Y = 1.50$	1.5000	1.5811	1.8028	2.1213	2.5000
$Y = 2.00$	2.0000	2.0616	2.2361	2.5000	2.8284

from 0 m to 2 m, the columns correspond to values of X varying from 0 m to 2 m, and each entry is the corresponding result of S.

From this data, we can map a "terrain" where the elevations are the values of the performances. Figure 5.19 on the left shows a map of the data from Table 5.7. Here, the elevations are the values of $\sqrt{X^2 + Y^2}$ at each X-Y coordinate. We'll provide more detail about drawing maps of design spaces later in the examples, but in this case, the shape of the terrain turns out to be a portion of a scooped-out, circular bowl.

Just as a topographic map of a physical region helps us recognize areas of high elevation, low elevation, or steepness of terrain, a design space map helps us visualize areas of high or low performance for a system with respect to the input variables in a model, as well as how changes in inputs lead to small or large changes in outputs. In the same way that the contours in a topographic map indicate areas at the same elevation, contours in a design map such as Figure 5.19 show the combinations of values for input variables that lead to identical values of output variables. For example, Figure 5.20 illustrates that there are many combinations of values of X and Y values that result in the same stretch length for the slingshot spring. Each of these combinations lie on the $S = 2$ contour line in Figure 5.19.

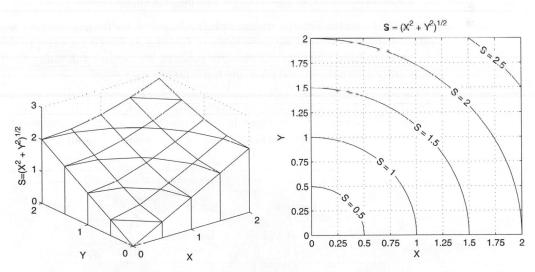

Figure 5.19 Plots of a design space for the stretch distance of a slingshot spring, where the elevation is $S = \sqrt{X^2 + Y^2}$. The plot on the right is a topographic map or *contour plot* of the bowl-shaped "terrain" on the left.

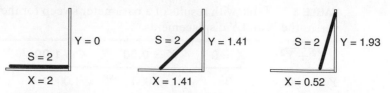

Figure 5.20 Different combinations of X and Y values that result in the same stretch length $S = 2$ for the slingshot spring.

Back in Chapter 2, we gave a cartoon view of an acceptable region, repeated here in Figure 5.21, as a hole cut out in the environment, bounded by constraints, and an acceptable design as one that fit inside the hole. Similarly, when we use maps in searching for an acceptable design, the acceptable region is an area in the map that is bounded by constraints, and an acceptable design is one that lies within this region.

To illustrate, let's first consider how an acceptable region might be defined on a topographic map. Suppose, for example, that a group wanted to locate a campsite on a mountain north of the valley at C4 in Figure 5.18, at an elevation of at least 3000 feet. Figure 5.22(a) shows the set of points that meet these constraints highlighted on the map.

Now suppose that as part of a launch problem, we need to find the set of X and Y slingshot settings where Y is at least 1 m and that will stretch the spring at least 2 m. Figure 5.22(b) illustrates the acceptable region, which is located above the line where $Y = 1$ and outside of the contour where $S = 2$. Any point in this region will satisfy the constraints and thus be an acceptable solution.

5.5.2 Problem Definition and Plan of Attack

Our initial problem statement is as follows:

> Build an empirical model for the slingshot and use it to determine settings for launching a softball within 2 m of a given target with minimal launch energy.

As we did with the handpump example in Chapter 4, we'll solve this problem following the framework that we introduced in Chapter 3. We give a verbose description in order to fully illustrate the thought process. We begin by *defining* the problem, *exploring* alternatives, and *planning* the attack.

Define   Figure 5.23 shows a schematic of the problem and lists the variables. The model has two design variables which are the X and Y settings and two behavioral variables, which are the distance D downrange where the ball lands after launch,

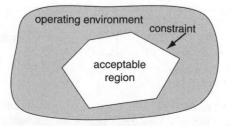

Figure 5.21   Cartoon view of an acceptable region, from Chapter 2.

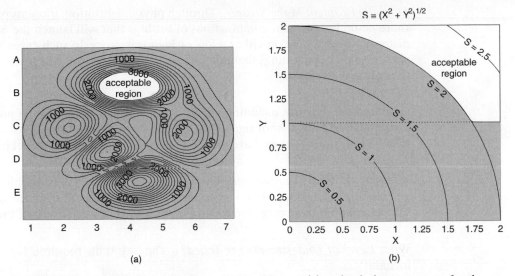

(a)   (b)

Figure 5.22   Acceptable regions in a topographic map (a) and a design space map for the slingshot stretch model (b). The acceptable region in the topographic is the set of points north of C4 and at an elevation greater than 3000 feet. The acceptable region in the design space map is the set of points where $Y \geq 1$ and $S \geq 2$.

and the energy E that is initially stored in the spring and then transferred to putting the ball in motion. For the sake of simplicity, we will not include any environmental variables in the model, but one logical choice would be the wind velocity, which could have a significant impact on the distance that the ball will travel.

Explore   Recall that the "explore" step is a pre-planning step where we answer several questions, listed below.

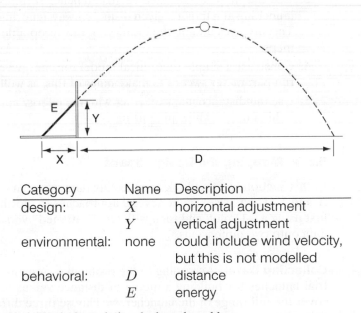

Category	Name	Description
design:	X	horizontal adjustment
	Y	vertical adjustment
environmental:	none	could include wind velocity, but this is not modelled
behavioral:	D	distance
	E	energy

Figure 5.23   Variables in formulating the launch problem.

Does the Problem Make Sense? Through physical intuition, it seems plausible that there could be different combinations of settings that will launch the ball the same distance, for example on either a low or high trajectory. In such cases, the problem asks us to pick the settings that minimize energy, which is related to the amount that the spring is stretched.

Assumptions By not including any environmental variables in the model, we are assuming that they are not significant. As mentioned earlier, wind is one factor that may be important, but we have decided not to include it in the model for simplicity.

What are the Key Concepts and Possible Approaches? In solving this problem we will use trial launches to build an empirical model of launch distance versus X and Y, and a simple theoretical model for energy, based on the stretch of the spring.

What Level of Understanding is Tested? The key skills required for this problem include:

- application of techniques for building an empirical model, including selecting experimental data points, graphing results, and fitting equations,
- analysis of the overall problem into component parts
- synthesis of a plan for solving the problem
- judgement in choosing appropriate launcher settings

Plan We will take the following steps toward solving this problem:

1. Collect data on launch distance versus the two launcher settings X and Y.
2. Map the space of launcher settings versus distance using 3-D plots, side-view cross-section plots, and contour plots.
3. Use the maps of distance versus launcher settings to determine settings for launching at a target a given distance away, ignoring energy for the time being. This involves finding the location of the acceptable region within the distance maps.
4. Construct a simple theoretical model for energy versus launcher settings and run parameter sweeps to make maps of this, as well.
5. Use the distance maps together with the energy maps to find launcher settings to hit a target while minimizing energy.

5.5.3 Mapping the Design Space

In this Section, we first look at collecting data using an experiment that varies both X and Y. Then, we examine several approaches to plotting three-dimensional data, first using a 3-D plot, and then with two "flattened" versions of this that project the data onto two dimensions.

Collecting Data According to the problem statement, we are allowed up to nine trial launches for building a model of distance versus launcher settings. To evenly cover the full range of the launcher, we choose three different X settings at each of three different Y settings. Table 5.8 shows the results of the nine trials.

TABLE 5.8   Results of trials for launch process with two adjustable settings.

Trial	x	y	Distance
1	0.5	1.00	4
2	1.0	1.00	18
3	1.5	1.00	36
4	0.5	1.25	10
5	1.0	1.25	26
6	1.5	1.25	45
7	0.5	1.50	16
8	1.0	1.50	34
9	1.5	1.50	52

3-D Plots   Figure 5.24 shows a 3-D plot of the trial data from Table 5.8. In this example, X and Y are assigned to the x- and y-axes, respectively, and D is assigned to the vertical z-axis. Whereas the data points defined a curve composed of line segments in the case of one design variable, in this case they define a *surface* composed of *patches*. Note that the height of the surface increases as we increase either X or Y, and that the maximum value of the surface occurs when both X and Y are maximized.

While a 3-D plot provides a good overview of the general trends in the design space, it is difficult to use to make actual predictions. With one design variable, in order to determine the distance travelled for a given X setting, we interpolated along the length of a line segment; now, we would need to interpolate into the interior of

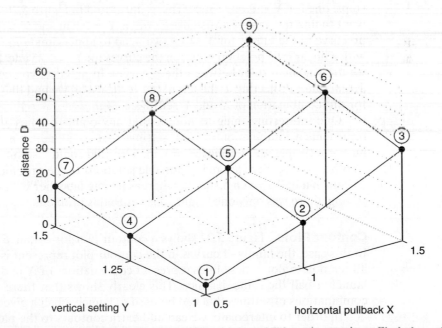

Figure 5.24   3-D plot of flight distance versus the X and Y launcher settings. Circled numbers refer to trials from Table 5.8.

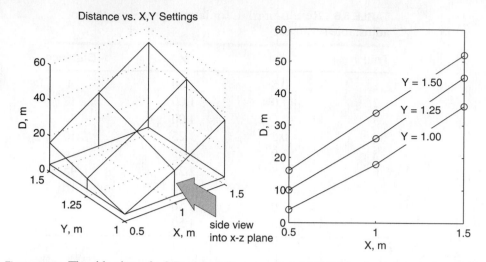

Figure 5.25 The side view of a 3-D plot of D versus X and Y, looking into the x-z plane, with a line-of-sight parallel to the y-axis produces a family of D versus X curves at cross-sections $Y = 1.00$, $Y = 1.25$, and $Y = 1.50$.

a patch. So instead of using the 3-D plots directly, we'll "flatten" them first into 2-D plots by considering either a side view or a top view.

Side View Cross-Section Plot First, we'll consider a side view of a 3-D plot, looking into the X-D plane, parallel to the y-axis. Figure 5.25 shows this view with cross-sections of the surface at $Y = 1.00$, $Y = 1.25$, and $Y = 1.50$.

In order to estimate a value between data points now requires two interpolation steps: one for X and one for Y. This is illustrated in Figure 5.26. In this example, we want to find the value for distance D when $X = 0.8$ m and $Y = 1.4$ m. Since there is no curve in the family for $Y = 1.4$, we need to approximate *its* location. We do this with a linear interpolation between the curves for $Y = 1.25$ and $Y = 1.50$. To do this, we drew a small ruler between these curves. In practice, you could do this with an actual ruler. Projecting to the left to the vertical axis, we find that the launch distance for this combination of X and Y settings is 24 m.

One interesting thing to note is that any point that lies along the line $D = 4$ represents a combination of X, Y settings that could launch a ball a distance of 24 m. For example, the settings $X = 1.18$, $Y = 1.00$ could also hit a target 24 m away. In other words, there are many possible solutions to the problem of finding settings to launch a ball a given distance, whereas for the launcher with only one adjustable setting, there was only one "right" answer to this problem.

Contour Plots Figure 5.27 shows a contour plot for distance D versus X and Y. In this case, the family of curves in the contour plot represent iso-D lines, such that all the points along these lines represent combinations of X and Y settings that will launch a ball the same distance. This clearly shows that there are infinitely many combinations of settings that can be used to launch a ball a given distance.

In order to interpolate, we can add extra contours to the plot if necessary. Suppose, for example, that we wanted to find the set of all possible settings for launching a ball a distance of 24. Figure 5.28 shows the addition of the contour line $D = 24$

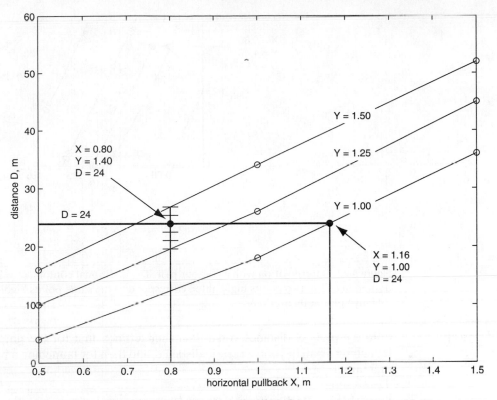

Figure 5.26   Interpolation using a side-view "family" plot.

interpolated between the $D = 20$ and $D = 30$ contour lines. The two points high-lighted correspond to the same points highlighted in the "family" plot in Figure 5.26.

5.5.4 Finding Settings to Satisfy Distance Constraints

Given plots of distance D versus the X-Y launcher settings, the next step is to find ac-ceptable settings for hitting a target. In this Section, we show how we map constraints

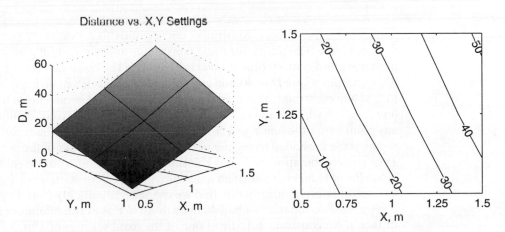

Figure 5.27   Contour plot for slingshot distance D versus X and Y.

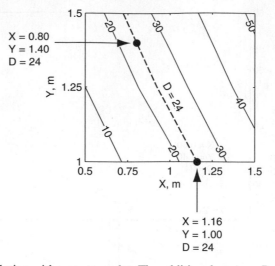

Figure 5.28 Interpolation with a contour plot. The additional contour $D = 24$ is shown as a dashed line. The two points highlighted correspond to the same points highlighted in the "family" plot in Figure 5.26.

onto the plots of distance versus launcher settings, first for the simpler example of launching a ball beyond a certain distance, and then for launching a ball within limits of a target.

Launching a Ball beyond a Given Distance   As a first example, suppose that we want to determine a pair of X and Y settings that will launch the ball a distance of at least 40 m. Mathematically, we can express this as a constraint of the form:

constraint 1: $D \geq 40$

Since the maximum X and Y settings of the launcher are both 1.5 m, there are two additional constraints on acceptable values for X and Y:

constraint 2: $X \leq 1.5$
constraint 3: $Y \leq 1.5$

Each of these three constraints corresponds to a boundary of the acceptable region. Figure 5.29 illustrates these constraints and the resulting acceptable region, drawn within a contour plot of D versus X and Y. The boundary for the first constraint is the contour where $D = 40$, and any point above that curve is acceptable with respect to that one constraint. The second constraint is satisfied by any point to the left of the line $X = 1.5$, which is the right side of the plot, and the third constraint is satisfied by any point below the line $Y = 1.5$, which is the top of the plot. The resulting acceptable region is the unshaded triangular area in the top-right corner of the plot, as indicated. Thus any combination of X and Y settings in this region would yield the acceptable behavior of launching a softball a distance greater than 40 m.

Another bit of information that we can see immediately from Figure 5.29 is that it is impossible to launch a ball 40 m or more if Y is at the minimum setting of 1 m. In fact, if we constrain Y to be at one of the fixed settings of 1 m, 1.25 m, or 1.5 m, then a launch of 40 m is only feasible at the two upper settings.

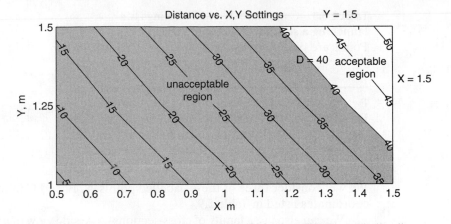

Figure 5.29   Acceptable region for values of X and Y for launching a softball a distance greater than 40 m.

Launching a Ball within Limits of a Target   In this next example, we consider determining X and Y settings for launching a ball close to a target. Suppose that we want to find settings to launch the ball to within 2 m of a target 20 m away. We can express this problem using four constraints:

constraint 1:	$D \geq 18$
constraint 2:	$D \leq 22$
constraint 3:	$X \leq 1.5$
constraint 4:	$Y \leq 1.5$

Figure 5.30 illustrates the acceptable region of X and Y settings. In this case, the acceptable region is a band centered on the $D = 20$ contour. Again, there are many possible combinations of X and Y settings that will yield an acceptable launch.

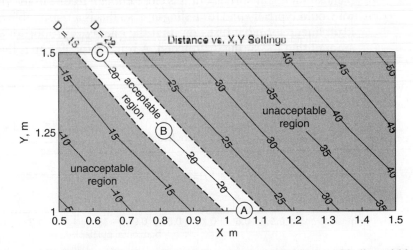

Figure 5.30   Acceptable region for values of X and Y for launching a softball to within 2 m of 20 m. A, B, and C denote three acceptable pairs of settings, whose coordinates are listed in Table 5.9

TABLE 5.9 Candidate X and Y settings for launching a softball at a target 20 m away. Points are plotted in Figure 5.30.

	X	Y
A	1.05	1.00
B	0.81	1.25
C	0.61	1.50

We will consider three of these, labelled points A, B, and C in the Figure 5.30, with coordinates listed in Table 5.9.

We can also use a family of cross-sections—the side view of a 3-D plot—to locate acceptable points for launching a ball at a target. Figure 5.31 illustrates the use of a family of cross-sections to find X and Y settings to hit a target at a distance of 20 m. The points A, B, and C are the same identified in Figure 5.30 and Table 5.9.

5.5.5 Minimizing Energy while Launching at a Target

Now that we have identified the acceptable region of X-Y slingshot settings that will launch a ball within a specified distance of the target, we're ready to find the points among those settings that minimize energy. Our strategy for doing so will be to make a map of energy versus launcher settings, just as we did earlier for distance versus launcher settings. Then, given the two maps, we first find the acceptable region in the distance map, and then use the energy map to find the location within that region where the energy is minimized.

Modelling Energy Versus Launcher Settings   As the basis for the energy model, we will assume that the spring is an ideal linear spring that follows Hooke's law, where the amount that it stretches is proportional to the applied force. A rubber spring, in fact, isn't a linear spring, but it is close enough to one in the range of operation that we would typically find for a slingshot, so this is a reasonable approximation. For an ideal linear spring, it turns out that the energy is proportional to the *square* of the amount that the spring is stretched. A rough explanation is that the energy stored in

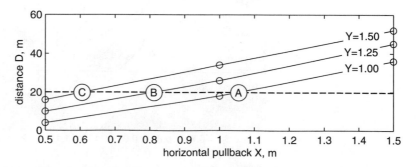

Figure 5.31   A family of cross-sections used to locate acceptable X and Y settings for launching a ball at a target 20 m away. Points A, B, and C are the same points indicated in Figure 5.30 and Table 5.9.

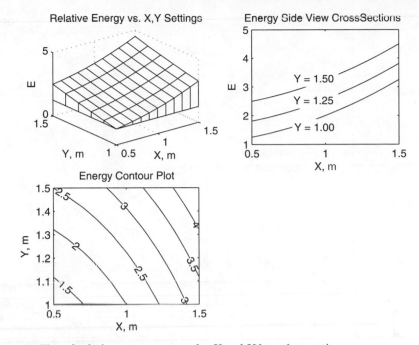

Figure 5.32   Plot of relative energy versus the X and Y launcher settings.

the spring is equal to the mechanical work of moving a force over a distance, and for an ideal spring, force is proportional to the amount that it stretches. Thus, the stretch distance factors into the energy calculation twice.

In order to determine the actual energy, we would need more information than is given in the problem description. Since we are only interested in monitoring the relative energy, we can use the value of the square of the stretch distance as an indicator. Neglecting the slack length of the spring, we may calculate this from the Pythagorean theorem. Thus, the relative energy is

$$E = X^2 + Y^2$$

Figure 5.32 shows a 3-D plot, side view cross-sections, and contour plots of relative energy versus X and Y. The surface has a parabolic "bowl" shape, and the cross-sections as well as contours have a more pronounced curve than they did for the distance plots. Note that in the contour plot, the lowest energy is in the bottom left corner of the plot, and it increases as we move towards the upper right corner.

Overlaying Distance and Energy Maps   Armed with plots of both distance and energy versus the X and Y launcher settings, we can now finally address the problem of finding settings that satisfy distance constraints while minimizing energy. To do this, we need to overlay the acceptable region from a distance plot on top of an energy plot. Figure 5.33 illustrates the process using two contour plots. A contour plot of distance versus X and Y is shown in the top of the Figure and directly below it is a contour plot of energy versus X and Y. Note that the X and Y axes of the two plots are identical, so that they line up perfectly. The region of acceptable distance settings is highlighted in gray in the top distance plot. Since the axes of the two plots line up, we can overlay the acceptable region onto the energy plot by projecting it downward.

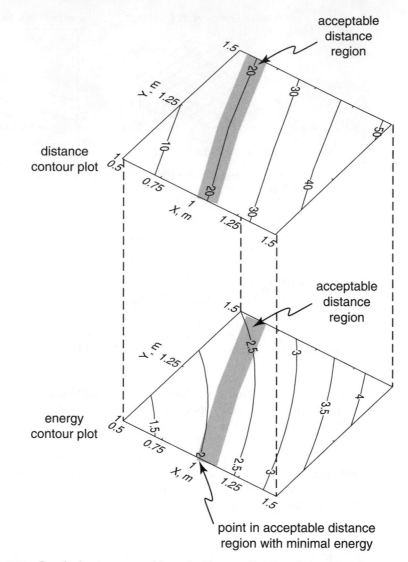

Figure 5.33 Overlaying an acceptable region from a distance plot on top of an energy plot.

Once we have the acceptable distance region superimposed on the energy plot, we can then find the point with minimal energy by locating the point in the acceptable distance region that lies on the lowest iso-energy contour line. For this example, that point is in the lower left corner of the acceptable region.

To illustrate the process of overlaying the distance and energy plots in further detail, we'll use the same example that we used in Section 5.5.4, where we found the combinations of launcher settings to hit a target 20 m away. Recall that in that example, we produced two versions of maps, a contour plot in Figure 5.30 and a family plot in Figure 5.31 and identified an acceptable region within those maps. Further, we identified three points A, B, and C that were found to be acceptable combinations. Table 5.10 shows the coordinates of these three points, together with

TABLE 5.10   Relative energy E of X and Y settings for launching a ball at a target 20 m away.

	X	Y	$E = X^2 + Y^2$	
A	1.05	1.00	2.10	lowest energy
B	0.81	1.25	2.22	
C	0.61	1.50	2.62	

the calculation of relative energy, $E = X^2 + Y^2$. The Table indicates that point A has the lowest relative energy of the three.

In order to superimpose the acceptable distance region onto the energy contour map, we can plot point A, B, and C on the energy map. Figure 5.34 show these results for both a contour plot and a family of cross-sections of energy versus X and Y settings. The dotted line connecting points A, B, and C corresponds to the iso-distance contour $D = 20$, which is the center of the acceptable region. The contour plot in particular clearly shows the changes in relative energy between the three points, where point A is just outside of the $E = 2$ contour, while point C is outside of the $E = 2.5$ contour.

More generally, we might ask if there are any points along this line that have a lower energy than point A. To answer this question, we scan along the length of the

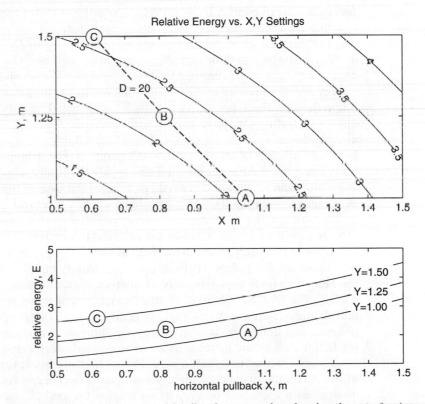

Figure 5.34   Energy contour map and family of cross-section showing the set of points were distance $D = 20$.

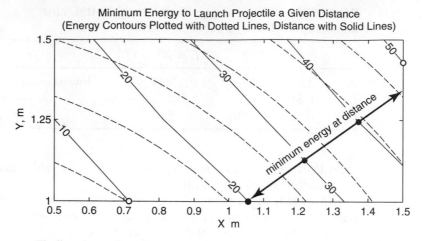

Figure 5.35   Finding the settings for launching a ball any distance with minimal energy.

$D = 20$ line to find the point closest to the $E = 2$ contour. Visually, the closest point does appear to be point A, or at least a point very close to it.

A General Solution for Minimal Energy   Based on the previous example, we may be tempted to conclude that the minimal energy launcher settings for hitting a target at any distance would involve using the minimal Y setting of 1 m. To see if this is in fact true, we can overlay more distance contours on top of the energy contour plot, and find the minimal energy point along each distance contour. Figure 5.35 shows this analysis. Just as we did for the single distance contour $D = 20$ in Figure 5.34, we visually scan the length of each distance contour and find the point with the lowest energy. Looking at the contours for $D = 30$ and $D = 40$, we observe that the minimal energy points—marked by filled dots—are *not* at the minimal Y setting. If we draw a line through these points, we get the set of X and Y points that use the minimal energy for hitting a target at any distance, *within the limits of the launcher*. For targets at some distances, we can't use the theoretical minimum settings defined by this line because the limits of the launcher won't permit it. For example, to hit a target 50 m away using the theoretical minimal energy would require a horizontal displacement X greater than 1.5 m, but this would go beyond the base of the launcher. So instead, the practical minimum energy point occurs when X is set to 1.5 m, as indicated by an unfilled dot. Similarly, the minimal energy point for hitting a target 10 m away would require a vertical setting Y below the practical minimum value of 1 m.

If we look closely at the theoretical minimal energy line in Figure 5.35, we note that these are the points where X is approximately equal to Y, or in other words, the launch angle is approximately 45 degrees. Figure 5.36 helps explain why this is so. Suppose that we ran a set of trial launches, where we gradually increased the launch angle from a very low angle to nearly vertical, but keeping the spring stretch distance and hence the launch energy the same with each trial. The distance that the ball travels would increase as we increase the launch angle until it reaches some maximum distance, and then it would begin to decrease. It turns out that the angle of the maximum distance is slightly less than 45 degrees; in fact, it would be exactly 45 degrees if there was no *drag* or air friction on the ball, but of course this is not the case in practice. You can also confirm this by varying the angle of a garden hose

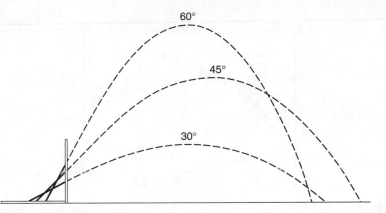

Figure 5.36 Trajectories of a softball launched at different angles, but with the same launch energies (same spring stretch distance). The maximum flight distance is achieved when the launch angle is approximately 45 degrees.

that is spraying water at constant pressure and see that the spray will travel furthest when the angle is approximately 45 degrees. Thus, the most efficient use of energy to launch a ball a given distance is to set the launch angle close to 45 degrees.

PROBLEMS

1. **Fitting Theory to Data for an Unknown Spring**
   A spring system consists of a plunger and a cylinder with an unknown mechanism inside, as pictured below, with experimental results of the force on the plunger versus its exposed length as given in the table.

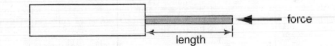

   (a) Which theory best describes the relationship between the two?
   - the length is inversely proportional to the force
   - the length is inversely proportional to the square root of the force
   - the length is inversely proportional to the square of the force

   Use a graphical technique that transforms the data and finds the best-fit line passing through it to answer this question.

length (cm):	0.0	1.0	2.0	3.0	4.0
force (N):	2.36	0.72	0.27	0.14	0.10

   (b) Give an equation based on the best-fit line for the force versus the length.

2. **Fitting a Model to Data**
   Label each graph with the mathematical model of the form $y = f(x)$ that best fits the data. Possible answers for $f()$ are logarithmic, exponential, linear, or quadratic.

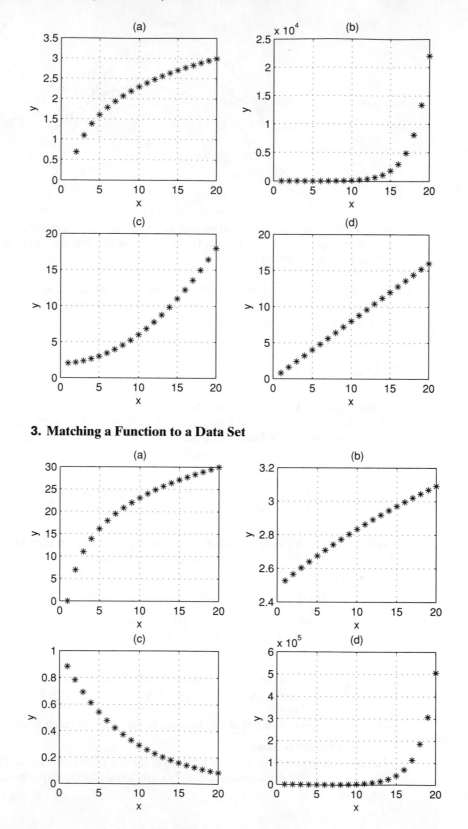

3. Matching a Function to a Data Set

Match the following functions to the data sets plotted above:

(a) $y = \ln(12 + x/2)$

(b) $23e^{0.5x}$

(c) $e^{-0.123x}$

(d) $10\ln(x)$

4. **Modeling Drag in a Wind Tunnel**

The drag D on a scale model of a rocket is measured for several different values of velocity v in a wind tunnel. The results of four measurements are:

Measurement, j:	1	2	3	4
Velocity, v_j (m/s):	20	25	30	40
Drag, D_j (N):	2.5	3.9	5.6	10.1

(a) Find the mean velocity.

(b) Find the standard deviation of the velocity.

(c) Plot D versus V.

(d) Plot the log of D versus the log of V.

(e) Someone claims that the mathematical model for drag versus velocity is

$$D = kV^c,$$

where k and c are constants. From your plots, would you tend to agree with that claim?

(f) Using the plot from (d), find the equation of the line that joins the first and the last points.

5. **Measurement Precision**

A meter stick is divided into centimeters, and each centimeter is divided into millimeters. Estimate the degree of precision with which a human observer can measure the length of a piece of string approximately 1/2 meter in length. Your answer should be of the form $\pm$ some numerical value, with units.

6. **Modeling Error**

Suppose that the following theoretical model specifies the position x versus time for an object of mass m set in motion against an opposing force F with an initial velocity v_0 from an initial position x_0:

$$x(t) = \frac{F}{2m}t^2 + v_0 t + x_0,$$

The three graphs below, (a), (b), and (c), show the measured values of x versus time for three different experiments, compared with the value for x predicted by the model. Which of the three graphs corresponds to each of the following sources of modeling error?

- the initial position x_0 of the object was improperly recorded;
- the opposing force on the object was less than the value for F used in the model;
- the actual initial velocity was greater than the value for v_0 used in the model.

Explain your choices.

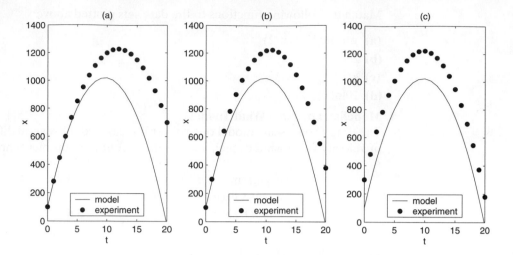

7. **Sources of Error in Building Empirical Models**
   Name possible sources of error in building empirical models for the following situations:
   (a) modeling the braking distance of an automobile as a function of its speed,
   (b) modeling the distance of a golf shot for each club in a set,
   (c) modeling the amount of time that it takes for a kettle of water to boil, as a function of the amount of water in the kettle.

8. **Measurement and Experimental Error**
   Random error can come from reading measured values. If the error in reading a sample's value stays constant regardless of the magnitude of the value, is it preferable to design the experiment so that larger or smaller sample values are obtained (assuming all else is constant)? For example, suppose that a 1000 mL graduated cylinder used for measuring liquids has a measurement error of $\pm$ 0.05 mL. Would it be preferable to design an experiment that requires a 50 mL volume of water, or a 500 mL volume of water? Explain your answer.

9. **Estimating the Mean of a Data Sample**
   In Section 5.4.2, the mean value of the launch distance was computed, using the data from Table 5.5. You will now compute the mean using a shortcut method (this method is especially useful if you are dealing with high numbers that are mostly in the same range):

   • come up with an estimate of the mean (integer value, to make it simpler)
   • write down each distance's offset from the estimated mean (i.e., the difference between them; sign is important)
   • add up those offsets, and find their mean
   • add that value to the estimated mean to get the correct mean

10. **Calculating Mean and Standard Deviation**
    Calculate the mean and standard deviation for the following data set:

Trial:	1	2	3	4	5	6
Value:	15.2	14.9	15.0	14.2	15.4	15.1

11. Thinking About Standard Deviation

At the end of Section 5.4.2, the standard deviation is explained to be the square root of the sum of the squares of the errors, noting that just adding the errors themselves would cancel them out and result in 0. If that were the only issue, though, why wouldn't we then take the sum of the absolute values of the errors? What might be other benefits of taking the squares of the errors? Explain.

12. Variations in the Speed of Graphics Chips

Suppose that a semiconductor company has developed a new type of graphics chip that can improve the quality of images and videos for a wide range of applications, from games consoles to handheld devices. The chip is designed to process images very fast, but because of variations in the manufacturing process, not all fabricated chips have the same speed. The table below lists the times that it takes for each of 20 random samples of the chip to process a reference image, measured in nanoseconds:

samples 1–5:	33.9	36.4	24.5	30.4	29.7
samples 6–10:	22.5	31.8	25.2	37.6	26.4
samples 11–15:	33.2	31.6	25.9	19.7	30.2
samples 16–20:	25.5	33.6	33.1	38.9	33.5

(a) What is the mean time that it takes to process the reference image?

(b) What is the standard deviation?

(c) Draw a histogram of the speed of the sample lot, with bins 1 ns wide.

(d) Suppose that the company can sell chips that are tested to be 25 ns or faster for $30 each, chips that are tested to be 30 ns or faster for $20 each, and chips that are tested to be 40 ns or faster for $15 each. What would be the expected average selling price of a chip, based on the distribution of speeds in the sample lot?

13. Evaluating the Goodness-of-Fit of an Empirical Model

When comparing different possible models, it is useful to have a single quantity that describes how well the model fits the data. One way to do this is to examine the average error between the values predicted by the model and the experimental data at each of the sample points. In this exercise, we'll look at the goodness-of-fit of the numerical model of the slingshot from Section 5.3.3.

(a) Three possible ways of averaging the error are by calculating:

- the mean of the error for all of the trials (mean error),
- the mean of the squares of the errors for all of the trials (mean square error),
- the square root of the mean of the squares of the errors for all of the trials (root mean square or *RMS* error).

Calculate each of these quantities for the data in Table 5.4. Explain why RMS error is generally the most used measure of goodness-of-fit of the three ways of averaging the error of a model.

(b) Suppose that one of your colleagues suggested using an alternative numerical model for the slingshot specified by the equation

$$D = 28.80X - 9.20$$

Using RMS error as a measure, which model better fits the data: the one used to generate Table 5.4 or your colleague's alternative?

14. Error Tolerance

Engineers often use the term "error tolerance" when designing systems. Explain this concept from a probabilistic perspective.

15. Probabilities of Coin Tosses

Find the probabilities of each of the following occurrences:

(a) Obtaining two consecutive heads when tossing a coin twice

(b) Rolling two dice and getting at least one 4

(c) Rolling two dice and getting two different numbers

16. Expected Gain from a Dice Game

Your friend proposes a game where you roll a pair of dice. If the sum of the die's faces is either seven or eleven, you earn $3 from him; otherwise, you owe him $1. Is the game favorable to you or to your friend?

17. Everyday Probability and Statistics

Give at least three examples where probabilities and statistics are used to make decisions in everyday situations. Describe the decision in terms of the "winning" and "losing" outcomes, as well as the nature of the statistics that are collected. Your examples may come from the following situations or from others:

- sports
- health care
- finance

18. Guessing on an Exam

Suppose that you are taking a multiple question test that has 50 problems, each of which has four choices (A B C D). The grading system is such that you will earn 2 points for a correct answer, and 0 points for an unanswered question, and will be penalized 1 point for every three incorrect answers. Come up with a strategy to help you determine when to answer a question based on how certain you are about it, or based on how many choices you can eliminate, so as to maximize your grade on the test.

19. Estimating the Size of a Population

An ecologist wants to estimate the frog population in a pond. One day she decides to catch a small number of frogs. She marks them, and returns them to the pond. A few days later, she returns (similar weather conditions) and catches a large number of frogs. Based on her observation of the frogs, how can she determine an estimate for the total frog population? Explain your answer with specific numbers.

20. Cumulative Probabilities for Slingshot Distances

Section 5.4.4 showed data for 20 trials for the slingshot, namely in Table 5.6 where the count and probability were shown for each distance (14 to 19) and in Fig. 5.14 where the data was shown in histograms. Whereas the above provides information for launch distances that are equal to given distances, you would now like to find similar data, but for launch distances that are up to those same distances. Generate a table as well as histograms for cumulative counts ($N \leq i$) and for cumulative probabilities, using the same cutoffs.

21. Percentiles

When a pediatrician weight a child and finds her height, the measures are plotted on a chart so that the doctor can determine the child's *percentile* in each category. Explain what a percentile is, and through some research on the web, explain how the doctor's charts were created in the first place.

22. Six-Sigma Quality

We stated that for a quantity whose values are distributed according to the Gaussian model, there is approximately a 70 percent chance that the value of a particular trial value will lie within one standard deviation of the mean. Doing some research on the web, answer the following:

(a) What is the probability that the value of a trial will lie within 2, 3, or 6 standard deviations (6-sigma) of the mean?

(b) If a company claims that it has better than 6-sigma quality, how many defective parts would it have per million units shipped?

23. Predicting Foot Size

How accurately can you predict the length of a person's foot if you know the distance between their wrist and their elbow? Write a short technical memo 2–3 pages in length that presents your solution, following the problem-solving framework from Chapter 3 and using analytic techniques from Chapter 5.

24. Picking a Door—Stay or Switch

This is a famous problem suggested by Marilyn vos Savant, who writes in the magazine *Parade*. Suppose that you are on a game show and are presented with three doors labeled A, B, and C, are told that there is a prize behind one of them, and are asked to choose a door. Suppose that you pick door A. The game host then opens door C and shows you that there is no prize behind it and asks if you want to keep your original choice or switch to door B. Should you stay, switch, or does it make no difference on the odds of you winning the prize? Explain your rationale.

25. Interpreting Slingshot Tradeoff Plots

The following are plots of the relative launch energy for the slingshot as described in Section 5.5.5.

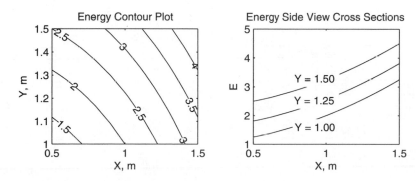

(a) Locate the point Q where $x = 0.75$ and $y = 1.25$ in both plots. What is the value of the relative energy at this point?

(b) Suppose that there are constraints on the launch process that the relative energy can be no more than 3 and that y must be fixed at 1.25. Locate the set of acceptable launch settings on both plots.

(c) Locate the point R where $E = 3$ and $X = Y$.

Modeling Interrelationships in Systems: Lightweight Structures

LEARNING OBJECTIVES

- to explain and give examples of the nature of interrelationships in systems, arising either from the interactions among parts or from viewing a system from multiple perspectives; to describe specifically the interrelationships in modeling a truss from the statics, materials, and geometric perspectives;
- to formulate the static equilibrium equations for a truss using the method of joints;
- to use Hooke's Law and Young's modulus to formulate equations describing the relationship between the internal force in a bar and its elongation;
- to discuss some of the mechanisms by which structures fail, and to be able to apply the concept of buckling strength;
- to explain the difference between a statically determinate and statically indeterminate truss, and to be able to formulate and solve a simple system of linear equations for either case;
- to explain some simple tradeoffs between strength and weight in truss design.

6.1 INTRODUCTION

Modeling the behavior of trivial systems that have only one "part" and one behavior of interest is as simple as finding the right law and applying it directly. Most engineering systems, however, have many parts and many behaviors of interest. This leads to a much more complex modeling task. The goal of Chapters 6 and 7 is to look

Figure 6.1 Trusses. The weight of a structure is especially important when it must be carried into space. Lower right photo courtesy of NASA.

in depth at modeling interrelationships in engineering systems, including how they arise, how we model them, and their importance in design.

To illustrate interrelationships in systems, we'll focus on the example of the design of lightweight structures. Specifically, we'll focus on a type of structure called a *truss*, which is built from relatively long and thin but strong bars connected together at joints, as Figure 6.1 illustrates. Trusses are important in the design of many types of structures, including bridges, roof supports, and heavy equipment. One way to look at a truss is to think of a solid slab of material that provides some basic support function, which you then cut as much away of the material as possible and still retain much of the strength of the original slab. The resulting form will be much lighter and cheaper than the original slab, but will still satisfy the performance constraints. Because most trusses aren't cut from a slab but put together from components, they have the additional virtue of being relatively easy to assemble on site. Construction toys such as Erector Sets, TinkerToys, and K'Nex are based on trusses, as are the common "pre-engineering" projects of building bridges out of popsicle sticks.

Modeling a truss brings in both the complexity of many parts *and* the complexity of multiple disciplines. We only need to introduce a few, fairly simple physical concepts—which you've likely already encountered in high school—in order to build accurate models of trusses. Within its operating environment, a truss is required to support a load and satisfy geometric constraints without overly deforming or breaking. Within the engineering environment, three different corresponding perspectives must be considered when designing and analyzing trusses:

Statics perspective: The statics perspective is the view of the system according to Newton's Laws, namely, since the truss is *static* or not moving, then the net force on each part of the truss must be zero. Models of a truss from the statics perspective consider the forces on each of the bars, joints, and supports.

Materials perspective: The materials perspective considers how the truss will deform when a load is applied to it. We can think of this as Hooke's view of the system, since from this perspective, we'll model the bars of a truss as very stiff springs that stretch or compress in proportion to the force across them. How much a bar deforms when a given force is applied to it depends upon both the material of the bar and its dimensions, and models from this perspective will consider these factors.

Geometric perspective: Geometry, and in particular trigonometry, plays several roles in the design and analysis of trusses. First, there's the basic requirement that the truss as a whole satisfy some spatial constraints, such as supporting a roof with some peak height over some distance. Second, we'll use trigonometry in our analysis of the forces acting on each joint. Finally, we'll again use geometry to determine the new positions of all the joints when the bars deform under load.

As we develop a model for a truss structure, we make several limiting assumptions that simplify calculations:

- First, we only consider *two-dimensional* problems. Real structures are—of course—three-dimensional, but with trusses, we're often most interested in forces that lie entirely within a plane. Many three-dimensional structures are assembled as space frames that consist of multiple planar trusses connected to form a box, where we can analyze each of the planar trusses mostly independently. Hence, this simplification isn't a serious limitation.

- We'll assume that all the joints in the truss are implemented as simple pins, such that the bars connected to them are free to rotate about them. In practice, many of the joint technologies used in real trusses don't allow the bars to rotate, but this simplification greatly simplifies the analysis of forces, and is widely used in structural engineering.

- We'll assume that all the structures we consider undergo small deformations when subjected to load, so that the overall geometry of the structure when subjected to load isn't too different from before the loads were applied. We'll use this assumption to greatly simplify the calculation of the new joint positions when bars are stretched or compressed under load.

If you're a student majoring in mechanical, civil, or aerospace engineering, this introduction to the analysis of lightweight structures has a direct connection to material you can expect to see in greater depth in later courses, although our treatment differs somewhat from the approach in some structures textbooks. If you're a student majoring in other branches of engineering, such as electrical or chemical engineering, while the physical concepts are different, the analysis method has much in common with the analysis of electrical circuits or networks of flows between chemical reactors. The most important point of this chapter, however, is to illustrate some of the general issues and approaches involved in the modeling of systems with many parts and multidisciplinary analysis, which is important for engineers of any discipline.

6.2   THE STATICS PERSPECTIVE

6.2.1   Force as a Vector

A single number can describe some of the physical quantities we deal with. Examples of such quantities are the mass or volume of an object. Such quantities are called *scalar* quantities. There are other quantities, however, where a single number doesn't give a complete picture, such as a force. Force is a *vector* quantity expressed in terms of both a magnitude—how much—and direction—which way. Remember that all forces are vectors, but not all vectors are forces! Velocity and acceleration are also vector quantities, and displacement can also be considered a vector quantity when viewed as a distance and direction from a fixed reference point. In electronics, vectors are used to describe certain characteristics of electrical signals. Here, we'll touch only on some of the very basics; you'll learn much more about vectors and vector quantities later in mathematics, physics, and engineering courses.

It's important to keep in mind that a force is a quantity with dimensions. In the SI system of units, forces are expressed using the unit Newton, N. In the engineering or English system of units, things are more complicated. You may see forces expressed in units of "pound," written as "lb" or sometimes pound-force "lb_f," or ounce, ton, etc. The proper use and understanding of units is critical in solving any engineering problem.

Forces—and other vector quantities—are often expressed as "directed arrows." Figure 6.2 illustrates a type of diagram called a *free-body diagram* that represents a body B with two forces **F** and **G** applied to it. The point at which a force is applied to a body is called the *point of application*. The arrow points along a *line of action* in the direction of the force and the length often indicates the magnitude of the force.

In many engineering applications, we must consider force vectors in three dimensions; after all, the real world is three-dimensional! But often, as we'll do here, engineers are able to simplify a problem by making certain assumptions about the behavior of a system or the nature of a problem. For the following discussions, we'll assume we can effectively model a real world problem by assuming an equivalent two-dimensional problem. Thus, all the forces and structures that we'll study will be limited to a single plane. As you'll see, this assumption makes some of the analysis— as well as drawing diagrams—easier and it allows us to study some important basic issues. Much of what we'll present concerning forces and structures can easily extend to three dimensions. We hope that by developing a good understanding of some of the basic concepts by using the easier two-dimensional problems, you'll be better prepared to deal with the added mathematical and notational complexity associated with the three-dimensional problems when you encounter them in your studies.

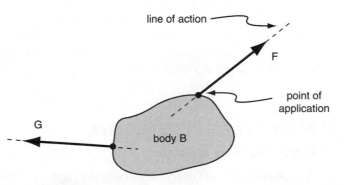

Figure 6.2   A free-body diagram.

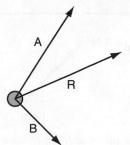

Figure 6.3 The resultant of two vectors added together.

6.2.2 Addition of Forces

If two forces are applied to a body at the same point, they may be replaced by a single force, called the *resultant*, that'll have the same net effect, as Figure 6.3 shows. Here, the two vectors **A** and **B** can be replaced by the single resultant vector **R**. In this text, we'll set vector quantities in boldface and scalar quantities in italics. The magnitude of vector **R** is the scalar quantity *R*.

Vectors add according to a rule called the *parallelogram law*, which Newton introduced in his *Principia Mathematica* [Newton1686] shown in Figure 6.4. To add two vectors **A** and **B** using the parallelogram law, first construct a parallelogram with **A** and **B** as adjacent sides. The resultant vector **R** is the diagonal of the parallelogram passing through the point where **A** and **B** meet. Another way to think of vector addition, which follows as a consequence of the parallelogram law, is to join the vectors head-to-tail, so that the tail of one vector is at the same point as the head of the other. The resultant vector is found by drawing a vector with its tail at the same point as the tail of the first vector, and with its head at the same point as the head of the second vector. This is also shown in Figure 6.4.

Given the parallelogram rule, we could dismiss the addition of vectors as a basic geometry problem for two reasons. Engineers typically use a common systematic approach for adding vectors that uses more steps than you may think necessary. First, it minimizes the chance of common errors, particularly sign errors. Second, it's easily automated, which is important if we write an algorithm for adding forces to program into a computer. The steps in this approach are as follows:

1. Select a convenient Cartesian (x y) reference system.
2. Resolve each force into its components along the reference axes.
3. Add all of the components along each reference axis.
4. Combine the two resultant components to form the desired vector.

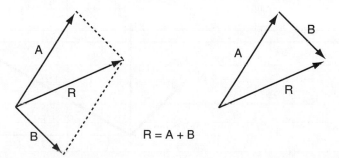

Figure 6.4 Addition of vectors; parallelogram law (left), "head-to-tail" rule (right).

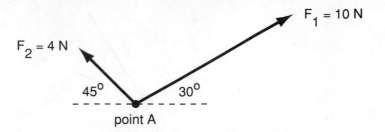

Figure 6.5 Example of addition of forces used throughout this section.

To illustrate this approach, we use the example problem shown in Figure 6.5. The goal of this problem is to determine the resultant force acting at point A when the two forces F_1 and F_2 act together at point A. In other words, we'll determine a single, *statically equivalent* force, which we'll call **R**, that could replace F_1 and F_2 and point A wouldn't know the difference. The directions of both forces are indicated with respect to the dashed horizontal reference line shown on the figure. Both forces act in the same plane—in this case, the plane of the paper you're reading. Given this problem definition, we now follow the steps listed above to add the forces.

Step 1: Select a Convenient Cartesian Reference System Since much of the information about the direction the vectors act is expressed relative to the dashed line in Figure 6.5, it makes sense to use that to our advantage in selecting the reference system. Figure 6.6 includes a simple right-handed Cartesian "*x-y*" coordinate system with the origin at point A. Note that we didn't have to define the coordinate system this way. For example, if we align the *x*-axis along the line of action of the force F_1, then we already know the component of the force F_1 along the *x*-axis. For now, however, we'll proceed using the axes shown in the Figure 6.6.

Step 2: Resolve Each Force into its Components Along the Reference Axes Just as we can add two vectors to produce a resultant, we can decompose a vector into two vectors that add to form the original. In particular, we commonly decompose or resolve vectors into components parallel to the reference axes of the coordinate system, which simplifies the bookkeeping when adding forces.

Let's see what this means for force F_1 in our running example. The process of resolving F_1 into its components involves constructing a parallelogram—in this case, a simple rectangle with the vector along the diagonal and the two sides of the rectangle

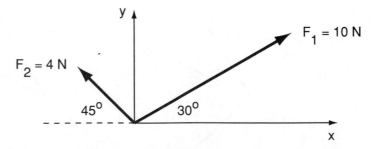

Figure 6.6 Definition of a Cartesian coordinate system.

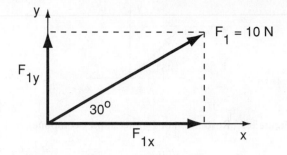

Figure 6.7 Application of the parallelogram rule in resolving $\mathbf{F}_1$ into components $\mathbf{F}_{1x}$ and $\mathbf{F}_{1y}$.

aligned with the coordinate axes—as shown in Figure 6.7. The x-component of the force $\mathbf{F}_1$ is a vector $\mathbf{F}_{1x}$, and from a geometric perspective, it is the base of a rectangle with magnitude:

$$F_{1x} = F_1 \cos(30°) = 10\text{N} \cos(30°) = 10\text{N}(0.866) = 8.66\text{N}$$

The y-component of the force $\mathbf{F}_1$, is another vector, $\mathbf{F}_{1y}$, which is the height of the rectangle, and its magnitude is:

$$F_{1y} = F_1 \sin(30°) = 10\text{N} \sin(30°) = 10\text{N}(0.500) = 5.00\text{N}$$

Thus, as illustrated in Figure 6.8, the action of force $\mathbf{F}_1$ on point A can be expressed either in terms of the resultant $\mathbf{F}_1$ or the two components $\mathbf{F}_{1x}$ and $\mathbf{F}_{1y}$.

The same process can be applied to the force $\mathbf{F}_2$, as illustrated in Figure 6.9, but some care must be taken to account for the fact that the x-component of the force $\mathbf{F}_2$ is directed in the negative x direction. Note that you must use the same reference system for both force vectors if you eventually want to add them together and not lose track of the direction associated with each of the vectors. The x-component of the force $\mathbf{F}_2$ is a vector $\mathbf{F}_{2x}$ and its magnitude is:

$$F_{2x} = -F_2 \cos(45°) = -4\text{N} \cos(45°) = -4\text{N}(0.707) = -2.83\text{N}$$

This force has a negative magnitude since it acts in the direction of the negative x axis. The y component of the force $\mathbf{F}_2$ is the vector $\mathbf{F}_{2y}$ and its magnitude is.

$$F_{2y} = F_2 \sin(45°) = 10\text{N} \sin(45°) = 4\text{N}(0.707) = 2.83\text{N}$$

Thus, as we showed before, the action of force $\mathbf{F}_2$ on point A can be expressed either in terms of the resultant $\mathbf{F}_2$ or the two components $\mathbf{F}_{2x}$ and $\mathbf{F}_{2y}$, as these two representations of the force $\mathbf{F}_2$ are statically equivalent.

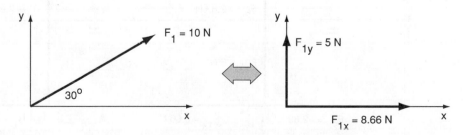

Figure 6.8 Resolution of $\mathbf{F}_1$ into x and y components.

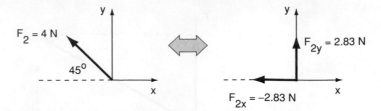

Figure 6.9 Resolution of $\mathbf{F}_2$ into x and y components.

Step 3: Add All of the Components Along Each Reference Axis Now that each of the two forces have been resolved into their components, it's simple to add vectors with the same point of application and line of action. This is shown in Figure 6.10.

If the resultant force is represented by the symbol $\mathbf{R}$, then the components of the resultant can be expressed as $\mathbf{R}_x$ and $\mathbf{R}_y$ where

$$R_x = F_{x1} + F_{x2} = 8.66\text{N} + (-2.83\text{N}) = 5.83\text{N}$$

and

$$R_y = F_{y1} + F_{y2} = 5.00\text{N} + (2.83\text{N}) = 7.83\text{N}$$

Figure 6.10 shows this result.

Step 4: Combine the Two Resultant Components to Form the Desired Vector The final step in this process is to combine these into a single vector, again using the parallelogram law, as Figure 6.11 shows. The magnitude and direction of the resultant

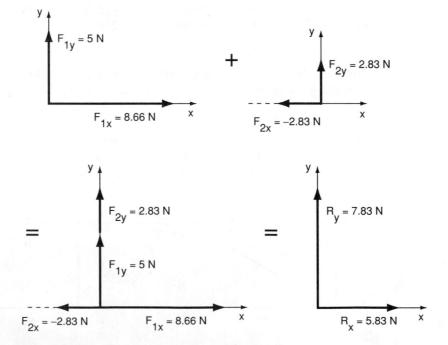

Figure 6.10 Addition of x- and y-components of vectors $\mathbf{F}_1$ and $\mathbf{F}_2$.

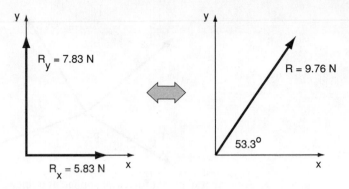

Figure 6.11   Recombining $\mathbf{R}_x + \mathbf{R}_y$ back into a single vector $\mathbf{R}$.

are determined from the components, where

$$\tan \theta = R_y/R_x = 7.83\text{N}/5.83\text{N} = 1.343$$
$$\theta \quad = 53.3°$$

and

$$R = \sqrt{R_x^2 + R_y^2} = \sqrt{(5.83\text{N})^2 + (7.83\text{N})^2} = 9.76\text{N}$$

The resultant $\mathbf{R}$ is then *statically equivalent* to the initial forces $\mathbf{F}_1$ and $\mathbf{F}_2$. Finally, we confirm that the resultant of adding $\mathbf{F}_1$ and $\mathbf{F}_2$ by breaking them down into x- and y-components is the same as adding them by either the head-to-tail or parallelogram rule, as shown in Figure 6.12.

6.2.3 Equilibrium of a Point or Particle

Figure 6.13 illustrates a scenario where several forces are all applied at a single point A. Such a point is sometimes referred to as a *particle*, which implies that it could have mass, but is infinitesimally small. Again, we restrict our discussion to cases where all the forces lie in a single plane.

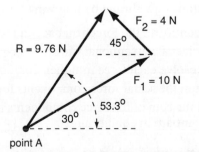

Figure 6.12   The resultant force $\mathbf{R}$ is sum of $\mathbf{F}_1$ and $\mathbf{F}_2$, as determined by the parallelogram rule.

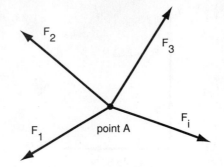

Figure 6.13 A number of planar forces all applied at point A.

As before, we can replace the set of applied forces with a single resultant force, sometimes called the net force, which is the sum of the individual forces:

$$\mathbf{R} = \mathbf{F}_1 + \mathbf{F}_2 + \mathbf{F}_3 + \cdots + \mathbf{F}_i = \sum_{j=1}^{i} \mathbf{F}_j$$

According to Newton's Laws, the net force on an object equals its mass times its acceleration. For the problems we'll address later in this chapter, point A will represent a joint or pin that ties together a simple structure that *we don't want to move*. Thus, the acceleration of each point in the structure must be zero and, according to Newton's Laws, the net force at each point must be zero, as well. Therefore,

$$\mathbf{R} = \mathbf{F}_1 + \mathbf{F}_2 + \mathbf{F}_3 + \cdots + \mathbf{F}_i = \sum_{j=1}^{i} \mathbf{F}_j = 0$$

at every point in the structure. This situation, in which the net force on a point or particle is equal to zero, is called *static equilibrium* because the particle will be static, or not moving.

As in Section 6.2.2, we can resolve each of the forces applied to a particle into their *x*- and *y*-components. Then the condition of equilibrium at a point must satisfy two independent algebraic equations:

$$F_{1x} + F_{2x} + F_{3x} + \cdots F_{ix} = \sum_{j=1}^{i} F_{jx} = 0$$

$$F_{1y} + F_{2y} + F_{3y} + \cdots F_{iy} = \sum_{j=1}^{i} F_{jy} = 0$$

Thus, we determine whether the forces acting on a particle such as a joint in a structure are in static equilibrium by following these steps:

1. Identify all the forces that act on the point.
2. Select a conventional Cartesian (*x-y*) reference system.
3. Resolve each force into their components in the *x*- and *y*-directions.
4. Sum the scalar force components for both directions.
5. If the sum of the force components in both directions is zero, the forces on the point are in equilibrium.

6.2.4 Equilibrium of Pinned Joints and Bars

As we described earlier, a truss is a lightweight structure made of long, thin bars connected together at joints. In actual truss designs, the joints can be fabricated from

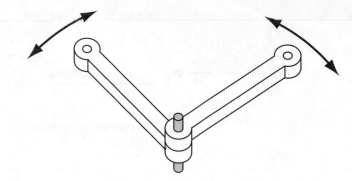

Figure 6.14   A pinned joint. The bars are free to rotate about the pin.

a variety of technologies, including welds and bolts that hold the bars rigidly in place. Throughout our analysis, however, we'll assume that all joints are a type of idealized joint called a *pinned joint*, illustrated in Figure 6.14. The main feature of a pinned joint is that its bars are free to rotate about the joint. Unlike a welded or bolted joint, this joint offers no resistance to the bars. This feature is significant in the orientation of forces in a truss, as we'll see shortly.

Whenever a pinned truss supports a load, the bars in the truss apply forces to the pins, and vice versa. If the truss as a whole is in static equilibrium, then each bar and pin must be in static equilibrium as well, or else that part of the truss would be moving. This means that the net force on each bar and each pin must equal zero. Another requirement, however, on the forces exerted on the bars is that they must be directed along the same line of action, parallel to the length of the bar, as illustrated in Figure 6.15. In the first case in the figure, even though the applied forces are equal and opposite in both magnitude and direction, and hence the net force is zero, the bar will have a tendency to rotate. Because pinned joints can't resist rotation, the bar is not in static equilibrium. In the next two cases, in which the forces are directed parallel to the length of the bar, there's no tendency for rotation and the bar is in static equilibrium, even with pinned joints. Objects such as a pin-ended bars that have exactly two forces acting upon them are known as *two-force elements*, and for

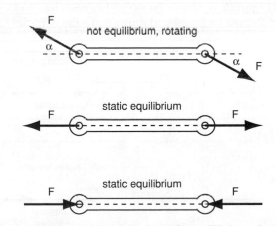

Figure 6.15   In order for a pin-ended bar to be in static equilibrium, the forces applied by the pins at either end must be of equal magnitude, in opposite directions, *and* be directed parallel to the length of the bar.

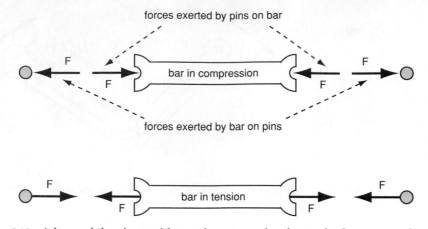

Figure 6.16 A bar and the pins at either end exert equal and opposite forces on each other.

any two-force element to be in static equilibrium, the two forces *must* be equal and opposite *and* directed along the same line of action.

Just as the pinned joints exert forces on the bars, the bars exert forces on the pins. According to Newton's Third Law, the force that a bar exerts on a pin must be equal and opposite to the force that the pin exerts on the bar. Furthermore, both of these forces must be directed parallel to the length of the bar for it to be in static equilibrium. The resulting arrangement of forces between a bar and the pins at either end is shown in Figure 6.16. If the forces exerted on the bar are directed toward the center of the bar such that the bar is being compressed, then the bar is in *compression*. If the forces exerted on the bar are directed away from the center of the bar so that the bar is being stretched, then the bar is in *tension*. Note that when a bar is in compression, it is pushing on the pins, while a bar in tension is pulling on the pins.

Whether a bar is in tension or compression, there are forces within the material of the bar itself that prevent it from coming apart. We call these forces *internal forces*. We discuss the nature of these forces in detail in Section 6.3, in which we consider the materials perspective. For now, we consider the relationship between the internal forces and the forces between a bar and pins. Suppose that we could take a bar in static equilibrium, slice it in half, and measure the internal forces exerted in each half, as Figure 6.17 shows. In order for each half to be in static equilibrium, the internal force must be equal and opposite to the external force that the pin exerts on the bar.

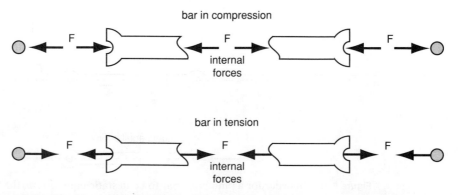

Figure 6.17 The internal forces within a bar exert the same force on each half of a bar that that half of the bar exerts on its pin.

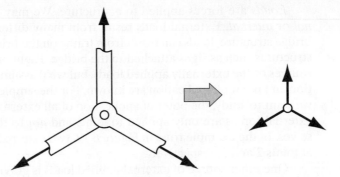

Figure 6.18 Modeling a joint as a particle with the attached bar internal forces applied to it.

This implies that the internal force is the same as the force the bar exerts on the pin. We repeat this point because it's very important to the analyses that we perform in later sections:

> The internal force within a bar is equivalent to the force that the bar exerts on the pinned joints at either end.

Finally, we consider the static equilibrium conditions of a pinned joint. Each pin joint in a truss has forces exerted upon it by the bars that it connects. These forces are equivalent to the internal bar forces. Because a pin is so small, we consider it as a particle or point, as Figure 6.18 shows. Thus for a joint to be in static equilibrium, the internal forces of the bars connected to it must sum to zero, as we discussed in Section 6.2.3. Note that because a joint is approximated as a point without dimension, we don't need to worry separately about the line of action as we did with the equilibrium conditions for a bar.

6.2.5 Loads, Supports, and Reaction Forces

Before we begin a complete static analysis of a truss, it's necessary to consider briefly how this structure is *loaded* and *supported*. This involves determining what forces are applied to the structure, and where and how they are applied. This includes how the structure is attached to the ground and how this influences the behavior of the structure. Throughout this discussion, we'll refer to the example truss shown in Figure 6.19.

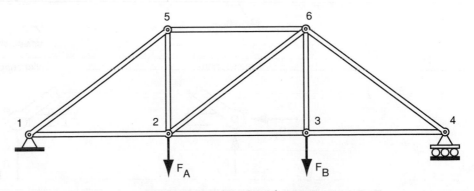

Figure 6.19 An example truss with external loads and supports.

Loads are forces applied to a structure. We may classify loads as either *external* or *internal*. External loads result from many different sources. In the case of a bridge structure, loads can result from traffic on the bridge, from wind, or from other structures such as signs attached to the bridge. Right now we won't worry about the sources of the externally applied loads, but we'll assume that their magnitude, direction and point of application are known. For the simple pinned-joint truss structures we want to study, the point of application of all external loads *must* be a joint. Thus, external forces are only applied to joints and *not* to the structural members themselves. In the example truss in Figure 6.19, two external loads $\mathbf{F}_A$ and $\mathbf{F}_B$ are applied at joints 2 and 3, respectively.

One other source of external applied loads is gravitational forces. The structure must be able to support its own weight along with any other forces that might be applied. Obviously, the weight is distributed across all parts of the structure and in reality is not only applied at the joints. In general, we'll assume for our examples that the weight forces are negligible as compared to other forces applied to the structure. If we include weight, however, we would approximate it as external loads distributed across all the joints.

Another important source of external forces is the supports that prevent the structure from moving. Figure 6.19 includes two different kinds of supports, one at joint 1 and the other at joint 4. The symbols for these supports are shown in detail in Figure 6.20.

The support at joint 1 is a fixed support that fully constrains the structure at that joint. This implies that no matter how large the forces applied to this structure are, we assume that this point in the structure never moves in any direction. In other words, a fixed support is capable of applying whatever load is necessary to the structure at this joint to stop it from moving. This further implies that at joint 1 there will be an unknown applied force that *reacts* in any manner needed to constrain the structure. Because of the nature of this force, we call it a *reaction force*, which can be represented as a force with unknown magnitude and direction. In keeping with our methodology for adding vectors, however, we represent the reaction force as two vectors of known direction—parallel to the coordinate axes—but with unknown magnitude. We denote the magnitudes of these forces as R_x and R_y.

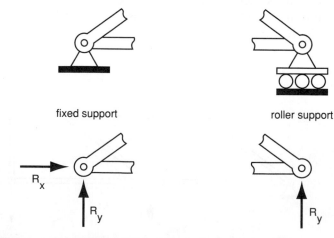

fixed support   roller support

Figure 6.20 Two types of supports and their equivalent reaction forces.

Another type of support was used at joint 4 in Figure 6.19 that's also shown in detail in Figure 6.20. This type of support, whose symbol looks like a little cart, represents a condition in which only a single unknown reaction or support force is applied to the structure at this joint. We refer to this type of support as a frictionless *roller* support. A roller support allows the structure to move or roll along the solid surface, but doesn't allow any motion perpendicular to this surface. Thus, at joint 4 a single, unknown vertical reaction or support force is introduced to the structure. Note that although the symbol for a roller support suggests you might be able to lift the little cart off the ground, you can't. The reaction will provide whatever force is needed to resist movement perpendicular to the solid surface, even if the cart is attached to a vertical wall or to the ceiling.

Reaction forces are pretty amazing. As the external loads applied to the structure change, the reactions change in an attempt to maintain equilibrium. They change directions as needed, sometimes they pull and under other conditions they push. By convention, we generally *assume* that the reaction forces point in the directions of the coordinate axes. If, after solving a static analysis problem we discover that a reaction force is *negative*, we know that we assumed the incorrect direction and the force actually acts in the opposite direction. Making and checking assumptions is, as we've discussed before, an important part of engineering problem solving.

6.2.6 Static Analysis of a Complete Truss

In order for a complete truss to be in static equilibrium, each of its components—each bar and joint—must be in static equilibrium. To illustrate the static analysis of a truss, we'll consider a three-bar truss as shown in Figure 6.21:

Definition of the Problem   In this problem, the structure of the truss, its supports, and external loading are known, and the internal bar forces and external reaction forces are unknown. The goal in solving the problem is to determine these unknown

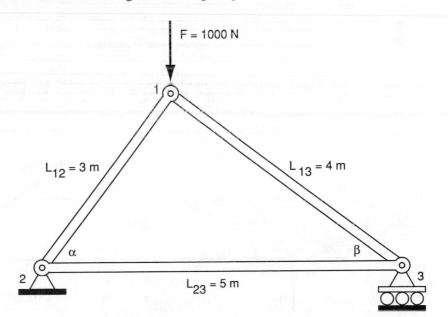

Figure 6.21   A three-bar truss.

forces. The truss has three joints, three bars, and two supports, and has a single external force, $\mathbf{F} = 1000\,\text{N}$, applied vertically at the top joint. The three joints are labeled 1, 2, and 3. The bars are named according to the joints at either end, namely bar 12, bar 13, and bar 23. The corresponding lengths of the bars are $L_{12} = 3\,\text{m}$, $L_{13} = 4\,\text{m}$, and $L_{23} = 5\,\text{m}$.

Assumptions We make the following assumptions in solving the problem:

- Each joint is a frictionless pinned joint, so that the bars can rotate freely about the joint. We'll model each joint as a particle or point, so that for a joint to be in static equilibrium, the sum of the forces applied to the joint must equal zero.
- Each bar is a two-force element in static equilibrium, where the joints at either end apply equal and opposite force to the bar, directed in line with the length of the bar.

Plan To solve the problem:

1. Draw a diagram that shows the forces acting on each joint and that includes the internal bar forces, the reaction forces at the supports, and the applied external force.
2. For each joint, write a pair of equations that state the static equilibrium conditions, namely, equating the sums of the forces in the x- and y-directions to zero, or $\sum F_x = 0$, $\sum F_y = 0$.
3. Solve the resulting system of equations symbolically for the unknown internal and reaction forces.

Implementation of the Plan

Diagram of Forces at Each Joint Figure 6.22 illustrates the forces at each of the joints in the truss. Each bar applies an internal force $\mathbf{N_{ij}}$ to the joints at either end

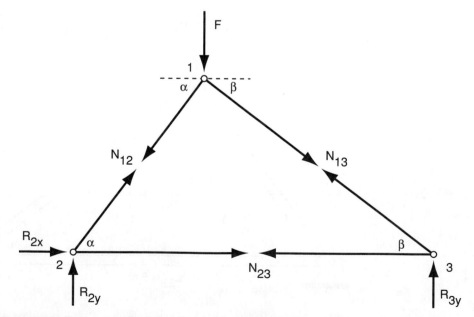

Figure 6.22 Force at each of the joints in the truss from Figure 6.21.

of the bar, where i and j are the joint numbers. We've adopted the convention of drawing the bar forces on the joints as pulling on the joints, thus assuming that the bars are in tension. If the values of these unknown forces turn out to be negative, then the bar is in compression.

The supports at joints 2 and 3 apply reaction forces to the truss. The support at joint 2 is a fixed support, so it applies reaction forces in both the x- and y-directions. The support at joint 3 is a rolling support, and it only applies a reaction force in the y-direction.

Static Equilibrium Equations at Each Joint   This is the step in solving the problem in which we impose Newton's Laws on each joint in the truss. Note that we've already applied Newton's Laws to the bars in stating that there are equal and opposite forces at either end. Given the diagram of forces from Figure 6.22, we can now sum the x- and y-components of the forces at each joint and set them equal to zero.

Joint 1 has three forces acting on it: the external force $\mathbf{F}$ and the two internal forces $\mathbf{N}_{12}$ and $\mathbf{N}_{13}$. The x- and y-components of each of these forces are as follows:

$$N_{12x} = -N_{12}\cos(\alpha) = -\frac{3}{5}N_{12} \qquad N_{12y} = -N_{12}\sin(\alpha) = -\frac{4}{5}N_{12}$$

$$N_{13x} = N_{13}\cos(\beta) = \frac{4}{5}N_{13} \qquad N_{13y} = -N_{13}\sin(\beta) = -\frac{3}{5}N_{13}$$

$$F_x = 0 \qquad F_y = -1000$$

Adding the forces in the x- and y-directions and setting the sums to zero, we get the following two equations:

$$\sum F_x = 0 \text{ at joint 1:} \qquad -\frac{3}{5}N_{12} + \frac{4}{5}N_{13} = 0 \qquad (6.1a)$$

$$\sum F_y = 0 \text{ at joint 1:} \qquad -\frac{4}{5}N_{12} + -\frac{3}{5}N_{13} - 1000 = 0 \qquad (6.1b)$$

Similarly, we determine the static equilibrium conditions at joint 2. Joint 2 has four forces acting on it, the two internal forces N_{12} and N_{23}, as well as the reaction forces R_{2x} and R_{2y}. We need only to decompose N_{12} into x and y components, since the other forces already point in the directions of one of the axes:

$$N_{12x} = N_{12}\cos(\alpha) = \frac{3}{5}N_{12} \qquad N_{12y} = N_{12}\sin(\alpha) = \frac{4}{5}N_{12}$$

Now, equating the sums of the forces in the x- and y-directions to zero, we get:

$$\sum F_x = 0 \text{ at joint 2:} \qquad \frac{3}{5}N_{12} + N_{23} + R_{2x} = 0 \qquad (6.2a)$$

$$\sum F_y = 0 \text{ at joint 2:} \qquad \frac{4}{5}N_{12} + R_{2y} = 0 \qquad (6.2b)$$

Finally, we determine the equilibrium conditions at joint 3. Joint 3 has three forces acting on it, the two internal forces N_{13} and N_{23} and the reaction force R_{3y}. Only N_{13} needs to be decomposed into x- and y-components.

$$N_{13x} = -N_{13}\cos(\beta) = -\frac{4}{5}N_{13} \qquad N_{13y} = N_{13}\sin(\beta) = \frac{3}{5}N_{13}$$

The equilibrium equations are then

$$\sum F_x = 0 \text{ at joint 3:} \quad -\frac{4}{5}N_{13} + -N_{23} = 0 \tag{6.3a}$$

$$\sum F_y = 0 \text{ at joint 3:} \quad \frac{3}{5}N_{13} + R_{3y} = 0 \tag{6.3b}$$

Solve the Resulting System of Equations for the Unknown Forces At this point, we have six equations in six unknowns that we can solve for the unknown forces. The six equations are the equilibrium conditions (6.1a,b), (6.2a,b), and (6.3a,b). The six unknowns are the three internal forces N_{12}, N_{13}, N_{23} and the three reaction forces R_{2x}, R_{2y}, R_{3y}.

The typical approach to solving a system of equilibrium equations such as this is to first find any equations with one unknown that we can solve trivially. In this case, there aren't any. Next, we look for two equations in two unknowns, which we find at joint one in equations (6.1a) and (6.1b). Solving these, we obtain:

$$N_{12} = -800 \text{ N} \quad \text{or} \quad 800 \text{ N compression}$$
$$N_{13} = -600 \text{ N} \quad \text{or} \quad 600 \text{ N compression}$$

Substituting the value of N_{13} into (6.3.a) and (6.3.b), we get:

$$N_{23} = 480 \text{ N} \quad \text{or} \quad 480 \text{ N tension}$$
$$R_{3y} = 360 \text{ N}$$

Finally, substituting the values of these forces into (6.2.a) and (6.2.b), we get:

$$R_{2x} = 0 \text{ N}$$
$$R_{2y} = 640 \text{ N}$$

Figure 6.23 illustrates these results.

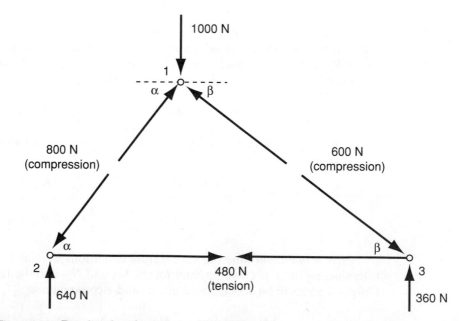

Figure 6.23 Results of static analysis of 3-bar truss.

Discussion There are several interesting points that arise from this analysis. First, you might be surprised that the reaction force R_{2x} is zero. In general, the values of all the unknown forces depend upon the magnitude and direction of the applied force **F**, a notion we'll explore later in problems at the end of the chapter. Second, we observe that the equilibrium equations depend upon the angles between the bars in the truss, not the lengths of the bars themselves. This implies that if we scale the truss up or down in size, keeping the angles and applied force the same, we would get the same results for the unknown forces. Finally, note that this analysis didn't require any information about the dimensions or material of the bars. This implies—for this particular truss design—that the forces would be the same regardless of the composition of the bars!

6.3   THE MATERIALS PERSPECTIVE

Whenever we apply a force to a solid object constrained in a way such that it can't move, that object will *deform* or change shape. The deformation may be so small we don't even notice it, but it happens nonetheless. In this section, we examine the relationships among the force applied to a bar, what it's made of, its dimensions, and how much it deforms.

The materials perspective on trusses is important for several reasons. First, given an understanding of how much each bar deforms under load, we can determine how much the truss as a whole changes shape, which can be critical to whether it performs its function acceptably. For example, a truss supporting a piece of scientific equipment or a bridge may have a requirement that it not deflect more than a certain amount under load. Second, a bar can only sustain up to a critical level of deformation before it'll break, and one of the most important considerations in modeling a truss is to determine if any of the bars exceed that level.

6.3.1   Bars as Springs: Hooke's Law and Young's Modulus

Suppose that we have a bar made of some material with length L and cross-sectional area A, as shown in Figure 6.24. If a force F is applied to either end of the bar, then the bar will change in length. This change is called *elongation*. We represent the elongation with the symbol ΔL, in which "Δ" means a "small but measurable change" in common engineering notation. By convention, if the applied force is positive, then

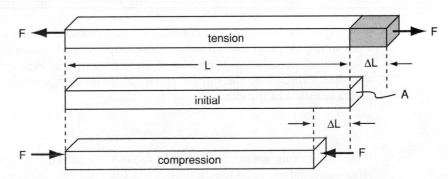

Figure 6.24 A bar of length L with cross-sectional area A stretches with some elongation ΔL when a force F is at either end.

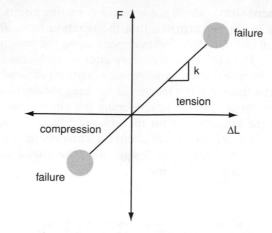

Figure 6.25 Force versus elongation of a bar.

the bar is in tension and the elongation is positive; if the applied force is negative, then the bar is in compression and the elongation is negative.

For most materials from which truss bars are made, including wood and metal, a plot of the force F versus the elongation ΔL would have a shape similar to that shown in Figure 6.25. For small elongations—either positive or negative—the elongation is roughly proportional to the force, and thus the bar obeys Hooke's Law,

$$F = k \cdot \Delta L,$$

in which k is the *spring constant* of the bar. As the magnitude of the force increases, however, this behavior changes and eventually for large enough forces, the bar will break or *fail*. In this section, we focus on the region Hooke's Law applies to, and consider failure mechanisms in Section 6.3.2.

Given a bar, we could determine its spring constant k by performing an experiment in which we apply a known force F to the bar and then measure its elongation ΔL. The spring constant would then be

$$k = \frac{F}{\Delta L}.$$

This is impractical to do, however, for every bar that we might use in building a truss. A much more useful approach is to find a way to determine the spring constant from the dimensions of the bar and some property of the material of the bar. To see how we might separate the spring constant into terms for the dimensions and a property of the material, let's consider an experiment in which we take a sample of a bar in the shape of a small cube, where length L and cross-sectional area A both equal 1, independent of the system of units, as shown in Figure 6.26. Now, suppose we apply a force F to the sample and measure its elongation ΔL to determine the spring constant of the cube:

$$k_{\text{cube}} = \frac{F}{\Delta L}$$

Rearranging this, we can also say that the force required to stretch a sample whose cross-sectional area and length both equal 1 is

$$\text{area} = 1, \text{length} = 1 \quad \Rightarrow \quad F = k_{\text{cube}} \cdot \Delta L$$

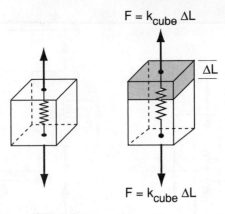

$$F = k_{\text{cube}} \, \Delta L$$

ΔL

$$F = k_{\text{cube}} \, \Delta L$$

Figure 6.26   A unit cube of material modeled as a spring.

Next, suppose we double the cross-sectional area of the cube. As illustrated in Figure 6.27, if we think of the original cube as a spring with a constant of k_{cube}, then a sample with a cross-sectional area of 2 is like having two springs in parallel, each with a spring constant of k_{cube}. In order to stretch this combination by the same elongation ΔL, we apply a force $k_{\text{cube}} \cdot \Delta L$ to *each* spring. Thus the total force we apply to the sample is

$$\text{area} = 2, \text{length} = 1 \quad \Rightarrow \quad F = 2k_{\text{cube}} \cdot \Delta L$$

If we increase the cross-sectional area of the sample to A, then this is like having A springs in parallel, and the force required to stretch the sample to an elongation of ΔL is

$$\text{area} = A, \text{length} = 1 \quad \Rightarrow \quad F = Ak_{\text{cube}} \cdot \Delta L$$

Now, suppose we go back to the original cube and double its length. This is like connecting two springs with a spring constant of k_{cube} in series, as shown in Figure 6.28. If we stretch this sample by ΔL, and assume the elongation is evenly distributed across the sample, then each half would stretch by $\Delta L/2$. Since it only takes half the

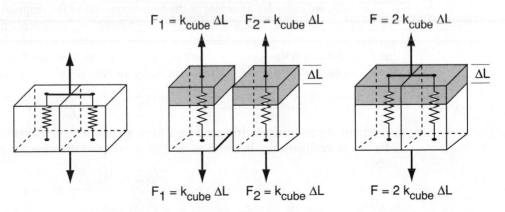

$$F_1 = k_{\text{cube}} \, \Delta L \quad F_2 = k_{\text{cube}} \, \Delta L \qquad F = 2\,k_{\text{cube}} \, \Delta L$$

ΔL ΔL

$$F_1 = k_{\text{cube}} \, \Delta L \quad F_2 = k_{\text{cube}} \, \Delta L \qquad F = 2\,k_{\text{cube}} \, \Delta L$$

Figure 6.27   Doubling the cross-sectional area of a unit cube can be modeled as connecting two springs in parallel.

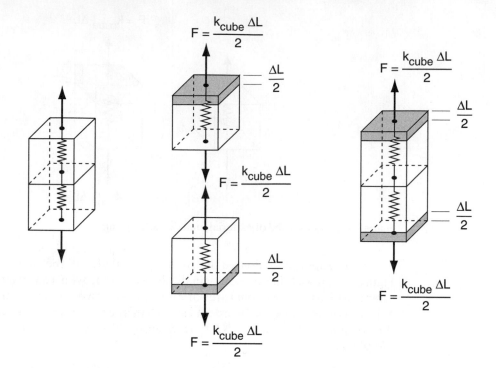

Figure 6.28 Doubling the length of a unit cube can be modeled as connecting two springs in series.

force to stretch the spring half the distance, the force required to stretch this sample ΔL is

$$\text{area} = 1, \text{length} = 2 \quad \Rightarrow \quad F = \frac{k_{\text{cube}}}{2} \cdot \Delta L$$

Generalizing this result, if we have a sample with length L and cross-sectional area of 1, this is like having L springs in series, and the force required to stretch the sample ΔL would be

$$\text{area} = 1, \text{length} = L \quad \Rightarrow \quad F = \frac{k_{\text{cube}}}{L} \cdot \Delta L$$

Finally, we consider a sample of the same material as the original unit cube with length L and cross-sectional area A. We could think of this sample as a network of springs with A springs in parallel and L springs in series. Since placing A springs in parallel multiplies the force by A and placing L springs in series divides the force by L, the force required to stretch this sample by an elongation ΔL would be

$$\text{area} = A, \text{length} = L \quad \Rightarrow \quad F = \frac{A k_{\text{cube}}}{L} \cdot \Delta L \qquad (6.4)$$

Looking at equation (6.4), we now have an expression for the spring constant of a bar of arbitrary size in terms of its dimensions and the spring constant of a unit cube:

$$k_{A \times L} = \frac{A k_{\text{cube}}}{L}.$$

In other words, if we know the spring constant k_{cube} of a unit cube made of some material, we can determine the spring constant of a bar of any dimensions made of

$$F = \frac{A\,k_{cube}\,\Delta L}{L}$$

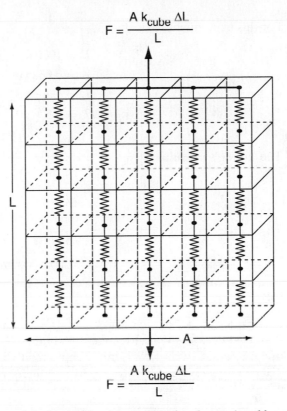

$$F = \frac{A\,k_{cube}\,\Delta L}{L}$$

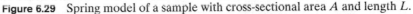

Figure 6.29 Spring model of a sample with cross-sectional area A and length L.

that same material. Thus k_{cube} is the material property we're looking for. Formally, we call k_{cube} *Young's modulus* or *modulus of elasticity* of a material, and represent it with the symbol E. Like density, Young's modulus is an intrinsic property of a material independent of the size of a sample made of that material.

So how do we determine the Young's modulus of a material in practice? One way is to carefully cut a sample of material into an exact cube, apply a force to it, and measure its elongation. Looking back at equation (6.4), however, we can determine k_{cube} from a sample bar of any dimensions. If we solve (6.4) for k_{cube}, we get

$$\text{Young's modulus:} \quad E = k_{cube} = \frac{F/A}{\Delta L/L}. \tag{6.5}$$

Thus, if we apply a force F and measure the elongation ΔL (or vice-versa) to a sample of material with length L and cross-sectional area A, we can determine the Young's modulus from (6.5).

Example 6.1	**Determining the Young's Modulus of Steel**

Determining the Young's Modulus of Steel

A steel rod with a cross-sectional area of 0.0625 square inches and a length of 1 inch stretches 5.5×10^{-4} inches when loaded with a weight of 1000 lb. What is the Young's modulus of this type of steel?

Solution For this experiment

$$F = 1000\,\text{lb}$$
$$A = 0.0625\,\text{in}^2$$
$$L = 1\,\text{in}$$
$$\Delta L = 5.5 \times 10^{-4}\,\text{in}$$

The Young's modulus is

$$E = \frac{F/A}{\Delta L/L}$$
$$= \frac{1000/0.0625}{5.5 \times 10^{-4}/1}$$
$$= 29,000,000\,\text{lb/in}^2$$

Example 6.2

Elongation of a Popsicle Stick
Popsicle sticks are typically made of white birch, which has a Young's modulus of approximately $15 \times 10^9\,\text{N/m}^2$. The nominal dimensions of a popsicle stick are 110 mm long, by 9.5 mm wide, by 2.1 mm thick. What would be the elongation of a popsicle stick if a tensile force of 20 N (approximately 4.5 lb) were applied to it?

Solution The tricky part of this problem is keeping the units straight. Since the Young's modulus is given in SI units, we'll convert the popsicle stick dimensions to meters to solve for the elongation. Thus

$$F = 20\,\text{N}$$
$$E = 15 \times 10^9\,\text{N/m}^2$$
$$A = (9.5 \times 10^{-3}\,\text{m})(2.1 \times 10^{-3}\,\text{m})$$
$$= 2.0 \times 10^{-5}\,\text{m}^2$$
$$L = 0.11\,\text{m}$$

From equation (6.5),

$$\Delta L = \frac{FL}{AE}$$
$$= \frac{(20)(0.11)}{(2.0 \times 10^{-5})(15 \times 10^9)}$$
$$= 7 \times 10^{-6}\,\text{m}$$
$$= 0.007\,\text{mm}$$

For comparison, the thickness of a human hair is typically between 0.05–0.10 mm.
You might wonder how it is possible to measure distances so small. Today, there are a number of inexpensive options for doing this. One approach is to use a

mechanical device called a dial indicator, as shown in Figure 6.30. A dial indicator uses a precise gear mechanism to convert a tiny linear motion into the rotation of a dial. Today, dial indicators accurate to 0.001 inches are available for under $20 and indicators accurate to 0.0001 inches are available for under $150.

6.3.2 Strength of Materials

As we mentioned briefly in the beginning of the last section, the typical materials used in trusses obey Hooke's Law for small elongations, but for large elongations, the relationship breaks down and the bars start to fail. In this section, we consider how we quantify the strength of materials, and look at some of the factors that contribute to it.

Whether a given elongation ΔL is considered "large" depends upon the length of the bar being stretched. As we begin to look at failure in materials, we want a measure of relative elongation independent of the bar dimensions, much as the Young's modulus is independent of dimensions. This relative elongation is called *strain*, ε, and is defined as

$$\text{strain} = \varepsilon = \Delta L/L.$$

Note that since both the numerator and denominator of strain have units of distance, the ratio has no units or is *dimensionless*.

Similarly, whether or not a given applied force is considered "large" depends upon the area over which it's distributed. The force relative to the cross-sectional area of a sample is called stress, σ, and is defined as

$$\text{stress} = \sigma = F/A.$$

Stress has units of *pressure*, which is force per unit area. In the SI system, this is the *pascal*, Pa, where $1\,\text{Pa} = 1\,\text{N/m}^2$. In the US system, pressure is measured in lb/in^2, pounds per square inch, or psi. Note that we can write the definition of Young's

Figure 6.30 A dial indicator is used in an experiment to measure the displacement of a K'NEX structure under load.

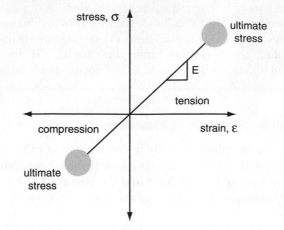

Figure 6.31 Stress versus strain plot for a material.

modulus from equation (6.5) in terms of stress and strain as

$$E = \frac{\sigma}{\varepsilon}.$$

Since strain in the denominator is dimensionless, we see that Young's modulus has the same units as stress, which is pressure.

In Figure 6.25, we illustrated the force versus elongation behavior for a particular sample of material with particular dimensions. To characterize the strength of a material without regard to the dimensions of a bar made from that material, engineers use plots of stress versus strain, as shown in Figure 6.31. A stress versus strain plot has the same basic shape as a force versus elongation plot. The slope of a stress versus strain plot in the region where the material obeys Hooke's Law is the Young's modulus, whereas it was the spring constant in the force versus elongation plot. Recall that the Young's modulus is simply the spring constant of a bar of material in the shape of a unit cube.

When the stress in a sample becomes very large, it begins to fail. Different materials fail in different ways and may have very different behaviors depending upon whether they're in tension or compression. Some materials fail abruptly when the stress reaches a certain level, while others gradually deform before they break. Independent of how a sample fails, the stress at which a sample of a material ultimately breaks is called the *ultimate stress* or *ultimate strength*, as indicated in Figure 6.31.

We can take two different perspectives on the strength and failure of materials: the *macroscopic* and the *microscopic*. The macroscopic perspective focuses on the gross characteristics of materials deforming under load such as we might identify in a stress versus strain plot. If we think of the macroscopic perspective as focusing on the symptoms of deformation, then the microscopic focuses on the causes in terms of the composition of the material. A hierarchy of viewpoints that spans from the

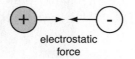

Figure 6.32 The electrostatic force between positively and negatively charged particles holds materials together.

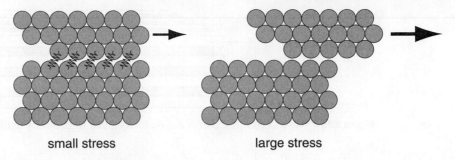

small stress   large stress

Figure 6.33 Crystalline structure of a metal.

macroscopic to microscopic considers materials in greater and greater detail all the way to the molecular and atomic levels.

At the atomic level, what holds materials together is the force between positively and negatively charged particles, called an *electrostatic* force, as shown in Figure 6.32. These electrostatic forces bind atoms together to form molecules with various shapes, and the shapes of these molecules and how they pack together significantly impact how materials behave at the macroscopic level. Below consider two of the most common construction materials, metal and wood, and how their composition affects their mechanical behavior.

Metal   The atoms in metals form in tightly packed regular structures called *crystals*, as shown in Figure 6.33. If a small stress is applied to a metal, the structure will stretch the interatomic bonds that hold the metal together, and the metal will behave according to Hooke's Law. If a large enough stress is applied, one part of the structure can "slip" relative to another part. This slippage typically begins at a location in the material where there's a defect, such as a missing atom. If this large stress continues to be applied, the slipping will continue until the structure slips apart and breaks.

This behavior appears in the stress versus strain curve for a metal and is shown in Figure 6.34. Once a critical stress called the *yield stress* is reached, the metal begins to stretch with little or no increase in stress. The yield stress is the point where slipping begins in the crystal structure. Eventually, the metal stretches to the point where the ultimate strength is reached and the material breaks. This stretching with little increase in stress is called *plastic* behavior, while stretching according to Hooke's

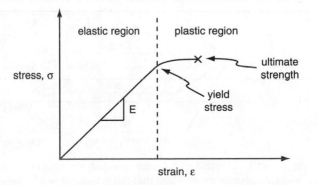

Figure 6.34 Stress versus strain curve for a metal. A metal will yield and exhibit plastic behavior when slipping begins in the crystal structure.

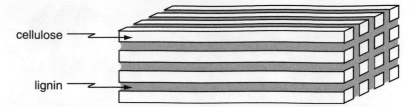

Figure 6.35 Wood is composed of cellulose fibers glued together by lignin.

Law is called *elastic* behavior. As an example of plastic behavior, think of chewing gum, which will stretch to a very long elongation before it breaks.

Wood The structure of wood is obviously quite different from that of a metal, and its mechanical behavior reflects the peculiarities of its structure. The substance of wood is the walls of cells of once-living plants. Structurally, these cell walls are mostly made up of long fibers of *cellulose*, bonded together by a glue-like substance called *lignin*, which gives wood its "grain," as shown in Figure 6.35.

Both cellulose and lignin are *polymers*. A polymer is a very large molecule composed of many simpler, repeating chemical units, called *monomers*, linked together. The shapes of these linked molecules can range from simple linear chains to complex three-dimensional meshes. Cellulose is simply a long, straight rigid chain of glucose monomers. Lignin has a much more complex molecular structure described as resembling a three-dimensional tangle of chicken wire. Figure 6.36 illustrates the structure of these two molecules.

Because it's a composite material, wood behaves differently depending upon the direction in which a forces is applied and whether it's in tension or compression. Figure 6.37 illustrates approximate values for the ultimate strength of wood, for stresses applied along various lines of action. Wood is strongest in compression parallel to the grain, and weakest when stresses are applied perpendicular to the grain.

Table 6.1 lists the density, Young's modulus, and ultimate stress for some common materials.

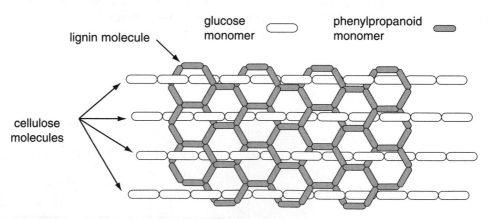

Figure 6.36 Molecular structure of wood, showing cellulose polymer chains held together by a lignin polymer mesh. Note that the actual shape of a lignin molecule is much more complex that this diagram suggests, and that the molecule actually has three different varieties of phenylpropanoid monomers.

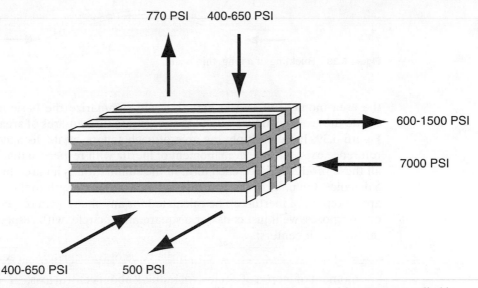

Figure 6.37 Approximate values for the ultimate strength of wood for stresses applied in different directions.

6.3.3 Buckling

Imagine taking a piece of uncooked spaghetti and compressing it along its length. Eventually it will fail, but not because the ultimate strength of pasta has been exceeded. Rather, the cause of failure is *buckling* or bending of the spaghetti, as shown in Figure 6.38. Swiss mathematician Leonhard Euler developed the theory of buckling during the mid-1700s. He showed that there was a critical load at which long thin beams in compression become unstable, and that the slightest force perpendicular to the beam will cause it to bend with an unlimited displacement—complete failure. Derivation of the buckling model requires calculus, but we can use Euler's results to state a formula for beams with certain dimensional relationships.

The critical buckling load of a beam depends upon its length, cross-sectional area, and Young's modulus, as we might expect. It also depends on the shape of its cross section. The important properties of a cross section is captured by a single quantity called *area moment of inertia, I*. We won't go into all the details of calculating

TABLE 6.1 Density, Young's modulus, and ultimate stress of some common engineering materials.

Material	Density kg/m^3	Density lb/in^3	Young's Modulus 10^9 N/m^2	Young's Modulus 10^3 lb/in^2	Ultimate Stress 10^6 N/m^2	Ultimate Stress lb/in^2
Steel	7860	0.2840	200	29008	400	58015
Aluminum	2710	0.0979	70	10153	110	15954
Glass	2190	0.0791	65	9427	50	7252
Concrete	2320	0.0838	30	4351	40	5802
Wood	525	0.0190	13	1885	50	7252
Bone	1900	0.0686	9	1305	170	24656
Polystyrene	1050	0.0379	3	435	48	6962

Figure 6.38   Buckling of a long, thin beam.

the area moment of inertia here, but will summarize the basic idea. Suppose we divide the cross-sectional area of the beam into tiny squares of area ΔA, as shown in Figure 6.39. If we multiply the area of each square by its distance squared from a reference axis, then the area moment of inertia with respect to that axis is the sum of all these products. Since each term in the summation is an area times the square of a distance, the units of the area moment of inertia is length to the fourth power. The area moment of inertia can be calculated for any shape with respect to any axis. For our purposes, we'll just consider a square and a circle, with respect to axes passing through their centers:

$$\text{moment of inertia of circle with radius } r \text{ about central axis:} \quad I = \frac{\pi r^4}{4} \qquad (6.6)$$

$$\text{moment of inertia of square with side } d \text{ about central axis:} \quad I = \frac{d^4}{12} \qquad (6.7)$$

One interesting aspect of the formula for the area moment of inertia is that the parts of the area that are further from the axis have a greater influence on the total value than the parts of the area that are close to the axis. Suppose, for example, that one of the tiny squares ΔA is 1 unit away from the axis, while another of the tiny squares is 2 units away. The contribution of the first square to I is $\Delta A(1)^2$, while the contribution of the second is $\Delta A(2)^2$, which is 4 times greater. We'll consider this point further in the problems at the end of the chapter, when we look at beams with cross-sections of different shapes.

Given this background, Euler's formula for critical buckling load for a beam of length L with area moment of inertia I and Young's modulus E is

$$\text{Euler's critical buckling load:} \quad F_{\text{buckling}} = \frac{\pi^2 EI}{L^2} \qquad (6.8)$$

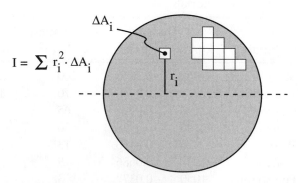

$$I = \sum r_i^2 \cdot \Delta A_i$$

Figure 6.39   Calculation of the area moment of inertia.

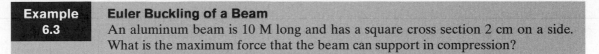

| Example 6.3 | **Euler Buckling of a Beam**
An aluminum beam is 10 M long and has a square cross section 2 cm on a side. What is the maximum force that the beam can support in compression? |

Solution We know of two possible ways that the beam could fail, either because its ultimate stress, σ_u, is exceeded or through buckling. We will calculate the force required to induce both of these and then see which is less. From Table 6.1, for aluminum

$$\sigma_u = 110 \times 10^6 \, \text{N/m}^2$$
$$E = 70 \times 10^9 \, \text{N/m}^2$$

The force at ultimate stress is

$$
\begin{aligned}
F_u &= A\sigma_u \\
&= (.0004 \, \text{m}^2)(110 \times 10^6 \, \text{N/m}^2) \\
&= 44,000 \, \text{N}
\end{aligned}
$$

The area moment of inertia for the beam is

$$
\begin{aligned}
I &- \frac{d^4}{12} \\
&= \frac{(0.02)^4}{12} \\
&= 1.6 \times 10^{-7} \, \text{m}^4
\end{aligned}
$$

The critical buckling load is

$$
\begin{aligned}
F_{\text{buckling}} &= \frac{\pi^2 EI}{L^2} \\
&= \frac{(3.14)^2 (70 \times 10^9)(1.6 \times 10^{-7})}{(10)^2} \\
&= 1104 \, \text{N}
\end{aligned}
$$

The critical buckling load is much lower than the ultimate stress, so the beam will fail at the critical buckling load of 1104 N.

6.4 PUTTING IT ALL TOGETHER

Having looked at the statics and materials perspectives on trusses independently, we now consider an example that constructs a complete model of a truss and looks at both at the same time. Consider the truss in Figure 6.40, which is made of aluminum bars with a square cross section 2 cm on a side. Our main concern is whether this truss can support the applied weight of 1000 N without failing; if it can, we'd also like to know the internal and reaction forces, the bar elongations, and the displacement of the top joint.

6.4.1 Statics Perspective

We recall that the model of the truss from the statics perspective expresses the condition that the sum of the forces on each of the joint pins must equal zero.

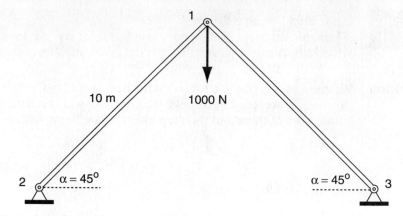

Figure 6.40   A statically determinate two-bar truss.

Figure 6.41 illustrates the forces at the joints. For this truss there are six unknown forces, the two internal bar forces and the four reaction forces.

As before, we break down each of the forces into its x- and y-components, and then write the force equilibrium equations at each joint in the x- and y-directions. This gives us a total of six equations:

$$\sum F_x = 0 \text{ at joint 1:} \qquad -N_{12}\cos(\alpha) + N_{13}\cos(\alpha) = 0 \qquad (6.9a)$$

$$\sum F_y = 0 \text{ at joint 1:} \quad -N_{12}\sin(\alpha) + -N_{13}\sin(\alpha) - F = 0 \qquad (6.9b)$$

$$\sum F_x = 0 \text{ at joint 2:} \qquad N_{12}\cos(\alpha) + R_{2x} = 0 \qquad (6.9c)$$

$$\sum F_y = 0 \text{ at joint 2:} \qquad N_{12}\sin(\alpha) + R_{2y} = 0 \qquad (6.9d)$$

$$\sum F_x = 0 \text{ at joint 3:} \qquad -N_{13}\cos(\alpha) + R_{3x} = 0 \qquad (6.9e)$$

$$\sum F_x = 0 \text{ at joint 3:} \qquad N_{13}\sin(\alpha) + R_{3y} = 0 \qquad (6.9f)$$

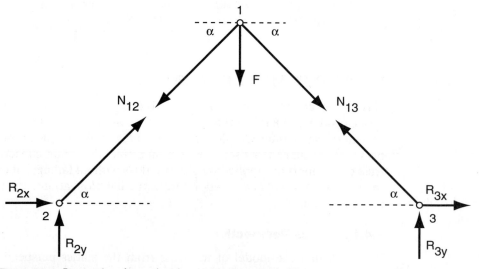

Figure 6.41   Internal and reaction forces for the truss in Figure 6.40.

Since there are six equations and six unknowns, we can solve for the forces at this point. From the x-components of the forces at joint 1, equation (6.9a), we find that

$$N_{13} = N_{12} \qquad (6.10)$$

Substituting this result into the equation for the y-components of forces at joint 1, equation (6.9b) and solving, we get

$$N_{12} = N_{13} = -\frac{F}{2 \sin \alpha}$$

$$= -\frac{1000}{2 \sin 45°} \qquad (6.11)$$

$$= -707 \, \text{N(compression)}$$

Now, substituting this result into the remaining equations (6.9c–f), we get the reaction forces

$$R_{2x} = R_{2y} = R_{3y} = 500 \, \text{N}$$
$$R_{3x} = -500 \, \text{N} \qquad (6.12)$$

6.4.2 Materials Perspective

The materials perspective describes the forces in each bar in terms of its elongation, dimensions, and Young's modulus. Since we determined the internal bar forces from the statics perspective, we'll first check if the forces are sufficient to induce buckling. The area moment of inertia for the two identical bars is

$$I = \frac{d^4}{12}$$

$$= \frac{(0.02)^4}{12}$$

$$= 1.6 \times 10^{-7} \, \text{m}^4$$

The critical buckling load is

$$F_{\text{buckling}} = \frac{\pi^2 EI}{L^2}$$

$$= \frac{(3.14)^2 (70 \times 10^9)(1.6 \times 10^{-7})}{(10)^2}$$

$$= 1104 \, \text{N}$$

Since the internal force of 707 N is less than the critical buckling load, the truss won't fail under this load. Note that the ultimate strength is much higher than the critical buckling load, so the truss won't fail by exceeding that, either.

Since the truss won't fail, we'll go on to find the elongations. The equations below describe the Hooke's Law relationships for the two bars:

$$\text{force in bar 12:} \quad N_{12} = \frac{EA_{12}}{L} \cdot \Delta L_{12} \qquad (6.13a)$$

$$\text{force in bar 13:} \quad N_{13} = \frac{EA_{13}}{L} \cdot \Delta L_{13} \qquad (6.13b)$$

As the two bars are identical and bear the same internal forces, their elongations will be the same:

$$\Delta L_{12} = \Delta L_{12} = \frac{N_{12}L}{EA_{12}}$$

$$= \frac{(-707\,\text{N})(10\,\text{m})}{(70 \times 10^9\,\text{N/m}^2)(0.0004\,\text{m}^2)} \tag{6.14}$$

$$= -0.25\,\text{mm}$$

6.4.3 Statically Determinate and Indeterminate Trusses

Curiously, for each of the trusses we've analyzed so far, the number of equations and unknowns in the static analysis were the same, so that we could solve the system of equations for a unique answer. However, this is not guaranteed to happen for *every* truss design.

The previous examples led to an equal number of equations and unknowns because the elements of the trusses had been arranged in a special way that makes the trusses *statically determinate*. In a statically determinate structure, it's possible to solve for the internal forces simply by solving the equilibrium equations at each joint. If a structure is statically determinate, then the number of unknowns and the number of equilibrium equations will be the same. Otherwise, the structure is *statically indeterminate*. In a statically indeterminate structure, the numbers of static equilibrium equations and unknowns are *not* the same, and it's necessary to consider both the materials and geometric perspectives simultaneously with the statics perspective in order to determine the forces. In the remainder of this section, we consider the special arrangement of elements that makes a structure statically determinate.

To understand the makeup of a statically determinate truss, look at the use of both triangles and supports in the design. Figure 6.42 shows a basic triangular or 3-bar truss. We support the truss in such a way that the location of the truss as a whole is fixed, but that each element is free to expand or contract when a load is applied. We achieve this by placing a fixed support at joint 1 and a rolling support at joint 2. This basic structure has three joints, three unknown internal forces (one in each member), and three unknown reactions (two at joint 1 and one at joint 2). Each joint provides two equilibrium equations—one for the sum of the *x*-components of

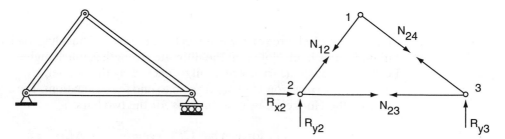

Figure 6.42 A static analysis of a single triangle with one fixed support and one rolling support leads to 6 equations in 6 unknowns.

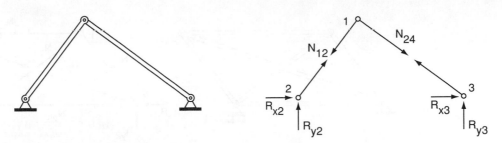

Figure 6.43 Removing one bar from a triangular truss while adding one reaction force maintains 6 equations in 6 unknowns in the static analysis.

the force at the joint and one for the sum of the y-components—so we will have a total of six equations and six unknowns. Thus, we can solve for the forces using the equilibrium equations alone. We note that we could also trade one truss element for an additional constraint on the support at joint 2, as shown in Figure 6.43, and still have six equations in six unknowns.

Next, we add a second triangle to the structure by introducing one more joint and more structural elements, as shown in Figure 6.44. This structure has 4 joints, which produce 8 equations, three unknown reactive forces and 5 unknown internal forces, for a system with 8 equations in 8 unknowns. If we continue to build up the structure by adding triangles in this manner, each time we add one joint and two structural elements to connect that joint to an already existing truss, we introduce two more unknown internal forces and two more equations, keeping the number of equations and unknowns in balance. Thus, structures with such an arrangement of triangles will be statically determinate.

Now, let's see what happens if we add an additional cross-brace to the structure as shown in Figure 6.45. The structure still has 4 joints that produce 8 equations, as well as 3 unknown reactive forces, but it now has 6 unknown internal forces, which leads to a system of 8 equations in 9 unknowns. As a result, we can't uniquely determine the internal forces in this truss through static analysis alone. In order to uniquely determine the forces, we need to consider how the structure will deform under load, which means considering both the materials and geometric perspectives together with the statics perspective. We won't go over this type of analysis here, but it's an important topic in courses on structural analysis.

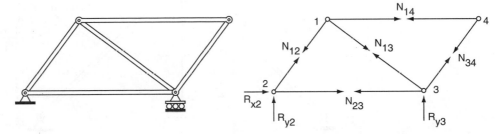

Figure 6.44 Adding one joint and 2 bars to add a second triangle leads to 8 equations in 8 unknowns.

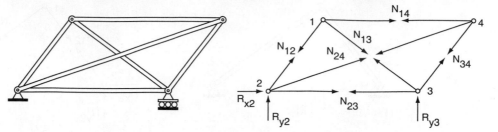

Figure 6.45 Adding another bar without adding another joint produces an extra unknown without adding more equations. Static analysis of this truss leads to 8 equations in 9 unknowns.

6.5 EXAMPLE: A TRADE STUDY OF STRENGTH VERSUS WEIGHT IN A TRUSS

One of the main advantages of trusses over other types of structures is that they're strong, yet lightweight. In this example, we look at the problem of designing the lightest truss possible that's capable of supporting a given load without failing. In solving the problem, we'll use the trade study approach we defined in Chapter 5 and used in solving the problem of launching a softball using a slingshot to hit a target in Section 5.5. As we'll see, even for a simple truss topology, a number of interrelationships in the system make this problem an interesting challenge.

6.5.1 Problem Definition and Plan of Attack

Popsicle stick bridge building contests have proved to be an excellent way of introducing high-school students to engineering problem solving. One of the goals of this example is to draw from the experience that many people have in building structures out of sticks and to add a bit to that through some elementary structural analysis.

Figure 6.46 illustrates the basic design problem. The goal of the problem is to design a two-bar truss, made of wooden sticks, that will span a distance of 24 in

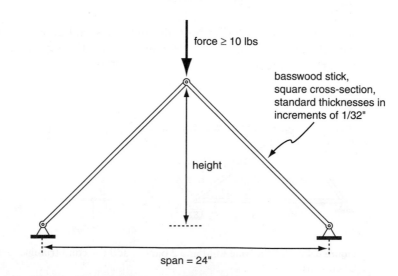

Figure 6.46 The two-bar truss design problem.

TABLE 6.2   Variables in the truss design problem.

Category	Name	Description	Known/Unknown
design:	h	truss height	unknown
design:	t	bar thickness	unknown
design:	material	type of wood	known, basswood $E = 1.46 \times 10^6$ psi $\sigma_u = 4730$ psi $\rho = 13.37 \times 10^{-3}$ lbs/in^3
environmental:	s	span of truss	known, 12 in
environmental:	F_{req}	required minimum applied force (weight) that the truss must support	known, 10 lbs
behavioral:	W	truss weight	unknown
behavioral:	F_{fail}	actual maximum weight that a given truss design will support at failure	unknown

while supporting a weight of at least 10 lbs. The sticks must be the same length and made of basswood—a common hardwood found in hobby stores—with a square cross section of standard dimensions in increments of 1/32 in. An additional objective is to minimize the weight of the design.

The critical three problem-solving steps are:

Define   The first step in solving the problem is to restate it in terms of variables, and to define the known and unknown. Table 6.2 defines the main variables in the problem. There are three design variables: the height of the truss h, the bar thickness t, and the bar material. Both the height and bar thickness are unknown, while the bar material for this problem is given as basswood. Note that the choice of material is a collection-type variable that comprises a set of properties. The table also contains the two important material properties relevant to this problem: the Young's modulus E and ultimate compression strength σ_u for basswood. There are two environmental variables, both of which are known. The first is the span of the truss, which is 24 in. The second is the minimum applied force or weight the truss is required to support, which is 10 lbs. Finally, there are two behavioral variables that describe the performance characteristics of a given truss design, both of which are unknown. The first is the actual truss weight, W, and the second is the actual maximum weight that the truss will support at the point of failure, F_{fail}.

Figure 6.47 illustrates the variables used in describing the dimensions of the truss. Note that we need only one of the three variables height h, bar length L, or bar angle α to describe the geometry uniquely, but we'll use each of these as convenient.

Explore   Before we define a plan for solving the problem, we'll first explore the problem in more detail.

Does the Problem Makes Sense?   The problem asks us to design a two-bar truss that'll satisfy two conditions: (1) to support at least 10 lbs without failing, and (2) to

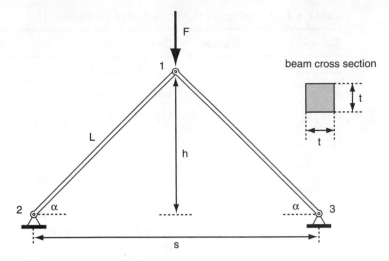

Figure 6.47 Variables for truss dimensions.

minimize the weight of the truss. We know two ways a truss might fail, either because the internal force in a bar exceeds the ultimate stress or because a bar buckles. Thus, the problem asks us to design the truss such that neither failure mechanisms occur with an applied load of 10 lbs.

There are only two unknown design variables in the problem, the height of the structure and the thickness of the bars, so the solution will be some combination of values for these. To get a better understanding of the problem, look at what each of these variables contributes to the solution of the problem and how they might be related.

The problem is in many ways similar to the slingshot problem in Chapter 5, in which we also were looking for a combination of values for two variables that would satisfy two conditions. Recall that in the slingshot problem, we were looking for a combination of horizontal and vertical settings for the slingshot that would satisfy the conditions of (1) hitting a target and (2) minimizing energy in doing so. In solving the slingshot problem, we first considered a simplified version of the problem with only one design variable. Suppose we did the same here, and fixed either the height of the structure or the bar thickness. If the height of the structure were fixed, it's clear what the role of the bar thickness is in solving the problem. If we make the bars thicker, we'll make the truss stronger in terms of either stress failure or buckling. This will also clearly make the truss heavier.

The role of the height of the structure, or equivalently the length of the bars, is less obvious. If we make the bars longer, the truss will become heavier. Also, if we make the bars longer, they'll buckle under a lower applied force; to see this, think about compressing a short piece of spaghetti versus a long piece of spaghetti. Both of these factors favor making the bars as short as possible. But what may not be obvious is that as the bars become shorter and the bar angle becomes smaller, the internal force in the bars increases. In fact, the internal bar forces can grow to be *much* larger than the applied force, and tend towards *infinite* as the bars move towards horizontal! This will become clearer as we perform a static analysis of the truss. As a result, if the bars are too short, the internal forces become so high that the bars will fail under a low applied load. Thus, there must be some intermediate value for the length that strikes a balance between strength and weight. Figure 6.48 illustrates these tradeoffs.

Assumptions As usual, we'll assume that the truss has ideal pinned joints. Also, we'll initially assume that the cause of failure will be buckling rather than a result

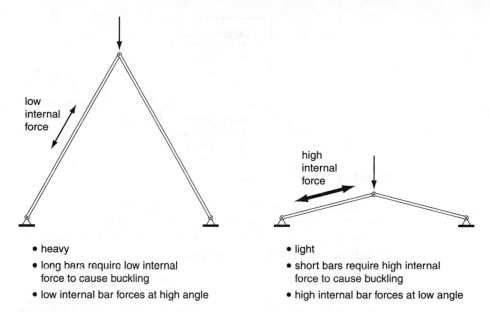

- heavy
- long bars require low internal force to cause buckling
- low internal bar forces at high angle

- light
- short bars require high internal force to cause buckling
- high internal bar forces at low angle

Figure 6.48   Tradeoff in bar length versus truss strength.

of ultimate stress, since the critical buckling load for a thin beam is generally much lower than its ultimate stress. We'll confirm this, however, before we choose a final design.

What Are the Key Concepts and Possible Approaches?   This problem involves two main groups of concepts. The first is modeling the interrelationships among components in a lightweight structure, from the static, materials, and geometric perspectives, which is the main focus of this chapter. It equally involves the ability to make tradeoffs in a design—the focus of Chapter 5.

What Level of Understanding is Tested?   While a two-bar truss appears to be a very simple structure, this problem is challenging for several reasons. The previous truss examples we've considered in this chapter focused on either application of a principle such as buckling or analysis of a known design. This example requires *evaluation* of a range of possible designs and selecting one that is optimal. A second factor that makes this problem complex involves the interrelationships in the analysis of the truss. When we performed the trade study of the launch system in Chapter 5, the distance model was a simple empirical model, and the energy model was simply the length of the stretched spring. This time, we need to build a model that considers the multiple parts of the truss—although this is only two bars—as well as multiple perspectives. Even though the individual equations that we'll produce aren't that complicated in isolation, we have many ideas to juggle in solving this problem.

Plan   Because solving this problem involves working with a number of concepts simultaneously, we'll be especially careful to make sure we have a clear plan before we start to write and solve equations. In developing the plan, we'll start with the goal and work backwards toward the primary inputs. We'll do this with three versions of the plan, each adding more detail than the one before it. Figure 6.49 shows a first version of the plan. The main goal is to find an acceptable design for the truss—one that supports at least 10 lbs—that also minimizes weight. To find this design, we'll

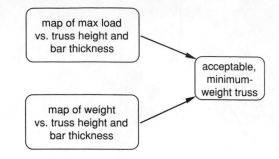

Figure 6.49 Version 1 of plan.

need maps of the two behavioral variables, maximum load and truss weight, each versus the two design variables, truss height and bar thickness.

The next version of the plan, shown in Figure 6.50, focuses on constructing the models we need for drawing the maps.

- The weight of the truss depends on the dimensions of the bars and the density of the wood.
- Assuming that the truss will fail by buckling, the critical buckling load for a bar depends upon the internal force in the bar, the length of the bar, and its moment of inertia.

In the third version of the plan, shown in Figure 6.51, we further break down the calculation of internal bar force and moment of inertia.

- The internal bar force comes from a static analysis of the truss; this will consider the applied force and the bar angles, which, in turn, depend on the span and height of the truss.
- The moment of inertia depends upon the bar thickness and Young's modulus.

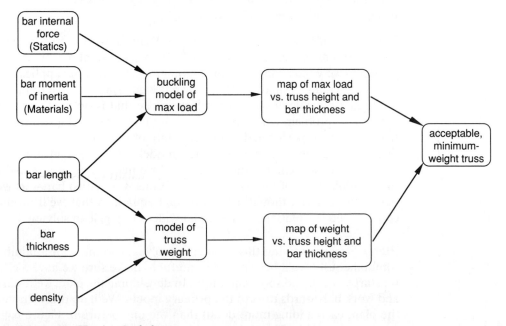

Figure 6.50 Version 2 of plan.

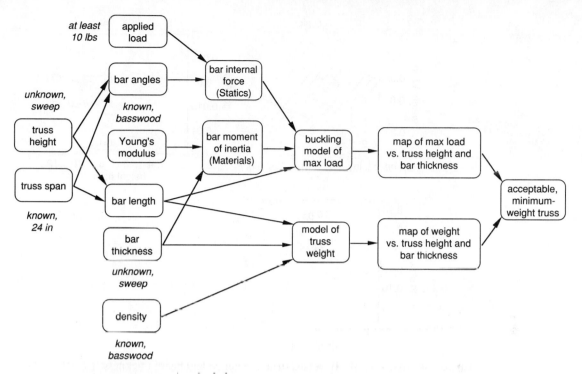

Figure 6.51 Version 3 of plan.

Figure 6.51 also annotates which quantities required by the plan are known, and which are unknown and that we'll sweep in drawing the maps.

Given the steps to solving the truss design problem in Figure 6.51, all that remains to defining a plan is picking an order for implementing them. We'll begin with the "easy part" of developing and mapping the weight model, then move on to the load model, and finally superimpose the maps—as we did earlier for the slingshot problem—to find the optimal design.

6.5.2 Implementation of the Plan

Truss Weight Model   The first part of the plan is to build a model for the weight of the truss and to map the weight versus the design variables height h and bar thickness t. The weight is simply the combined volume of the two bars times the density of the wood:

$$
\begin{aligned}
W &= 2 \times \text{bar volume} \times \text{density of basswood} \\
&= 2(t^2 L)\rho \\
&= 2(t^2\sqrt{h^2 + (s/2)^2})\rho \\
&= 2(t^2\sqrt{h^2 + (12)^2})(13.37 \times 10^{-3}) \text{ lbs}
\end{aligned}
\tag{6.15}
$$

Given this formula for weight in terms of h and t, we can now construct a map for the weight by sweeping the values of these two variables. Figure 6.52 shows the resulting plots, in which we vary h from 1 inch to 12 inches, and t from 1/8 inch to 1/4 inch in increments of 1/32 inch.

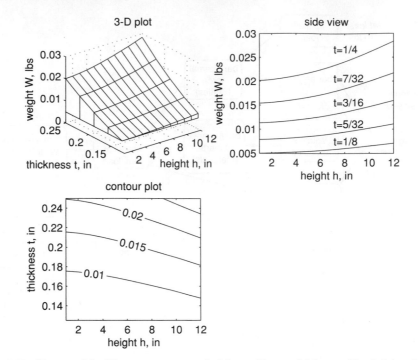

Figure 6.52   Truss weight W versus structure height and beam thickness. Top left is a 3-D plot, with h on the x-axis, t on the y-axis, and W on the z-axis. Top right is a side view looking into the y-z plane, with cross sections at t ranging from 1/8 inch to 1/4 inch in increments of 1/32 inch. Bottom left is a top-view contour plot, looking down onto the x-y plane, showing the iso-weight lines for different combinations of h and t.

Truss Strength Model   The truss strength model describes the maximum load the truss can sustain before buckling, versus the height h and bar thickness t. To develop this model, we need to consider the truss from two perspectives: the statics perspective to determine the bar internal forces, and the materials perspective to determine the critical buckling load.

Statics Perspective   The two-bar truss in this example is statically determinate. As a result, we only need a static analysis to solve for the forces. Furthermore, since we're only concerned about the bar internal forces, and not the reaction forces at the supports, we only need to solve the static equilibrium equations at the free joint, joint 1. Summing the forces in the x- and y-directions at joint 1, we obtain the following two equations:

$$\sum F_x = 0 \text{ at joint 1:} \qquad -N_{12}\cos(\alpha) + N_{13}\cos(\alpha) = 0 \qquad (6.16a)$$

$$\sum F_y = 0 \text{ at joint 1:} \qquad -N_{12}\sin(\alpha) - N_{13}\sin(\alpha) - F = 0 \qquad (6.16b)$$

Solving these equations, we obtain

$$N_{12} = N_{13} = -\frac{F}{2\sin\alpha}$$
$$= -\frac{FL}{2h} \qquad (6.17)$$

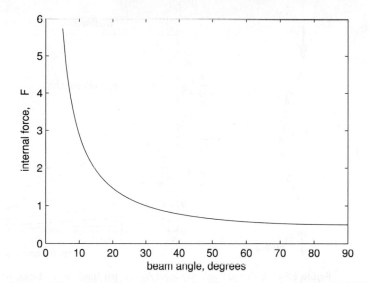

Figure 6.53   Internal force N_{12} versus beam angle α for the truss in Figure 6.46. The internal force is measured as a fraction of the applied force F.

From this result, we see that as the truss height h, or equivalently the angle α, approaches zero, the bar internal force grows very large, approaching infinity. Figure 6.53 shows a plot of the bar internal force versus α. For a very tall truss, as the bars approach vertical, each bar carries an internal load of approximately half the applied force F. When the angle is lowered to $45°$, the internal force in each bar is still less than the applied force. At $30°$, the internal force equals the applied force and grows rapidly after that, so that at $10°$, the internal force is nearly three times the applied force and at $5°$ it's nearly six times the applied force. Thus, while shortening the length of the bar reduces the weight of the truss, it can dramatically increase the internal forces.

Materials Perspective   The materials perspective on the strength model looks at the critical buckling load for a bar versus its length, thickness, and Young's modulus. In general the critical buckling force is

$$N_{\text{buckle}} = \frac{\pi^2 EI}{L^2} \tag{6.18}$$

where, for a bar with a square cross section,

$$I = \frac{t^4}{12} \tag{6.19}$$

Thus, for basswood sticks,

$$
\begin{aligned}
N_{\text{buckle}} &= \frac{\pi^2 E t^4}{12 L^2} \\
&= \frac{\pi^2 (1.46 \times 10^6) t^4}{12 L^2} \text{ lbs}
\end{aligned}
\tag{6.20}
$$

Figure 6.54 shows a plot of N_{buckle} as we vary both the length and thickness of basswood sticks. Note that in this calculation, N_{buckle} is a compression force applied directly to the bar and parallel to its length; it's *not* the force F applied to joint 1 of

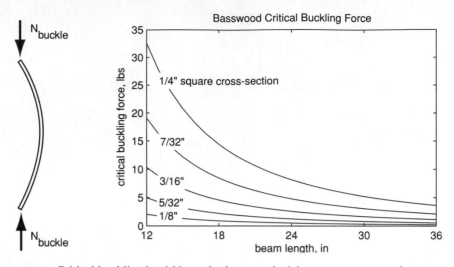

Figure 6.54 Critical buckling load N_{buckle} for basswood sticks, square cross section, versus bar length and thickness.

the truss. The plot clearly shows how the critical buckling load decreases with bar length and increases with thickness.

Combining the Statics and Materials Perspectives Given the formula for the internal bar force obtained from statics perspective in Equation (6.17) and the formula for critical buckling load in a bar obtained from the materials perspective in Equation (6.20), we can combine these results to determine a formula for the force F_{buckle} that we'd need to apply to the truss at joint 1 in order to cause the bars to buckle. At $F = F_{\text{buckle}}$, $N_{12} = N_{\text{buckle}}$, so we can rewrite equation (6.17) as

$$N_{\text{buckle}} = \frac{F_{\text{buckle}}L}{2h}$$

$$F_{\text{buckle}} = \frac{2hN_{\text{buckle}}}{L}$$

(6.21)

Substituting equation (6.20) into equation (6.21), we get

$$F_{\text{buckle}} = \frac{2hN_{\text{buckle}}}{L}$$

$$= \frac{2h\pi^2 Et^4}{12L^3}$$

$$= \frac{2h\pi^2(1.46 \times 10^6)t^4}{12L^3}$$

And since

$$L = \sqrt{h^2 + (s/2)^2}$$

$$= \sqrt{h^2 + (12)^2}$$

we finally get the following formula for F_{buckle} versus h and t:

$$F_{\text{buckle}} = \frac{2h\pi^2(1.46 \times 10^6)t^4}{12(h^2 + 144)^{3/2}}$$

(6.22)

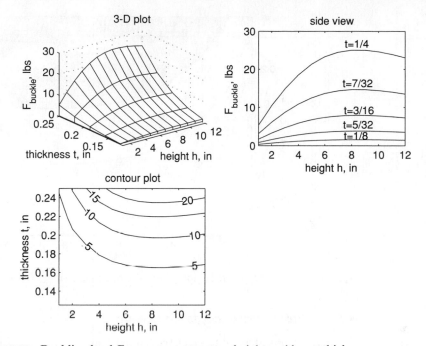

Figure 6.55   Buckling load F_{buckle} versus structure height and beam thickness.

Figure 6.55 presents the plots of the strength model, including a 3-D plot, side view cross sections, and a top-view contour plot. As a step towards understanding the results in these plots, let's begin with just the $t = 1/4$ inch curve in the side-view plot and follow what happens as we increase the height h from 1 inch to 12 inches. When h is very low, as we saw from equation (6.17) and Figure 6.53, the internal force in the bars is much greater than the applied force and thus F_{buckle} is low. As we gradually increase the truss height, the internal force becomes smaller relative to the applied force, until at around 7 inches or an angle of 30°, they become approximately equal. Throughout this region, F_{buckle} steadily increases as the ratio of the internal load to the applied load goes down. All the while, however, as the bars get longer the critical buckling load of a bar in isolation is decreasing, as we saw from equation (6.20) and Figure 6.54. When the truss height becomes greater than approximately 8 inches, the fact that long beams buckle at lower loads begins to dominate, and F_{buckle} starts going down again. This trend continues as the height increases.

Next, we consider the curves for the other thicknesses in the side-view plot in Figure 6.55. Each of these curves has the same basic shape at the $t = 1/4$ inch curve, where F_{buckle} increases as the height increases from 1 to 8 inches, and then decreases after that. As we decrease the thickness from $t = 1/4$ inch down to $t = 1/8$ inch, each curve lies beneath the one before it, showing how F_{buckle} decreases as the bar thickness decreases. We can see the relationship between F_{buckle} and t more clearly by looking at the 3-D plot. If we trace a line along the surface where the truss height h is constant, say $h = 12$ inches, we see that F_{buckle} increases rapidly with t, since F_{buckle} is proportional to t^4. In other words, a small increase in the bar thickness significantly improves the strength of the truss.

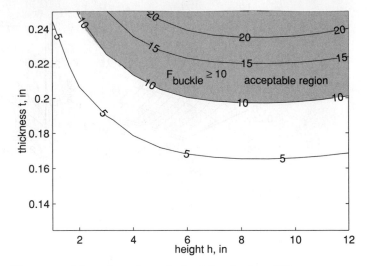

Figure 6.56 The acceptable region where the buckling load $\geq$ 10 lbs.

6.5.3 Finding an Acceptable Design

Given the maps of weight versus truss height and bar thickness in Figure 6.52 and the maps of strength versus truss height and bar thickness in Figure 6.55, we're ready to find the design for the lightest truss that'll support at least 10 lbs. We'll do so by first taking the strength contour plot, locating the acceptable region in that plot where $F_{buckle} \geq 10$ lbs, and then superimposing the acceptable region for F_{buckle} onto the weight contour plot, to determine the point in the region with the lowest weight.

Figure 6.56 shows the acceptable region where $F_{buckle} \geq 10$ lbs within the strength contour plot. Recall that a contour plot is a top-view of a 3-D plot, and the lines in the plot are contours where the dependent variable—in this case F_{buckle}—have the

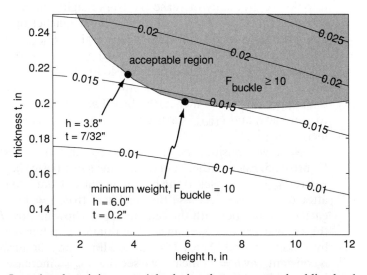

Figure 6.57 Locating the minimum-weight design that supports a buckling load $\geq$ 10 lbs.

same value. The acceptable region in Figure 6.56 is thus the set of all combinations of h, t points that lie on or above the $F_{buckle} = 10$ contour.

Figure 6.57 shows the acceptable region for strength superimposed onto the the weight contour plot. We see that the minimum-weight point lies just below the $W = 0.015$ lbs contour, and has coordinates $h = 6$ inches, $t = 0.2$ inch. Since we're required to use stock basswood sticks with standard thicknesses in multiples of 1/32 inch, the lightest feasible design that supports 10 lbs is $h = 3.8$ inches and $t = 7/32$ inch.

PROBLEMS

1. Vector Components
Determine the x and y components of the vector F in each of the cases below.

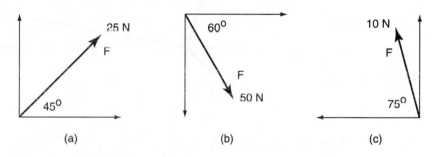

(a)   (b)   (c)

2. Resultant of Vector Addition
Find the vector sum of the two vectors **A** and **B**

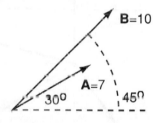

(a) using the trigonometric ("magnitude/direction") method
(b) using Cartesian components.

3. Resultant of Two Vectors
Consider the following two vectors:

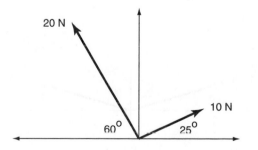

Find their resultant (magnitude and direction) using

(a) the parallelogram rule,

(b) the "head-to-tail" rule,

(c) the Cartesian components approach.

4. Resultant of Three Vectors

Find the resultant (magnitude and direction) of the three vectors shown using the Cartesian components approach.

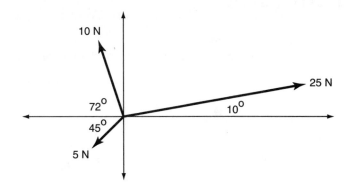

5. Resultant of Three Vectors

Find the resultant (magnitude and direction) of the three vectors shown using the Cartesian components approach.

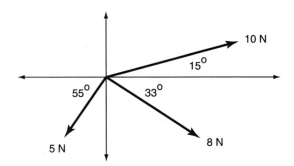

6. Satisfying Static Equilibrium

For each of the two situations below, determine the additional force required so that the system will be in static equilibrium.

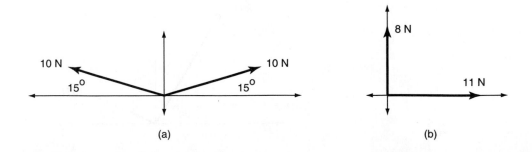

7. Determining an Unknown Force for a Particle in Static Equilibrium

A particle, Point A, is subjected to four planar forces, three of which are known and the fourth is unknown. If the particle is in equilibrium, determine the magnitude of the force **F** and the direction of its line of action, relative to the 200 N force, θ.

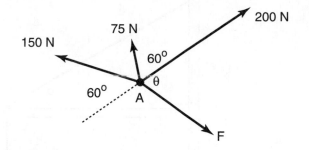

8. Adding Vectors

The vector **R** (with unknown magnitude) is known to be the vector sum of two forces: one, **A**, making an angle of 30° with **R**, and with a magnitude of 25 lb, and the other, **B**, with magnitude 16.8 lb.

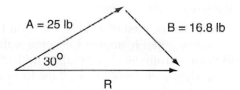

Find

(a) the direction of **B**, in degrees measured from **R**

(b) the magnitude of the resultant vector **R**

Note that there are two possible answers to this. Why?

9. Velocity Vector

A projectile is moving with velocity **V**. If the magnitude of the x-component of **V**, $V_x = 3$ m/s and the magnitude of the y-component of **V**, $V_y = 4$ m/s, find

(a) the magnitude of **V**

(b) the direction of **V**, measured in degrees from the horizontal x-axis.

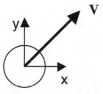

10. Forces on the Joints of a Truss

Draw each joint of the truss in Figure 6.19 as a particle with all the internal bar forces and external forces acting on that particle. Assume all bars are in tension. Assume all reaction forces are pushing on the truss.

11. Supporting a Shelf

A shelf is attached to a wall with a hinge and a cable as illustrated below.

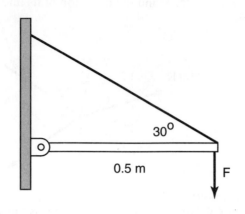

- If a force with magnitude $F = 100$ N is applied to the edge of the shelf, what is the tension in the cable?
- Suppose that instead of a cable, the shelf is supported by 50 lb test fishing line tied to a nail hammered into the wall. If it takes a force of 45 lbs to pull the nail from the wall, by what mechanism will the shelf fail if the force F is gradually increased from 0 lbs to the point of failure? Explain your result.

Now suppose that instead of being supported by a cable from above, the shelf is supported from beneath by a single wooden bar whose cross section is 1 cm by 1 cm, as shown in the figure below.

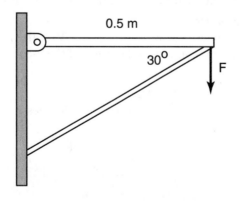

- What is the maximum weight F that the shelf can support?
- What would be the minimum diameter of a cylindrical aluminum rod that could support the same weight as the wooden bar?

12. Analysis of a Five-Bar Truss

Consider the 5-bar truss below.

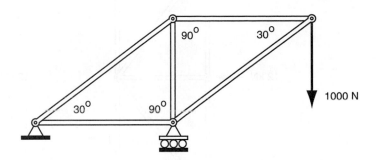

(a) Determine the internal forces for each of the five elements of the truss.

(b) If you doubled the height of the truss but kept its width the same, would the magnitudes of the internal forces increase or decrease for each bar? Explain your reasoning.

13. Zero-Force Members

A zero-force member is an element of a truss that has no internal force—it is in neither tension nor compression—for a given applied load. For each of the load configurations below, determine which members of the truss are zero-force, as well as determine the reaction forces at points C and D. The height of the truss is 16 m and the width is 12 m.

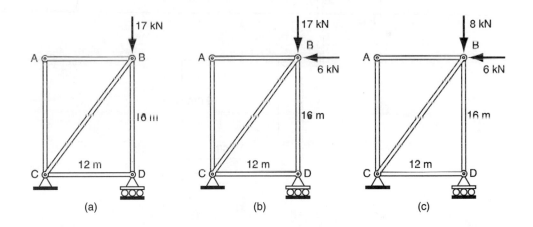

(a)   (b)   (c)

14. Alternative Truss Designs

The following two trusses are being considered for an application with a horizontal load. The truss is 10 m wide and 26 m tall, with each individual bay being 13 m tall. The three point loads are each 12 kN. Find the internal bar forces and the reaction forces for these two trusses. Are there any benefits of one design over the other?

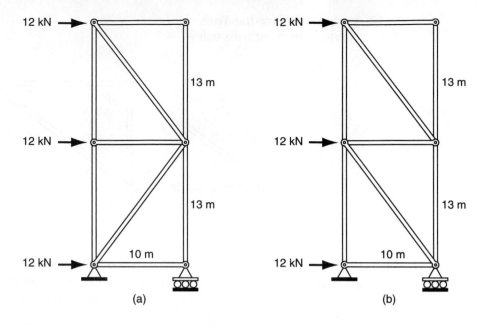

(a)   (b)

15. Catenary Curves and the St. Louis Arch

Look up the definition of a catenary curve. Explain why the St. Louis Arch was designed as an inverted catenary, as opposed to, say, a parabola.

16. Spring Constant of a Material Sample

Find the spring constant of a material sample 10 cm by 10 cm by 2 cm that elongates 0.1 mm when the sample is subjected to a tensile force of 24 kN along the axis shown below. What might this material be?

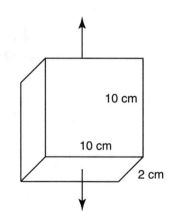

17. Compression of a Pipe

Find the elongation of an aluminum pipe subjected to a compressive force of 200 lbs. The inner diameter of the pipe is 2 in and the outer diameter is 2.125 in. The initial length of the pipe is 4 ft.

18. Strength of a Polyethelene Cylinder

Find the stress, strain, and Young's modulus of a polyethylene cylinder 1/2″ in diameter and 6″ in length. When loaded with a tensile force of 475 lb, the sample elongates 1/8″.

19. Deformation of an Aluminum Alloy Sample

A certain aluminum alloy has the stress vs. strain characteristics shown below. If a cylindrical sample of the material has a length of 10 cm and a diameter of 0.5 cm, what force is required to produce an elongation of 2 mm? Would this elongation be elastic?

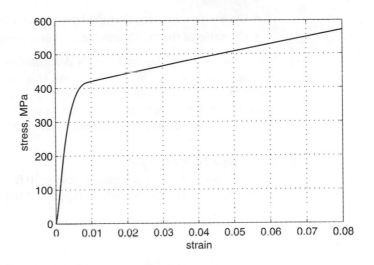

20. Critical Buckling Load of a Rectangular Beam

In Section 6.3.3 we provided the background needed to find the critical buckling load of a beam with a square cross section.

(a) What do you think would be the effect on the critical buckling load of doubling the width of the beam but leaving the thickness the same (so that it now has a rectangular cross section). Explain. Hint: you might look up information on area moment of inertia online to see if your theory is correct.

(b) Find the force required to buckle a glass plate 1 m × 1 m × 5 mm, if the force is applied to one of the faces with the smallest area.

21. Why Use an I-Beam?

An I-beam is a beam whose cross section is shaped like the letter I, as shown below. What do you think is the rationale behind designing a beam shaped like this?

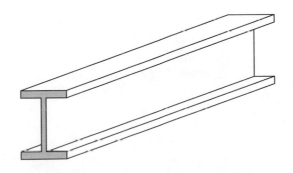

22. Solid Versus Hollow Beams

Consider three cylindrical aluminum beams in compression, each 5 m long. The first is a solid rod with a diameter of 10 mm, the second is a solid rod with a diameter of 6 mm, and the third is a hollow pipe with an outer diameter of 10 mm and a wall thickness of 2 mm. For each of the beams, find:

- its mass
- the maximum force that the beam will withstand before failing
- the ratio of the maximum force over the mass

The tricky part of this problem is determining the area moment of inertia of the pipe—you should be able to figure this out from the definition of the area moment of inertia and the formula for the area moment of inertia for a cylinder given in the chapter, along with a bit of reasoning. (You can also look it up online to confirm your theory.) In general, what conclusions can you draw about solid versus hollow beams?

23. Climbing a Telephone Pole

Suppose that a wooden telephone pole is 30 ft tall and 1 ft in diameter. Approximately how many adults could it support at the top before it starts to buckle?

Modeling Interrelation- ships in Systems: Digital Electronic Circuits

LEARNING OBJECTIVES

- to describe the difference between the logical and physical views of a digital circuit, and to discuss some of the history of the evolution of both;
- to express simple logic statements using Boolean equations and networks of switches;
- to use the concepts of current, voltage, and power, together with Ohm's Law and Kirchhoff's Laws, to analyze a simple circuit;
- to discuss some of the tradeoffs between the logical and physical views of a digital circuit, such as power versus size considerations.

7.1 INTRODUCTION

It's difficult to overstate the extent to which computers are woven into society, and even leaders in the industry have grossly underestimated their impact. In 1943, IBM chairman Thomas J. Watson, Sr. predicted that "there's a total world market for maybe five computers." In 1977, president of Digital Equipment Corporation Ken Olsen stated that "nobody would want to buy a computer for their home."[1] To-day, computers are nearly everywhere, and are used for applications as diverse as

---

[1] Digital Equipment Corporation was acquired by Compaq Computer in the mid-1990s. Now part of Hewlett-Packard, Compaq was the number-two manufacturer of personal computers worldwide in 2004.

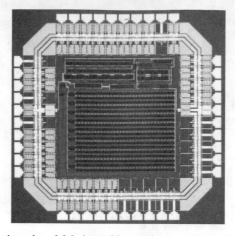

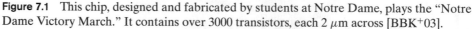

Figure 7.1 This chip, designed and fabricated by students at Notre Dame, plays the "Notre Dame Victory March." It contains over 3000 transistors, each 2 μm across [BBK+03].

forecasting the weather, controlling the timing of an automobile engine, and animating the latest video game. Fundamentally, a computer is a machine that processes information. Modern computers—in their various forms—follow a chain of inventions that extends back thousands of years. Today's portable MP3 players, with their songs stored with tiny magnets on a hard drive or as electrical charge in a flash memory, descend from music boxes, with their songs stored as tabs on metal cylinders. Even the word "calculator" pays homage to one of the very first tools that people devised for working with numbers: the pebble, or *calculus* in Latin.

Over the span of the twentieth century, the substrate of a computer has come full circle to a small rock, in this case a chip of silicon such as the one shown in Figure 7.1, containing tiny electrical switches called *transistors*. The purpose of this chapter is to take an introductory look at digital electronic circuits, and to examine how we use them to perform calculations and make decisions. Similar to the lightweight structures called trusses we examined in Chapter 6, a digital circuit is a network of interrelated parts that work together to provide a common function. Moreover, just as a truss has to obey different mechanical laws, such as Newton's Laws for the balance of forces and Hooke's Law for springs, so must a circuit obey different electrical laws, such as Ohm's Law and Kirchhoff's Laws. For either example, the analysis leads to a system of equations that must be solved simultaneously.

In this chapter, we first give an overview of some of the key issues common to the design of any computing machine and take a brief look at their history. Next, we look at digital circuits from the logical perspective, and see how new ideas in mathematics during the late 19th and early 20th centuries opened the door to modern computation. Then, we shift our focus to the electronics perspective and provide an introduction to the analysis of electrical circuits. Finally, we pull both perspectives together and look at tradeoffs in the design of a simple digital logic circuit called an *inverter*.

7.2 COMPUTING MACHINES

7.2.1 The Logical and Physical Views

As we've done with other artifacts, we can analyze a computing machine from the perspectives of its developers and its consumers. On top of this, however, there are

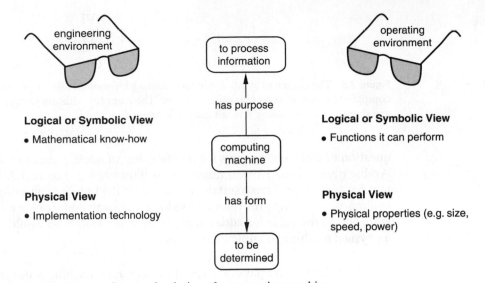

Figure 7.2 Perspectives on the design of a computing machine.

two other important viewpoints: the *logical or symbolic* viewpoint and the *physical* viewpoint—or the logical machine and the physical machine, for short—as shown in Figure 7.2.

Logical or Symbolic View We can think of the logical or symbolic view of a computing machine as the mathematician's or computer scientist's view, disembodied from the physical construction of the machine. From the consumer's perspective, the logical view defines the functions the computer performs. For example, a consumer evaluating a handheld device from the logical view might ask the following questions:

- Can I use it to do arithmetic calculations?
- Can it store my address book and calendar?
- Will it play music?
- Can it make phone calls?

From the producer's perspective, the resources available for developing the logical machine is essentially mathematical knowledge. This includes how to represent different kinds of information and how to operate on this information to get the desired result. In mathematics and logic, we represent information using an "alphabet" of *symbols*. According to the dictionary, a symbol is

> **symbol** n. **1** Something chosen to stand for or represent something else. **2** A character, mark, abbreviation, or conventional sign, or letter indicating something. (Funk and Wagnalls)

Examples of symbols abound. The letters of the alphabet are symbols, as are the digits 0, 1, 2, 3, 4, 5, 6, 7, 8, 9, the card suits ♣, ♢, ♡, and ♠, and all the other "marks" we can make with a keyboard. Using an alphabet of symbols, we can encode all different kinds of information, including text and numbers, as well as images and sound.

From the logical viewpoint, a system of mathematical and logical rules defines the operations of a computer. Choices made in the design of the logical machine can profoundly impact the design of the physical machine. Consider, for example, the

$$56 \qquad\qquad LVI$$
$$+\ 44 \qquad\qquad +\ XLIV$$
$$\overline{100} \qquad\qquad \overline{C}$$

Figure 7.3   The choice of symbols for representing information has a major impact on the complexity of building a computing machine. The rules for addition using Arabic numerals are simple, the rules using Roman numerals are not.

question of choosing symbols for numbers for an adding machine. Two choices are Arabic numerals or Roman numerals, as illustrated in Figure 7.3. If we represent numbers using the Arabic system, we can apply the familiar rules of adding numbers column-wise, carrying the ones, and so forth. We can't add Roman numerals column-wise, and so the rules for addition are more complicated, as would be the design of a physical machine to implement them.

Physical View   The physical view of a computing machine is its tangible construction. From the consumer's perspective, the characteristics of the physical machine are its physical properties. A consumer evaluating our hypothetical handheld device, for example, might ask the following questions:

- What does it weigh? How big is it?
- Is it energy-efficient? How long do the batteries last? Does it get warm when you operate it?
- How fast is it?
- How reliable is it? Will it always give the correct answer?

From the producer's perspective, the resources available for developing the physical machine are the design and fabrication technology. Over the centuries, this technology has advanced from pebbles, to the beads on a string of an abacus, through precision metal parts, on to today's semiconductor technology and beyond.

We say that a physical computing machine is an *implementation* of a logical machine, and to build one, we require two important things. First, we need a way to represent information such that the physical machine can "understand" it. A CD or DVD, for example, encodes symbolic information as tiny pits and bumps on the surface of the disk, while a microprocessor encodes information through voltages at points in an electrical circuit. Second, we require some "trick" for using physical components to interpret encoded symbols, and for processing them by strictly physical means, in a manner that *emulates* arithmetic or logical operations. This "trick" depends on the implementation technology, be it mechanical parts such as gears, electrical devices, or even molecular or biological components.

Example: Burroughs Adding Machine Vs. Sony PlayStation 3   To illustrate these ideas, we consider the design of a mechanical adding machine and compare its performance with the Cell Broadband Engine, the processor inside the Sony PlayStation 3. Figure 7.4 shows the Burroughs Class 1/Model 9, built between 1905 and 14.[2] From the logical perspective, the machine could perform a single operation—tabulating a running sum—on nine-digit integers. An upgraded version of the

[2]In 1986, Burroughs merged with Sperry-Univac to form Unisys Corporation.

Figure 7.4 The Burroughs Class 1 adding machine, 1905–14. Copyright David G. Hicks.

machine, the Model 12, supported 12-digit integers. From the physical perspective, each input digit is represented by a column of 10 buttons labeled '0' through '9.' To enter a number, the user presses one button in each digit column. When the crank is pulled, the machine adds this encoded integer to a running sum. Inside, the "trick" for emulating addition through physical means is counting the rotation of gears. As shown in Figure 7.4, each digit has a ten-tooth counting gear that rotates when the crank is pulled, and the buttons set stop pins that limit the gears' rotation.

Size: At 19 inches deep, over a foot tall, and weighing 63 pounds, the Burroughs Class 1/Model 9 is considerably larger than a 2006 state-of-the-art "adding machine" such as the Cell Broadband Engine [Hof05], the chip at the heart of the Sony PlayStation 3, shown in Figure 7.5.

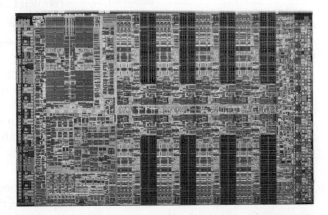

Figure 7.5 The Cell Broadband Engine was developed jointly by IBM, Sony, and Toshiba for games and other high-performance computing applications. It measures approximately 10 mm by 20 mm. It can perform over 256 billion arithmetic operations per second and consumes approximately 80 watts of power [Hof05]. Photo courtesy IBM Corporation.

Speed: A skilled operator could perhaps enter a nine-digit number and pull the crank on the Burroughs adding machine once every 5 seconds. In contrast, the Cell Broadband Engine can perform up to 256 billion arithmetic operations per second.

Energy and Power: In one respect, the Burroughs adding machine is similar to the handpump we analyzed in Chapter 4: it produces a result each time the user pulls on the handle. Thus, we can analyze the energy requirements of the Burroughs machine in a similar way as we did for the pump. Recall from Chapter 4 that work is equal to force times distance. If we assume that a user exerts approximately 1 lb to pull the crank over a distance of approximately 6 inches, then the work performed per addition is

$$\text{work per addition} = 1\,\text{lb} \times 0.5\,\text{ft}$$
$$= 0.5\,\text{ft} - \text{lb}$$
$$= 0.7\,\text{J (joules)}$$

Power is the rate at which energy is consumed over time, and has units of watts (W), where 1 watt is equal to 1 Joule per second. Thus, if the operator can enter a number and pull the crank once every 5 seconds, the power consumption of the Burroughs adding machine is:

$$\text{power} = \frac{\text{energy per addition}}{\text{time per addition}}$$
$$= \frac{0.7\,\text{J}}{5\,\text{s}}$$
$$= .14\,\text{W}$$

By comparison, the power consumption for the Cell chip has been estimated at 80 W. If it can perform 256 billion additions per second, then the energy per addition is

$$\text{energy per addition} = \frac{\text{power} \times \text{time}}{\text{number of additions}}$$
$$= \frac{(80\,\text{W})(1\,\text{s})}{256 \times 10^9}$$
$$= 0.3 \times 10^{-9}\,\text{J}$$

which is over 2 billion times lower than the energy per addition for the Burroughs adding machine. The Burroughs machine isn't likely to get very warm during operation at maximum speed, as long as it's kept well oiled. The Cell Broadband Engine, although low power for a chip of its complexity, can get very hot. As illustrated in Figure 7.6, the Sony PlayStation 3 provides sophisticated cooling technology in order to keep the Cell chip from overheating.

Reliability: Because it has moving parts, the Burroughs adding machine is subject to wear. If gears became stripped or if springs broke, it would not produce a correct answer. Generally speaking, the greater the number of parts and the finer their tolerances, the more difficult it is to build a machine of any type that will function reliably. The Cell chip has 234 million transistors, each as small as 90 nanometers long.

Figure 7.6 Dissection of a Sony PlayStation 3, courtesy of Kristopher Kubicki and Marcus Yam at *DailyTech* [KY06]. The outer casing removed (top left). A 160 mm fan provides active cooling from below (top right). A massive heat sink attached to the fan has heat pipes that attach directly to the two most powerful chips on the motherboard, the Cell Broadband Engine, and an NVIDIA RSX "Reality Synthesizer" (bottom left). The packaged Cell chip on the motherboard, with the NVIDIA RSX to its left (bottom right).

7.2.2 History and Background

From Counting to Computing The path between the Burroughs Class 1/Model 9 adding machine and the IBM/Sony/Toshiba Cell Broadband Engine covers as great an advance in mathematical knowledge as it does in implementation technology. When the Burroughs Class 1 was built at the turn of the 20th century, the engineering environment for computing machines hadn't changed much since the 17th century. Blaise Pascal built an adding/subtracting machine using the principles of interlocking gears in 1642, and Gottfried Leibniz suggested a way of building a machine that could also perform multiplication and division in 1694. Although it took until the 1800s for manufacturing techniques to advance to the point that such machines could be mass-produced, it wasn't just implementation technology that held machines such as the Burroughs Class 1 back. In 1900, developers of computing machines still focused on the mathematics of *counting*, and it would take several decades for innovations from a new generation of mathematicians to begin to change this.

In 1854, the British mathematician George Boole published a monograph that—once engineers realized its implications—would have tremendous impact on the design of computing machines. Entitled *An investigation into the Laws of Thought, on Which are founded the Mathematical Theories of Logic and Probabilities*, Boole's work for the first time made the connection between algebra and logic. This effectively introduced a way to determine the conditions under which a combination of statements would be true or false by juggling symbols, in the same way that we juggle symbols to solve for unknowns that represent numbers in "common" algebra. Other mathematicians, including Alfred North Whitehead and Bertrand Russell, Kurt Gödel, Alan Turing, Alonzo Church, and John von Neumann, built the bridge between logic and *programming*, which forms the theoretical basis for the computer as we know it today.

In 1937, more than 80 years after Boole's *Laws of Thought*, a 21-year-old graduate student at MIT wrote what has been called the most important Master's thesis of the twentieth century [Sha37]. In it, Claude Shannon, of Petosky, Michigan, described a way to build a computing machine out of electromechanical switches called *relays*, based on Boole's system of logic. Moreover, Shannon made the connection between Boolean logic and binary arithmetic—arithmetic using just 1s and 0s—that's used almost exclusively in computers today. Claude Shannon wasn't the only person to see the potential in using switches for building computing machines: around the same time Shannon was formulating his ideas, Konrad Zuse in Germany, George Stibitz at Bell Labs in New York, and John Atanasoff at Iowa State College were building prototypes of machines that performed binary arithmetic using switches. Shannon, however, was the first to write up a rigorous approach that made it possible to both use switches to solve complex logic problems, as well as to use mathematics to simplify the design of computing machines. The modern programmable digital computer, with memory and automatic control as well as arithmetic and logic functions, took shape over the course of a series of independent projects in the early 1940s. These include Zuse's Z3, demonstrated in 1941, the ABC computer by Atanasoff and his student Clifford Berry in 1941, the Colossus, developed in 1943 during World War II by British engineer Tommy Flowers and a team of mathematicians to break German codes, the IBM/Harvard Mark I, designed in 1944 by a team led by Howard Aiken, and the ENIAC, designed by John Mauchly and J. Presper Eckert at the University of Pennsylvania in 1944. Each of these used switching circuits, some in combination with gears or other mechanical components, to implement complex logic operations.

Digital Circuits and Reliability The arrangement of switches that Claude Shannon described in his Master's thesis remains the basis for designing logic circuits today, and it's difficult to imagine designing a modern electronic computer with billions of transistors without using this approach. This is not only because of the relationship between switches and the mathematics of computing, but also because the approach is well suited to building computers that will function *reliably*. As Figure 7.7 illustrates, we might encode information using electronic signals in two general ways: the *analog* approach, which lets the signal take on any value over a continuous range, or the digital approach, which restricts the range of values to a discrete number of values, typically only two. In the analog system, the value of the electrical signal directly corresponds with or is *analogous* to the real-world information. For example, a signal value of 2.33 volts might correspond to a temperature of 23.3°C from an analog temperature sensor. In the digital approach, the two possible values of the signal, called "high" and "low" or 1 and 0, form the "alphabet" of symbols or *digits* with which we "spell out" information, be it numbers, text, music, or other kinds of data. A single binary digit—a 1 or a 0—is called a *bit*, a term popularized by Shannon.

When used in logic circuits, switches naturally encode two distinct values: they're either open or closed, conducting electricity or not. From the physical perspective, the main advantage of using digital encoding over analog encoding is that it's *easier* to build a machine that can reliably distinguish between two different states than it

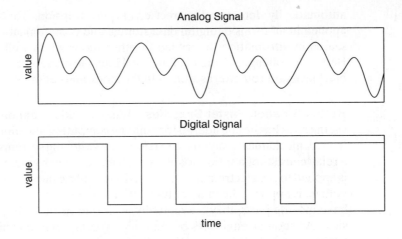

Figure 7.7 Analog signals take on a continuous range of values, while digital signals take on a discrete number of values, typically only two.

is to build one that can accurately determine a value from within a closely spaced range. Consider, for example, the pits and bumps that encode information on a CD or DVD. In a digital scheme, we only need to determine if a bump or pit is present, whereas in an analog scheme we would need to be able to precisely determine their sizes. Moreover, even if there are some variations or inaccuracies in either the disk itself—such as scratches—or in the player—such as the alignment of the laser—there can still be a high likelihood that it'll be readable. As illustrated in Figure 7.8, the pits and bumps on an ideal DVD would have sharp edges and flat surfaces, whereas realistically there will always be some imperfections. Still, as long as the imperfections aren't too severe, it's still possible to tell a pit from a bump.

Engineers call imperfections in a signal—such as the fluctuations in the pits and bumps of a DVD—*noise*. There are many possible sources of noise in an electronic system, including anything that might cause static on a radio or television such as an electrical storm or interference from an appliance. In the design of a computer, it's important that it interprets ones as ones and zeros as zeros even when there's noise: imagine if the digital circuitry in an artificial insulin pump made a wrong decision whenever lightning struck outdoors! In general, as the separation between the high and low values increases, a system stands a better chance of tolerating noise than if they are close together. Still, there's always some chance of data being corrupted, especially during communication. In 1948, while working at Bell Laboratories, Shannon published another revolutionary paper entitled *A Mathematical Theory of Communication* [Sha48], where he used probability and statistics to analyze the likelihood of bits of data being lost during communication, and thus laid the groundwork for schemes of cleverly sending extra bits with a message that can be used to

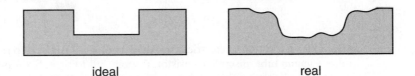

ideal   real

Figure 7.8 Ideal pits and bumps on the surface of a CD or DVD versus realistic ones. One of the greatest advantages of digital encoding over analog encoding of information is that it can still be correctly interpreted even when there are imperfections in the recording of data.

automatically detect and correct errors on reception. This basic idea is routinely applied in all forms of digital data storage and communication today. Shannon's research in information theory led to other successes as well: together with his wife Betty and colleagues from MIT and Bell Labs, he made a fortune in Las Vegas playing blackjack and roulette, as well as in the stock market on Wall Street [Pou05].

Better, Smaller, Faster Switches   When Claude Shannon wrote his master's thesis, the switches he considered for building computing machines were electrically controlled, mechanical *relays* used in telephone switching systems. Pictured in Figure 7.9, a relay consists of a spring-loaded switch and an electromagnet. When a high voltage is applied to the electromagnet, it turns it on, pulling the switch closed, and when a low voltage is applied, the magnet turns off and the switch springs open. Because relays have moving parts, they'll eventually wear out; further, they're large and relatively slow. As soon as engineers began building computing machines in the early 1940s, they sought better, faster, and smaller switches, an evolution shown in Figure 7.9.

Originally used for amplifying signals in radios, *vacuum tubes* followed relays as the next advance in switch technology. Shown in Figure 7.9, a vacuum tube looks like a light bulb, with a filament sealed in a glass tube, except that it has multiple pins on the bottom. When the filament is heated by passing a current through it via some of the pins, it controls the flow of current between other pins. The Atanasoff-Berry Computer from Iowa State, the British Colossus, and the ENIAC from the University of Pennsylvania all used vacuum tubes in their logic design. Like light bulbs, tubes

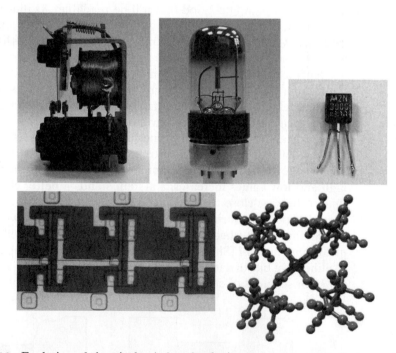

Figure 7.9   Evolution of electrical switch technologies. Top, left to right: electromechanical relay, vacuum tube, discrete transistor. Bottom, left to right: Six transistors wired together in an integrated circuit, fabricated by undergraduates at Notre Dame (image courtesy Greg Snider). A proposed molecular switch, made from 4 ruthenium atoms arranged in a box (image courtesy Olaf Wiest).

burn out and have to be replaced. The ENIAC contained over 17,000 vacuum tubes, and according to a 1989 interview with Presper Eckert, "we had a tube fail about every one or two days, and we could locate the problem within 15 minutes" [Bos06]. Tubes are also as large as relays and burn a lot of power: the ENIAC weighed nearly 30 tons and consumed 150 kW of power, and could perform approximately 5,000 additions on 10-digit numbers per second.

In December 1947, a few months before Shannon published *A Mathematical Theory of Communication* [Sha48], another team of researchers at Bell Labs developed a new kind of switch that would revolutionize electronics. John Bardeen and Walter Brattain, working in the laboratory of William Shockley, succeeded in producing a device made of a piece of germanium with gold contacts attached called a *transistor* [RH97]. Germanium is a type of element called a *semiconductor*, that depending upon its temperature or upon other elements added to it, can act either as a conductor or as an insulator. The first commercial transistors were packaged individually and had three wires: one to effectively control the flow of electricity through the other two, as shown in Figure 7.9. The first computers to use transistors were built in the early 1950s, including the Transistor Computer prototype at the University of Manchester in England in 1953, IBM Model 604 in 1953, and the TRADIC at Bell Labs in January 1954.

During 1958 and 1959, Jack Kilby at Texas Instruments and Robert Noyce at Fairchild Semiconductor Corporation were both working independently on the same idea: how to integrate multiple electronic components on a single semiconductor chip. Kilby got his prototype working first, but Noyce received the first patent. Today, both men are recognized as the inventors of the integrated circuit. Kilby later invented the portable electronic calculator in 1967. Noyce's version of the design, made of silicon rather than germanium, with glass as an insulator and aluminum interconnecting wires, became the prototype for future manufacturing, as pictured in Figure 7.9. Noyce left Fairchild and in 1968 co-founded Intel Corporation with Gordon Moore and Andy Grove.

What comes next? As we discussed in Chapter 1, the number of transistors on an integrated circuit has approximately doubled every two years, as the devices themselves have become both smaller and faster, as Gordon Moore predicted in 1965. People have repeatedly expected this trend to falter, and each time new innovations have kept it on track. Now, however, there are significant signs of the trend slowing, and most experts don't expect for it to continue much beyond 2015–2020. In recent years, much research, time and money have been invested in nanoelectronics, or building computing machines at the molecular scale. Figure 7.9 shows one possible switch proposed by Marya Lieberman, Craig Lent, and colleagues at the University of Notre Dame [Len00] [LIL03]. The switch consists of ruthenium atoms arranged in a box around a central pivoting axis. Depending upon its surrounding charges, the molecule can spin into one orientation or another. Much as the computing machines of a century ago were made from rotating metal gears, the computers of the next century may be made from rotating molecules.

7.3 DIGITAL CIRCUITS FROM THE SYMBOLIC AND LOGICAL PERSPECTIVE

The symbolic or logical perspective on a computer is the mathematician's view of the machine. In this section, we look at how digital computers use switches to do arithmetic. We begin with an introduction to the system of mathematical logic developed by George Boole in the mid-1800s, called *Boolean logic*. Next, we examine the system for implementing logic expressions using switches devised by Claude Shannon. After this, we introduce *binary arithmetic*, which uses only the digits 1 and 0, and describe the connection between this and Boolean logic. Finally, we look at the design of a simple switching circuit that adds two binary numbers.

7.3.1   Boolean Logic

There was a widely held belief in Britain during the mid-1800s that if a person is ill, then the cure should resemble the cause. When her husband took sick after walking to Queen's College in the rain and teaching all day in wet clothes, Mary Everest Boole—the niece of the surveyor for whom the world's tallest mountain is named—put him in bed and dumped buckets of cold water on him, thus quickening the death of the inventor of mathematical logic. When George Boole published *The Laws of Thought* in 1854, he certainly had no idea his system of algebra would provide the underpinnings for machines that one day would beat grandmasters at chess. The elementary algebra we study in high school focuses on how to use symbols to reason about numeric quantities. We use symbols called *variables* to represent numbers, symbols called *operators* to represent ways of combining numbers, and symbols called *relations* to represent ways of comparing them. Together with the symbols is a system of rules, properties or *axioms* that pertain to statements formed from these symbols. For example, the commutative property of addition states that we can change the order of the variables on either side of the '+' symbol and still obtain the same result. Boole's system, called *Boolean algebra*, involves rules for manipulating symbolic statements that represent the logic values "true" or "false" and operations on these values. By convention, we use the symbol "1" to denote "true" and the symbol "0" to denote "false."

Variables and Operators   Just as elementary algebra uses variables to represent a quantity in the real world, Boolean algebra uses variables to represent the truth of statements. For example, in elementary algebra, we might represent the statement that "the voltage of a battery is 1.5 volts" as

$$V = 1.5$$

Similarly, in Boolean algebra, we could represent the statement that "the lights are on" as

$$L = 1$$

or "the lights are off" as

$$L = 0$$

Under elementary algebra, we build expressions for combining quantities using operations such as addition ('+'), multiplication ('×'), and negation ('−'). Under Boolean algebra, we build compound expressions using the operators AND OR and NOT. These operations also have mathematical symbols associated with them—"∧," "∨," and "¬," respectively—but for simplicity, we'll stick with their text equivalents.

The NOT Operation: The NOT operator signifies logical negation. Suppose, for example, we represent the assertion, "somebody is home" with the logic statement

$$H = 1$$

Using the NOT operator, we can express that, "nobody is home" as

$$\text{NOT } H = 1$$

If A represents a logic variable whose value is "true," then the value of NOT A is false, and vice versa. We can represent all the cases of the NOT operation in a kind of table called a *truth table*, as shown below:

A	NOT A
0	1
1	0

The AND Operation: If A and B are both logic variables, then the expression

$$A \text{ AND } B$$

will be true only if both A and B are true. The truth table for the AND operation covers all four combinations of values for A and B:

A	B	A AND B
0	0	0
0	1	0
1	0	0
1	1	1

Using the AND operation, for example, we can write the expression "the lights are on and nobody's home" as

$$(L = 1) \text{ AND } (\text{NOT } H = 1)$$

Generally, logic variables have a "default" value of "true," and so when we write expressions that combine variables, we don't bother to explicitly equate them to "1." Thus we can simplify the above expression as

$$L \text{ AND } (\text{NOT } H)$$

The OR Operation: The expression

$$A \text{ OR } B$$

will be true if either A, or B, or both are true. The truth table for the OR operation is:

A	B	A OR B
0	0	0
0	1	1
1	0	1
1	1	1

Thus, we can write the expression "either the lights are on or nobody is home" as

$$L \text{ OR } (\text{NOT } H)$$

Properties   As with elementary algebra, Boolean algebra has a number of laws or properties that apply to operations. To familiarize ourselves with the concept, let's first review some of the basic properties under elementary algebra.

commutative property of addition:	$A + B = B + A$
commutative property of multiplication:	$A \times B = B \times A$
associative property of addition:	$(A + B) + C = A + (B + C)$
associative property of multiplication:	$(A \times B) \times C = A \times (B \times C)$
additive inverse:	$A + (-A) = 0$
distributive property:	$A \times (B + C) = (A \times B) + (A \times C)$

Under Boolean algebra, the AND and OR operations are both commutative and associative. That is,

commutative property of AND:	$A \text{ AND } B = B \text{ AND } A$
commutative property of OR:	$A \text{ OR } B = B \text{ OR } A$
associative property of AND:	$(A \text{ AND } B) \text{ AND } C = A \text{ AND } (B \text{ AND } C)$
associative property of OR:	$(A \text{ OR } B) \text{ OR } C = A \text{ OR } (B \text{ OR } C)$

The AND and OR operations also have inverse properties:

inverse property of AND:	$A \text{ AND } (\text{NOT } A) = 0$
inverse property of OR:	$A \text{ OR } (\text{NOT } A) = 1$

To illustrate, consider the example where L represents the statement "the lights are on." $L \text{ AND } (\text{NOT } L)$ would thus represent the statement "the lights are on and the lights are not on," which clearly can't be true. On the other hand, $L \text{ OR } (\text{NOT } L)$, which represents the statement "either the lights are on or the lights are not on," must always be true, since it covers all possible situations.

A common way to demonstrate that a given property holds is to construct the truth table. For example, below is a truth table showing all cases for the expression $A \text{ OR } (\text{NOT } A)$:

A	$\text{NOT } A$	$A \text{ OR } (\text{NOT } A)$
0	1	1
1	0	1

There are only two possible cases: where $A = 0$ and where $A = 1$. In each of these cases, the table shows the value of $A \text{ OR } (\text{NOT } A)$ to be a 1, so the statement must *always* be true. We'll use this same technique to demonstrate another property under Boolean algebra, the distributive property of AND over *OR*:

distributive property: $A \text{ AND } (B \text{ OR } C) = (A \text{ AND } B) \text{ OR } (A \text{ AND } C)$

Since the property involves three variable, A, B, and C, the truth table will have 2^3 or 8 cases. We build the table with three parts: the first three columns enumerate the combinations of values for the variables, the next two columns calculate the value of A AND $(B$ OR $C)$ for each case, and the final three columns calculate the value of $(A$ AND $B)$ OR $(A$ AND $C)$.

A	B	C	B OR C	A AND $(B$ OR $C)$	A AND B	A AND C	$(A$ AND $B)$ OR $(A$ AND $C)$
0	0	0	0	0	0	0	0
0	0	1	1	0	0	0	0
0	1	0	1	0	0	0	0
0	1	1	1	0	0	0	0
1	0	0	0	0	0	0	0
1	0	1	1	1	0	1	1
1	1	0	1	1	1	0	1
1	1	1	1	1	1	1	1

Observe that the values in the fifth and eighth columns are the same for each case, showing that the statements are equivalent.

Example: Voicemail   Suppose that as part of the design of a cell phone network, we wanted to express the logic for when to route a call to a person P's voicemail. One possible rule might be that a call should go to P's voicemail if the call is to P's number and there's no answer. We can represent this rule as:

$$V = P \text{ AND } N$$

where

- V: condition (true/false) to route call to P's voicemail
- P: condition (true/false) that call is to P's phone number
- N: condition (true/false) that there is no answer

Further, there might be several conditions that would lead to the "no answer," $N =$ true condition, such as if there were 4 rings without a pickup, or if the phone is offline. We can express this as an OR statement:

$$N = R \text{ OR } X$$

where

- R: condition (true/false) that there were 4 rings without pickup
- X: condition (true/false) that the phone is off-line

We can combine these statements into a single statement as:

$$V = P \text{ AND } (R \text{ OR } X)$$

Using the distributive property, we could rewrite the conditions for forwarding a call to P's voicemail as

$$V = (P \text{ AND } R) \text{ OR } (P \text{ AND } X)$$

In other words, this states that the call should be forwarded to P's voicemail if the call is to P's number and there are 4 rings, or, if the call is to P's number and the phone is offline. In this example, we didn't really simplify the condition for forwarding a call to voicemail, but for more complex examples, we would typically use the rules of algebra to do just that.

7.3.2 Building Computing Machines Out of Switches

Claude Shannon's Master's thesis demonstrated that Boolean logic expressions could be implemented as electronic machines built from switches [Sha37]. For all their power, the basic ideas behind Shannon's arrangement of switches to implement Boolean operations are really very simple. First, let's consider the computing machine shown in Figure 7.10. The machine has a single input push-button that represents a variable A, where if the button is depressed then A = true, and if not, then A = false. The machine also has a single output lamp that represents a variable Y, where if the lamp is on, then Y = true, and if not, then Y = false. The purpose of the machine is to implement the—trivial—Boolean statement

$$Y = A,$$

or in other words, Y indicates true (lamp on) when A is true (button pressed), and Y indicates false (lamp off) when A is false (button not pressed).

How might we implement such a machine using switches? Figure 7.11 shows a solution. When the button is pressed, it closes a switch, which causes "electricity" to flow through the circuit and turn on the lamp. Next, suppose we want to build a machine to implement the Boolean statement

$$Y = \text{NOT } A,$$

such that Y is false (lamp off) when A is true (button pressed) and vice versa. Figure 7.12 shows one possible implementation. Here, the switch is reversed, so that pushing the button opens the switch, while not pushing it leaves the switch closed.

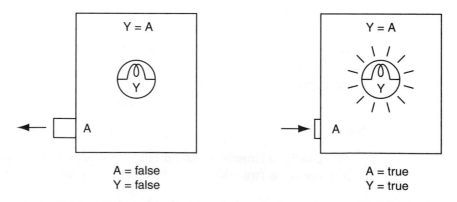

Figure 7.10 A computing machine that implements the logic statement $Y = A$. The input variable A is represented by a push-button that indicates a value of "true" when pressed. The output variable Y is represented by a lamp that indicates "true" when lit.

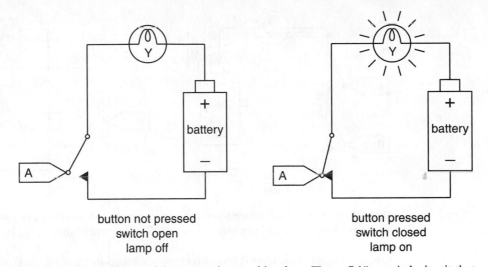

Figure 7.11   The "insides" of the computing machine from Figure 7.10: a switch circuit that implements the logic statement $Y = A$.

Now suppose we wanted to build machines to implement the Boolean AND and OR operations:

$$Y = A \text{ AND } B$$
$$Y = A \text{ OR } B$$

Both these machines would have two input buttons, one for variable A and the other for variable B, in addition to a single output lamp for Y. The switch circuit on the left in Figure 7.13 shows an implementation of an AND operation. In this circuit, two switches—one controlled by variable A and the other controlled by variable B—are wired together in series. In order for the lamp to turn on, *both* switches must be pressed. The circuit on the right in Figure 7.13 shows an implementation of an

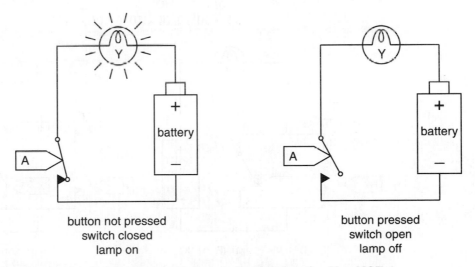

Figure 7.12   A switch circuit that implements the logic function $Y = \text{NOT } A$.

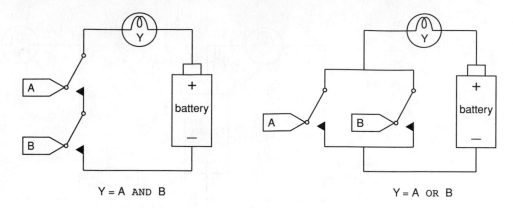

$$Y = A \text{ AND } B \qquad\qquad Y = A \text{ OR } B$$

Figure 7.13 Two switches wired in series can implement an AND operation (left), while two switches wired in parallel can implement an OR operation (right).

OR operation. This circuit has two switches wired in parallel, so that pushing either button A, or button B, or both will complete the circuit and turn on the lamp.

Combinations of switches can be wired together in series and parallel connections to implement more complex Boolean statements. As a final example, Figure 7.14 shows a switch circuit that implements the two equivalent rules for forwarding a call to voicemail,

$$V = P \text{ AND } (R \text{ OR } X)$$
$$V = (P \text{ AND } R) \text{ OR } (P \text{ AND } X)$$

7.3.3 Binary Representation of Numbers

Much of how we characterize the world—especially in engineering—is described in quantitative terms, and thus computing machines need symbols for *numbers*. The representation of numbers we learn in grade school—and continue to use throughout our lives—is based on a decimal, or base-10, formulation. With this formulation, the number 142, for instance, is interpreted as

$$(1 \times 10^2) + (4 \times 10^1) + (2 \times 10^0)$$

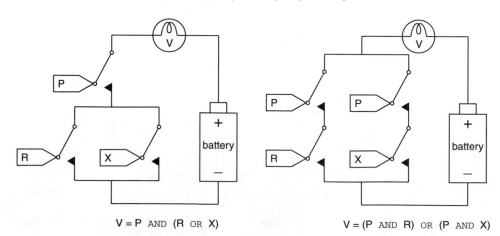

$$V = P \text{ AND } (R \text{ OR } X) \qquad\qquad V = (P \text{ AND } R) \text{ OR } (P \text{ AND } X)$$

Figure 7.14 A combination of switches in series and in parallel implements the logic statement $V = P \text{ AND } (R \text{ OR } X)$

Here, "10" refers to the number ten, that is, the number of fingers and thumbs most of us have on our two hands. So to use this representation, we need symbols called "digits" to represent ten different integers, namely, the numbers 0 through 9. These digits indicate the coefficients used as factors of the various powers of 10, and the location of each digit indicates which power of 10 that digit should be multiplied by. More generally, the decimal number $a_n a_{n-1} \ldots a_1 a_0$ represents the number

$$(a_n \times 10^n) + (a_{n-1} \times 10^{n-1}) + \cdots + (a_1 \times 10^1) + (a_0 \times 10^0),$$

where each a_i is a digit from 0 to 9.

The base-10 representation of numbers, however, is not the only possible representation. If we had evolved with only six fingers rather than ten, we might well have developed a base-6 representation for numbers. To employ such representation, we would need symbols for six different integers—namely, the numbers 0 through 5— and the position of each digit would indicate what power of 6 that coefficient should be multiplied by. For instance, the base-6 number $(51)_6$ represents[3] the decimal number $(31)_{10}$, because

$$(5 \times 6^1) + (1 \times 6^0) = (3 \times 10^1) + (1 \times 10^0)$$

The principle behind representing numbers in the binary system is identical to that of the decimal system: each bit in a number is a coefficient to be multiplied by a power of 2, and the position of the bit indicates that power of two. For instance:

$$(110011)_2 = (51)_{10}$$

because

$$(1 \times 2^5) + (1 \times 2^4) + (0 \times 2^3) + (0 \times 2^2) + (1 \times 2^1) + (1 \times 2^0)$$
$$= 32 + 16 + 0 + 0 + 2 + 1$$
$$= (51)_{10}.$$

All the operations one can carry out in a decimal system—addition, subtraction, multiplication, and division—can be executed in a completely analogous fashion using base-two arithmetic. For instance, in base-10 addition we "carry" a power of ten when a column of numbers sums to 10 or more; in an analogous fashion, in base-2 addition a power of two is "carried" whenever the column sums to two or more. This is shown below for decimal and binary addition of the same two quantities.

decimal (base 10) addition	binary (base 2) addition
1	1111
27	11011
+ 14	+ 1110
41	101001

Binary-to-Decimal Conversion   Once you understand base-2 representation of numbers, it's easy to convert binary numbers to their decimal equivalent. Simply writing out the "meaning" of a binary string and then carrying out the operations using the familiar decimal notation quickly yields the desired results; all that's required for quick conversion is that you be able to come up with the required powers of two. For

---

[3]When there is potential confusion as to which base a number is written in, we will use parentheses and a subscript — e.g., the number of months in the year is $(12)_{10} = (20)_6$.

instance, $(1110110)_2$ represents the decimal number $64 + 32 + 16 + 4 + 2 = (118)_{10}$. The conversion is "automatic" provided you know the powers of two; and if you don't, you can always use a calculator. (For example you may not have memorized what 2^{13} is, but a calculator will quickly reveal that it is 8192 in the decimal system.)

Decimal-to-Binary Conversion   This is a bit trickier than the conversion in the opposite direction because we didn't grow up using binary representation. There is, however, an algorithm we can use to carry out this process, which we describe below. Let x be an integer in decimal format. Our job is to find the binary representation of x, or in other words, to find the bits $(b_n b_{n-1} \ldots b_1 b_0)$ such that

$$x = (b_n \times 2^n) + (b_{n-1} \times 2^{n-1}) + \cdots + (b_1 \times 2^1) + b_0.$$

Before presenting the algorithm, there are a few key observations that explain why it works:

- $b_0 = 0$ if x is even and $b_0 = 1$ if x is odd. Put another way, b_0 is the remainder you get when you divide x by 2.
- Once you find b_0 you can find b_1 by applying the above step to $(x - b_0)/2$.
- You can continue similarly until you're done.

Given these observations, here is the algorithm

1. Given an integer x, set $i = 0$ and $x_0 = x$.
2. Divide x_i by two. Call the quotient x_{i+1} and the remainder b_i. Thus, $x_i = 2x_{i+1} + b_i$.
3. If $x_{i+1} = 0$ stop.
4. Set $i = i + 1$ and go to step 2.

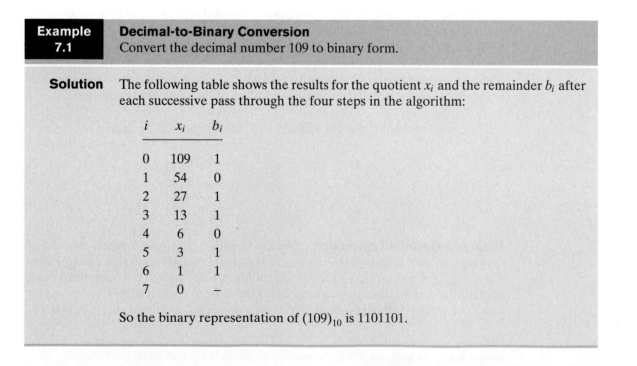

| Example 7.1 | **Decimal-to-Binary Conversion**
Convert the decimal number 109 to binary form. |

Solution   The following table shows the results for the quotient x_i and the remainder b_i after each successive pass through the four steps in the algorithm:

i	x_i	b_i
0	109	1
1	54	0
2	27	1
3	13	1
4	6	0
5	3	1
6	1	1
7	0	–

So the binary representation of $(109)_{10}$ is 1101101.

7.3.4 Adding Numbers with Switches

The "trick" to performing addition in digital computers today is fundamentally different from the "trick" used in the Burroughs adding machine: rather than counting to add numbers, they use logic to determine the result. In this section, we'll first show how to use Boolean logic to add two single-bit numbers, called a *half adder*, and design a switching circuit to implement the logic. Then, we'll generalize this result to design a circuit for adding numbers of any length, called a *ripple-carry adder*.

Half Adder   Suppose A and B are each single-bit numbers, that is, just 0 or 1. When we add A and B together, we'll need up to two bits to hold the result, since 2 in base-10 is 10 in base-2. We'll call the least significant bit of the result the "sum" bit and the most significant bit the "carry" bit, as illustrated below:

	case 1	case 2	case 3	case 4
$\leftarrow A$	0	0	1	1
$+ \quad \leftarrow B$	+ 0	+ 1	+ 0	+ 1
	00	01	01	10

carry   sum

Each of the four different cases produces a value for the sum bit and a value for the carry bit. We can gather these values into truth tables for each:

A	B	carry
0	0	0
0	1	0
1	0	0
1	1	1

A	B	sum
0	0	0
0	1	1
1	0	1
1	1	0

The next step is to find logic expressions for the sum and carry bits from the truth tables. The carry bit is a 1 only when both A and B are 1s. Thus,

$$\text{carry} = A \text{ AND } B$$

The logic for the sum bit is slightly more complicated. The sum bit will be a 1 when either A or B, but not both, is a 1. This relationship is called an *exclusive or*, XOR, which we can express in terms of AND, OR, and NOT operators as follows:

$$\text{sum} = A \text{ XOR } B$$
$$= [(\text{NOT} A) \text{ AND } B)] \text{ OR } [A \text{ AND } (\text{NOT } B)]$$

The final step is to construct switching circuits that implement the sum and carry logic. Figure 7.15 shows the result.

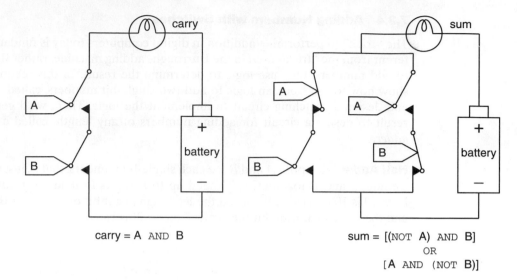

carry = A AND B

sum = [(NOT A) AND B]
OR
[A AND (NOT B)]

Figure 7.15 Switch implementation of carry and sum logic circuits for a half-adder.

Ripple-Carry Adder By itself, a half adder isn't very useful. What we really want is a circuit that'll let us add two values that are any number of bits in length. Figure 7.16 illustrates the *interface* to a 4-bit adder, which shows its inputs and outputs, but not the details of its implementation. The A and B inputs to the adder are each 4 bits wide, where A_0 and B_0 are the least significant bits and A_3 and B_3 are the most significant bits. The output sum S is five bits wide, since there may be a final carry in the last column.

The most common implementation of a multibit adder is the *ripple-carry adder*, shown in Figure 7.17. The ripple-carry adder is a modular design that works much the same way that we add numbers column by column, from least significant to most significant, carrying a 1 to the next column as needed. In order to build a ripple-carry adder, we need a module called a *full adder* for each column that accepts 3 inputs—the A and B bits in that column, and the carry passed from the previous column—as well as sum and carry outputs. Since the least significant column doesn't have a carry into it, we can use the half-adder we designed earlier, which only has 2 inputs. The ripple-carry adder gets its name from the way that the carry bits "ripple" from module to module.

Now that we have the basic structure of a ripple-carry adder, the next step is to figure out the details of the full adder. We begin by constructing a truth table

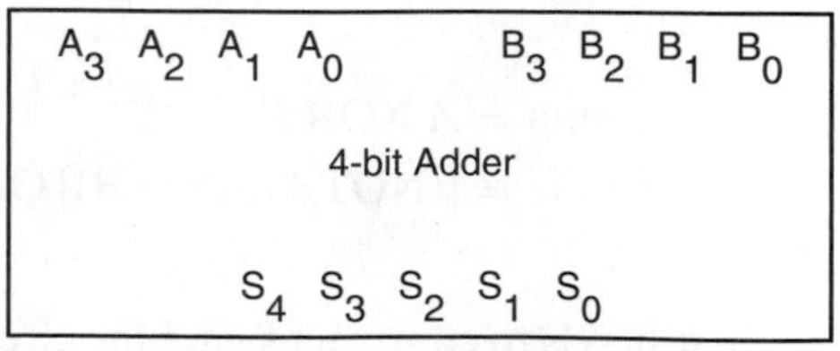

$$A_3 \quad A_2 \quad A_1 \quad A_0 \qquad B_3 \quad B_2 \quad B_1 \quad B_0$$

4-bit Adder

$$S_4 \quad S_3 \quad S_2 \quad S_1 \quad S_0$$

Figure 7.16 A 4-bit adder.

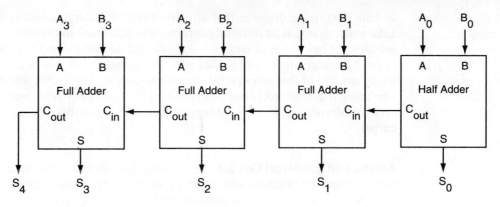

Figure 7.17 Ripple-carry adder.

that expresses the values of the sum S and carry output C_{out} for each of the eight combinations of A, B, and the carry input C_{in}:

A	B	C_{in}	C_{out}	S
0	0	0	0	0
0	0	1	0	1
0	1	0	0	1
0	1	1	1	0
1	0	0	0	1
1	0	1	1	0
1	1	0	1	0
1	1	1	1	1

From the truth table, we can determine Boolean logic expressions for both C_{out} and S:

$$C_{out} = (A \text{ AND } B) \text{ OR } (A \text{ AND } C_{in}) \text{ OR } (B \text{ AND } C_{in})$$

$$\begin{aligned} S &= A \text{ XOR } B \text{ XOR } C \\ &= [(\text{NOT } A) \text{ AND } (\text{NOT } B) \text{ AND } C_{in}] \\ &\quad \text{OR } [(\text{NOT } A) \text{ AND } B \text{ AND } (\text{NOT } C_{in})] \\ &\quad \text{OR } [A \text{ AND } (\text{NOT } B) \text{ AND } (\text{NOT } C_{in})] \\ &\quad \text{OR } [A \text{ AND } B \text{ AND } C_{in}] \end{aligned}$$

We leave it as an exercise to verify that these expressions are correct. Given these expressions, the final step is to design a switching circuit that implements them. We leave this as an exercise as well.

7.4 DIGITAL CIRCUITS FROM THE ELECTRONICS PERSPECTIVE

7.4.1 Electricity

The ancient Greeks observed that if they rubbed petrified resin or amber—called *elektron* in Greek—with fur, then it would attract or repel lightweight objects such as hair. Today, we describe that attraction or repulsion in terms of *electric charge*. Like mass, charge is an intrinsic property of nature; we can't really say what it *is*, but we identify two kinds of charge—positive and negative—and say that bodies with opposite charge attract and bodies with like charge repel, with a force proportional to the amount of charge and the distance between the bodies. We also say that charge, like mass, is conserved; it can be neither created nor destroyed, only moved around and transferred, such as by rubbing amber with fur or by scuffing our feet on the carpet.

Atoms and Electrical Charge   According to modern theory, all matter in the universe is made up of atoms, and the atoms themselves contain charged particles: a nucleus with positively charged protons (and neutrally charged neutrons), surrounded by orbiting, negatively charged electrons. The unit of charge is the *coulomb* (C), and one proton has a charge of 1.6×10^{-19} C, while one electron has a charge of -1.6×10^{-19} C. When atoms join together to form molecules, they share their outer electrons. In some materials, such as glass, the outer electrons are tightly bound between a pair of neighboring atoms and can't move freely. In metals, the outer electrons are loosely bound, and surround all the atoms in a fluid sea. Because electrons move easily through metals, they are good *conductors* of electricity. Materials like glass, on the other hand, do not conduct electricity and are called *insulators*.

In 1746 at the University of Leyden in Holland, Pieter van Musschenbroek found that he could store electricity—he didn't understand it as charge the way we do now—in an early form of capacitor called a Leyden jar. When Ben Franklin flew his kite in a thunderstorm in 1752 and stored the jolt in a Leyden jar, he discovered that lightning was the same electricity others had produced in the lab using friction. Franklin envisioned electricity as a kind of fluid that has a preferred level in any material. If a material charged to a "positive" level was brought into contact with a material charged to a "negative" level, electricity would flow between them until both were charged to a neutral level. While our understanding today is different, Franklin's terminology has stuck, and it's still useful to use the analogy of a fluid in explaining the concepts of electrical circuits, current, and voltage.

Figure 7.18 shows two simple systems: an hydraulic or water-powered system on the left and an electrical system on the right. In the hydraulic system, water from a tank flows through a pipe to turn a paddle wheel. In the electrical system, charge from a battery flows through a wire to light up a lamp. In the analogy, positive charge is analogous to the mass of water. When making analogies, it's important to recognize limitations: after all, charge and mass are different physical quantities and there's no such notion as "positive" and "negative" mass. Nevertheless, we identify enough similarities in the basic concepts that our tangible experience with flowing water helps us understand the more abstract ideas of electricity.

Energy and Voltage   Both systems use stored energy to perform a useful task, namely, turning a paddle wheel in the hydraulic system or lighting a bulb in the electrical one. As we discussed in Chapter 4, energy—like mass and charge—is a conserved physical quantity that can neither be created nor destroyed. We can add energy to a system by attaching a battery to a circuit; likewise, energy can leave a system through radiation as heat or light. However, if we carefully account for the energy that enters, leaves, or stays within a system before and after some change, the net total will always be the same.

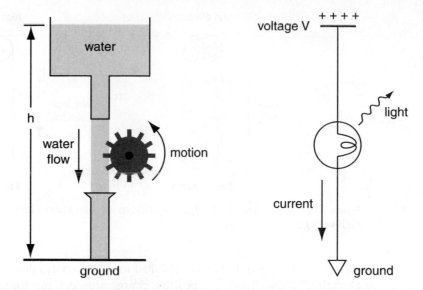

Figure 7.18 Two circuits: water flowing through a pipe turns a wheel, electricity flowing through a wire lights up a lamp.

In the hydraulic system, energy is initially stored as *gravitational energy* in the form of water in the tank. Gravitational potential energy is the energy of a weight W held at some height above the ground. If the mass of water in the tank is m and the height of the tank is h, then the gravitational potential energy E_g is

$$E_g = Wh$$
$$= (mg)h$$

where g is the acceleration due to gravity.[4]

In the electrical system, energy is stored as *electrical potential energy* in the form of charge stored at a *voltage*, V. If Q is an amount of charge, then the electrical potential energy E_e of charge Q at voltage V is

$$E_e = QV$$

It's difficult to find an exact mechanical analog to voltage, and the assortment of alternative names for voltage adds to the confusion.[5] We'll use the analogy that voltage is similar to *elevation*, which is directly related to the height of a water tank. The greater the height of the tank, the greater the flow at the outlet of the pipe.

Both elevation and voltage are relative quantities, and elevation and voltage measurements are always made with respect to some reference point. Elevation is often measured with respect to sea level. In an electrical circuit, voltages are typically made with respect to *ground*. In most electrical systems, ground is defined with respect to the voltage at the "ground;" imagine simply attaching a wire to the ground point in your circuit and burying the other end in the earth.

---

[4]Note that we're ignoring the variation of height of the water in the tank.

[5]Voltage has been called "electromotive force," or EMF, but it is not a force. Voltage is also sometimes called "electrical potential," which is not the same as electrical potential energy.

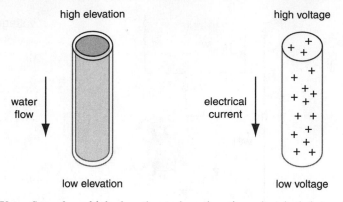

Figure 7.19 Water flows from high elevation to low elevation, electrical charge flows from high voltage to low voltage.

Current As shown in Figure 7.19, just as water always flows from an area of high elevation to low elevations, positive charge always flows from high voltage to low voltage (and negative charge flows the other way). In a pipe, the flow of water can be specified by the *mass flow rate*, which has units of kg/s as well as a direction, which in this case is from top to bottom. In a wire, the flow of charge is called *electrical current* or simply current. Current is measured in units of coulombs/s or *amperes* (A), or just "amps" for short. Current also has a direction, which is also from top to bottom in this example. Note that we're using the common convention that the direction of current is the direction in which *positive* charge flows, which is opposite to the direction in which negative charge or electrons flow.

Power, Resistance, and Heat Loss As we start a flow of water from a tank or a flow of charge from a battery, we convert stored potential energy into kinetic energy. When the flowing water hits the paddle wheel, some of the energy is converted to rotating the wheel (which could in turn drive other machinery). Similarly, when the electrical current flows through the bulb, it heats up the filament, causing it to radiate light. If we let either system run for a while, we'll find that we'll have drained some amount of energy from the supply, in the form of water from the tank or charge from the battery. The rate at which we drain energy from the supply is called *power consumption*. We can evaluate power consumption by measuring the amount of energy drained, and dividing this by the time interval:

$$\text{power consumption} = \frac{\text{energy drained from supply}}{\text{time interval}}$$

In the electrical system, energy is equal to charge times voltage. Thus power consumption would be:

$$\text{electrical power consumption} = \frac{(Q \text{ in coulombs})(V \text{ in volts})}{(\text{time in seconds})}$$

and has units of watts (W), where 1 watt equals 1 Joule per second. Since the flow of charge over time, however, is current, electrical power is equal to current times voltage:

$$\text{electrical power consumption} = (I \text{ in amperes})(V \text{ in volts}) \tag{7.1}$$

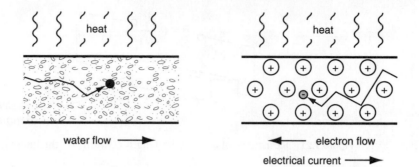

Figure 7.20 Resistance. In a pipe filled with gravel, water molecules collide with the gravel, producing friction that resists the flow of water and gives off heat. In a wire, negatively charged electrons collide with positively charged metal ions that resist the electrical current and also gives off heat. Note that the direction of electrical current is defined as the direction opposite to the flow of electrons.

Part of the energy drained from the power supply goes into "useful" work, either rotations of the paddle wheel or visible light from the bulb. The rest of the energy, however, is wasted. Friction is created in the hydraulic system when the wheel turns about its axle. This friction produces heat, which is a form of lost energy. Similarly, not all of the energy consumed by the bulb results in visible light; in fact, for a typical incandescent light bulb, only about 10 percent of the energy turns into visible light and the rest is lost as heat.

The friction in the paddle wheel and the heat of the lamp aren't the only sources of wasted energy in these two systems. In any real hydraulic system, the pipes themselves have friction that resists the flow of water. The total friction depends on the diameter of the pipe and its length. The friction would be even greater if something were blocking the pipe, such as gravel inside it. Similar to pipes, real wires also have a kind of "friction," called *resistance*, that resists the flow of charge. The unit of electrical resistance is the ohm (Ω). As shown in Figure 7.20, similar to the way that friction in a pipe filled with gravel results from collisions between water molecules and the gravel, resistance in a wire results from collisions between electrons and the positively charged metal ions—what's left behind when the electrons are separated from metal atoms.

Just as the friction of a pipe depends on its dimensions, the resistance of a wire depends on its dimensions as well as the material of which it is made. In general, the longer the wire, the greater its resistance, and the greater the cross-sectional area of a wire, the smaller the resistance. Therefore, thick, short wires have low resistance, and long, thin wires have high resistance, as is the case with pipes. Under most circumstances, the resistance of a wire is proportional to its length and inversely proportional to its cross-sectional area. If a wire has length L, width W, and thickness t, the resistance R along its length is

$$R = \rho \frac{L}{Wt}, \tag{7.2}$$

where ρ is a constant of proportionality called the *resistivity*, which depends on the material. Figure 7.21 illustrates this relationship, and Example 7.2 shows sample calculations of the resistance of a wire in an integrated circuit.

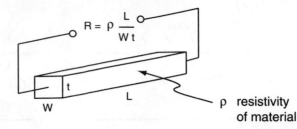

Figure 7.21 Resistance of a wire with length L, width W, thickness t, and resistivity ρ.

Example 7.2	**Resistance of a Wire in an Integrated Circuit** A metal wire 15 mm long runs the full length of an integrated circuit chip. The wire is 150 nm wide and 200 nm thick. What would be the resistance of the wire if it were made of aluminum? Of copper? The resistivity of aluminum is 2.8×10^{-8} Ω-m and the resistivity of copper is 1.7×10^{-8} Ω-m.

Solution   For this problem

$$L = 15 \times 10^{-3} \quad \text{m } W = 150 \times 10^{-9} \quad \text{m } t = 200 \times 10^{-9} \text{ m}$$

$$\rho_{\text{Al}} = 2.8 \times 10^{-8} \, \Omega - \text{m } \rho_{\text{Cu}} = 1.7 \times 10^{-8} \, \Omega - \text{m}$$

If the wire is made of aluminum, then the resistance is

$$R_{\text{Al}} = \rho_{\text{Al}} \frac{L}{Wt}$$

$$= 2.8 \times 10^{-8} \frac{15 \times 10^{-3}}{(150 \times 10^{-9})(200 \times 10^{-9})}$$

$$= 14 \, \text{k}\Omega$$

If the wire is made of copper, then the resistance is

$$R_{\text{Cu}} = \rho_{\text{Cu}} \frac{L}{Wt}$$

$$= 1.7 \times 10^{-8} \frac{15 \times 10^{-3}}{(150 \times 10^{-9})(200 \times 10^{-9})}$$

$$= 8.5 \, \text{k}\Omega$$

Note that these are large values for wire resistance in an integrated circuit. A typical integrated circuit would be laid out so that wires this long are avoided.

Voltage Drops and Ohm's Law   As water flows through a pipe, any friction or blockage causes a drop in pressure. To see this, think of what happens when you open a water faucet a small amount; even though there's water at high pressure in the pipes behind the faucet, the flow coming out of it is at low pressure. Similarly, when electrical current flows through a resistive material, there's a drop in voltage. The relationship among voltage, current, and resistance is called Ohm's Law, after German physicist Georg Ohm, who published this relationship in 1827. Ohm's Law

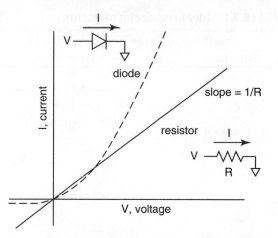

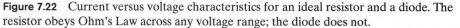

Figure 7.22 Current versus voltage characteristics for an ideal resistor and a diode. The resistor obeys Ohm's Law across any voltage range; the diode does not.

states that the voltage drop across the resistance is equal to the product of the current and the resistance, or

$$\text{Ohm's Law: } V = IR$$

Ohm's Law is an example of a "narrow law." It *doesn't* state that the current through *any* electrical device is proportional to the voltage across it; it only says that this is true for a device that's purely resistive. In fact, another way of looking at Ohm's Law is as the definition of a circuit element called a resistor: a device in which the current through it is proportional to the voltage across it, with a constant of proportionality R. Laws such as Ohm's Law that define the behavior of a particular device are known as *constitutive relationships*.

It's important to recognize that there's a wide variety of commonly used electrical devices whose behaviors do *not* follow Ohm's Law. One such device is a diode, whose behavior is like a valve that lets current flow through it in one direction but not the other. Figure 7.22 shows a plot of the current versus voltage characteristics for a resistor and a diode, and also shows the symbols for those two devices. The current versus voltage or IV characteristic for the resistor is a straight line with slope $1/R$. The IV characteristic for the diode is far from straight: when the voltage is negative, the current is almost zero, and when the voltage is positive, the current increases exponentially—not linearly—with voltage.

7.4.2 Electronic Devices

Thus far, we've seen a variety of electrical devices, including the wire, battery, lamp, resistor, and diode. Each of these devices serves a different purpose, and has different behaviors when a voltage is applied across it or a current is passed through it. There are many different kinds of electrical devices, but in our analysis, we'll only need to consider three *idealized* components: a resistor, a voltage source, and a switch. We call these idealized components because the basic "laws" that describe their behavior are simple theoretical models. Further, we can use these devices either alone or in combination to represent other devices. For example, from the perspective of calculating currents and voltages in a circuit, we can model a light bulb as a resistor.

TABLE 7.1 Idealized circuit elements.

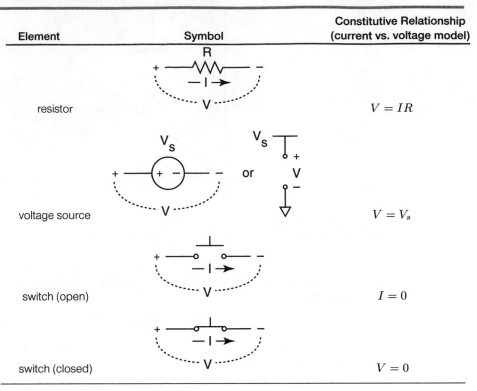

Element	Symbol	Constitutive Relationship (current vs. voltage model)
resistor		$V = IR$
voltage source		$V = V_s$
switch (open)		$I = 0$
switch (closed)		$V = 0$

Similarly, we'll model the type of switch used in digital integrated circuits—called a transistor—as an ideal switch connected to an ideal resistor.

Resistor, Voltage Source, and Switch   Table 7.1 shows the schematic symbols and the constitutive relationships (current versus voltage models) for each of the idealized circuit elements. We've already encountered the resistor; its constitutive relationship is Ohm's Law. A voltage source is like an ideal battery, and its constitutive relationship is that the voltage across it is a constant. A switch can be either open or closed. If it's open, then no current flows through it and if it's closed, then it behaves as an ideal wire with no resistance, so the voltage drop across it is zero.

Switches for Integrated Circuits: The MOSFET   The first digital logic circuits built in the late 1930s and early 1940s used electromechanical relays from the telephone industry. As shown in Figure 7.23, a relay consists of a spring-loaded switch and an electromagnet, and has three connecting points or terminals, which we'll call the "source," "drain," and "gate." When a positive voltage is applied to the gate, it turns the electromagnet on, which pulls the switch closed. This enables current to flow from the "source" to the "drain." When a low voltage is applied to the gate, the electromagnet turns off and the spring pulls the switch open.

Within 20 years, vacuum tubes replaced relays; in turn, transistors replaced vacuum tubes. Since Bardeen and Brattain's 1947 prototype made from germanium and gold, many different forms of transistors have been devised. Today, the vast majority of digital integrated circuits use a device called a *MOSFET*

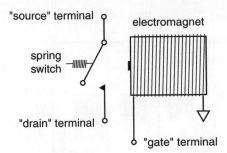

"source" terminal

electromagnet

spring
switch

"drain" terminal

"gate" terminal

Figure 7.23 A electromechanical relay.

(metal-oxide-semiconductor, field-effect transistor), fabricated in silicon. Figure 7.24 shows the symbol for a MOSFET. Like an electromechanical relay, it has three terminals—a gate, source, and drain—where the gate controls the flow of current between the source and drain.

Figure 7.25 illustrates the layout of a MOSFET. As the cross section shows, the middle part of a MOSFET is a metal-oxide-semiconductor sandwich. It consists of polysilicon—a form of silicon "doped" to conduct like a metal—on top, silicon dioxide—common glass—in the middle, and silicon on the bottom. On either side of the MOS sandwich, another material such as phosphorus is implanted into the silicon. Like an electromechanical relay, MOSFETs have three terminals—a gate, source, and drain—connected to metal wires, typically made from either aluminum or copper. The MOSFET has two dimensions of interest: its length L—the distance between the source and the drain, and its width W—the width of the source and drain areas.

The physics of how precisely these materials form a switch is beyond the scope of this discussion, but we can explain the operation of a MOSFET qualitatively using the simple model illustrated in Figure 7.26. In both pictures in Figure 7.26, the drain terminal of the MOSFET is connected to a high voltage and the source is connected to ground. In the picture on the left, the gate is connected to ground, and the region of silicon between the source and the drain acts as an insulator so that no current can flow between the source and drain. When the gate is connected to a high voltage, as shown on the right, electrons are drawn into this region, forming a conducting channel between the source and the drain. This effectively closes the switch, enabling electrons to flow from the source to the drain (or current, which is defined in terms of the flow of positive charge, flows from drain to source).

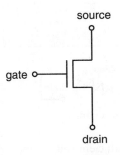

source

gate

drain

Figure 7.24 Symbol for a MOSFET.

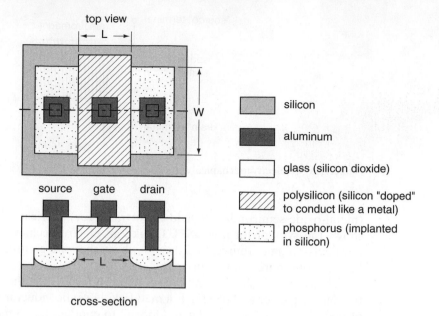

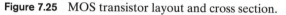

Figure 7.25   MOS transistor layout and cross section.

The channel between the source and the drain in a MOSFET is not a perfect conductor. The physics of what's actually going on in the channel are more complex than we need to discuss here, but for our purposes, we can model the channel with a resistor in the path of the switch, as shown in Figure 7.26. As we did for a wire in Figure 7.21 and equation (7.2), we can represent the channel as a bar, in which the length of the bar is the distance between the source and drain L, the width is the width of the source or drain W, and the thickness t is the depth of the conducting

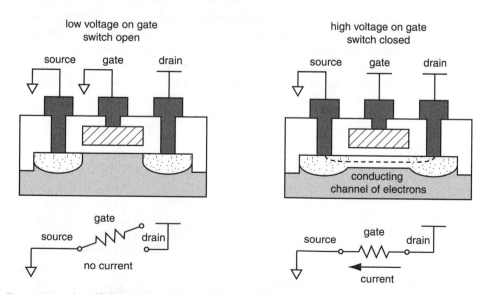

Figure 7.26   A sufficient voltage applied to the gate of a MOS transistor causes current to flow between the source and drain.

channel into the silicon. The channel thickness t is as small as 5 nm, and the resistivity ρ for this channel model depends upon the gate voltage. Example 7.3 shows a sample calculation for the resistance of a MOSFET using this simple model.

Example 7.3	**Resistance of a MOSFET**

Resistance of a MOSFET

Suppose that a MOSFET has a length from source to drain of 180 nm and a width of 900 nm as shown below. If the channel has a depth of 5 nm when the MOSFET is turned on, and the effective resistivity of the channel is 3×10^{-5} Ω-m, what is the resistance of the channel?

If we wanted to change the resistance of the MOSFET to 600 Ω, while keeping the length the same, what would the new width have to be?

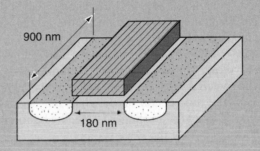

900 nm

180 nm

Solution   Applying Equation (7.2), we calculate the resistance of the channel from its dimensions and resistivity:

$$R = \rho \frac{L}{Wt}$$
$$= (3 \times 10^{-5}) \frac{180 \times 10^{-9}}{(900 \times 10^{-9})(5 \times 10^{-9})}$$
$$= 1.2 \, \text{k}\Omega$$

In order to change the resistance to 600 Ω, we would need to double the width to 1800 nm.

7.4.3 Electrical Circuits

By connecting together electronic devices in various ways, we can build electrical systems that perform many different functions, from receiving radio signals to adding numbers. Networks of electrical elements are called *circuits*. Figure 7.27 shows a simple circuit diagram called a *schematic*, which we'll use to introduce some terminology. Fundamentally, a circuit consists of a set of *branches*, connected together at a set of points called *nodes*. You might recognize this as an example of a *graph* we defined in Chapter 2. In this case the edges of the graph are the branches. Each branch in the circuit represents an electrical device with two ideal wires (no resistance) for making connections, shown as lines coming out of either end of a box. In Figure 7.27, we've drawn the nodes as dots. One rule for constructing a circuit is that every node must have at least two branches connected to it. Oftentimes in a schematic, the dots for the nodes are not drawn; in this case, a node is simply any point in the circuit that connects two or more branches. The example circuit in Figure 7.27 contains five

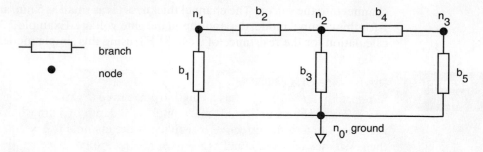

Figure 7.27   A circuit schematic.

branches, labeled $b_1 \ldots b_5$, and four nodes, labeled $n_0 \ldots n_3$. One node in a circuit is always designated as ground and if the nodes are numbered, the ground node is always labeled as node 0.

Figure 7.28 shows two equivalent schematics for a flashlight. It consists of a switch, a voltage source, and a light bulb modeled by a resistor connected in a loop. When the switch is closed, current will flow through the circuit, causing the light bulb to glow. While the version on the right may not look like a loop, note that in the alternative symbol for a voltage source, the bar on top and the ground triangle at the bottom are all part of the same symbol.

Kirchhoff's Laws and the Dependencies Between Currents and Voltages in a Circuit   A key step in almost any circuit design problem is analyzing the circuit to determine the voltages at the nodes and the currents through the branches. The branch constitutive relationships—such as Ohm's Law for resistors—give us some, but not all, of the information we need to do this. To illustrate, consider the circuit in Figure 7.29. Suppose we want to find the voltage at node n_2. If we knew the current I_2 through R_2, we could calculate the voltage drop across it using Ohm's Law, and then determine the voltage at n_2, because the other end of R_2 is at ground, or 0 V. The problem is, we don't know the current through R_2, and furthermore, since current passes through R_1 before it gets to R_2, the current I_2 depends on both resistors. In other words, in this circuit, "everything depends upon everything" and we need some additional information to capture these dependencies.

We've neglected two important laws so far in our analysis: conservation of charge and conservation of energy. The law of conservation of charge tells us that charge can

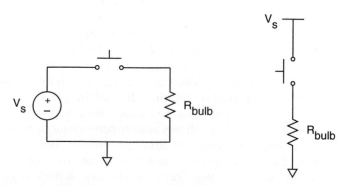

Figure 7.28   Two equivalent circuit schematics for a flashlight.

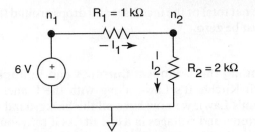

Figure 7.29 An example circuit with two resistors in series. The current I_1 flowing through R_1 must somehow affect the current I_2 flowing through R_2.

neither be created nor destroyed. With this in mind, let's look again at the situation at node n_2. The current I_1 flows into the node and I_2 flows out of it. Suppose that I_1 were smaller than I_2. This would imply that charge was flowing out of n_2 faster than charge flowing into it. For this to be true, we'd have to somehow create charge at n_2, but according to the law of conservation of charge, this is impossible. Similarly, if I_1 were greater than I_2, we would appear to be destroying charge at the node, which is also impossible. The only way that either of these scenarios would be possible is if we could store excess charge at a node and then siphon it off later. But nodes don't work this way; think of the analogy with a water pipe. If water flowed into one end of a pipe faster than it flowed out the other end, the pressure would build up and the pipe would burst. Thus, by the law of conservation of charge, the current I_1 flowing into the node *must* equal the current I_2 flowing out of the node.

This observation, which is a consequence of the law of conservation of charge, is known as *Kirchhoff's Current Law*. Specifically, it states:

Kirchhoff's Current Law (KCL): The sum of all currents flowing into a node through branches must equal the sum of all currents flowing out of a node through branches. Alternatively, if we consider currents flowing out of a node to be negative currents flowing in, the sum of all currents flowing into a node must equal zero.

Kirchhoff posed a second law, which is a consequence of the law of conservation of energy, known as Kirchhoff's Voltage Law. It states:

Kirchhoff's Voltage Law (KVL): In tracing any closed-path "loop" through a circuit, the sum of the voltage drops over branches in the loop must equal the sum of the voltage gains over branches. Alternatively, if we consider voltage drops to be negative voltage gains, the sum of all voltage drops around a circuit must equal zero.

To see how KVL works, let's again consider the circuit in Figure 7.29. The circuit contains only one closed-path loop, which we trace. Starting at the ground node and proceeding clockwise, the first branch we cross is the voltage source, which brings us to n_1. This produces a voltage gain of 6 V. Next we cross R_1 to get to n_2 and then cross R_2 to get back to ground and complete the loop. Since there is (unknown) current flowing through the loop, there will be voltage drops across each of the resistors. We don't yet know their individual values, but by KVL, their sum must be 6 V. To see how this is a consequence of the law of conservation of energy, imagine a blob of charge Q traveling around the loop. With the voltage gain across the voltage source, the potential energy of Q increases and with each voltage drop across the resistors, the potential energy of Q decreases. Because we can neither create nor destroy energy, the net total of the increases and drops around the loop and back to the starting point must be zero.

Finding the Unknown Currents and Voltages Using Nodal Analysis   Armed with Kirchhoff's Laws, along with the branch constitutive relationships (such as Ohm's Law), we now have all the background we need to determine the unknown currents and voltages in a circuit. As it turns out, we can use *either* Kirchhoff's Current or Voltage Law, along with the branch constitutive relationships—we don't have to use both of Kirchhoff's laws. In beginning circuits courses, students are taught to use both approaches, as well as to recognize which might lead to a simpler set of equations when analyzing a given circuit by hand. When using a computer to solve the equations, however, it's not nearly as important to find the simplest approach for a *given* circuit, as it is to apply a method that can reliably generate a set of equations for *any* circuit. Generally, it's more straightforward to formulate a set of equations using Kirchhoff's Current Law, because it's easier to identify the nodes in a circuit than it is to identify the loops. This formulation of circuit equations is known as *nodal analysis*. The basic formulation of the circuit equations is as follows:

1. The unknown variables are the node voltages and the branch currents. If a circuit has n nodes (excluding ground) and b branches, then there is a total of $n + b$ unknowns.
2. For each node except ground, we apply Kirchhoff's Current Law to set the sum of the currents into the node equal to zero, which gives us n equations.
3. For each branch, we use its constitutive relationship to express the relationship between the node voltages on either end and the current flowing through it. This gives us b more equations for a total of $n + b$ equations, which is the same as the number of unknowns.

To illustrate, we'll analyze the same circuit with two resistors in series, repeated in Figure 7.30 showing the unknown quantities. Note that we've drawn the arrows for the currents pointing in arbitrary directions—if the analysis shows that the currents flow in the opposite direction, then they'll turn out to have negative values. First, we list the unknown variables:

$$\text{unknown node voltages:} \quad v_1 \quad v_2$$
$$\text{unknown branch currents:} \quad I_s \quad I_1 \quad I_2$$

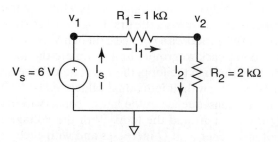

Figure 7.30   Example circuit for nodal analysis.

Next, we write the KCL equations for each node as the sum of the currents into the node equal to zero:

$$\text{KCL at node 1:} \quad I_s - I_1 = 0$$
$$\text{KCL at node 2:} \quad I_1 - I_2 = 0$$

Then, we write the branch constitutive relationships (BCR) for each of the branches:

$$\text{BCR for } V_s: \quad v_1 - 0 = V_s$$
$$\text{BCR for } R_1: \quad v_1 - v_2 = I_1 R_1$$
$$\text{BCR for } R_2: \quad v_2 - 0 = I_2 R_2$$

Thus we have 5 equations in 5 unknowns. One way to solve this system of equations is to arrange them so that we can have a computer software package solve them automatically, and we'll show how to do this with MATLAB in Chapter 11. This particular system, however, is easy to solve manually.

First, from the node 1 equation, we see that $I_1 = I_s$ and from the node 2 equation, $I_2 = I_1$. Thus, we can express all of the branch currents in terms of I_s:

$$I_1 = I_2 = I_s$$

From the BCR for the voltage source V_s, we get $v_1 = V_s$. If we substitute this, together with the expressions for I_1 and I_2 in terms of I_s, into the remaining two BCRs, we end up with 2 equations in 2 unknowns, I_s and v_2:

$$V_s - v_2 = R_1 I_s$$
$$v_2 \quad = R_2 I_s$$

Substituting the expression for v_2 into the first equation and solving, we get

$$I_s = \frac{V_s}{R_1 + R_2}$$
$$= \frac{6}{1000 + 2000}$$
$$- 2\,\text{mA}$$

and plugging this result back into the second equation, we get

$$v_2 = \frac{V_s R_2}{R_1 + R_2}$$
$$= \frac{(6)(2000)}{1000 + 2000} \tag{7.3}$$
$$= 4\,\text{V}$$

Thus, the values of all the node voltages and branch currents are:

$$\text{node voltages:} \quad v_1 = 6\,\text{V} \quad v_2 = 4\,\text{V}$$
$$\text{branch currents:} \quad I_s = 2\,\text{mA} \quad I_1 = 2\,\text{mA} \quad I_2 = 2\,\text{mA}$$

The Voltage Divider The very simple circuit with two resistors in series is actually one of the most useful in electronics. The final step in our analysis of this circuit, equation (7.3), provides a formula for determining the voltage between the two

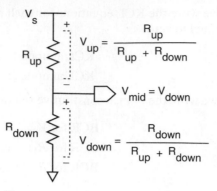

$$V_{up} = \frac{R_{up}}{R_{up} + R_{down}}$$

$$V_{mid} = V_{down}$$

$$V_{down} = \frac{R_{down}}{R_{up} + R_{down}}$$

Figure 7.31 Voltage divider.

resistors in terms of their values. By carefully choosing values for each of the resistors, we can set this voltage to be some fraction of the supply voltage V_s, and hence the circuit is known as a *voltage divider*.

Figure 7.31 illustrates an analysis for a voltage divider. The resistor between the output and V_s is called the *pull-up resistor*, R_{up}, and the one between the output and ground is called the *pull-down resistor*, R_{down}. The voltage between the two resistors relative to ground, V_{mid}, is equal to the voltage drop over the pull-down resistor, V_{down}. Generalizing equation (7.3), this voltage is

$$V_{mid} = V_{down} = V_s \frac{R_{down}}{R_{up} + R_{down}} \tag{7.4}$$

The voltage across the pull-up resistor, V_{up} is

$$V_{down} = V_s \frac{R_{down}}{R_{up} + R_{down}} \tag{7.5}$$

In other words, the voltage across each resistor is equal to the supply voltage times the fraction of the total resistance, $R_{up} + R_{down}$, in that resistor.

Designers often use voltage divider circuits to generate a lower voltage from a higher voltage. Figure 7.32 shows a device called a *potentiometer* that was used as a volume control knob for a radio. It consists of a curved strip of a resistive carbon film with terminals attached to either end, one of which may be connected to ground and the other to a supply voltage. A third terminal is attached to a wiper that sweeps along the resistive strip as the knob is turned. The wiper terminal serves as the middle node of a voltage divider, where the lengths of the carbon strip to either side serve

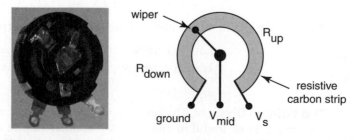

Figure 7.32 Potentiometer.

as the resistors. As the position of the wiper varies, so do the values of the resistors, as does the voltage at the wiper terminal.

7.5 PUTTING IT ALL TOGETHER: DESIGN OF AN INVERTER

7.5.1 Background

Now that we've looked at digital circuits from both the logical and electronic perspectives, we'll pull them together to design an inverter, which is a circuit that implements the NOT operation, using a MOSFET as the switch. Figure 7.33 illustrates the symbol for an inverter, as well as its logical and electrical behavior. From the logical perspective, the inputs and outputs are 1s and 0s, while from the electrical perspective, they are voltages. Ideally, the voltage value for a 0 should be the lowest voltage in the circuit, which is ground, and the value for a 1 should be the supply voltage, which by convention is known as V_{DD}. In practice, however, the output voltages may not reach these extremes, because of voltage drops across resistances in the circuit—remember that in our model, MOSFETs are switches with resistors. The actual output low voltage, which may be higher than ground, is called V_{OL} and the actual output high voltage, which may be less than V_{DD}, is called V_{OH}.

We'd like V_{OL} to be as low as possible and V_{OH} to be as high as possible so that the inverter will produce logically correct results even when there's noise on the input or output signals, as shown in Figure 7.34. As we discussed in Section 7.2.1, random fluctuations on signals caused by static or other sources affect the reliability of any real system. If V_{OH} and V_{OL} are widely separated, a system stands a much better chance of tolerating noise than if they're close together. Since the values of V_{OH} and V_{OL} depend upon resistances in the inverter, and since the resistances are tied to the dimensions of the resistor and the MOSFET, there's a tradeoff between reliability and the size of the inverter.

There are several different ways of building an inverter using transistors. For this problem, we'll use a design that's similar to the approach in Section 7.3.2 where we used an ideal switch and a light bulb, except we'll replace the switch with a MOSFET and the light bulb with a resistor, as shown in Figure 7.35. Qualitatively, the operation of the inverter is as follows: when IN is low, the MOSFET is turned off and the resistor pulls OUT high, and when IN is high, the MOSFET is conducting and pulls OUT low. How high and how low depend upon the lengths and widths of the MOSFET and the resistor, which are labeled in Figure 7.35.

We note that this *isn't* the design most widely used in integrated circuits today, and it has some shortcomings, but those shortcomings will serve to make our example more interesting. This design *is* used, however, in some important situations, and we'll introduce the most common design style—which uses two MOSFETS instead of one MOSFET and a resistor—at the end of the section.

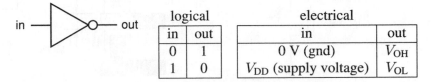

logical		electrical	
in	out	in	out
0	1	0 V (gnd)	V_{OH}
1	0	V_{DD} (supply voltage)	V_{OL}

Figure 7.33 Inverter symbol, logical behavior, and electrical behavior.

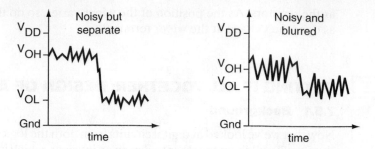

Figure 7.34 A digital circuit can better tolerate noise when the high and low output voltage levels are widely separated.

7.5.2 Problem Definition and Plan of Attack

Given this background, we're now ready to state the design problem:

For the inverter in Figure 7.35, determine values for length and width of the resistor and MOSFET, such that $V_{OL} \leq V_{DD}/4$ and $V_{OH} \geq 3V_{DD}/4$, where $V_{DD} = 2$ V. Model the MOSFET as a resistor and a switch and assume the following technology parameters:

polysilicon thickness:	t_{poly}	120 nm
polysilicon resistivity:	ρ_{poly}	$1.5 \times 10^{-4} \Omega - m$
MOSFET channel thickness:	$t_{channel}$	5 nm
MOSFET channel resistivity:	$\rho_{channel}$	$3 \times 10^{-5} \Omega - m$

Design the inverter to minimize the size of the MOSFET and resistor, subject to the following constraints:

- the minimum length or width of a polysilicon region is 180 nm
- the minimum length or width of a phosphorus implant region is 360 nm

Determine the power consumption for the final design.

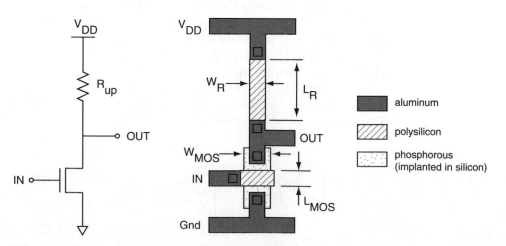

Figure 7.35 Schematic and layout of an inverter with a MOSFET and pull-up resistor.

TABLE 7.2 Variables in the inverter design problem.

Name	Description	Type	Value
V_{DD}	supply voltage	environmental	2 V
V_{OL}	output low voltage	behavioral	0.5 V or less
V_{OH}	output high voltage	behavioral	1.5 V or more
W_R	resistor width	design	at least 180 nm
L_R	resistor length	design	at least 180 nm
W_{MOS}	MOSFET width	design	at least 360 nm
L_{MOS}	MOSFET length	design	at least 180 nm
R_{up}	pull-up resistance	design	unknown
R_{MOS}	MOSFET resistance	design	unknown
t_{poly}	polysilicon thickness	environmental	120 nm
ρ_{poly}	polysilicon resistivity	environmental	$1.5 \times 10^{-4}\ \Omega - m$
$t_{channel}$	MOSFET channel thickness	environmental	5 nm
$\rho_{channel}$	MOSFET channel resistivity	environmental	$3 \times 10^{-5}\ \Omega - m$

Define The problem asks us to determine sizes for the resistor and the MOSFET in the inverter design to ensure that the output voltage for a 1 is sufficiently high and that the output voltage for a 0 is sufficiently low. The sizes of these devices determine the resistances in the circuit, which in turn control the output voltage levels. Table 7.2 lists the known and unknown variables in the problem.

Explore Since we're modeling the MOSFET as a switch with a resistor, when the MOSFET is turned on, the inverter becomes two resistors with the output in between them: *a voltage divider*. Now we can see more clearly why the output may not pull all the way up to V_{DD} or all the way down to ground, since for a voltage divider, the output voltage depends on the relative sizes of the pull-up and pull-down resistors. Looked at another way, the inverter has something in common with the potentiometer: turning on the MOSFET in the inverter is like turning the knob on the potentiometer from Figure 7.32, in that it changes the value of the pull-down resistor, lowering the output voltage.

Plan Now that we see that the inverter works as a voltage divider, we're ready to plan our solution to the problem:

1. Determine the dimensions of the MOSFET and pull-up resistor.

   (a) Redraw the circuit with the MOSFET modeled as a switch and a resistor. This will transform the inverter into a voltage divider when the switch is closed.

   (b) Using the voltage divider equations, determine V_{OH} and V_{OL} in terms of the pull-up and pull-down resistance values.

   (c) Determine the resistance values in terms of the dimensions of the pull-up resistor and the MOSFET.

   (d) Set the length of the MOSFET and the width of the resistor to their minimum values, and then solve for the remaining dimensions, satisfying the constraints on V_{OH} and V_{OL}.

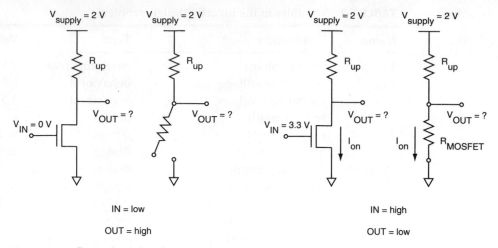

Figure 7.36   Operation of an inverter.

2. Calculate the power consumption for the circuit design from the current through the resistors and the voltages across them.

7.5.3   Choosing Device Sizes

The first step is to redraw the circuit, modeling the MOSFET with a switch and a resistor. As shown in Figure 7.36, there are two cases: the first in which V_{IN} is low and the switch is open, and the second in which V_{IN} is high and the switch is closed. Below, we analyze each case separately to determine the output voltage V_{OUT}.

Analysis When V_{IN} is Low   The schematic on the left in Figure 7.36 illustrates the case where V_{IN} is low and V_{OUT} is high. Here, V_{OUT} is connected to V_{DD} through the pull-up resistor R_{UP}. Since the switch is open, there's no connection to ground and thus there's no current flowing out of the output node OUT. According to Kirchhoff's Current Law, the current flowing into the output node OUT must equal the current flowing out of it, and therefore there cannot be any current flowing into the output node through R_{UP}. Because there's no current flowing through it, there's no voltage drop across R_{UP}, and thus V_{OUT} must equal V_{DD}. Therefore, the output high voltage is:

$$V_{OH} = V_{DD}$$

Note that this result is independent of the resistor values. Thus, for this inverter design, the output high voltage will equal the supply voltage V_{DD}, regardless of the size of either the pull-up resistor or the MOSFET.

Analysis When V_{IN} is High   The schematic on the right in Figure 7.36 illustrates the case in which V_{IN} is high and V_{OUT} is low. In this case, the output is the middle node of a voltage divider, connected to V_{DD} through R_{UP} and to ground through R_{MOS}, and the voltage at this node will depend on the ratio of the values of these two resistances. From Equation (7.4), the value for V_{OUT} is

$$V_{OUT} = V_{OL} = V_{DD}\frac{R_{MOS}}{R_{MOS} + R_{UP}}$$

From the requirement that $V_{OL} \leq V_{DD}/4$,

$$\frac{V_{DD}}{4} \geq V_{DD}\frac{R_{MOS}}{R_{MOS} + R_{UP}}$$

Solving, we get

$$R_{UP} \geq 3R_{MOS} \tag{7.6}$$

Thus, to meet the requirement that $V_{OL} \leq V_{DD}/4$, the resistance of R_{UP} must be at least three times as large as the resistance of R_{MOS}.

Determining Device Sizes   There are infinitely many combinations of sizes for the pull-up resistor and the MOSFET that'll satisfy equation (7.6). From the problem statement, we are to pick the combination that minimizes the sizes of these two devices. We'll use the following approach to do so:

1. determine the resistance of a minimum-sized MOSFET,
2. calculate the minimum value for the pull-up resistor to satisfy Equation (7.6),
3. set the width of the resistor to its minimum allowed value, and calculate the required length.

MOSFET Size and Resistance   From Figure 7.35 and Table 7.2, the length and width of a minimum-sized MOSFET are:

$$l_{MOS} = 180\,\text{nm}$$
$$W_{MOS} = 360\,\text{nm}$$

From equation (7.2), the MOSFET resistance is

$$R_{MOS} = \rho_{channel}\frac{l_{MOS}}{W_{MOS}t_{channel}}$$
$$= (3 \times 10^{-5})\frac{180 \times 10^{-9}}{(360 \times 10^{-9})(5 \times 10^{-9})}$$
$$= 3\,\text{k}\Omega$$

where the values for $\rho_{channel}$ and $t_{channel}$ also come from Table 7.2.

Pull-up Resistor Value and Size   From Equation (7.6), the minimum resistance for R_{UP} is

$$R_{UP} = 3R_{MOS}$$
$$= 9\,\text{k}\Omega$$

R_{UP} will be fabricated as a thin strip of polysilicon. The minimum width of this strip (Table 7.2) is

$$W_R = 180\,\text{nm}$$

From equation (7.2), the resistance of the polysilicon resistor R_{UP} is

$$R_{UP} = \rho_{poly}\frac{L_R}{W_R t_{poly}}$$

Solving for L_R, we get

$$L_R = \frac{1}{\rho_{poly}} R_{UP} W_R t_{poly}$$

$$= \frac{1}{1.5 \times 10^{-4}} (9000)(180 \times 10^{-9})(120 \times 10^{-9})$$

$$= 1300 \text{ nm}$$

7.5.4 Calculating Power Consumption

In Section 7.4.1, we saw that power consumption is equal to the current flowing from the power supply times its voltage. In the inverter, when the output is low, there's current flowing from the power supply through the circuit, but when the output is high, there's no current. Thus, the power consumption of the inverter depends upon the stream of data that it processes. We can reasonably assume that for a large data output stream, half of the bits are 0s and half are 1s, so the power consumption is the average of both cases:

$$P_{avg} = \frac{P_{\text{output low}} + P_{\text{output high}}}{2}$$

$$= \frac{0 + V_{DD} I_{ON}}{2}$$

$$= \frac{V_{DD} I_{ON}}{2}$$

where I_{ON} is the current flowing through the circuit when the MOSFET is turned on.

From Kirchhoff's Current Law, the current flowing through the circuit from the power supply equals the current flowing through the resistor, which also must equal the current flowing through the MOSFET. From Ohm's Law, thus

$$I_{ON} = \frac{V_{OL}}{R_{MOS}}$$

$$= \frac{0.5 \text{ V}}{3000 \text{ }\Omega}$$

$$= 167 \text{ }\mu\text{A}$$

Substituting the value of I_{ON} back into the formula for average power, we get

$$P_{avg} = \frac{(2 \text{ V})(167 \times 10^{-6} \text{ A})}{2}$$

$$= 167 \text{ }\mu\text{W}$$

Thus, on average, a single inverter of our design will consume 167 μW of power. Is this a lot? To put this number in perspective, let's consider not just one, but the average number of inverters we might find on a computer chip. Figure 7.37 shows a photo of an Intel Core Duo processor, designed for low-power mobile computing. More than half of the area of the chip is composed of a type of fast memory called *cache* that holds the data in a program most likely to be accessed frequently by the processor. In this version of the Core Duo, there are more than 16 million bits of memory in the cache, and each bit contains two inverters. Thus, the power consumption of this part of the chip is at least 32 million times the power consumption of a single inverter. If each inverter consumed 167 μW, then the total power consumption of

Figure 7.37   An Intel Core Duo module. Half of the area of the chip is the cache memory. Photo courtesy Intel Corporation.

the cache would be over 5,000 W. This is a *huge* number: about four times the power consumption of an average household in the U.S. packed into an area of roughly one square centimeter—more power per unit area than a nuclear reactor! In reality, the power consumption of this version of the Core Duo is approximately 21 W with a supply voltage of approximately 1.3 V, so clearly this chip is using a far more energy-efficient inverter design than the one that we presented here.

On the positive side, our inverter design didn't draw current when the input was low because the MOSFET functioned as an open switch and there was no path through the circuit between V_{DD} and ground. If we could replace the pull-up resistor with a switch that's open when the input is low, then it wouldn't draw current in that case, either. This style of logic circuit, which uses a combination of "normally open" and "normally closed" MOSFET switches, is known as *complementary MOS* or *CMOS*, and is the most common approach to building digital integrated circuits today. Figure 7.38 illustrates a CMOS inverter. The "normally open" switch that connects the output to the negative terminal of the power supply or ground is called an n-type MOSFET, while the "normally closed" switch that connects the output

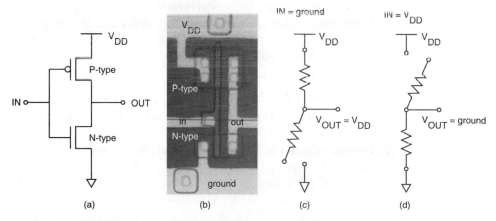

Figure 7.38   CMOS inverter. (a) Schematic showing symbols for p-type and n-type MOSFETs. (b) Photomicrograph of CMOS inverter on a chip fabricated by Notre Dame undergraduates, courtesy Greg Snider. (c) When the input is low, the n-MOSFET is open and the p-MOSFET is closed. (d) When the input is high, the p-MOSFET is open and the n-MOSFET is closed.

to the positive terminal or V_{DD} is called a p-type MOSFET. CMOS circuits only draw current when the output is in the process of switching between a high and low voltage—not when the output has a stable value of one or zero—so the power consumption is extremely low.[6] The tradeoffs are that supporting both p-type and n-type MOSFETs on the same chip requires a more complex fabrication process and that the design will have twice as many switches.

PROBLEMS

1. **Evaluating Logic Expressions**
   Given that $A = 1$, $B = 0$, and $C = 1$, evaluate the following logic expressions:

   (a) $(A$ AND NOT $B)$ AND NOT C

   (b) $[(A$ OR B OR $C)$ AND A AND $C]$ OR $(B$ AND $C)$

   (c) $(A$ AND $B)$ OR (NOT C AND $A)$ OR A

2. **DeMorgan's Law**
   DeMorgan's Law states that

$$\text{NOT}(A \text{ AND } B) = (\text{NOT } A) \text{ OR } (\text{NOT } B)$$
$$\text{NOT}(A \text{ OR } B) = (\text{NOT } A) \text{ AND } (\text{NOT } B)$$

   Using a truth table, show that both statements in DeMorgan's Law are true.

3. **Equivalence of Logic Expressions**
   Show that the following two logic statements are equivalent, first by constructing a truth table, and then by using De Morgan's Law.

$$(\text{NOT } A) \text{ OR } (\text{NOT } B) \text{ OR } [(\text{NOT } A)\text{AND}(\text{NOT } B)]$$
$$\text{NOT}(A \text{ AND } B) \text{ OR NOT}(A \text{ OR } B)$$

4. **Equivalent Boolean Logic Statements**
   Using the properties of Boolean logic, prove that the following equality holds:

$$((\text{NOT } A) \text{ AND } (\text{NOT } B) \text{ AND } C) \text{ OR}$$
$$((\text{NOT } A) \text{ AND } B \text{ AND } C) \text{ OR}$$
$$(A (\text{NOT } B) \text{ AND } (\text{NOT } C)) \text{ OR}$$
$$(A \text{ AND } B \text{ AND } (\text{NOT } C)) = ((\text{NOT } A) \text{ AND } C) \text{ OR } (A \text{ AND } (\text{NOT } C))$$

5. **Design a Circuit to Detect Values Greater Than 5**
   Consider a box that has three buttons that represent a 3-digit binary number as input and a red lamp as output. Design a circuit using switches that will light the lamp if the value of the input is greater than 5.

---

[6]They also consume power as a result of "leaky" switches that allow a tiny amount of current to flow when they are supposed to be turned off—a problem that is becoming more significant as billions of switches are placed on a single chip.

6. Decoder Logic

Suppose that you are designing a decoder circuit that takes a 4-bit binary number as input and produces a single bit as output. The decoder should output a 1 when the input equals any of the values 2, 3, 6, 7, 10, 11, 14, 15, and should output a 0 otherwise. Write a Boolean logic equation that defines the controller output in terms of the input.

7. Design of a Multiplexor

A multiplexor is a logic component that acts as a switch for selecting one input from many. The component selects which input to route to the output based on a select signal S, such that when $S = 0$ it selects the A input and when $S = 1$ it selects the B input. The symbol and truth table for a multiplexor are given below.

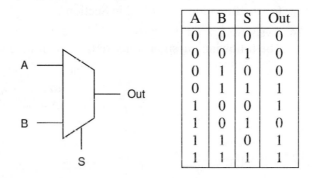

A	B	S	Out
0	0	0	0
0	0	1	0
0	1	0	0
0	1	1	1
1	0	0	1
1	0	1	0
1	1	0	1
1	1	1	1

(a) Given the truth table, write a Boolean logic expression for a multiplexor.

(b) If you were to construct the multiplexor from AND gates, OR gates, and inverters, how many switches would it take to implement this logic equation?

(c) Can you think of a way to implement a multiplexor using far fewer switches than the answer to the previous question? Explain your solution.

8. Conversion of Binary Numbers

Convert the following binary numbers to hexadecimal and decimal numbers.

(a) 101

(b) 1010

(c) 1100101011111110

9. Decimal-to-Binary Conversion

Convert the following decimal numbers to binary.

(a) 32

(b) 73

10. How Many Bits?

How many bits are needed to represent the decimal number 1234 as a binary number? Briefly describe a general approach for determining how many bits are needed to represent any positive integer as a binary number.

11. Binary Addition

Calculate the following sums of binary numbers:

(a) $10 + 10$

(b) $111010 + 10111$

(c) $1111 + 1101$

12. Logic Equations for a Full Adder

Using a truth table, demonstrate that the following logic equations for the carry-out C_{out} function and sum function S are correct:

$$C_{out} = (A \text{ AND } B) \text{ OR } (A \text{ AND } C_{in}) \text{ OR } (B \text{ AND } C_{in})$$
$$S = A \text{ XOR } B \text{ XOR } C$$

13. Full Adder Circuits

Design circuits using switches that implement the sum and carry-out functions for the full adder described in Section 7.3.4.

14. Resistive Circuit Analysis

The following questions refer to the circuit below:

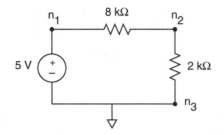

(a) What is the voltage between nodes n_1 and n_2?

(b) What is the voltage between nodes n_2 and n_3?

(c) What is the voltage between nodes n_3 and n_1?

(d) What is the sum of the voltage *drops* around the complete circuit?

15. Calculating an Unknown Resistance

What is the value of R_2 in the circuit below if $I_2 = 12$ mA?

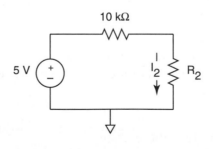

16. Power Dissipation Through a Wire

As illustrated in Figure 7.21 (repeated below), the resistance of a particular wire is given by the expression $R = \rho L/(Wt)$, where L is the length, W is the width, t is the thickness, and ρ is the resistivity of the material. Suppose that a wire with these properties is used to connect two components on either side of an integrated circuit.

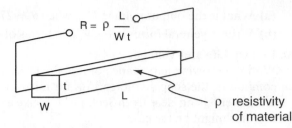

$$R = \rho \frac{L}{W\,t}$$

ρ resistivity of material

(a) Suppose that the voltage drop across the wire is V volts. Write out an expression for the power dissipation by the wire in terms of its dimensions, resistivity, and V.

(b) In order to fit more wires on a chip, and to save on materials, suppose that your fellow engineers have proposed scaling the width and thickness of wires to make them smaller by a factor s. Using the equation you derived earlier, explain the impact of this proposal in terms of power dissipation.

(c) What would the voltage drop across a scaled wire have to be in order for it to have the same power dissipation as an original unscaled wire, assuming that the length of both wires is the same?

17. Resistance and Power

Do you agree or disagree with the following statement? Explain.

> As the resistance in a circuit increases, so does the power dissipation, because a larger resistance gets hotter when current flows through it than does a smaller resistance.

18. Light Bulb Current and Power

The resistance of a certain light bulb increases from about 10 Ω when it is cold to about 144 Ω when it is hot. Suppose that a 120 V source is applied to the light bulb.

(a) What is the current passing through the bulb the instant after it is switched on?

(b) What is the current passing through the bulb once it heats up?

(c) What is the power dissipation of the bulb once it heats up?

19. Using a Potentiometer

Consider a potentiometer whose resistance varies linearly with the angle θ of the wiper, where the resistance is 0 Ω when $\theta = 0°$ and 50 Ω when $\theta = 270°$. The potentiometer is wired up in a circuit as shown in the schematic below.

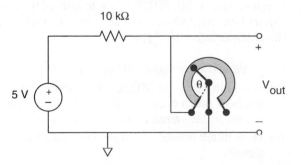

(a) What is the output voltage V_{out} when $\theta = 270$?

(b) Write a general formula for V_{out} in terms of θ.

20. Battery Life and Voltage

With the growing importance of mobile devices such as cell phones and laptop computers, the supply voltage V_{DD} for microprocessors has been getting lower and lower. Using ideas from Section 7.4.1, give at least two reasons why you think that this might be the case.

21. NMOS Logic Circuits

Construct a truth table and write a logic expression for each of the following circuits.

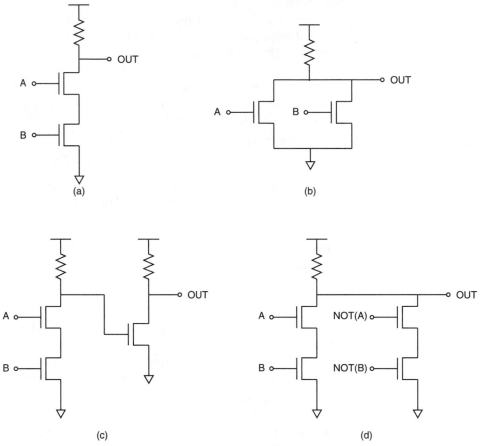

22. MOSFET Power Scaling

Suppose that a MOSFET has a length of 90 nm and a width of 500 nm. The channel has a depth of 2 nm when it is turned on, and the effective resistivity of the channel is 3×10^{-5} Ω-m.

(a) What is the resistance of the channel?

(b) Assuming that the MOSFET behaves as a simple resistor when it is turned on, how much current would flow through the MOSFET when the voltage between the source and drain is 1.8 V?

(c) How much power would be dissipated by the MOSFET under these conditions?

(d) By how much would you have to scale the width of the transistor to reduce the power by 50 percent?

(e) By how much would you have to scale the voltage across the transistor to reduce the power by 50 percent, assuming that the dimensions are kept the same?

23. A 2-Input NAND Gate

A NAND gate implements a "NOT AND" function, that is,

$$A \text{ NAND } B = \text{NOT}(A \text{ AND } B)$$

(a) Sketch a schematic and circuit layout for a 2-input MOS NAND gate, similar to that shown for an inverter in Figure 7.35. Try to make the layout as compact as you can, while keeping the resistance values the same.

(b) Determine the output voltage for each of the four possible input combinations. Assume that the supply voltage is 2 V.

(c) Is the power dissipation the same in all cases? Explain.

(d) Suppose that during the operation of the circuit, any input has a 50 percent change of being a logical 1 (or 0) at any point in time. What would the average power dissipation be?

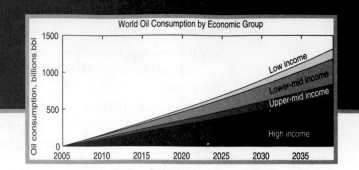

World Oil Consumption by Economic Group

Modeling Change in Systems

LEARNING OBJECTIVES

- to explain what engineers mean by the *state* of a system, to recognize that the state of many important physical systems *changes with time*, and to give examples of why engineers need to model such *dynamic systems*;
- to apply a simple approach to modeling the rate of change in a system as a ratio of small, finite changes;
- to apply a simple tabular approach to modeling the accumulation of change in a system by adding up finite changes over discrete steps in time;
- to apply the tabular approach to simulating changing systems such as the trajectory of a moving object or a liquid flowing out of a tank.

8.1 INTRODUCTION

Engineers are often called upon to predict the future. Sometimes these predictions are on a grand scale, such as when planning a system of levees in New Orleans that will withstand extreme weather conditions during an age of climatic uncertainty. Most of the predictions engineers make, however, are less publicized, and being able to model changes in systems is a routine part of engineering practice. Many engineering problems involve modeling how some quantity builds up or depletes over some interval. The quantity may be a material entity such as a fluid or electrical charge, a physical concept such as speed or distance, or a mathematical abstraction such as the probability of some event occurring. Similarly, the interval itself may be of many forms, such as a time interval, a region of space, or over the elements of some collections. The following problems all involve modeling an accumulation of change.

- What is the distance covered by a person walking at a certain speed over a given time interval?

- How much energy is stored in a spring as we stretch it across some distance?
- How fast can we run the clock on a microprocessor so that signals traveling over wires on a chip will arrive at different destinations on time?
- What is the depth of water in a tank over time as streams of given rates flow into and out of the tank?

One of the most common ways of representing changes in a system is as a series of "snapshots" taken at regular intervals. This view of the behavior of a system is similar to watching a moving object illuminated by a strobe light. Each time the light flashes, we get a brief look at the state of the system at that instant, but have no direct knowledge of what's happening to the system when it's in the dark between flashes. A model that represents time in discrete "flashes," with "darkness" in between the flashes, is called a *discrete* model.

The remainder of this chapter presents an approach for building discrete models for change in systems. After introducing the basic approach, we then apply it in two detailed examples:

- modeling the motion of a falling object
- predicting when we'll run out of oil

8.2   PREDICTING THE FUTURE: ACCUMULATION OF CHANGE

8.2.1   The State of a System

The first issue in modeling change in a system is deciding on change of *what*. Engineers usually talk about change in the *state* of a system; the state is determined by the combined values of its *properties*. A property, in turn, is any characteristic of a system we could, at least in principle, measure. There are many possible properties we could ascribe to a system, but usually we're only interested in a few of these. To illustrate the way that engineers look at change, we return to the example in Chapter 2 of lifting a weight onto a shelf. In this example, we'd certainly be interested in the weight of the box and its height off the floor. We wouldn't be interested in its color, however, since this is irrelevant to the problem. Properties that are part of the definition of the state are called *state variables*. Figure 8.1 illustrates two states for the box-and-shelf problem, *on-floor* and *on-shelf*. The example contains a single state variable, Y—the height of the box above the floor. $Y_{\text{on-floor}}$ is the value of the state variable Y when the box is in the state *on-floor*, which is 0, and $Y_{\text{on-shelf}}$ is the value of the state variable Y when the box is in the state *on-shelf*, which is 6.

The state of a system is determined only by the values of its properties and has nothing to do with how the system came to be that way. If a box is sitting on the shelf, and we come back a week later and the box is in the exact same spot on the shelf— and it's weight is exactly the same—then the state of box is the same at both points in time, regardless of what activity the box may have seen in the interim. Conversely, saying that a system is in a given state is equivalent to fixing the values for each of its properties.

Process and Path   A *process* is the activity of transforming a system from one state to another. A simple example of a process is moving the box from the floor to the shelf. A *path* is the series of intermediate states a system passes through in moving from one state to another. If someone lifts the box onto the shelf using a

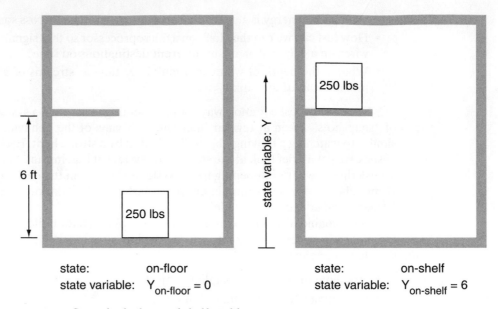

Figure 8.1 States in the box and shelf problem.

block and tackle, it follows one path between the two states. If someone loads the box onto a truck, drives it around the block, and then returns and lifts it onto the shelf, it has followed a different path. The beginning and end states in either case, however, are the same.

Usually when we talk about modeling how a system changes over time, we mean to estimate the path the state variables of the system follow as the system changes. Figure 8.2 illustrates a discrete view of the box falling from the shelf and also introduces some notation. The figure shows a series of six states of the box during the process of falling, labeled states 0 through 5, where state 0

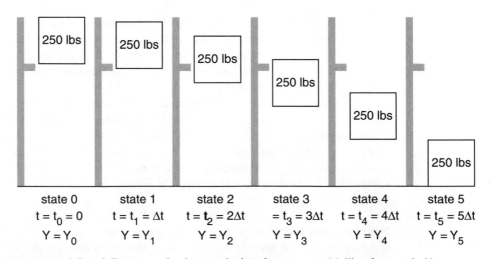

Figure 8.2 A box follows a path of states during the process of falling from a shelf.

is the initial state with the box at the height of the shelf. Each of the snapshots is taken at evenly spaced points in time t_0 through t_5, with a *time step* or interval of Δt between them, where the Greek letter Δ (delta) denotes a measurable change. The corresponding heights of the box at each of these time points are Y_0 through Y_5.

We also use Δ notation to signify changes in the values of state variables between states. If X is a state variable and i and j are two states, then the change in the value of X in the process of changing from state i to state j is

$$\Delta X_{i \to j} = X_j - X_i$$

Thus, for the box-and-shelf problem, for the states shown in Figure 8.1,

$$\Delta Y_{\text{on-floor} \to \text{on-shelf}} = 6$$
$$\Delta Y_{\text{on-shelf} \to \text{on-floor}} = -6$$

For a process such as that pictured in Figure 8.2, where a system follows a path through a series of sequentially numbered steps, we'll adopt the shorthand of just identifying the first step during a change from one step to the next. Thus

$$\Delta Y_0 = \Delta Y_{0 \to 1} = Y_1 - Y_0,$$
$$\Delta Y_1 = \Delta Y_{1 \to 2} = Y_2 - Y_1,$$

$$\dots$$

Rates of Change   A rate of change is a measure of how much one quantity changes with respect to changes in another quantity. In the discrete time model, the rate of change of a state variable at a given step is equal to the change in value of the variable divided by the duration of the time step. Thus, if X is a state variable and i is the step number in a series of steps, then the rate of change of X with respect to time at step i is

$$\frac{\Delta X_i}{\Delta t} = \frac{X_{i+1} - X_i}{\Delta t} \tag{8.1}$$

For the falling box, the rate of change of Y with respect to time is the *velocity* of the box. Example 8.1 examines the velocity of the box at various points in time.

Example 8.1	**Velocity of a Falling Box**					

Velocity of a Falling Box

The following table lists the height Y of a falling box at a series of points in time.

step	0	1	2	3	4	5
time, s	0	0.1	0.2	0.3	0.4	0.5
Y, ft	6.00	5.84	5.36	4.55	3.42	1.98

What is the initial velocity of the box at step 0? At 0.1 s? At 0.45 s?

Solution   The duration of the time step is $\Delta t = 0.1$ s. The velocity of the box at step i is

$$v_i = \frac{\Delta Y_i}{\Delta t} = \frac{Y_{i+1} - Y_i}{\Delta t}$$

The initial velocity at step 0 is

$$v_0 = \frac{Y_1 - Y_0}{\Delta t} = \frac{5.84 - 6.00}{0.1} = -1.6 \, \text{ft/s}$$

0.1 s corresponds to step 1, where the velocity is

$$v_0 = \frac{Y_2 - Y_1}{\Delta t} = \frac{5.36 - 5.84}{0.1} = -4.80 \, \text{ft/s}$$

0.45 s lies between steps 4 and 5. In the discrete model, the value of Y is undefined between steps—recall that the model goes "dark" between flashes. To *estimate* the value between steps, we can use a variety of methods. One method is to simply pick the value at step 4, another is to pick the value at step 5, and a third is to calculate the average of these two values.

8.2.2   Euler's Method: Predicting Change from One State to the Next

As we discussed earlier, engineers often make predictions about how systems will behave. In this section, we show how we can use a discrete time model to make such predictions. The basic logic of using a discrete time model to "predict the future" is as follows:

> If we know the state of a system now, and we know the current rate of change of the state variables, we can estimate the state of the system a short time in the future.

We must answer two key questions before we can apply this logic to modeling a changing system.

- Assuming we know the current state and the current rate of change, how do we use this information to make a prediction?
- We can determine the current state of a system by taking measurements, but how can we determine the current rate of change without knowing the future?

Predicting Values of State Variables at the Next Step   The answer to the first question is straightforward. Suppose X_i is the current value of state variable X and r_i is the current rate of change of X. Since

$$r_i = \frac{\Delta X_i}{\Delta t} = \frac{X_{i+1} - X_i}{\Delta t},$$

we can estimate the next value of X, X_{i+1} as

$$X_{i+1} = X_i + r_i \Delta t \qquad (8.2)$$

In other words, the value of the state variable at the next time step is equal to its current value plus its rate of change times the duration of the time step. Example 8.2 illustrates the application of this rule.

Now suppose we want to predict the behavior of a system further into the future. A good approach is to chain a series of predictions over small steps together. In other words, just as we can use the current state and current rate of change to predict the

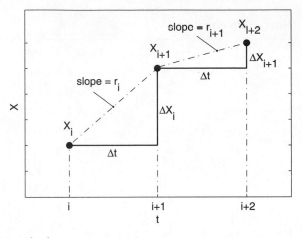

Figure 8.3 Euler's method.

next state, we can apply the same process again to predict the state after that, and so on. Mathematically, this involves repeatedly applying Equation (8.2) in order to predict values for X_{i+1}, X_{i+2}, X_{i+3}, and further values in the series. If we begin at step 0, the resulting series of predictions is

$$X_1 = X_0 + r_0 \Delta t$$
$$X_2 = X_1 + r_1 \Delta t$$
$$X_3 = X_2 + r_2 \Delta t$$

(8.3)

$$\ldots$$

This approach of predicting the behavior of a system as a series of discrete steps in time, evaluating the change at each step and accumulating the changes, is known as *Euler's Method*. It was developed by the Belgian mathematician Leonhard Euler—the same man who developed the theory of beam buckling we discussed in Chapter 6—in the mid-1700s. Figure 8.3 provides a graphical illustration of Euler's Method. Given a point at time t_i where the value of X is X_i, the rate of change r_i is the slope of a line segment that connects the point at t_i to the next point in the series at t_{i+1}. Along the horizontal axis, points are evenly spaced at intervals of Δt, while along the vertical axis, the spacing between successive points is ΔX.

Example 8.3 illustrates a simple application of Euler's Method for modeling a box moving at a constant velocity.

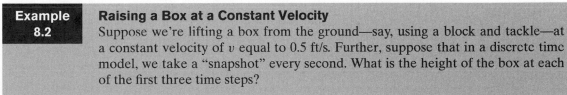

| Example 8.2 | **Raising a Box at a Constant Velocity** Suppose we're lifting a box from the ground—say, using a block and tackle—at a constant velocity of v equal to 0.5 ft/s. Further, suppose that in a discrete time model, we take a "snapshot" every second. What is the height of the box at each of the first three time steps? |

Solution First, we define the variables:

$$Y_0 = 0 \, \text{ft} \qquad v = \frac{\Delta Y_i}{\Delta t} = 0.5 \, \text{ft/s} \qquad \Delta t = 1 \, \text{s}$$

Next, from equation (8.2),

$$Y_1 = Y_0 + v\Delta t$$
$$= 0 + 0.5(1)$$
$$= 0.5 \text{ ft}$$
$$Y_2 = Y_1 + v\Delta t$$
$$= 0.5 + 0.5(1)$$
$$= 1 \text{ ft}$$
$$Y_3 = Y_2 + v\Delta t$$
$$= 1 + 0.5(1)$$
$$= 1.5 \text{ ft}$$

Figure 8.4 shows a plot of the results.

Laws of Nature and Rates of Change In response to the second question, we might know the rate of change through several methods. Sometimes, the rate of change may be given as part of the problem statement, as it was in Example 8.2. More generally, however, we need a model of how a system is *expected* to behave to predict the rate of change. As we first saw in Chapter 4, a model can be either theoretical or empirical, and in the case of a theoretical model, we predict behavior using laws of nature.

To illustrate, consider the example of predicting the velocity of a box falling off a shelf. In Example 8.1, we determined the velocity of the box at different time steps *empirically* from experimental data on the height of the box at each step. Galileo, Newton, and others performed similar experiments during the 1600s and 1700s that led to the *theoretical* result of Newton's Universal Law of Gravitation. From this, we now know that objects on earth will fall—if we ignore the effects of air resistance—with

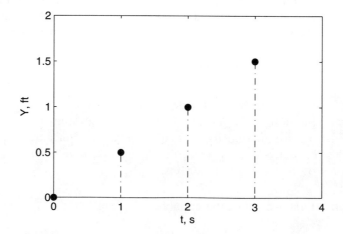

Figure 8.4 Plot of height versus time for a discrete model of raising a box at constant velocity, from Example 8.2.

an acceleration of 32.2 ft/s^2 or 9.81 m/s^2. Since acceleration is the rate of change of velocity, we can use the laws of gravity to estimate the velocity after a time step, as illustrated by Example 8.3.

Example 8.3	**Using the Laws of Gravity to Predict Velocity of a Falling Object**
	A box weighing 250 lbs falls off a shelf at a height of 6 ft. Using a discrete time model with a step size of 0.1 s, estimate the velocity of the box after one time step.

Solution From the laws of gravity, the acceleration of the box will be 32.2 ft/s^2. Assuming that the box is initially at rest:

$$v_0 = 0 \, \text{ft/s} \qquad a = 32.2 \, \text{ft/s}^2 \qquad \Delta t = 0.1 \, \text{s}$$

Since acceleration, a, is the rate of change of velocity, v,

$$a = \frac{\Delta v_0}{\Delta t} = \frac{v_1 - v_0}{\Delta t} = -32.2$$

Solving for v_1, we get

$$v_1 = v_0 + a \Delta t$$
$$= 0 - 32.2(.1)$$
$$= -3.22 \, \text{ft/s}$$

Thus the predicted velocity at 0.1 s using the theoretical discrete model is −3.22 ft/s. It is interesting to compare this with the experimental result from Example 8.1, which showed the velocity at 0.1 s to be −4.80 ft/s. Why does the box seem to be moving faster from the experimental model than from the theoretical model? Hint: Look at the values for v_0 under both models.

8.3 LAUNCHING A SOFTBALL

In Chapter 4, we looked at the problem of finding settings for a giant slingshot to launch a softball at a target. We built an *empirical* model of the launch process, using experimental data from trial launches, in order to make our prediction. In this section, we consider some of the basic elements we'd need to construct a *theoretical* model for the launch system. Specifically, we address the following problem:

A softball is launched straight up with an initial velocity of 20 m/s outdoors on a windless day. What is the maximum altitude that it'll reach, when will it hit the ground, and how fast will it be going when it hits?

This problem appears to be similar to a common problem a student might see in a beginning physics class. In order to make the problem tractable, these classes usually assume that the only force acting on the ball after launch is gravity and ignore other forces such as air resistance, also know as *drag*. These conditions are called "vacuum conditions," as if the launch were performed in an airless chamber. In our statement of the problem, however, the launch is performed "outdoors on a windless day"—clearly not a vacuum.

We'll build two versions of a discrete model for the trajectory of the ball, one that ignores air resistance and one that includes it, as shown in Figure 8.5. What

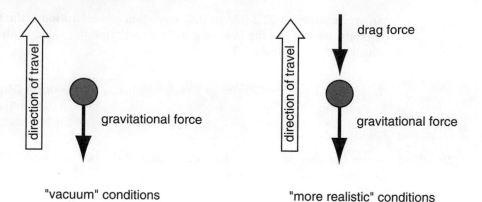

"vacuum" conditions   "more realistic" conditions

Figure 8.5 Forces acting on a ball immediately after launch. Beginning physics courses typically assume "vacuum conditions," where the only force acting on the ball is gravity. A more realistic model includes drag forces caused by air resistance that oppose the direction of travel.

we'll find is that even though a discrete model is only an approximation of the "real" trajectory, it enables us to include important effects such as air resistance that would be very difficult to model otherwise, and hence obtain a more realistic solution.

8.3.1 Problem Definition and Plan of Attack

As usual, we'll begin to solve this problem using our standard framework, first *defining* the problem, then *exploring* approaches, and *planning* an attack, before *implementing* the plan.

Define The first step in solving this problem is to make sure we understand it. This includes identifying the known and unknown quantities in the problem, drawing a diagram and defining the problem in terms of variables, and restating the problem in clearer or more specific terms. Fundamentally, this problem is about modeling the path—the series of intermediate states—that a softball will follow during the process of launching it straight up in the air with a given initial velocity. Once we've modeled the path, we can answer specific questions about it, such as the maximum altitude of the ball, when it returns to the ground, and how fast it's moving when it does.

The state of a system is defined by the values of properties of interest, and those properties are represented by state variables. For an object in motion, three properties of interest determine its state: its position, velocity, and acceleration. Figure 8.6 illustrates an example path of the ball using a discrete model in terms of these variables, and Table 8.1 lists the variables and their values, when known. We define y_{apogee} as the maximum height the ball reaches, and t_{impact} and v_{impact} as the time and velocity of the ball when it returns to the ground.

Given this understanding of the problem, we can restate it as:

Develop a discrete model for the path of a softball that is launched vertically with an initial velocity $v_0 = 20$ m/s and evaluate the model from time 0 through the time that the ball returns to earth. From the resulting series of steps, determine the maximum altitude y_{apogee} and the time and velocity at impact, t_{impact} and v_{impact}.

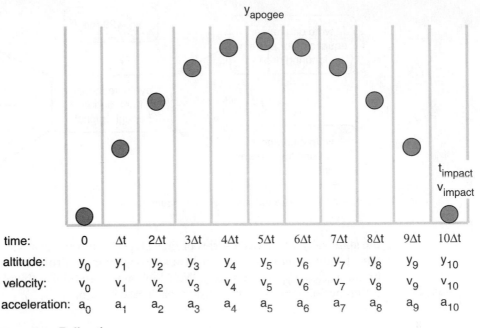

y_{apogee}

time:	0	Δt	$2\Delta t$	$3\Delta t$	$4\Delta t$	$5\Delta t$	$6\Delta t$	$7\Delta t$	$8\Delta t$	$9\Delta t$	$10\Delta t$
altitude:	y_0	y_1	y_2	y_3	y_4	y_5	y_6	y_7	y_8	y_9	y_{10}
velocity:	v_0	v_1	v_2	v_3	v_4	v_5	v_6	v_7	v_8	v_9	v_{10}
acceleration:	a_0	a_1	a_2	a_3	a_4	a_5	a_6	a_7	a_8	a_9	a_{10}

t_{impact}
v_{impact}

Figure 8.6 Ball path.

Explore This step is the pre-planning step in which we check that the problem makes sense, note any assumptions that must be made, identify key concepts and possible approaches, and assess the level of understanding required in solving the problem. Solving this problem requires *analysis* of the forces acting on the ball and the *application* of Newton's Laws and Euler's Method, define a model using difference equations to describe the motion.

We'll assume that the ball is subject to two different forces during flight: the force of gravity and the force of air resistance, called *drag*. In general, the drag force depends upon velocity and we'll assume a common empirical model where the drag is proportional to the square of the velocity.

Plan Figure 8.7 summarizes the plan for solving this problem. We'll construct two models, one ignoring drag and one including drag, and then compare the results. In either case, we begin by writing difference equations that describe the changes in

TABLE 8.1 State variables for modeling the trajectory of a softball.

Name	Description	Type	Value
t	time	behavioral	unknown
t_{impact}	time at impact	behavioral	unknown
$v(t)$	velocity (vs. time)	behavioral	unknown
v_0	initial velocity	environmental	20 m/s
v_{impact}	velocity at impact	behavioral	unknown
$y(t)$	height (vs. time)	behavioral	unknown
y_0	initial height	environmental	0 m
y_{apogee}	maximum height	behavioral	unknown

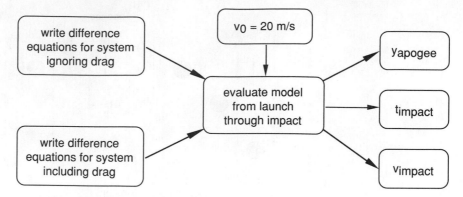

Figure 8.7 Plan for modeling a softball launch.

the state variable across a series of time steps. Next, we'll evaluate the models from launch with an initial velocity of 20 m/s through impact. From the results, we can then obtain y_{apogee}, t_{impact}, and v_{impact}. We'll look more carefully at the plan for writing the difference equations as we consider each case in turn.

8.3.2 Modeling the Softball Trajectory Without Drag

Before developing a more accurate model that takes drag into consideration, we'll first develop a simplified model that considers only the effects of gravity on the trajectory of the ball. First, we'll write the state equations for the model, and then evaluate it from launch through impact.

Setting up the State Equations   The plan for the formulation of the state equations has three steps:

1. Given the altitude of the softball at a given time step, we can determine its altitude at the next step if we know its velocity, which is the rate of change of position.
2. Similarly, we can determine the velocity at the next step from the current velocity and the acceleration, which is the rate of change of velocity.
3. Finally, we can determine the acceleration from Newton's Laws, which describe the relationship between the acceleration of an object and the forces acting upon it.

First, we write the state equation for the altitude.

$$v_i = \frac{\Delta y_i}{\Delta t} = \frac{y_{i+1} - y_i}{\Delta t}$$

$$y_{i+1} = y_i + v_i \Delta t$$

Next, we write the state equation for the velocity.

$$a_i = \frac{\Delta v_i}{\Delta t} = \frac{v_{i+1} - v_i}{\Delta t}$$

$$v_{i+1} = v_i + a_i \Delta t$$

$$F_g = mg$$

Figure 8.8   The only force acting on the ball between launch and impact is gravity.

Finally, we need to predict the velocity at the current state. The only force acting on the ball after launch is gravity, as illustrated in Figure 8.8. From Newton's Second Law of Motion, we know that for an object with constant mass, the net force equals its mass times its acceleration, and for a gravitational force, the acceleration is constant and equal to g, or 9.81 m/s^2 towards the ground.

$$F = ma = mg$$

Thus, a_i at every time step is simply equal to the constant g.

Evaluating the State Equations Between Launch and Impact   In summary, our model now consists of three state equations to evaluate at every time step:

$$y_{i+1} = y_i + v_i \Delta t$$
$$v_{i+1} = v_i + a_i \Delta t$$
$$a_{i+1} = g$$

A convenient way to evaluate the state equations is to arrange them in a table in which the columns are the state variables and each row is a time step. Table 8.2 shows the first few steps in the evaluation. Each row in the table contains both the formula version of the state equations and the numerical calculations. The first row in the table lists the initial state of the ball at step 0, in which

$$y_0 = 0 \qquad v_0 = 20 \qquad a_0 = g = -9.81$$

The second row of the table calculates the state of the ball at step 1, assuming a time step size Δt equal to 0.1 s. As expected, the altitude is increasing and the velocity is

TABLE 8.2   Evaluation of the state equations of the softball trajectory for the first few time steps.

Step	Time	Altitude	Velocity	Acceleration
0	t_0	y_0	v_0	a_0
	0	0	20	g
	0	0	20	-9.81
1	$t_1 = t_0 + \Delta t$	$y_1 = y_0 + v_0 \Delta t$	$v_1 = v_0 + a_0 \Delta t$	$a_1 = g$
	$0 + 0.1$	$0 + 20(0.1)$	$20 - 9.81(0.1)$	-9.81
	0.1	2.0	19.02	-9.81
2	$t_2 = t_1 + \Delta t$	$y_2 = y_1 + v_1 \Delta t$	$v_2 = v_1 + a_1 \Delta t$	$a_2 = g$
	$0.1 + 0.1$	$2.0 + 19.02(0.1)$	$19.02 - 9.81(0.1)$	-9.81
	0.2	3.90	18.04	-9.81

decreasing. The third row in the table continues the process at step 2. Note that as we fill in the table, the calculations in a given row only depend on values in the row immediately above it.

Conceptually, it would be simple to continue adding rows to the table, evaluating the state of the softball at successive time steps until the ball reaches its maximum altitude, begins falling, and then finally returns to an altitude of zero. In practice, however, this would be very tedious to do by hand. This is precisely the type of calculation ideally suited for a computer—nothing but lots of repetition of simple arithmetic. One way of performing these calculations on a computer is to use a spreadsheet, which arranges formulas in a table much like Table 8.2. A second approach is to write a program in a language such as MATLAB—or one of many others—that uses a "loop" to sequentially evaluate each row in the table.

Table 8.3 shows selected time steps from a complete evaluation of the softball trajectory with no drag from launch through impact with a time step length of Δt equal to 0.1 s. In the transition between steps 20 and 21, the sign of the velocity switches from positive to negative, meaning that the ball has reached its apogee and is headed back down. The maximum value of y in the table is at step 21,

$$y_{\text{apogee}} = 21.40\,\text{m}$$

The altitude continues to decrease after step 21, until it finally reaches zero between steps 41 and 42. Taking step 42 as the time of impact,

$$t_{\text{impact}} = 4.2\,\text{s}$$
$$v_{\text{impact}} = -21.20\,\text{m/s}$$

One of the curious results of the model is that the velocity at impact, -21.2 m/s, is *faster* than the initial velocity of 20 m/s. Physically, this is impossible; the model produces this value, however, because of errors that accumulate because time is broken into discrete steps. This error will decrease if we use a smaller time step

TABLE 8.3 Calculations.

Step	t Time, s	y Altitude, m	v Velocity, m/s	a Acceleration, m/s^2	
0	0.0	0.00	20.00	−9.81	
1	0.1	2.00	19.02	−9.81	
2	0.2	3.90	18.04	−9.81	
...	...	...	...	...	
19	1.9	21.22	1.36	−9.81	
20	2.0	21.36	0.38	−9.81	
21	2.1	21.40	−0.60	−9.81	*apogee*
...	...	...	...	...	
40	4.0	3.48	−19.24	−9.81	
41	4.1	1.56	−20.22	−9.81	
42	4.2	−0.46	−21.20	−9.81	*impact*

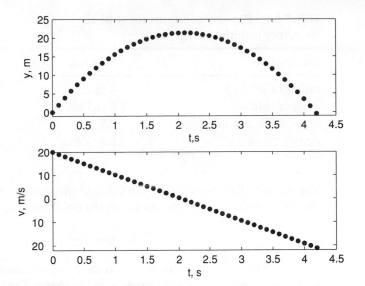

Figure 8.9 Plots of altitude and velocity versus time for the softball trajectory without drag.

length. For example, if we reduced Δt to 0.001 s, the ball would impact the ground between steps 4079 and 4080 and the new values at apogee and impact would be

$$y_{\text{apogee}} = 20.40 \text{ m}$$
$$t_{\text{impact}} = 4.08 \text{ s}$$
$$v_{\text{impact}} = -20.01 \text{ m/s}$$

If we continue to shrink the time step length, v_{impact} would eventually converge to 20 m/s.

8.3.3 Modeling the Softball Trajectory with Drag

Now, we turn to a more accurate model of the softball trajectory that includes the effects of air resistance or drag. The basic logic of the changes from state to state is nearly the same as the case without drag; we'll still calculate the next altitude from the current altitude and the velocity, and calculate the new velocity from the current velocity and the acceleration. The difference is that the acceleration is no longer constant, but rather is calculated at each time step from the net force on the ball—which includes both gravity and drag—and Newton's Second Law of Motion.

Air Resistance and Drag Force   Drag force is the force that opposes the motion of a projectile as a result of air resistance. As anyone who has held their hand out the window of a moving vehicle knows, the magnitude of this force is somehow related to the velocity of the object, as well the shape of the profile that the object presents to the air stream. Experimentation shows that the drag force is proportional to the *square* of velocity, so an *empirical* model for drag is

$$F_{\text{drag}} = k_d v^2$$

TABLE 8.4 Physical constants for softball trajectory including drag.

Description	Value
radius of softball	4.85 cm
mass of softball	0.185 kg
drag coefficient of sphere	0.5
density of air	1.29 kg/m^3

The constant k_d depends upon the size and shape of the object, and the properties of the fluid—such as air—through which the object moves. k_d is usually expressed as:

$$k_d = \frac{C_D}{2}\rho A \tag{8.4}$$

where ρ is the density of the fluid, A is the cross-sectional area of the object, and C_D is the *drag coefficient*, which depends on the shape. The drag coefficient for a sphere has been computed to be approximately 0.5. Table 8.4 lists other physical constants needed to calculate the drag force on a softball. Substituting values into Equation (8.4), we obtain the following result:

$$
\begin{aligned}
F_{\text{drag}} &= k_d v^2 \\
k_d &= \frac{C_D}{2}\rho A \\
&= \frac{0.5}{2}(1.29)(\pi)(.0485)^2 \\
&= .0024 \text{ kg/m}
\end{aligned}
\tag{8.5}
$$

Setting Up the State Equations As before, we begin with the definition of velocity as the rate of change of altitude and the definition of acceleration as the rate of change of velocity:

$$v_i = \frac{\Delta y_i}{\Delta t} = \frac{y_{i+1} - y_i}{\Delta t}$$

$$a_i = \frac{\Delta v_i}{\Delta t} = \frac{v_{i+1} - v_i}{\Delta t}$$

From these, we obtain the state equations for altitude and velocity at the next step.

$$y_{i+1} = y_i + v_i \Delta t$$
$$v_{i+1} = v_i + a_i \Delta t$$

Now, we need to determine the acceleration at each state, which we'll do by finding the net force on the ball and applying Newton's Second Law. As shown in Figure 8.10, two forces act on the ball, the gravitational force, F_g and the drag force, F_{drag}. F_g is always directed towards the ground, and F_{drag} always opposes the direction of motion. Thus as the ball rises, F_{drag} points down and as the ball falls, F_{drag} points up. Taking this into consideration, we restate Equation 8.5 as

$$F_{\text{drag}} = -\text{sign}(v)\, k_d v^2$$

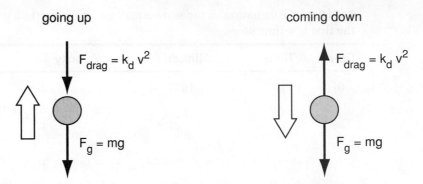

going up   coming down

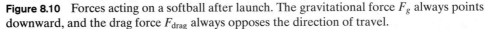

Figure 8.10 Forces acting on a softball after launch. The gravitational force F_g always points downward, and the drag force F_{drag} always opposes the direction of travel.

where sign(v) equals $+1$ when the direction of travel is up and equals -1 when the direction of travel is down. The net force F is the sum of F_g and F_{drag}:

$$F = F_g + F_{\text{drag}}$$
$$= mg - \text{sign}(v)\, k_d v^2$$

Finally, we calculate the acceleration from Newton's Law:

$$F = ma$$
$$a = \frac{F}{m}$$

Thus, the acceleration at step $i + 1$ is

$$a_{i+1} = \frac{mg - \text{sign}(v_{i+1})\, k_d v_{i+1}^2}{m} \tag{8.6}$$

Looking at equation (8.6), we note that the acceleration at a given state only depends on the velocity at that state and some physical constants. Also note that if k_d is set to zero, then the acceleration reduces to "vacuum" conditions, where $a = mg$.

Evaluating the State Equations Between Launch and Impact As before, we have three state equations—one for altitude, one for velocity, and one for acceleration.

$$y_{i+1} = y_i + v_i \Delta t$$
$$v_{i+1} = v_i + a_i \Delta t$$
$$a_{i+1} = \frac{mg - \text{sign}(v_{i+1})\, k_d v_{i+1}^2}{m}$$

Table 8.5 shows steps 0, 1, and 2 for an evaluation of the state equations with the following constants:

$$\Delta t = 0.1\,\text{s} \qquad v_0 = 20\,\text{m/s}$$

$$m = 0.185\,\text{kg} \qquad g = -9.81\,\text{m/s}^2 \qquad k_d = .0024\,\text{kg/m}$$

TABLE 8.5 Evaluation of the state equations of the softball trajectory with drag for the first few time steps.

Step	Time	Altitude	Velocity	Acceleration
0	t_0	y_0	v_0	$\frac{mg - \text{sign}(v_0)\, k_d v_0^2}{m}$
	0	0	20	$\frac{(0.185)(-9.81) - (.0024)(20)^2}{0.185}$
	0	0	20	-14.96
1	$t_1 = t_0 + \Delta t$	$y_1 = y_0 + v_0 \Delta t$	$v_1 = v_0 + a_0 \Delta t$	$\frac{mg - \text{sign}(v_1)\, k_d v_1^2}{m}$
	$0 + 0.1$	$0 + 20(0.1)$	$20 - 14.96(0.1)$	$\frac{(0.185)(-9.81) - (.0024)(18.50)^2}{0.185}$
	0.1	2.0	18.50	-14.22
2	$t_2 = t_1 + \Delta t$	$y_2 = y_1 + v_1 \Delta t$	$v_2 = v_1 + a_1 \Delta t$	$\frac{mg - \text{sign}(v_2)\, k_d v_2^2}{m}$
	$0.1 + 0.1$	$2.0 + 18.50(0.1)$	$18.50 - 14.22(0.1)$	$\frac{(0.185)(-9.81) - (.0024)(17.08)^2}{0.185}$
	0.2	3.85	17.08	-13.57

Note that the calculations for a given row depend only on values in the row above or in the same row.

Figure 8.11 shows plots of the altitude, velocity, and acceleration of a softball from just after launch through impact. The plots were made with a time step size of Δt equal to 1 millisecond, so the error related to step size is very small. Using this

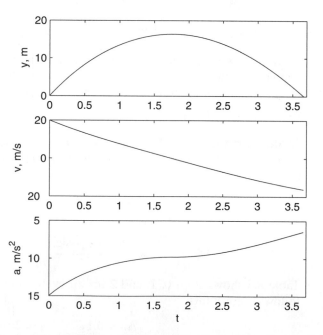

Figure 8.11 Altitude, velocity, and acceleration of a softball versus time, including the effects of drag. The model assumes that the ball was launched with an initial velocity of 20 m/s and Δt between evaluations is 1 millisecond.

smaller step size, the values for the maximum altitude and the time and velocity at impact are:

$$y_{apogee} = 16.39 \, m$$
$$t_{impact} = 3.66 \, s$$
$$v_{impact} = -16.20 \, m/s$$

All of these values are smaller in magnitude than their corresponding values when we ignored drag. Specifically, y_{apogee} is 20 percent smaller, t_{impact} is 7 percent smaller, and v_{impact} is 14 percent smaller.

The key to understanding the behavior of the system when we include drag is to look at the plot of acceleration versus time at the bottom of Figure 8.11. In the case with no drag, the acceleration was constant and equal to g or 9.81 m/s². When we consider drag, the acceleration changes with each time step. In the middle of the plot, when the ball is at apogee and its velocity is zero, there's no drag and the acceleration is equal to g. When the ball is on its way up, the acceleration is *more negative* than g, because the air resistance further "drags" the ball towards the ground. When the ball is on its way down, however, the acceleration is *less negative* than g. This is because the air resistance tends to "drag" the ball upwards, slowing its descent. An extreme case of an object with a lot of drag is a parachute—its drag coefficient is so high that it falls with constant low velocity. We can begin to see the "parachute effect" in the plot of velocity versus time in Figure 8.11, in which we observe that as time goes on, the velocity starts to gradually flatten out toward a constant value.

The effects of drag are much more pronounced if we launch the softball with a greater initial velocity. Figure 8.12 compares the results of launching the ball with an initial velocity of 50 m/s, with and without drag. When we include the effects of

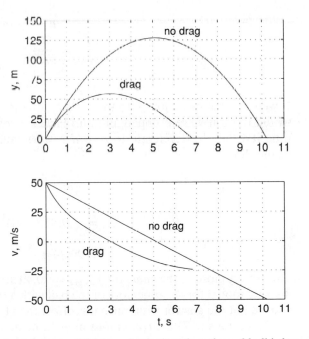

Figure 8.12 Effects of drag on altitude and velocity when the softball is launched with an initial velocity of 50 m/s.

drag, the maximum altitude is less than half as high as when we ignore drag. Also, the model predicts that the ball would be in the air for approximately 30 percent less time when drag is considered. Most striking, however, is the "parachute effect." In the plot of velocity versus time with drag, we see that the velocity of the ball never gets faster than 25 m/s. As the ball falls, the drag force increases—in proportion to the *square* of the velocity—until the drag force is the same as the gravitational force. At this point, the net force on the ball will be zero and it'll stop accelerating. The value of this velocity is known as the *terminal velocity*.

In summary, if we look back at the slingshot problem in Chapter 4, in which the goal was to determine launcher settings to shoot a softball at a target downrange, it's clear that we'd need to consider drag to build an accurate theoretical model.

8.3.4 Continuous Versus Discrete Models

You may remember some formulas from a physics class for the final velocity and distance traveled by an object with acceleration a over time t:

$$v = at \tag{8.7a}$$

$$y = \frac{1}{2}at^2 \tag{8.7b}$$

If such simple formulas existed, why didn't we use them to find the maximum altitude of the softball? Before explaining why we didn't, let's see how we could have used them. First, these formulas only pertain to vacuum conditions—they don't consider drag. The first step in calculating the altitude is to find the time t that the ball is in motion. Assuming that the initial velocity of the ball is 20 m/s, from the first equation, we get:

$$t = \frac{v}{a}$$

$$= \frac{20}{9.81}$$

$$= 2.04 \, \text{s}$$

If we plug this value for t into the second equation, this will tell us how far the ball would fall if it were dropped, until it reaches a velocity of 20 m/s. This is the same time it'd take to decelerate to zero if it were launched with an initial velocity of 20 m/s. So,

$$y = \frac{1}{2}at^2$$

$$= \frac{1}{2}(9.81)(2.04)^2$$

$$= 20.39 \, \text{m}$$

This is very close to the value for $y_{\text{apogee}} = 20.40$ m that we obtained when we used a discrete model for the trajectory of the ball with a time step size Δt of 1 millisecond, and we got the answer with much less effort than before.

Equations (8.7) are *continuous* models, meaning that they're defined for every point in time, as compared to discrete models that are only defined at intervals. A continuous model of a changing system can be very useful, provided that we can find one. This is one of the important topics calculus addresses in an area called *differential equations*. We won't say much about differential equations here, except to note that Equations (8.7) are solutions to the two differential equations

$$\frac{dy}{dt} = v \qquad \frac{d^2y}{dt^2} = a \qquad (8.8)$$

The process of *formulating* a differential equation is very similar to the process of formulating and setting up the difference equations we used in our models. The process of solving them, however, is very different. While Equations (8.8) have simple solutions, some differential equations that appear simple can have very complex solutions that are extremely difficult to find. For example, a differential equation for the velocity of a softball with drag is:

$$\frac{dv}{dt} = \frac{mg - k_d v^2}{m}$$

Note how similar this is to the difference equation (8.6). A continuous solution to this differential equation, however, is much more complicated:

$$v = \frac{\sqrt{mg}\,\tan\left(\sqrt{gk_d}\left[-\left(\frac{t}{\sqrt{m}}\right) + \frac{\arctan\left(\frac{v_0\sqrt{k_d}}{\sqrt{mg}}\right)}{\sqrt{gk_d}}\right]\right)}{\sqrt{k_d}}$$

Undergraduate calculus and physics courses will typically simplify or "idealize" problems — such as by neglecting the effects of drag — in order to make them solvable by the kinds of techniques that undergraduates learn. The tradeoff is that the results from the simplified version of the problem may be very different from the results from the original problem. For example, if we neglected the effects of drag, a parachutist would hit the ground at the same velocity as if the chute never opened! While discrete models are approximations of continuous models and have the problem of errors arising from the discrete step size, they can include effects that are difficult to solve analytically, and usually require only basic arithmetic to implement.

8.4 RUNNING OUT OF GAS

In Chapter 1, we looked at the world as a vast network of interconnections. Of all the things that tie us together, few have greater significance than the limited natural resources we all must share to survive. One such resource is safe drinking water. Another is fuel for producing energy. In this section, we'll apply some of the same modeling techniques we used to predict how long it would take for a falling ball to hit the ground, to predict when our supply of petroleum will run dry.

Of all the types of fuel society uses to produce energy, petroleum—oil—is the most important today. Figure 8.13 shows energy consumption in the United States by type, between 1949 and 2005. Petroleum, by far, makes up for the largest part of this, followed by the other "fossil fuels," natural gas and coal, which together accounted for 86 percent of the energy consumed in 2005. Nuclear power comes next, with 8 percent in 2005. Renewable energy sources, which include hydroelectric, biomass (ethanol), geothermal, solar, and wind power, together accounted for only 6 percent

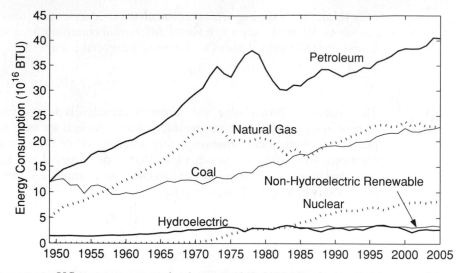

Figure 8.13 U.S. energy consumption by type, 1949–2005. Non-hydroelectric renewable energy resources include geothermal, biomass, solar, and wind power. Source: U.S. Energy Information Administration [Ene].

of energy consumption in 2005. Of all the petroleum consumed, 66 percent in 2005 was used for transportation, while 98 percent of all energy for transportation came from petroleum.

For the remainder of this section, we first provide some background on the problem and look at some basic patterns of oil supply and consumption in different parts of the world. Next, we define the problem we're going to solve—modeling the world's oil supply as a large tank that is being drained at different rates by different countries. Then, we'll develop our three main models:

- a model based on the Law of Conservation of Mass that relates the flow of oil to different countries to the remaining volume of the supply,
- a model for the average amount of oil that a person consumes per year,
- a model for population growth in different parts of the world.

By combining these three models together, we'll then predict how much oil we consume year by year, until the contents of the tank are depleted.

8.4.1 Background

Our basic model for oil consumption is to view the world's oil supply as a large tank, with each country pumping oil out of the tank at some flow rate, and with new discoveries adding oil to the tank. If oil is being pumped out at a rate faster than it's discovered—which is the case—then the tank will run dry. The question is, when.

How Much Oil is in the Tank?   We don't know exactly how much oil lies under the ground, and of the sources we know, some are more easily recovered than others. For our model, we'll assume that the starting volume of oil in the tank is the world's *proved oil reserve*. According to the Energy Information Administration of the U.S. Department of Energy, "proved reserves are estimated quantities that analysis of geologic and engineering data demonstrates with reasonable certainty are recoverable

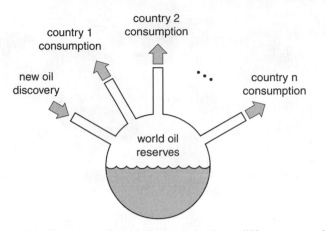

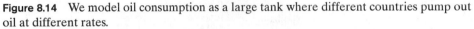

Figure 8.14 We model oil consumption as a large tank where different countries pump out oil at different rates.

under existing economic and operating conditions" [Ene]. There are different estimates for this quantity, and different views on what is "economically recoverable." Three reliable sources, the *BP Statistical Review* [bp:06], *Oil and Gas Journal* [oil05], and *World Oil Magazine*, place the proved reserves at the end of 2005 at between 1.1 and 1.3 trillion barrels, where 1 barrel is 42 US gallons.[1] For our model, we'll optimistically assume the upper end of this range of 1.3 trillion barrels. Of course, the beauty of having a model is that this quantity can be changed easily.

Figure 8.15 shows the geographical distribution of the proved reserves, according to the *BP Statistical Review 2006* [bp:06]. More than 60 percent of the reserves are in the Middle East, while less than 10 percent are in North America and the Asia-Pacific regions combined. As we'll see in the next section, this distribution is very different from where oil is consumed.

Where Does the Oil Go? Where oil lies is strictly a matter of geology; where it goes is a matter of economics. Roughly speaking, the quantity of oil that each country consumes depends upon two factors: the size of its population and the amount that each person, on average, uses. Behind these figures is the backdrop of the distribution of income around the globe. In short, most of the world's oil consumption is concentrated in a small number of wealthy countries. Most people, however, live in much poorer nations and use far less oil per person. Because the total population of these "developing nations" is so large, however, they still account for a large part of oil consumption. Further, as their population and standard of living increases, so will their oil usage.

The World Bank, which provides financial and technical assistance to developing countries around the world, classifies national economies by their *gross national income per capita* (GNI per capita)—the total income of the country from all sources, divided by the size of the population. They use four groups listed below:

- **High Income** (GNI per capita $10,726 or more, 56 countries): This group includes the United States and Canada, Western Europe, Japan, South Korea, Australia,

---

[1] The abbreviation for barrels of petroleum is "bbl," which stands for "blue barrel." During the 1860s, there was no standard container for oil. By the 1870s, the 42-gallon blue barrel Standard Oil used was adopted as the standard quantity [Ene].

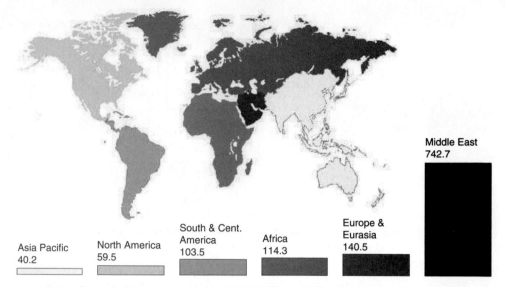

Figure 8.15   Proved oil reserves at end of 2005, billions of barrels. Source: *BP Statistical Review 2006* [bp:06].

New Zealand, and Singapore, Israel and some of the oil-producing nations in the Middle East.

- **Upper Middle Income** (GNI per capita $3,466–$10,725, 40 countries): This group includes Mexico and the wealthier nations in South and Central America, the wealthier nations in Eastern Europe and Eurasia (including Russia), Malaysia, South Africa, and small island nations with significant tourist economies.

- **Lower Middle Income** (GNI per capita $876–$3,465, 58 countries): This group includes the world's most populous country, China, as well as much of South America and the former Soviet Union. It also includes most of the Middle East and a few African nations, such as Morocco, Tunisia, and Angola.

- **Low Income** (GNI per capita $875 or less, 54 countries): The poorest group of countries accounts for nearly 40 percent of the world's population. It includes most of Africa, and South and Southeast Asia, including India, the world's second most populous nation. It also includes Haiti, a few hundred miles off the coast of the United States.

Figure 8.16 compares population size and oil consumption across each of these four economic groups for 2005. As the first plot in the figure shows, the high and upper middle income groups together accounted for only 25% of the world's population, yet consumed nearly 75% of the oil. The second plot in Figure 8.16 compares the per capita or average oil consumption per person for each group, obtained by dividing the national oil consumption by the size of the population. This reveals the dramatic difference in oil use between people in the wealthiest and poorest countries. On average, people in the high income nations use *more than 25 times* as much oil per year as people in the low income nations. The United States, which is the biggest oil consumer in the world, consumes over 25 barrels of oil per person per year, while China, which is the most populous, uses less than 2.

TABLE 8.1 The World Bank classification of national economies.

High Income (GNI per capita $10,726 or more)

Andorra	Denmark	Italy	Portugal
Antigua & Barbuda	Faeroe Islands	Japan	Puerto Rico
Aruba	Finland	Korea, Rep.	Qatar
Australia	France	Kuwait	San Marino
Austria	French Polynesia	Liechtenstein	Saudi Arabia
Bahamas, The	Germany	Luxembourg	Singapore
Bahrain	Greece	Macao, China	Slovenia
Belgium	Greenland	Malta	Spain
Bermuda	Guam	Monaco	Sweden
Brunei Darussalam	Hong Kong, China	Netherlands	Switzerland
Canada	Iceland	Netherlands Antilles	United Arab Emirates
Cayman Islands	Ireland	New Caledonia	United Kingdom
Channel Islands	Isle of Man	New Zealand	United States
Cyprus	Israel	Norway	Virgin Islands (U.S.)

Upper Middle Income (GNI per capita $3,466–$10,725)

American Samoa	Equatorial Guinea	Mauritius	Seychelles
Argentina	Estonia	Mayotte	Slovak Republic
Barbados	Gabon	Mexico	South Africa
Belize	Grenada	No. Mariana Islands	St. Kitts & Nevis
Botswana	Hungary	Oman	St. Lucia
Chile	Latvia	Palau	St. Vincent & Grenadines
Costa Rica	Lebanon	Panama	Trinidad & Tobago
Croatia	Libya	Poland	Turkey
Czech Republic	Lithuania	Romania	Uruguay
Dominica	Malaysia	Russian Federation	Venezuela, RB

Lower Middle Income (GNI per capita $876–$3,465)

Albania	Cuba	Jordan	Samoa
Algeria	Djibouti	Kazakhstan	Serbia & Montenegro
Angola	Dominican Republic	Kiribati	Sri Lanka
Armenia	Ecuador	Lesotho	Suriname
Azerbaijan	Egypt	Macedonia, FYR	Swaziland
Belarus	El Salvador	Maldives	Syria
Bolivia	Fiji	Marshall Islands	Thailand
Bosnia & Herzegovina	Georgia	Micronesia	Tonga
Brazil	Guatemala	Moldova	Tunisia
Bulgaria	Guyana	Morocco	Turkmenistan
Cameroon	Honduras	Namibia	Ukraine
Cape Verde	Indonesia	Nicaragua	Vanuatu
China	Iran, Islamic Rep.	Paraguay	West Bank & Gaza
Colombia	Iraq	Peru	
Congo, Rep.	Jamaica	Philippines	

(continued)

TABLE 8.6 The World Bank classification of national economies (continued).

Low Income (GNI per capita $875 or less)

Afghanistan	Gambia, The	Mauritania	Somalia
Bangladesh	Ghana	Mongolia	Sudan
Benin	Guinea	Mozambique	Tajikistan
Bhutan	Guinea-Bissau	Myanmar	Tanzania
Burkina Faso	Haiti	Nepal	Timor-Leste
Burundi	India	Niger	Togo
Cambodia	Kenya	Nigeria	Uganda
Cent. African Republic	Korea, Dem. Rep.	Pakistan	Uzbekistan
Chad	Kyrgyz Republic	Papua New Guinea	Vietnam
Comoros	Lao PDR	Rwanda	Yemen
Congo, Dem. Rep.	Liberia	Sao Tomé & Principe	Zambia
Cte d'Ivoire	Madagascar	Senegal	Zimbabwe
Eritrea	Malawi	Sierra Leone	
Ethiopia	Mali	Solomon Islands	

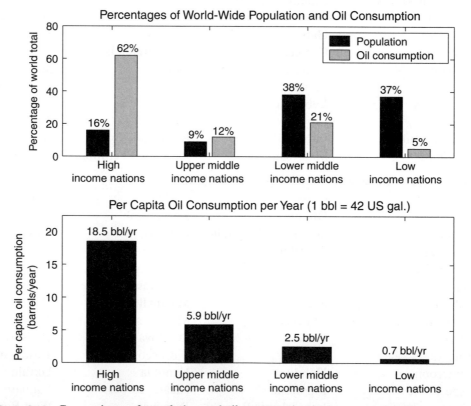

Figure 8.16 Comparisons of population and oil consumption by income group. Sources: *World Bank World Development Indicators* [Ban06], *BP Statistical Review* [bp:06], *CIA World Factbook* [U.S06b].

Historic Trends, 1965–2005   While Figure 8.16 provides a snapshot of the "big picture" of oil consumption in 2005, it's important to recognize that there are many individual stories behind these statistics, unique to the natural and geopolitical circumstances of each country. The uncertainty of events such as war, famine, and natural disasters makes it impossible to predict exactly how both population and oil consumption will change in the future. To illustrate, we'll look at past trends for three different situations: examples of high income industrialized nations, lower income developing nations, and Eastern Europe and the former Soviet Union.

Industrialized Nations: U.S., Japan, France, and Australia   Figure 8.17 shows 40 years of data for population and per capita oil consumption between 1965 and 2005 for four industrialized nations in different parts of the globe: the United States, Japan, France, and Australia. As the top plot shows, the populations in each country have grown steadily. The U.S. population has grown the fastest in terms of number of people per year, partly because the population is higher to start with, but also because of its percentage increase per year; the U.S. and Australia grew at rates of approximately 1.2% per year, while Japan and France grew at rates of approximately 0.2% and 0.4% per year, respectively.

The per-capita oil consumption trends have remarkably similar shapes over the forty-year period, with peaks and dips at the same times, although the consumption in the U.S. was at a significantly higher level than the other three countries. Per-capita

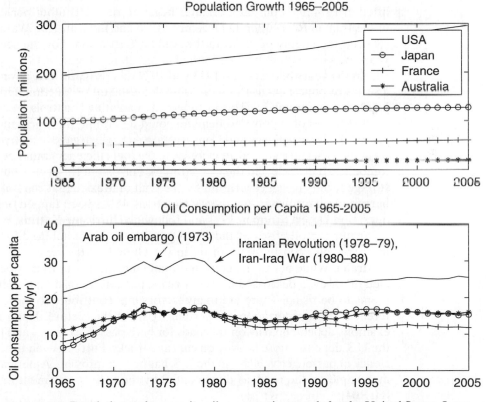

Figure 8.17   Population and per-capita oil consumption trends for the United States, Japan, France, and Australia, 1965–2005. Sources: *BP Statistical Review* [bp:06] and U.S. Census Bureau [U.S06a].

Figure 8.18 Lines at a U.S. gas station, June 15, 1979. Photo from U.S. Library of Congress.

consumption was rising sharply—reaching a peak of over 30 barrels per year in the U.S.—until 1973, when several Arab OPEC nations stopped selling oil to the United States and other nations in protest of their support for Israel in the Arab-Israeli "Yom Kippur" War. The U.S. government enforced gasoline rationing—where cars with even license plates could only purchase gas on even dates and cars with odd license plates on odd dates—for the first time since World War II. The embargo was lifted in 1974 and consumption later began to rise to another peak, when the start of the Iranian Revolution in December 1978 and the Iran-Iraq War in 1980 severely reduced the supply of oil from the Middle East again. For the second time in six years, cars lined up to buy gas in the United States (Figure 8.18).

In the years between the 1973 and 1979 oil crises, the U.S. government enacted a series of policies aimed at increasing domestic oil production and decreasing dependence on foreign oil. This included "deregulating" controls on oil prices, so that market forces of supply and demand could determine the price. In 1975, Congress passed the Energy Policy and Conservation Act, which gave incentives for producing oil in the U.S. and required an increase in the fuel efficiency (miles per gallon) for automobiles. It also created the U.S. Strategic Petroleum Reserve, a huge underground storage facility in salt caverns under the Gulf of Mexico that can hold up to a billion barrels of oil for use in emergency situations. It has been tapped only twice: during operation Desert Storm in 1991 and following Hurricane Katrina in 2005.

After the oil crises of the 1970s passed, there are subtle differences between the per-capita consumption of oil in the United States versus Japan, France, and Australia. While per-capita consumption in the latter countries appears to have leveled off or even decreased in recent years, per-capita consumption in the U.S. again seems to be rising. There are many factors that contribute to these trends, but differences in transportation are certainly part of the picture. Japan and France, for example, depend heavily on railroads for both passenger travel and freight, whereas the U.S. depends more heavily on cars and trucks. Further, even though the fuel efficiency of passenger vehicles in the U.S. has been improving, Americans are spending more time on the road, as documented by the U.S. Department of Transportation [HrR04]:

- the average number of vehicles per household has grown from close to 1 in 1969 to close to 2 in 2001;

- the average number of miles traveled per person each day in private vehicles grew from 30.8 in 1990 to 35.5 in 2001, while the miles traveled on public transportation shrank from 0.74 to 0.47 over the same period;

- the average time spent driving private vehicle grew from 49 to 62 minutes per day between 1990 and 2001.

Developing Nations: China, India, and Africa   Both the population growth and oil consumption trends for developing nations have been markedly different from those of the industrialized nations in the previous example. Figure 8.19 illustrates these trends from 1965 through 2005 for China, India, and the average of all nations in Africa. The populations in all three cases have grown dramatically. The annual growth rate in China is 0.9%, which is less than the growth rate in the U.S. (1.2%), but its population is more than 4 times as big. The growth rate in Africa, however, is more than twice that in the U.S. at approximately 2.5%. This is despite the fact that more than 7% of the population in sub-Saharan Africa is infected with the HIV virus and the prevalence in some individual nations is over 30%, as compared with 0.4% across all high income nations [Ban06].

The crises that shook oil consumption in the industrialized nations in the 1970s had little or no impact in developing nations that consume far less oil. As Figure 8.19 shows, however, the trends in oil usage in China, India, and Africa are very different. Although Africa has far more oil than either, per-capita consumption in China, and to a lesser degree India, is growing rapidly, while growth in Africa has stagnated. This is a direct reflection of the burgeoning industrial economies in both of these nations.

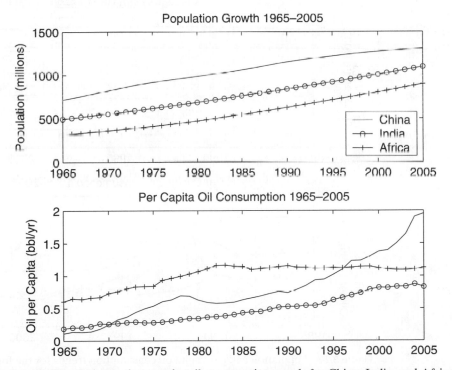

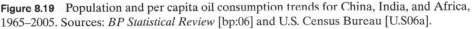

Figure 8.19 Population and per capita oil consumption trends for China, India, and Africa, 1965–2005. Sources: *BP Statistical Review* [bp:06] and U.S. Census Bureau [U.S06a].

In particular, per-capita consumption in China passed India in 1970 and Africa in 1997, and continues to rise sharply.

The Former Soviet Union: Russia, Romania, and Kazakhstan By the time that Mikhail Gobachev was appointed Secretary General in 1985, the economy of the Soviet Union had reached a breaking point. Decades of a rampant black market, an arms race with the United States capped by increased pressures from the Reagan administration, and a war in Afghanistan all contributed to pushing the Soviet Union to its limits as popular discontent boiled. When Gorbachev announced his policies of *glasnost* or "political openness" and *perestroika* or "economic restructuring" in 1986, he triggered a chain of events that began with a wave of revolution against the Communist governments of Soviet allies in Eastern Europe, and that ultimately led to the official breakup of the Soviet Union in 1991.

The strain of political and economic upheaval on Eastern Europe and the nations of the former Soviet Union led to drops in both population and per capita oil consumption over the 1990s, as shown in Figure 8.20. Romania, which was occupied by the Soviet army following World War II, saw an uprising in 1989 that led to the overthrow and execution of leader Nicolae Ceauşescu in 1989, and free elections the following year. Kazakhstan was conquered by Russia in the late 1700s and became a Soviet Republic in 1936. *Glasnost* gave rise to student protests against the communist system in 1986 and Kazakhstan formally declared independence in 1991. Proved oil reserves in Russia today are three times that of the United States, yet per capita oil consumption is less than one-third that of the U.S.

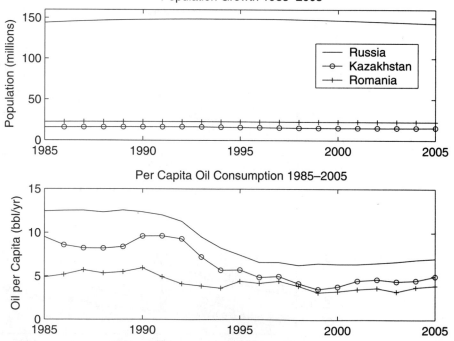

Figure 8.20 Population and per-capita oil consumption trends for the former Soviet Union countries Russia and Kazakhstan, 1985–2005. Sources: *BP Statistical Review* [bp:06] and U.S. Census Bureau [U.S06a].

TABLE 8.7   Assumed values for population and oil consumption growth.

	Population		Oil Consumption per Capita	
	2005 millions	Annual Growth Rate	Barrels per Year	Annual Growth Rate
High Income	1,011	0.4%	18.5	0%
Upper Middle Income	599	0.6%	5.9	1%
Lower Middle Income	2,475	0.8%	2.5	2%
Low Income	2,353	1.7%	0.7	3%

8.4.2   Problem Definition and Plan of Attack

As the scenarios from the last section showed, there are many factors that affect oil consumption, making it very difficult to forecast the future. Despite uncertainty in the details, however, oil consumption will inexorably increase and then decline. The tank may not actually run dry, but eventually oil will become prohibitively expensive to produce as proved reserves are depleted. The goal of our analysis, however, isn't to predict exactly when the last drop of oil is pumped. Rather, we want to get a sense of the urgency of the problem: is it years, decades, or centuries away? Most importantly, the point of this exercise isn't to forecast our doom; on the contrary, we need to understand the issues so that we can plan sufficiently to engineer our way out of a crisis.

Assumptions   Table 8.7 summarizes the assumed values for the main parameters of our model. The population figures come from both the U.S Census Bureau [U.S06a] as well as the World Bank *World Development Indicators* [Ban06]. The oil consumption figures are derived from information from the *BP Statistical Review* [bp:06], and the U.S. Central Intelligence Agency *World Factbook* [U.S06b]. Of all of the numbers in the table, the values for the growth rate of per-capita oil consumption called for the greatest judgment. For the high income nations, for example, we assumed that there's no increase in per-capita oil consumption beyond 2005 levels; this seems to be the trend in Western Europe and Japan, but is not the case for the United States.

Plan   The basic idea behind the model is to estimate the rate at which oil flows from the "tank" of the worldwide proved oil reserves into separate "buckets" for each economic group of nations. The amount of oil consumed by each group is the product of two factors: the per-capita oil consumption and the size of the population. In order to model the change in the oil supply, we'll have to model the change in both population and per-capita consumption. The next few sections will work through this model in several steps:

1. Develop a model for flows between the tank and the buckets, year by year, assuming given flow rates. The main idea behind the model will be conservation of mass—the mass of all the oil remains the same throughout the simulation, it just moves between the tank and the buckets.

2. Develop a model for population growth over time. In this model, the *percentage change* from year to year will be the same, but for a growing population, more people will be added each year than in the year before it.

3. Develop a model for the change in per-capita oil consumption. We'll use a growth model very similar to the one used for population growth.

4. Finally, pull each of the submodels together to build a comprehensive model for oil consumption over time. Then, use this model to calculate the remaining supply of oil from year to year until the volume of the tank goes to zero.

8.4.3 Flow Rates and Conservation of Mass

Figure 8.21 illustrates the process of filling a tank. The most important property of the tank over time is its volume, described by the state variable V. Just as we did with a falling object, we can represent the change in the state of the tank with a series of snapshots at regular time intervals Δt, where V_0 is the initial tank volume, V_1 is the volume Δt later, and so on. The change in tank volume from one time step to the next is ΔV, where

$$\Delta V_i = V_{i+1} - V_i$$

We can picture V_i as a drop of liquid falling into the tank (but not yet in it) at step i. Finally, the *accumulation rate* of the tank is the change in tank volume over time, or

$$\text{accumulation rate} = \frac{\text{change in tank volume}}{\text{change in time}}$$

$$= \frac{\Delta V_i}{\Delta t}$$

Now, suppose that water is pumped into the tank through a pipe, as shown in Figure 8.22. The *flow rate* of water through the pipe is q_{in}, and has units of volume per time, such as liters per second or gallons per minute. Since the pipe is the only supply of water, the flow rate out of the pipe must be equal to the accumulation rate in the tank. Thus,

$$\text{accumulation rate in tank} = \text{flow rate into tank}$$

$$\frac{\Delta V_i}{\Delta t} = q_{in}$$

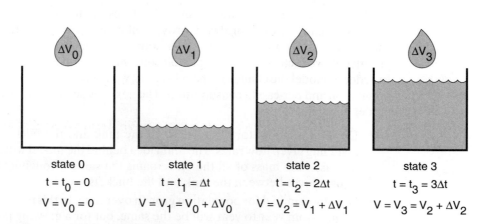

state 0	state 1	state 2	state 3
$t = t_0 = 0$	$t = t_1 = \Delta t$	$t = t_2 = 2\Delta t$	$t = t_3 = 3\Delta t$
$V = V_0 = 0$	$V = V_1 = V_0 + \Delta V_0$	$V = V_2 = V_1 + \Delta V_1$	$V = V_3 = V_2 + \Delta V_2$

Figure 8.21 Filling a tank in discrete "drops."

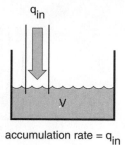

accumulation rate = q_{in}

Figure 8.22 A tank with one input stream.

Given this result, we can determine the change in volume in the tank from one time step to the next in terms of the flow rate:

$$\frac{V_{i+1} - V_i}{\Delta t} = q_{in}$$

$$V_{i+1} = V_i + q_{in}\Delta t \tag{8.9}$$

Example 8.4	**Filling a Tank** A tank initially contains 1.4 L of water. If water is flowing into the tank at a rate of 0.2 L/s, what will be the volume of water in the tank 5 s later?

Solution We define the following known variables for the problem:

initial tank volume: $V_i = 1.4\,\text{L}$

flow rate: $q_{in} = 0.2\,\text{L/s}$

time step: $\Delta t = 5\,\text{s}$

and the following unknown:

final tank volume: V_{i+1}

From equation (8.9), we find

$$V_{i+1} = V_i + q_{in}\Delta t$$

$$= 1.4\,\text{L} + \left(0.2\,\frac{\text{L}}{\text{s}}\right)(5\,\text{s})$$

$$= 2.4\,\text{L}$$

Thus 1 L flows into the tank during the time step and the final volume is 2.4 L.

Next, we'll add a second pipe to the tank to pump water out at a flow rate q_{out}, as shown in Figure 8.23. Now, the accumulation rate in the tank is the difference

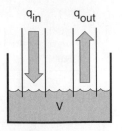

accumulation rate = $q_{in} - q_{out}$

Figure 8.23   A tank with one input stream and one output stream.

between the rate of water flowing in and the rate of water flowing out:

$$\text{accumulation rate in tank} = \text{flow rate in} - \text{flow rate out}$$
$$= q_{in} - q_{out}$$

In general, if there were more streams of water into (or out of) the tank, the accumulation rate would be the sum of the flow rates into the tank (with the rates negative for streams flowing out of the tank). This simple statement is actually a consequence of the Law of Conservation of Mass, that matter can be neither created nor destroyed. While this rule may seem obvious here, it becomes a very useful tool for analyzing more complicated flow problems, where there are mixtures of chemicals in the streams and reactions taking place in the tank, as one might find in a chemical processing plant.

8.4.4   Growth at a Constant Rate: Population and Per-Capita Oil Consumption

When a bucket is placed under a dripping faucet, it fills at a constant rate. Every second—drip, drip, drip—the volume of water in the tank increases by the same amount. Populations, however, don't grow this way. In an increasing population, the number of people added each year is typically larger than the number of people added the year before. The reason for this is that the more people in a population, the more people there are to reproduce. If we assume that the average number of children born per woman per year remains the same, and ignore changes in the death rate or migration, then the population growth rate will be proportional to the size of the population. We can write this growth rule as

$$\frac{\text{change in population}}{\text{year}} = \frac{\text{percentage growth}}{\text{year}} \times \text{population}$$

or

$$\frac{\Delta P_i}{\Delta t} = gP_i$$

where P_i is the size of the population in year i and g is the *annual growth rate*, measured as a percentage of the current population. As we've done before, we can

TABLE 8.8 Evaluation of the first few time steps for world population growth. The time step Δt is 1 year.

Step	Year	Population (billions)
0	t_0 1950	P_0 2.56
1	$t_1 = t_0 + \Delta t$ $1950 + 1$ 1951	$P_1 = P_0 + gP_0\Delta t$ $2.56 + (0.018)(2.56)(1)$ 2.60
2	$t_2 = t_1 + \Delta t$ $1951 + 1$ 1952	$P_2 = P_1 + gP_1\Delta t$ $2.60 + (0.018)(2.60)(1)$ 2.65

write this rate equation as a difference equation in the form:

$$\frac{P_{i+1} - P_i}{\Delta t} = gP_i$$
$$P_{i+1} = P_i + gP_i\Delta t \tag{8.10}$$

To illustrate, we'll use the model to predict the growth in total world population between 1950 and 2000 and compare this with the actual figures. In 1950, according to the U.S. Census Bureau, the actual world population was 2,556,518,868. The approximate population growth rate over that period was 0.18% per year. Thus,

$$P_0 = 2.56 \text{ billion}$$
$$g = 0.018$$

As we did for a falling softball, we can evaluate the state equations for population growth by setting up a table, as shown in Table 8.8. The third column of the table evaluates equation (8.10) at each time step to get the next value of P. See how the equation for each row only depends on the value of P in the row immediately above it.

Figure 8.24 shows the values for world population predicted by the model, versus actual values reported by the U.S. Census Bureau. In general, the population growth predicted by the model is quite close to the actuals. Note, however, that the model underpredicts the population slightly from the mid-1960s through the mid-1990s, and then overpredicts the population from the mid-1990s onward. The reason for this is that the growth rate hasn't really been constant at 1.8% over the entire period. During the mid-1960s, the actual population growth rate climbed to just over 2%, and through 2000 to around 1.25%. Experts expect the growth rate to continue to drop to approximately 0.5% by 2050. A more detailed model could take this "deceleration" of the growth rate g into account with another difference equation, much as we used equations for position in terms of velocity, and velocity in terms of acceleration, for a falling object.

As we saw in Section 8.4.1, changes in per-capita oil consumption are very sensitive to world events that are hard to predict, such as a war in the Middle East or the collapse of the Soviet Union. Over a long period of time, however, we can still model the trend for per-capita oil consumption for each economic group as

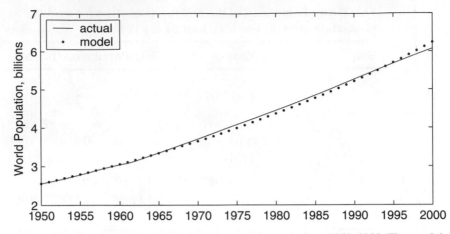

Figure 8.24   Predicted versus actual values for world population, 1950–2000. The model assumes an annual growth rate of 1.8%

changing at some constant rate, in the same way we did for population. We can express per-capita oil consumption as a difference equation:

$$\frac{C_{i+1} - C_i}{\Delta t} = rC_i$$
$$C_{i+1} = C_i + rC_i \Delta t \qquad (8.11)$$

where, C_i is the per-capita oil consumption at time step i and r is the rate of change.

8.4.5   Putting It All Together

Figure 8.25 shows a diagram of the oil flow for our model. We represent the proved world oil supply as a large tank that is drained into tanks for each of the four

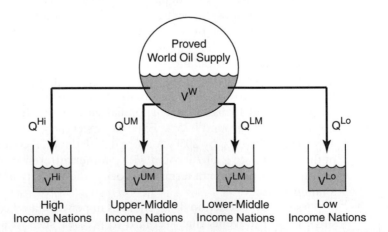

Figure 8.25   Modeling world oil consumption as a tank with oil flowing into four buckets, one for each economic group of nations.

TABLE 8.9 Variables for modeling world oil consumption.

Initial values:

Name	Description	2005 Value
V^W	remaining volume of proved world oil reserves	1.3 trillion bbl
P^{Hi}	population, high income nations	1.011 billion
P^{UM}	population, upper mid income nations	0.599 billion
P^{LM}	population, lower mid income nations	2.475 billion
P^{Lo}	population, low income nations	2.353 billion
g^{Hi}	population growth rate, high income nations	0.4%/year
g^{UM}	population growth rate, upper mid income nations	0.6%/year
g^{LM}	population growth rate, lower mid income nations	0.8%/year
g^{Lo}	population growth rate, low income nations	1.7%/year
C^{Hi}	per-capita oil consumption, high income nations	18.5 bbl/year
C^{UM}	per-capita oil consumption, upper mid income nations	5.9 bbl/year
C^{LM}	per-capita oil consumption, lower mid income nations	2.5 bbl/year
C^{Lo}	per-capita oil consumption, low income nations	0.7 bbl/year
r^{Hi}	per-capita oil consumption growth rate, high income nations	0%/year
r^{UM}	per-capita oil consumption growth rate, upper mid income nations	1%/year
r^{LM}	per-capita oil consumption growth rate, lower mid income nations	2%/year
r^{Lo}	per-capita oil consumption growth rate, low income nations	3%/year
Δt	time step	1 year

Other variables:

Name	Description	Units
V^{Hi}	accumulated oil volume, high income nations	bbl
V^{UM}	accumulated oil volume, upper mid income nations	bbl
V^{LM}	accumulated oil volume, lower mid income nations	bbl
V^{Lo}	accumulated oil volume, low income nations	bbl
Q^W	world oil consumption rate	bbl/year
Q^{Hi}	group oil consumption rate, high income nations	bbl/year
Q^{UM}	group oil consumption rate, upper mid income nations	bbl/year
Q^{LM}	group oil consumption rate, lower mid income nations	bbl/year
Q^{Lo}	group oil consumption rate, low income nations	bbl/year
P^W	world population	people

World Bank income groups of nations at different flow rates. Table 8.9 gives a listing of all of the variables in the model, along with initial values.

Flow Rates and Mass Balance Equations The first part of the model are the mass balance equations that relate the flow rates to tank volumes. Each of the four groups of nations is represented by a tank with oil flowing into it at some rate. We denote the rates as Q^{Hi}, Q^{UM}, Q^{LM}, and Q^{Lo} for the flows into the high income, upper middle income, lower middle income, and low income tanks, respectively. The accumulated volumes in each of these tanks at any point in time are respectively V^{Hi}, V^{UM}, V^{LM}, and V^{Lo}. Applying the mass-balance law from Equation (8.9), we get the following state equations:

$$\text{High income:} \quad V_{i+1}^{\text{Hi}} = V_i^{\text{Hi}} + Q_i^{\text{Hi}} \Delta t$$

$$\text{Upper middle income:} \quad V_{i+1}^{\text{UM}} = V_i^{\text{UM}} + Q_i^{\text{UM}} \Delta t$$

$$\text{Lower middle income:} \quad V_{i+1}^{\text{UL}} = V_i^{\text{UL}} + Q_i^{\text{UL}} \Delta t$$

$$\text{Low income:} \quad V_{i+1}^{\text{Lo}} = V_i^{\text{Lo}} + Q_i^{\text{Lo}} \Delta t$$

The total flow rate *out* of the proved world oil reserves is equal to the sum of the flow rates *into* the economic group tanks. If V^{W} and Q^{W} are the remaining volume and flow rate out of the "world" tank, then the mass balance equation is:

$$\text{World total:} \quad V_{i+1}^{\text{W}} = V_i^{\text{W}} - (Q_i^{\text{Hi}} - Q_i^{\text{UM}} - Q_i^{\text{LM}} - Q_i^{\text{Lo}}) \Delta t$$

$$= V_i^{\text{W}} - Q_i^{\text{W}} \Delta t$$

Population Growth   The next part of the model is population growth. This follows the model described by Equation (8.10). For each of the four economic groups, we thus have the following four state equations:

$$\text{High income:} \quad P_{i+1}^{\text{Hi}} = P_i^{\text{Hi}} + g^{\text{Hi}} P_i^{\text{Hi}} \Delta t$$

$$\text{Upper middle income:} \quad P_{i+1}^{\text{UM}} = P_i^{\text{UM}} + g^{\text{UM}} P_i^{\text{UM}} \Delta t$$

$$\text{Lower middle income:} \quad P_{i+1}^{\text{LM}} = P_i^{\text{LM}} + g^{\text{LM}} P_i^{\text{LM}} \Delta t$$

$$\text{Low income:} \quad P_{i+1}^{\text{Lo}} = P_i^{\text{Lo}} + g^{\text{Lo}} P_i^{\text{Lo}} \Delta t$$

The total world population at a given time point is simply the sum of the populations of each of the four groups:

$$\text{World total:} \quad P_i^{\text{W}} = P_i^{\text{Hi}} + P_i^{\text{UM}} + P_i^{\text{LM}} + P_i^{\text{Lo}}$$

Per-Capita and Group Oil Consumption   As we did for population, we assume that per-capita oil consumption grows at a constant rate, as described by Equation (8.11). For each of the four economic groups, we thus have the following four state equations:

$$\text{High income:} \quad C_{i+1}^{\text{Hi}} = C_i^{\text{Hi}} + r^{\text{Hi}} C_i^{\text{Hi}} \Delta t$$

$$\text{Upper middle income:} \quad C_{i+1}^{\text{UM}} = C_i^{\text{UM}} + r^{\text{UM}} C_i^{\text{UM}} \Delta t$$

$$\text{Lower middle income:} \quad C_{i+1}^{\text{LM}} = C_i^{\text{LM}} + r^{\text{LM}} C_i^{\text{LM}} \Delta t$$

$$\text{Low income:} \quad C_{i+1}^{\text{Lo}} = C_i^{\text{Lo}} + r^{\text{Lo}} C_i^{\text{Lo}} \Delta t$$

The oil consumption or flow rate for each of the economic groups at any point in time is equal to the size of the population times the per-capita oil consumption.

This also gives us four equations:

$$\text{High income: } \quad Q_i^{\text{Hi}} = P_i^{\text{Hi}} C_i^{\text{Hi}}$$

$$\text{Upper middle income: } \quad Q_i^{\text{UM}} = P_i^{\text{UM}} C_i^{\text{UM}}$$

$$\text{Lower middle income: } \quad Q_i^{\text{LM}} = P_i^{\text{LM}} C_i^{\text{LM}}$$

$$\text{Low income: } \quad Q_i^{\text{Lo}} = P_i^{\text{Lo}} C_i^{\text{Lo}}$$

Calculation Table Given the state equations for flow rates and mass balance, population growth, and per-capita and group oil consumption growth, we can now assemble them into a table and calculate values for each time step. Conceptually, this isn't any more difficult than our previous examples, but the table will be much larger because of the many variables. As before, the table is arranged in rows and columns, where each column is a series of values for each state variable at each time step. This table will have 20 columns, labeled A–T, which we'll organize according to income group as follows:

Year (Column A)

High Income Group (Columns B–E): population (P), per-capita oil consumption rate (C), group oil consumption rate (Q), group oil accumulation (V)

Upper Middle Income Group (Columns F–I): population (P), per-capita oil consumption rate (C), group oil consumption rate (Q), group oil accumulation (V)

Lower Middle Income Group (Columns J–M): population (P), per-capita oil consumption rate (C), group oil consumption rate (Q), group oil accumulation (V)

Low Income Group (Columns N–Q): population (P), per-capita oil consumption rate (C), group oil consumption rate (Q), group oil accumulation (V)

World Totals (Columns R–T): population (P), oil consumption (Q), volume of oil remaining (V).

The first two rows of the complete table are given below. The first row of the table contains the initial values given in Table 8.9. The second row is calculated from values in the first row, using the state equations. Each remaining row will follow the pattern of the second row, with its values calculated from the values in the previous row, according to the state equations.

.	Column A	Column B	Column C	Column D	Column E
		High Income Nations (Hi)			
step	Year	Population, P^{Hi} (billions)	Per Capita Oil Flow Rate, C^{Hi} (bbl/year)	Group Oil Flow Rate, Q^{Hi} (billion bbl/year)	Group Oil Accumulation, V^{Hi} (billion bbl)
0	t_0 2005	P_0^{Hi} 1.011	C_0^{Hi} 18.50	$Q_0^{\text{Hi}} = P_0^{\text{Hi}} C_0^{\text{Hi}}$ (1.011)(18.50) 18.70	V_0^{Hi} 0
1	$t_1 = t_0 + \Delta t$ 2005 + 1 2006	$P_1^{\text{Hi}} = P_0^{\text{Hi}} + g^{\text{Hi}} P_0^{\text{Hi}} \Delta t$ 1.011 + (0.004)(1.011)(1) 1.015	$C_1^{\text{Hi}} = C_0^{\text{Hi}} + k^{\text{Hi}} C_0^{\text{Hi}} \Delta t$ 18.50 + (0.00)(18.50)(1) 18.50	$Q_1^{\text{Hi}} = P_1^{\text{Hi}} C_1^{\text{Hi}}$ (1.015)(18.50) 18.78	$V_1^{\text{Hi}} = V_0^{\text{Hi}} + Q_0^{\text{Hi}} \Delta t$ 0 + (18.70)(1) 18.70

.	Column F	Column G	Column H	Column I
	Upper-Middle Income Nations (UM)			
step	Population, P^{UM} (billions)	Per Capita Oil Flow Rate, C^{UM} (bbl/year)	Group Oil Flow Rate, Q^{UM} (billion bbl/year)	Group Oil Accumulation, V^{UM} (billion bbl)
0	P_0^{UM} 0.599	C_0^{UM} 5.90	$Q_0^{\mathrm{UM}} = P_0^{\mathrm{UM}} C_0^{\mathrm{UM}}$ $(0.599)(5.90)$ 3.53	V_0^{UM} 0
1	$P_1^{\mathrm{UM}} = P_0^{\mathrm{UM}} + g^{\mathrm{UM}} P_0^{\mathrm{UM}} \Delta t$ $0.599 + (0.006)(0.599)(1)$ 0.603	$C_1^{\mathrm{UM}} = C_0^{\mathrm{UM}} + k^{\mathrm{UM}} C_0^{\mathrm{UM}} \Delta t$ $5.90 + (0.01)(5.90)(1)$ 5.96	$Q_1^{\mathrm{UM}} = P_1^{\mathrm{UM}} C_1^{\mathrm{UM}}$ $(0.603)(5.96)$ 3.59	$V_1^{\mathrm{UM}} = V_0^{\mathrm{UM}} + Q_0^{\mathrm{UM}} \Delta t$ $0 + (3.53)(1)$ 3.53

.	Column J	Column K	Column L	Column M
	Lower-Middle Income Nations (LM)			
step	Population, P^{LM} (billions)	Per Capita Oil Flow Rate, C^{LM} (bbl/year)	Group Oil Flow Rate, Q^{LM} (billion bbl/year)	Group Oil Accumulation, V^{LM} (billion bbl)
0	P_0^{LM} 2.475	C_0^{LM} 2.50	$Q_0^{\mathrm{LM}} = P_0^{\mathrm{LM}} C_0^{\mathrm{LM}}$ $(2.475)(2.50)$ 6.19	V_0^{LM} 0
1	$P_1^{\mathrm{LM}} = P_0^{\mathrm{LM}} + g^{\mathrm{LM}} P_0^{\mathrm{LM}} \Delta t$ $2.475 + (0.008)(2.475)(1)$ 2.495	$C_1^{\mathrm{LM}} = C_0^{\mathrm{LM}} + k^{\mathrm{LM}} C_0^{\mathrm{LM}} \Delta t$ $2.50 + (0.02)(2.50)(1)$ 2.55	$Q_1^{\mathrm{LM}} = P_1^{\mathrm{LM}} C_1^{\mathrm{LM}}$ $(2.495)(2.55)$ 6.36	$V_1^{\mathrm{LM}} = V_0^{\mathrm{LM}} + Q_0^{\mathrm{LM}} \Delta t$ $0 + (6.19)(1)$ 6.19

.	Column N	Column O	Column P	Column Q
	Low Income Nations (Lo)			
step	Population, P^{Lo} (billions)	Per Capita Oil Flow Rate, C^{Lo} (bbl/year)	Group Oil Flow Rate, Q^{Lo} (billion bbl/year)	Group Oil Accumulation, V^{Lo} (billion bbl)
0	P_0^{Lo} 2.353	C_0^{Lo} 0.70	$Q_0^{\mathrm{Lo}} = P_0^{\mathrm{Lo}} C_0^{\mathrm{Lo}}$ $(2.353)(0.70)$ 1.65	V_0^{Lo} 0
1	$P_1^{\mathrm{Lo}} = P_0^{\mathrm{Lo}} + g^{\mathrm{Lo}} P_0^{\mathrm{Lo}} \Delta t$ $2.353 + (0.017)(2.353)(1)$ 2.393	$C_1^{\mathrm{Lo}} = C_0^{\mathrm{Lo}} + k^{\mathrm{Lo}} C_0^{\mathrm{Lo}} \Delta t$ $0.70 + (0.03)(0.70)(1)$ 0.72	$Q_1^{\mathrm{Lo}} = P_1^{\mathrm{Lo}} C_1^{\mathrm{Lo}}$ $(2.393)(0.72)$ 1.73	$V_1^{\mathrm{Lo}} = V_0^{\mathrm{Lo}} + Q_0^{\mathrm{Lo}} \Delta t$ $0 + (1.65)(1)$ 1.65

.	Column R	Column S	Column T
	World Totals (W)		
step	Population, P^{W} (billions)	Flow Rate, Q^{W} (billion bbl/year)	Oil Remaining, V^{W} (billion bbl)
0	$P_0^{\mathrm{W}} = P_0^{\mathrm{Hi}} + P_0^{\mathrm{UM}} + P_0^{\mathrm{LM}} + P_0^{\mathrm{Lo}}$ $1.011 + 0.599 + 2.475 + 2.353$ 6.438	$Q_0^{\mathrm{W}} = Q_0^{\mathrm{Hi}} + Q_0^{\mathrm{UM}} + Q_0^{\mathrm{LM}} + Q_0^{\mathrm{Lo}}$ $18.91 + 3.53 + 6.19 + 1.65$ 30.27	V_0^{W} 1,300
1	$P_1^{\mathrm{W}} = P_1^{\mathrm{Hi}} + P_1^{\mathrm{UM}} + P_1^{\mathrm{LM}} + P_1^{\mathrm{Lo}}$ $1.015 + 0.603 + 2.495 + 2.393$ 6.505	$Q_1^{\mathrm{W}} = Q_1^{\mathrm{Hi}} + Q_1^{\mathrm{UM}} + Q_1^{\mathrm{LM}} + Q_1^{\mathrm{Lo}}$ $18.98 + 3.59 + 6.36 + 1.73$ 30.66	$V_1^{\mathrm{W}} = V_0^{\mathrm{W}} - Q_0^{\mathrm{W}} \Delta t$ $1,300 - (30.27)(1)$ 1,270

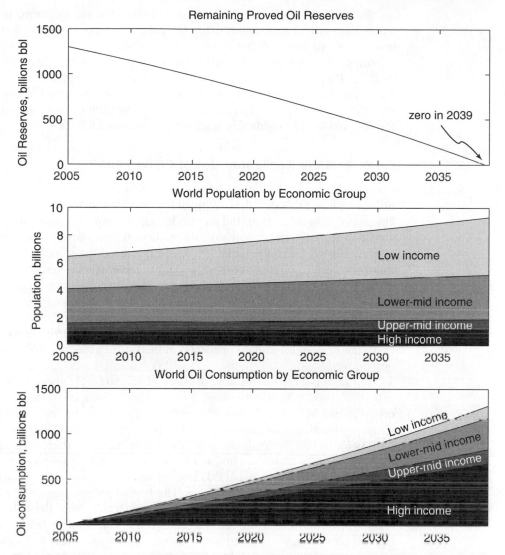

Figure 8.26   Oil reserves, world population, and oil consumption over time, according to our model.

Results   Figure 8.26 shows the results of the simulated model. As the top plot in the figure shows, the remaining volume of the worldwide proved oil reserves V^W goes to zero in 2039. Does this really mean we're going to run out of oil then? Before addressing this question, let's first look more closely at a breakdown of the result.

The bottom two plots in Figure 8.26 are *stacked area plots* of the group populations P^{Hi}, P^{UM}, P^{LM}, and P^{Lo} and of the group accumulated oil volumes V^{Hi}, V^{UM}, V^{LM}, and V^{Lo}. The advantage of a stacked area plot is that we can see both the total values and the relative contributions of each group in a single plot. By 2039, the total world population will be over 9 billion people. The percentage of the world population from low income nations continues to increase; by 2039, 45% of the world's population will be in low income nations while only 12% will be in high income nations.

In 2039, the high income nations will still account for more than half of the world's annual oil consumption, but the percentage consumed by lower middle and low income nations will have increased. Whereas these lower income nations only consumed approximately 25% of the oil output in 2005, this figure increases to 36% by 2039. This is mostly due, however, to the increase in population, and while higher than in 2005, the per capita oil consumption in the lower middle income nations is still under 5 bbl/year and in low income nations under 2 bbl/year—much less than the 18.5 bbl/year consumed in high income nations in 2005.

8.4.6   Will We Really Run Out of Oil by 2040?

Our model predicts that we will run out of oil in 2039, but will this really happen? In order to answer this question, we need to consider shortcomings of the model in two key areas: how accurately did we model supply and how accurately did we model consumption? On the supply side, the main deficiency of the model is that there are no pipes flowing into the tank—we only considered the world's proved oil reserves, those that could be recovered with reasonable certainty under the economic and operating conditions as of 2006. Undoubtedly, there's some quantity of oil as yet undiscovered that could be recovered with today's technology, and we could model this either by adding to the starting quantity in the tank or by adding a "discovery" input pipe. Other sources of oil are those reserves too costly to recover today, but that may be economically worthwhile in the future at higher prices, such as *oil sand* and *oil shale*.

Oil sand is a mixture of sand or clay and an extremely viscous form of petroleum called *bitumen*. Whereas most petroleum today is pumped from wells in liquid form, oil sands would have to be mined and then separated from the surrounding minerals either by heating it or diluting it with other hydrocarbons. Worldwide, there's perhaps twice as much petroleum to be found in oil sands as there is in liquid form, with the largest known reserves in Canada and Venezuela. Oil sand reserves in Alberta, Canada cover more than 141,000 km^2 of land and have been in production using open pit mining since the mid-1960s. The Canadian government estimates that approximately 174 billion barrels of oil are recoverable from the fields under current economic conditions using today's technology. The total estimated reserves are as high as 1.8 trillion barrels, but would be very costly to extract and would have severe environmental impact. *Oil shale* is a sedimentary rock that contains a mix of organic chemical compounds called *kerogen* that have not been buried long enough to develop into liquid petroleum. Oil shale can be processed into a fuel for transportation by heating it to over 300° C until it gives off a vapor, and then condensing the vapor to a liquid. Two of the main concerns with oil shale are the cost of the energy required to refine it, and its environmental impact. Still, estimates of the worldwide supply of oil shale exceed 3 trillion barrels of recoverable oil, with more than 2 trillion barrels in the United States, primarily in Colorado, Utah, and Wyoming.

On the consumption side, our model assumes that the demand for oil for the next 40 years will continue to follow historic trends. What the model doesn't consider is the economics of how demand changes with cost, and in particular, what would happen if oil became much more expensive. Our model, for example, shows oil being consumed at a steady pace until the tank runs dry. A more realistic scenario would show that as the supply becomes tighter, the cost of oil would rise, and that since fewer people would be able to afford to consume as much oil as they do today, the supply would last longer. On the other hand, if the price of oil increases beyond what people can afford to meet their energy needs, then a crisis could arise well before the supply of liquid petroleum dwindles to very low levels.

In short, a more accurate model for the production and consumption of oil would need to consider a number of important factors that our simple model does not. Nevertheless, the model does demonstrate that there'll likely be very significant changes in the supply and demand for oil worldwide over the next generation. And regardless of the actual amount of oil available or the rate of consumption, the following is certain: society will *need* a new generation of engineers and scientists who can plan for these changes, and more importantly, develop the innovations needed to keep our energy supply flowing.

PROBLEMS

1. **Changes in State**
   Each of the following scenarios describe a situation where something is changing. For each, describe at least three "state variables" that may be part of a model for a system that covers this change.

   (a) weather forecasting
   (b) plant and animal life in a pond
   (c) the cost of housing in a city

2. **Velocity and Distance**
   A car has been measured as having the following velocities at corresponding points in time. How far did the car travel in the first 12 s since the measurements started?

time, s	0	5	10	15
velocity, m/s	10	20	15	25

3. **Average Velocity**
   A car has been measured as having the following positions versus time. What is the average velocity at 12 s?

time, s	0	5	10	15
distance, m	0	20	50	75

4. **Euler's Method and Difference Equations**
   The following three difference equations describe three different ways in which the function $P(t)$ can change as a function of time, t, where b and c are constants.

   (a) $\Delta P(t)/\Delta t = c$
   (b) $\Delta P(t)/\Delta t = bt$
   (c) $\Delta P(t)/\Delta t = bt + c$

   Suppose that the value of $P(t)$ at time $t - 0$ is P_0. Use Euler's method to approximate $P(\Delta t)$ and $P(2\Delta t)$ for each of the three equations.

5. **Spreadsheet Model of a Softball Trajectory Without Drag**
   Develop a spreadsheet that implements a simulation of the trajectory of a softball launched vertically, ignoring the effects of drag, based on the model in Table 8.2. Compare your results with the values in Table 8.3 and correct any errors until they match. Use your simulation to answer the following questions.

   (a) If the initial velocity is changed to 10 m/s, how high with the ball fly and how long will it take to hit the ground? What if the initial velocity is changed to 30 m/s? Note that you will need to add rows to your spreadsheet to handle longer flights.

   (b) Determine how high the softball would fly if launched with an initial velocity of 20 m/s on the moon.

   (c) Run a series of experiments with your spreadsheet model, varying the initial velocity from 5 m/s to 50 m/s in increments of 5 m/s.

      i. Plot the maximum altitude versus initial velocity.

      ii. According to Galileo's theory as discussed in Chapter 4, what should be the relationship between the maximum altitude and initial velocity? Using techniques from Chapter 5, similar to the way that we validated Boyle's Law, determine how well the results of your simulation fit Galileo's theory.

6. **Spreadsheet Model of a Softball Trajectory with Drag**
   Develop a spreadsheet that implements a simulation of the trajectory of a softball launched vertically that includes the effects of drag, based on Table 8.5. Plot your results, compare them with Figure 8.11, and correct any errors until they match. Then use your simulation to answer the following questions.

   (a) The terminal velocity of a falling object is the maximum speed that it will reach when the drag force balances the force of gravity, and as a result the object stops accelerating and falls at a constant speed. Estimate the terminal velocities of a bowling ball, a basketball, a golf ball, and a ping pong ball.

   (b) Suppose that a basketball and a golf ball were dropped from the top of the Leaning Tower of Pisa at the same time. How far apart would the two balls be when the first one hits the ground?

   (c) Run a series of experiments with your spreadsheet model, varying the initial velocity from 5 m/s to 50 m/s in increments of 5 m/s.

      i. Plot the maximum altitude versus initial velocity.

      ii. Using Galileo's theory of the relationship between velocity and height as your basis, use techniques from Chapter 5 to determine a best-fit empirical model for maximum altitude versus initial velocity.

7. **Energy Stored in a Spring**
   In much the same way that we modeled a falling object as an accumulation of distance over time, we can model the energy stored in a spring as an accumulation of the work performed in pulling a force over a distance. In this exercise, we look at developing a model for the energy stored in a spring.

   (a) Just as the velocity of an object changes as it falls, so does the force across a spring change as it stretches. Assuming that it obeys Hooke's Law, write an expression for the rate of change of the force across a spring with respect to its change in length.

(b) Using Euler's method, develop a discrete model for the force across a spring as it stretches from some initial to final length in a series of regularly spaced steps.

(c) Develop a spreadsheet model for the force in terms of the stretch distance. Assume that the spring constant is 20 N/m, and that the stretch varies from 0 to 1 m in increments of 0.1 m. Plot the results.

(d) Write an expression for the work performed in stretching the spring across one step.

(e) Modify your spreadsheet model to include the accumulated value of the work performed at each step as the total energy stored in the spring. Plot energy versus stretch distance.

(f) Using techniques from Chapter 5, develop an empirical model for spring energy versus stretch distance. In doing so, first guess at a theory based on the shape of the energy versus stretch distance plot, and then transform the plot into a straight line. Be sure to note the values for any important constants related to the slope or intercept of the line.

8. Spring Energy

(a) Estimate the amount of internal mechanical energy stored in the spring whose force-elongation data is given by the table below.

Total Force (lbs)	Elongation (in)
5	2.2
10	4.0
15	5.3
20	6.2
25	7.0

(b) If you launched a 2 kg projectile vertically using this spring, estimate its takeoff speed if the spring was extended 6 inches prior to the launch. Assume that 100 percent of the spring energy is converted to kinetic energy.

9. Work in Stretching a Nonlinear Spring
The table below gives the forces and corresponding displacements for a certain (nonlinear) spring.

force, N	0	40	80	120
displacement, m	0	0.4	1.6	2.4

(a) Estimate how much work is done in stretching the spring from zero displacement to a displacement of 1 m.

(b) How much additional work is required to stretch it an additional 1 m?

(c) Estimate the launch velocity of a 0.2 kg projectile if the spring were pulled back 1.5 m, assuming that the conversion of spring energy to kinetic energy of the projectile is 100 percent efficient?

(d) Suppose that the measured velocity of the 0.2 kg projectile was 20 m/s when launched with the spring pulled back 1.5 m. What is the efficiency of the launcher in converting spring energy to kinetic energy of the projectile?

(e) Given the launcher efficiency calculated in **(b)**, how far would you have to pull the spring back to launch a 0.2 kg projectile with a velocity of 25 m/s?

10. Simulation of a Mass-Spring System

In this problem, we'll examine a discrete model for the motion of a mass bouncing on a spring, as illustrated by the figure below. Assume that the mass m is placed gently on a spring with spring constant k, so that as it begins to fall it compresses the spring. Let y be the height of the mass with respect to its initial position, where $y = 0$ when the mass is first placed on the spring and y becomes negative as the spring is compressed.

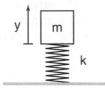

(a) Describe what you expect would happen if you actually performed the experiment with a real mass and spring. Sketch a plot of what you think y should look like versus time.

(b) Write an expression for the total force on the mass at any time in terms of m, y, k, and g.

(c) From the previous result, write an expression for the acceleration of the mass.

(d) Following the pattern used for modeling the motion of a softball in Sections 8.3.2 and 8.3.3, use Euler's method to write expressions for the position, velocity, and speed of the mass on the spring for a discrete series of time steps.

(e) Implement the model in a spreadsheet. Assume that the mass is 1 kg and that the spring constant is 20 N/m. Simulate the motion of the mass for 3 s with a step size of 0.05 s. Did your simulated results match your expectations? What are the differences? Do you think that the differences are something that would really happen in nature, or do you think that they are somehow related to the simulation?

(f) Repeat the simulation with steps sizes of 0.1 s and 0.01 s, still over a total simulation time of 3 s. What do you observe? Give an explanation.

11. Translation Vectors

Similar to force, as discussed in Section 6.2, we can also view the position or *translation* of an object as a vector quantity that has both a magnitude and direction. Further, we can add translation vectors using the head-to-tail method, the parallelogram method, or by decomposing the vectors into x- and y-components (for movement in two dimensions) and adding these separately, just as we did with force vectors. If you need to, review Sections 6.2.1 and 6.2.2 on the definition and addition of vectors before proceeding with this problem.

Suppose that an object undergoes two successive translations, $\mathbf{T}_1$ and $\mathbf{T}_2$, as pictured below. $\mathbf{T}_1$ has a magnitude of 2 m and a direction of 60° from horizontal, while $\mathbf{T}_2$ has a magnitude of 3 m and a direction of 30° from horizontal.

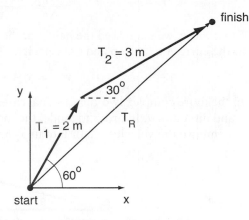

(a) The diagram above illustrates the resultant vector $\mathbf{T}_R$ as the results of the two translations $\mathbf{T}_1$ and $\mathbf{T}_2$ added together in "head-to-tail" fashion. Sketch a diagram that illustrates $\mathbf{T}_1$ and $\mathbf{T}_2$ added according to the parallelogram rule and show that the final position is the same.

(b) Determine the x- and y-components of the two vectors $\mathbf{T}_1$ and $\mathbf{T}_2$.

(c) What are the x- and y-components of the resultant vector $\mathbf{T}_R$?

(d) What are the magnitude and direction of $\mathbf{T}_R$?

12. **Velocity Vectors**

Like force and translation, velocity is also a vector quantity that has both magnitude and direction. If you need to, review Sections 6.2.1 and 6.2.2 on the definition and addition of vectors before proceeding with this problem.

(a) A person is rowing a boat across a river. If the person is rowing at a speed of 3 km/h perpendicular to the bank and the river is flowing at a rate of 4 km/h, what is the speed of the boat in the direction of travel? At what angle is the boat traveling relative to the riverbank?

(b) A ball thrown from a tower is traveling with a velocity of 20 m/s at an angle of 15° below horizontal. What are the horizontal and vertical components of its velocity?

13. **Airplane Takeoff**

A jet aircraft has taken off from a runway with its nose up at an angle of 12° and with a constant velocity of 260 km/h in the direction of travel.

(a) How many meters of elevation does it gain per second?

(b) How long will it take to travel a horizontal distance of 1 km?

14. **Using Euler's Method to Simulate an Airplane Climbing at Constant Velocity**

In Example 8.2, we used Euler's method to simulate the trajectory of a box being raised vertically at a constant velocity. In this exercise, we'll extend the model to consider an object moving at a constant velocity at an arbitrary angle, such as an airplane climbing. The explanation below will help guide you through the process.

In the simpler case of an object moving vertically at a constant velocity, we modeled the velocity as the change in height per change in time:

$$v = \frac{\Delta y}{\Delta t} = \frac{y_{i+1} - y_i}{\Delta t}$$

From there, we expressed the height at the next time step in terms of the height at the current time step and the velocity:

$$y_{i+1} = y_i + v\Delta t$$

In the more complex two-dimensional case, velocity is a *vector* with a magnitude v and direction θ. In order to model the motion of an object in this case, we can decompose the velocity vector into its x- and y-components as pictured below:

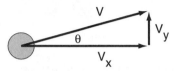

where

$$v_x = \frac{\Delta x}{\Delta t} \qquad\qquad v_y = \frac{\Delta y}{\Delta t}$$

$$= \frac{x_{i+1} - x_i}{\Delta t} \qquad\qquad = \frac{y_{i+1} - y_i}{\Delta t}$$

With velocity broken down into its x- and y-components, we can then use Euler's method to model the change in position from one time step to the next:

$$x_{i+1} = x_i + v_x\Delta t \qquad y_{i+1} = y_i + v_y\Delta t$$

Given this background, complete the table below to simulate the first few time steps of scenario from Problem 8.13: an airplane climbing at an angle of 12°, with a constant velocity of 260 km/h in the direction of travel. Use a time-step size $\Delta t = 1$ s.

	step	t	x	y	v_x	v_y
formula:	i	t_0	0	0		
value:	0	0				
formula:	i	$t_0 + \Delta t$				
value:	1					
formula:	i	$t_1 + \Delta t$				
value:	2					

15. Spreadsheet Model of an Object Moving at Constant Velocity at an Angle
Develop a spreadsheet model for the trajectory of an object moving at constant velocity at a fixed angle. The spreadsheet should be patterned after the table

completed in Problem 8.14, and include at least 20 rows (time steps). The spreadsheet should have the following primary inputs:

Name	Description	Units
Δt	time step size	m/s
x_0, y_0	initial position	m
v_0	initial velocity	m/s
θ	angle	degrees

Test the spreadsheet by simulating the scenario from Problems 8.13 and 8.14, namely an airplane climbing at an angle of 12°, with a constant velocity of 260 km/h in the direction of travel. Use a time step size $\Delta t = 1$ s. According to your spreadsheet, how long does it take for the airplane to cover a horizonal distance of 1 km? Did you get the same answer as for Problem 8.13?

16. Forces on a Tennis Ball in Flight

Shortly after being struck, a tennis ball is traveling at a speed of 25 m/s at an angle of 10° above horizontal. The ball has a mass of 57 g and a diameter of 6.5 cm. Solve each of the following problems *twice*: first, ignoring the effects of drag, and then second, including the effects of drag. Assume that the ball is a perfect sphere when calculating the effects of drag.

(a) Draw a free-body diagram of the ball, showing all (considered) forces acting on it.

(b) What are the horizontal and vertical components of the net force acting on the ball?

(c) What are the horizontal and vertical components of the acceleration of the ball?

Comment on the significance of the effects of drag.

17. Euler Model of a Ball Launched at an Angle, Vacuum Conditions

In this problem, we'll build on the results of Section 8.3.2, Problem 8.14, and Problem 8.16 to develop a model for the trajectory of a ball launched at an angle, ignoring the effects of drag, using Euler's method. You should study Section 8.3.2 and solve the two earlier problems before attempting this one.

To recap, in Section 8.3.2, we modeled the trajectory of a ball in flight, ignoring drag, traveling in the vertical direction only. At the heart of the model were three difference equations: one for altitude, one for velocity, and one for acceleration. To evaluate the model, we constructed a table—Table 8.2—where the columns corresponded to the state variables and the rows corresponded to their values at successive time steps. Then, in Problem 8.14, we modeled the motion of a object traveling in two dimensions, but at a constant velocity. Finally, in Problem 8.16, we determined expressions for the horizontal and vertical components of the acceleration of an object traveling at a given velocity. Now, we'll combine these results to build a complete model for the motion of an object in two dimensions ignoring drag (will include the effects of drag in a later exercise). Complete the table below to simulate the first few time steps of the flight of the tennis ball from

Problem 8.16, where at time $t = 0$, the ball has a speed of 25 m/s at an angle of $10°$ above horizontal. Use a time step size of 0.1 s.

	step	t	x	y	v_x	v_y	a_x	a_y
formula:	i	t_0	0	0				
value:	0	0						
formula:	i	$t_0 + \Delta t$						
value:	1							
formula:	i	$t_1 + \Delta t$						
value:	2							

18. Spreadsheet Model of a Tennis Ball in Flight, Ignoring Drag

Develop a spreadsheet model for the trajectory of a ball launched at an angle, ignoring the effects of drag. The spreadsheet should be patterned after the table completed in Problem 8.17. The spreadsheet should have the following primary inputs:

Name	Description	Units
Δt	time step size	m/s
x_0, y_0	initial position	m
v_0	initial velocity	m/s
θ	angle	degrees

Test the spreadsheet by simulating the trajectory of a tennis ball launched at an angle of $10°$ above horizontal with an initial speed of 25 m/s, with a time step size of 0.1 s. Include enough rows in the spreadsheet to determine how long it takes for the ball to return to the ground. From the data in your spreadsheet, generate the follow plots:

- distance versus time
- height versus time
- height versus distance

Finally, answer the following questions:

- What is the maximum altitude of the ball?
- How long does it take for the ball to return to the ground?
- How far does the ball travel horizontally before it returns to the ground?

19. Euler Model of a Ball Launched at an Angle, Including Effects of Drag

Redo Problem 8.17, this time including the effects of drag.

20. Spreadsheet Model of a Tennis Ball in Flight, Including Effects of Drag

Redo Problem 8.18, this time including the effects of drag, patterned after the table from Problem 8.19. The spreadsheet should have the following primary inputs:

Name	Description	Units
m	mass of ball	kg
r	radius of ball	m
Δt	time step size	m/s
x_0, y_0	initial position	m
v_0	initial velocity	m/s
θ	angle	degrees

21. Launching a Tennis Ball, Soccer Ball, Golf Ball, and Ping Pong Ball

Using the spreadsheet that you implemented for Problem 8.20, simulate how far a tennis ball, soccer ball, golf ball, and ping pong ball would travel when launched at an angle of 10° degrees above horizontal with an initial speed of 25 m/s.

22. Spreadsheet Model of Oil Consumption

Implement a spreadsheet simulation based on the model for oil consumption discussed in Section 8.4.5. Build up the simulation piece by piece, carefully checking the results for each part such as population growth before moving on to the next step. When you're convinced that your implementation is correct and matches the results in the text, come up with a list of five "what if" scenarios that change input conditions to the model and simulate them. Discuss your findings and their implications.

PART III

PROBLEM SOLVING
WITH MATLAB

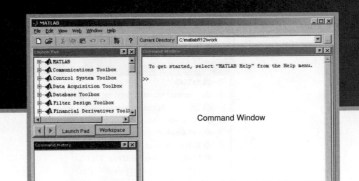

Getting Started with MATLAB

LEARNING OBJECTIVES

- to explain what MATLAB is and what advantages it holds over a pocket calculator in solving engineering problems;
- to explain what a variable is, and to use MATLAB to assign scalar values to variables and to evaluate arithmetic expressions using scalar variables and constants;
- to write a simple MATLAB *script*, with comments, that executes a series of commands.

9.1 YOUR FIRST MATLAB SESSION

9.1.1 Interpreting Simple Arithmetic Expressions

In this section, we'll take you through the basic steps of writing your first MATLAB program. You should read this section while sitting at the computer, and type in the examples as we present them.

When you first start MATLAB, a window similar to the one in Figure 9.1 appears. For the first few examples, we'll type commands directly at the prompt >> in the command window.

The central part of MATLAB is the MATLAB command interpreter. A typical interaction with the MATLAB interpreter consists of the following dialog:

- you type a command into the MATLAB interpreter's command window at the command prompt,
- MATLAB invokes the command and displays the results.

The very simplest form of command is to have MATLAB evaluate a primitive arithmetic expression, and one of the simplest primitive expressions is a number. If you enter the number 1.5 at the command prompt,

```
>> 1.5
```

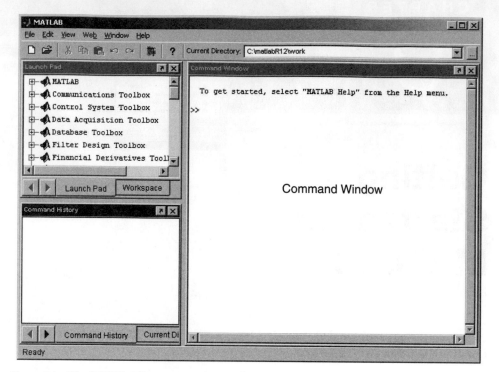

Figure 9.1   The MATLAB programming environment main window.

MATLAB responds by printing

```
ans =
    1.5000
```

What just happened is that you entered a command to evaluate the arithmetic expression 1.5, MATLAB executed this command, and replied with the answer 1.5.

You can compose more complex arithmetic expressions by combining numbers with primitive arithmetic operators such as + (addition), – (subtraction), * (multiplication), / (division), and ^ (exponentiation). For example, if you enter the arithmetic expression

```
>> 2 + 1.5
```

MATLAB responds by performing the addition and displaying the result

```
ans =
    3.5
```

Here are some other examples of arithmetic expressions. Try the first one with and without parentheses. What is the *precedence*, or the order in which operations are performed, of addition relative to multiplication in MATLAB?

```
>> (2 + 1.5) * 3
ans =
   10.5000

>> 3^2 + 4^2
ans =
   25
```

TABLE 9.1   Most common MATLAB mathematical functions and operators.

Basic Arithmetic

Function	Description	Function	Description	Function	Description
+	add	−	subtract	*	multiply
\	divide	^	exponent		

Trigonometry, Logarithms, and Exponents

Function	Description	Function	Description	Function	Description
sin	sine	cos	cosine	tan	tangent
asin	arcsine	acos	arccosine	atan	arctangent
log	natural logarithm	log2	$\log_2$	log10	$\log_{10}$
sqrt	square root	exp	e^x		

Rounding and Remainder

Function	Description	Function	Description	Function	Description
abs	absolute value	mod	modulus	rem	remainder
round	round to integer	floor	round down	ceil	round up

To use scientific notation, include an c or E to denote the exponent on 10.

$$-1.5\text{e}2 \ \ = \ -1.5 \times 10^2$$
$$2.75\text{e}\text{-}3 \ = \ 2.75 \times 10^{-3}$$

MATLAB offers a very large assortment of mathematical functions you can use in expressions. Table 9.1 lists some of the most common. A complete listing is available in MATLAB's online documentation under MATLAB > Functions by Category > Mathematics > Elementary Math.

9.1.2 Variables

Thus far, we've seen examples in MATLAB of two of the basic things we'd expect to find in any programming language: a way to express primitive ideas such as numbers and arithmetic operators, and a way to combine them into compound expressions. Now we'll look at some of the mechanisms that MATLAB provides for abstraction. The most basic form of abstraction used in any programming language is a way to assign names to either data objects or procedures. Named data objects are traditionally called *variables* in most programming languages. In MATLAB, we use the *assignment operator*, represented by a single equal sign (=), to assign the value of an expression to a variable with a given name. Entering the following at the command prompt:

```
>> radius = 4
radius =
     4
```

causes MATLAB to assign the value 4 to the variable radius. To get the value of a variable, simply enter the name of the variable at the prompt.

```
>> radius
radius =
     4
```

Once we define variables, we can use them in expressions, including expressions that define other variables. For example, we can define a variable that represents the area of a circle and uses both the variable `radius` as well as the built-in MATLAB variable `pi`.

```
>> area = pi * radius^2
area =
    50.2655
```

MATLAB stores all named variables in a kind of memory called the *workspace*. MATLAB, like many programming languages, lets you use different workspaces for different parts of your program; for now, we will store all our variables in the *global* workspace. To obtain a listing of all variables defined in the current workspace, use the command who:

```
>> who

Your variables are:

ans        area       radius
```

From the result of this command, we see listed the two variables that we explicitly created, area and radius, and a third variable, ans. ans is a built-in MATLAB variable that always contains the value of the last expression evaluated.

When you begin a new MATLAB session, it starts with an empty workspace, and when you quit a session, the workspace memory is cleared. You can explicitly clear the workspace at any time using the clear command. If you want to continue to work with the data in a workspace in a later session, you can write the contents of a workspace out to a file and then reload it using the save and load commands.

```
>> save
Saving to: matlab.mat

>> clear
>> who

>> load
Loading from: matlab.mat

>> who
Your variables are:
ans        area       radius
```

9.1.3 Scripts

When we define a variable, we associate a name with a data object so that we can later refer to that object by name. In a similar manner, we can associate a name with a collection of commands, so when we enter that name, MATLAB will execute each of the commands in the collection in sequence. In MATLAB, such a collection of commands is called a *script*.

A MATLAB script is a form of computer program, that is, a set of instructions for a computer to perform a particular procedure, written in a specialized programming language. Before we write our first script, we need to come up with a task to perform. As an example, let's write a script to calculate the mass of water in a cylindrical tank. The first step in writing a program in any programming language is to write an outline for the program in English (or another natural language, if more comfortable). In

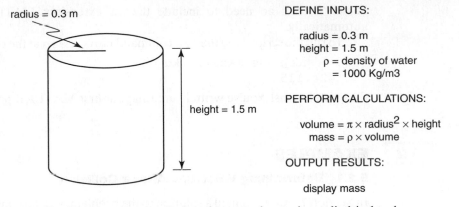

DEFINE INPUTS:

radius = 0.3 m
height = 1.5 m
ρ = density of water
= 1000 Kg/m3

PERFORM CALCULATIONS:

volume = $\pi \times$ radius$^2 \times$ height
mass = $\rho \times$ volume

OUTPUT RESULTS:

display mass

Figure 9.2   Outline of a script to calculate the mass of water in a cylindrical tank.

determining the mass of water in a cylindrical tank, one possible outline of a program is shown in Figure 9.2. Before encoding the program as a MATLAB script, let's observe a few things about the outline. First, the outline is not overly polished: it simply serves as notes for the programmer. A sketch helps to define the problem. The outline of the program itself has three parts: first, to define the input; second, to perform the calculations; and third, to display the results. Many of the programs we write will follow this basic structure of input, calculate, and display.

Given an outline for the program, we're ready to edit and save the MATLAB script. For MATLAB to locate and run a script, it must be saved with a .m extension, called an "M-file." From the MATLAB main menu, select File > New > M-file, or click on the blank page icon. This will open a MATLAB editor window. Below is a MATLAB script that follows the outline for calculating the mass of water in cylindrical tank. Type this script into the editor.

```
% tank_mass: procedure to find the mass of water in a tank

radius = 0.3;    % radius of tank base, m
height = 1.5;    % tank height, m
rho = 1000;      % density of water, kg/m^3

volume = pi * radius^2 * height;
mass = rho * volume;

disp('mass in kg');
disp(mass);
```

As you enter the script, note that any text following a percent sign turns green. These comments have no effect on the computation, but improve the readability of the program. The MATLAB command disp is used to display text or values to the terminal. When disp is invoked on a string of characters enclosed by single quotes, it displays that string literally, and when it's invoked on a variable name, it displays the value of the variable. By contrast, the semicolon at the end of each command suppresses output to the terminal; without the semicolons, each command would print a result as it's executed, resulting in a cluttered output.

Before you can run the script, you need to save and name it. Select File > Save As... from the editor main menu and enter tank_mass as the file name. There's no need to include the .m extension; MATLAB will add this automatically.

To run the script, type the new command `tank_mass` at the command prompt:

```
>> tank_mass mass in kg
   424.1150
```

Congratulations! You've written and run your first MATLAB program.

9.2 EXAMPLES

9.2.1 Determining Velocities After a Collision

In Section ??, we examined a solution to the problem of determining the velocities of two bodies after a collision. If two bodies have masses m_1 and m_2 and corresponding initial velocities v_{1i} and v_{2i}, then their final velocities after a perfectly elastic collision are:

$$v_{1f} = \frac{m_1 v_{1i} - m_2 v_{1i} + 2m_2 v_{2i}}{m_1 + m_2}$$

$$v_{2f} = \frac{m_2 v_{2i} - m_1 v_{2i} + 2m_1 v_{1i}}{m_1 + m_2}$$

Given these formulas, we'll use MATLAB to calculate the final velocities for specified masses and initial velocities. Assuming that m_1 and m_2 are both 5 kg and that v_{1i} is 4 m/s and v_{2i} is -6 m/s as they were in Example 4.3, we write the MATLAB script `collision.m` as follows:

```
% collision: calculate velocities after a perfectly
% elastic collision

m1 = 5;     % mass of body 1, kg
m2 = 5;     % mass of body 2, kg
v1i = 4;    % initial velocity of body 1, m/s
v2i = -6;   % initial velocity of body 2, m/s

% calculate final velocities

v1f = (m1*v1i - m2*v1i + 2*m2*v2i)/(m1+m2)
v2f = (m2*v2i - m1*v2i + 2*m1*v1i)/(m1+m2)
```

Running the script, we get

```
>> collision
v1f =
    -6
v2f =
     4
```

A few observations regarding this script:

- The very beginning of the script has a simple comment with the name of the script and a brief description.
- The next part of the script defines the input variables, with comments giving a description of each along with units.

- To display the results, we simply left off the semicolons from the ends of the lines with the formulas for the final velocities v1f and v2f, so that MATLAB will display their values when it evaluates them. Since the names of the variables are descriptive enough, there's no need for an explicit disp command. It's a very common trick to add or remove semicolons from a script to control which variable values are displayed on a given run.

9.2.2 Mass of CO_2 Produced by a Car

In this example, we use a MATLAB script to perform the calculations for the mass of CO_2 produced by a typical passenger car, first presented in Section 3.6. The script is typical of the most common use of MATLAB: first, it defines variables and then it expresses formulas in terms of these variables.

Summary of Variables and Equations The following table summarizes the variables in the calculation of the mass of CO_2 produced by a typical passenger car per year, repeated from Section 3.6:

Symbol	Description	Value
distance	typical distance driven per year	29370 km
fuel economy	typical gasoline milage	10.6 km/L
ρ_{gas}	density of gasoline	0.75 kg/L
AW_H	atomic weight of H	1 g/mol
AW_C	atomic weight of C	12 g/mol
AW_O	atomic weight of O	16 g/mol
AW_{gas}	atomic weight of gasoline	unknown
AW_{CO2}	atomic weight of CO_2	unknown
$mass_{gas}$	mass of gasoline	unknown
$mass_{CO2}$	mass of CO_2	unknown
mol_{gas}	moles of gasoline	unknown
mol_{CO2}	moles of CO_2	unknown

For convenience, we also repeat the key equations that resulted from the analysis in Section 3.6. First, the atomic weights of gasoline, C_8H_{18}, and carbon dioxide, CO_2, are:

$$AW_{gas} = 8 \times AW_C + 18 \times AW_H$$
$$AW_{CO2} = AW_C + 2 \times AW_O$$

Next, the fuel economy and the mass of gasoline burned are:

$$V = \frac{distance}{fuel\ economy}$$

$$mass_{gas} = \rho_{gas} \times V$$

From the mass of gasoline and its atomic weight, we calculated the number of moles of gasoline:

$$mol_{gas} = \frac{mass_{gas}}{AW_{gas}}$$

Given the reaction $C_8H_{18} + 12.5O_2 \rightarrow 8CO_2 + 9H_2O$, we calculated the moles of CO_2 as:

$$mol_{CO2} = 8 \times mol_{gas}$$

and finally, the mass of CO_2 produced as:

$$mass_{CO2} = mol_{CO2} \times AW_{CO2}$$

MATLAB Script   The script below, `annual_co2_mass.m`, shows a MATLAB program to calculate the mass of CO_2. The line numbers are included for reference.

```
 1 % annual_co2_mass.m
 2 % Calculate the mass of CO2 produced by a typical passenger
 3 % car in a year.
 4
 5 dist = 29370;              % distance/year, km
 6 fuel_ec = 10.6;            % fuel economy, km/l
 7 rho = 0.75;                % density of gasoline, kg/l
 8 aw_H = 1;                  % atomic weight H, g/mol
 9 aw_C = 12;                 % atomic weight C, g/mol
10 aw_O = 16;                 % atomic weight O, g/mol
11
12 aw_gas = 8*aw_C + 18*aw_H;      % atomic weight gas, g/mol
13 aw_CO2 = aw_C + 2*aw_O;         % atomic weight CO2, g/mol
14 vol_gas = dist/fuel_ec;         % volume of gas/year, l
15 mass_gas = rho * vol_gas;       % mass of gas/year, kg
16 mol_gas = 1000 * mass_gas/aw_gas;  % moles of gas/year
17 mol_CO2 = 8 * mol_gas;          % moles of CO2/year
18 mass_CO2 = mol_CO2 * aw_CO2/1000   % mass of CO2/year, kg
```

Line 1–3 provide beginning comments that give the name of the script and a brief description. Lines 5–10 in the script define MATLAB variables from the table of variables above. To improve readability, each variable has a meaningful name, and a comment that provides a definition of the variable and its units. The remainder of the script repeats the calculations from above. Note that the last line of the script leaves off the semicolon at the end, so that execution of this line will print its result to the screen—we didn't need to use the `disp` command. Running the program gives the following result—the same we determined earlier in Section 3.6.

```
>> annual_co2_mass
mass_CO2 =
   6.4165e+003
```

PROBLEMS

1. Evaluating MATLAB Expressions

Evaluate the following MATLAB expressions:

(a) `3.4 * (4.1 - 2.4)`

(b) `sin(pi/7)`

(c) `5.2 * exp(-0.14)`

2. Simple Expressions in MATLAB
Write a MATLAB script that evaluates each of the following expressions. Let $A = 1, B = 1.5, C = -10, D = 2.5$.

(a) $4B + C$

(b) $D^2 + 1.5D - 10$

(c) $\dfrac{-B + \sqrt{B^2 - 4AC}}{2A}$

(d) $\cos^2(B\pi) + \sin^2(B\pi)$

3. MATLAB Expressions: Roots and Exponents
Write a MATLAB script that evaluates the following expressions, where $A = 4$, $B = 7, C = 21$, and $D = 0.5$.

$$A^2 + \sqrt{\dfrac{B}{C}}$$

$$\sqrt{(B-D)^2 - (B^2 - D^2)}$$

$$\dfrac{B^3 - C}{A} + \sqrt[3]{D}$$

4. What Does This Do?
If (x_1, y_1) and (x_2, y_2) are the coordinates of two different points, what does the following MATLAB script find?

```
sqrt((abs(x1 - x2))^2 + (abs(y1 - y2))^2)
```

5. MATLAB Evaluation of a Polynomial
Write a MATLAB script that evaluates the following polynomial, where $x = 1.2$:

$x^5 + 7x^4 - 3x^3 + 8x^2 + 12^x - 7$

6. Trigonometry with MATLAB

(a) Write a MATLAB expression that finds the sine of $30°$.

(b) Write a MATLAB expression that finds the angle in degrees whose tangent is equal to 1.

7. Roots of a Quadratic Expression
The roots of a quadratic expression of the form

$$ax^2 + bx + c$$

are:

$$\dfrac{-b \pm \sqrt{b^2 - 4ac}}{2a}$$

Write a MATLAB script that finds the roots of a quadratic, where

(a) $a = 1$ $b = 5$ $c = 6$

(b) $a = 2$ $b = 10$ $c = 12$

(c) $a = 1$ $b = 0$ $c = 1$

In your script, write the expressions for the roots in terms of the variables a, b, and c. Note that the simplest way to write the complete script involves "cutting and pasting" these expressions several times.

8. Boolean Expressions

A Boolean expression is an expression that takes on a value of either "true" or "false." Write a script that evaluates the following MATLAB expressions. Following this experiment, in your own words, describe what the MATLAB '&' and '|' operators do.

```
true
false
false & false
false & true
true & false
true & true
false | false
false | true
true | false
true | true
```

Figure out what the values of the following expressions are, and then confirm your results by evaluating them with MATLAB.

(a) `true | false`

(b) `true & (true | false)`

(c) `true | (true & false)`

9. Problem Solving with MATLAB

Solve each of the following problems. The solution that you turn in should follow the homework problem format suggested in Chapter 3 (Given—Find—Diagram—Plan—Analysis—Comments), but use a MATLAB script to implement the "Analysis" part of the solution.

(a) Determine the energy and momentum of an object with a mass of 12.4 kg moving at a velocity of 5.8 m/s.

(b) Determine the force exerted by water pressure on a plug at the bottom of a bucket filled with 0.4 m of water, where the diameter of the plug is 1.5 cm.

(c) Find the maximum volume of water that could be boiled with 7 metric tons of coal, assuming that the water is already at boiling temperature, that the energy content of the coal is 25 MJ/kg, and that the heat of vaporization of water is 2260 kJ/kg.

10. Volume of a Pyramid in MATLAB

Find the volume of a pyramid with a square base of 10 units by 10 units and a height of 15 units using MATLAB.

11. Area of a Sphere in MATLAB

Calculate the surface area of a sphere 36″ in diameter in MATLAB.

12. Processing User Input

The following MATLAB expression prompts the user for a value and then assigns that value to the variable x:

```
x = input('Enter my input: ');
```

Using the input function, write a MATLAB script that prompts the user for a radius and then calculates:

(a) the circumference of a circle with that radius

(b) the area of a circle with that radius

(c) the volume of a sphere with that radius

(d) the surface area of a sphere with that radius

Be sure that the output from your script clearly indicates which result is which.

13. The Drake Equation

Dr. Frank Drake, president of the SETI Institute, developed an equation for estimating the number of "communicative civilizations" that exist in the Milky Way galaxy. The Drake equation states that:

$$N = R \times F_P \times N_E \times F_L \times F_I \times F_C \times L$$

where

N = the number of communicative civilizations within the Milky Way today

R = the number of suitable stars that form in our galaxy per year

F_P = the fraction of these stars that have planets

N_E = the number of Earth-like planets within each planetary system

F_L = the fraction of Earth-like planets where life develops

F_I = the fraction of life sites where intelligent life develops

F_C = the fraction of intelligent life sites where communication develops

L = the lifetime (in years) of a communicative civilization

Look up Drake's equation online and find estimates for the values of the parameters above. Write a MATLAB script that sets variables to the estimated values for each of these parameters and then evaluates Drake's equation.

14. Painting a Water Tower

Write a MATLAB program to calculate how many gallons of paint are needed to paint the water tower described in Example 3.1. The water tower consists of a tank on top of a tower. The tank is a sealed cylinder. The tower consists of eight support tubes arranged symmetrically around a wider drain tube. Two octagonal braces made from L-beams (that is, beams shaped like the letter "L") surround the set of support tubes at levels one-third and two-thirds up the height of the tower. The figure and table below give the dimensions of the tower.

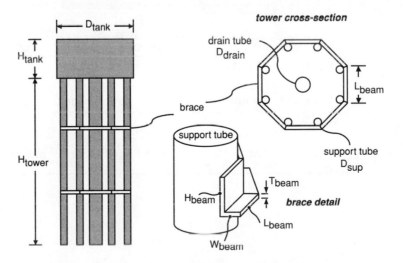

D_{tank}	tank diameter, ft	60	H_{beam}	L-beam height, ft	1.5
H_{tank}	tank height, ft	30	W_{beam}	L-beam width, ft	0.8
H_{tower}	tower height, ft	270	t_{beam}	L-beam thickness, ft	0.1
D_{sup}	support tube diameter, ft	4	L_{beam}	L-beam length, ft	15
D_{drain}	drain tube diameter, ft	8	C_{paint}	paint coverage, ft^2/gal	650

A skeleton of the MATLAB program is provided below.

```
% tower_paint.m
% Calculate the amount of paint required to
% coat the water tower.
%
% Your name and date

% define dimensions and other inputs
C_paint = 650;   % paint coverage, ft^2/gal
D_tank = 60;     % tank radius, ft
% and so on . . .

% surface area of tower structure, ft^2
A_brace = % fill in the blanks
A_sup =
A_drain =
A_tower = 2*A_brace + 8*A_sup + A_drain;

% surface areas of tank, ft^2
A_tank =

% total surface area of water tower, ft^2
A_total = A_tank + A_tower;

% total amount of paint needed, gal
paint =

disp(sprintf('Total is %f gallons', paint));
```

15. Surface Area of a House

Write a MATLAB script that finds the surface area of a house of the shape illustrated below. The script should prompt the user for dimensions and then display the result.

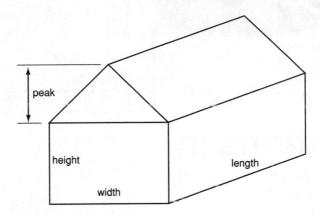

Test the script with the following values:

length (m)	width (m)	height (m)	peak (m)
20	10	3	2
25	15	6	2
15	12	9	3

CHAPTER 10

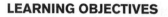

Vector Operations in MATLAB

LEARNING OBJECTIVES

- to explain what a vector is, and to use MATLAB to define vector variables and to access individual scalar elements of a vector;
- to be able to evaluate simple element-wise arithmetic expressions using vectors, and to use simple statistical functions to calculate averages and variations in a data set;
- to produce two-dimensional plots using vectors in MATLAB;
- to perform simple statistical operations using vectors, including calculating averages and generating histograms.

10.1 INTRODUCTION

In every MATLAB example we've considered thus far, all of the data we processed consisted of simple, numerical values, such as the radius of a circle, the mass of a cylinder, or the density of a liquid. Individual isolated variables, however, are very limiting in the kinds of ideas they can express. In addition to representing individual objects, we also need a way to represent *compound* objects that have many parts. Some examples of compound objects include:

- a *series* of positions versus time that make up the flight path of an aircraft,
- a *set* of samples of a chemical product processed in various batches,
- a *sequence* of stripes that make up a barcode,
- a *network* of beams and joints that make up a bridge.

In this section, we'll take a first look at one of the basic features in MATLAB for representing collections of data as a unit, called a *matrix*. Matrices are absolutely fundamental to MATLAB—in fact, MATLAB is an abbreviation for MATrix

432

LABoratory. Quite simply, a matrix is a table of values. Generally, matrices have two dimensions: rows and columns. For now, we'll just consider 1-dimensional matrices, that is, a single row of values, which is called a *vector*. As we'll see, this simple structure for organizing data is flexible enough to accommodate a wide variety of engineering models. For now, we'll consider two applications of vectors:

- using a *series* of data values to represent some aspect of a system that varies in an orderly manner, either over time, or over some region of space, or with respect to some other condition. This includes examples such as a series of distances traveled by a vehicle at regular time intervals, or the amount a spring stretches when increasing amounts of force are applied to it. The order of data values in a series is important in interpreting its meaning.

- representing a *set* of values for some aspect of a system where the order isn't important. This includes examples such as the weights of a random sample of rocks collected from a riverbed.

We'll also present some of the basic graphing and plotting utilities that MATLAB provides for visualizing collections of data, as well as some of MATLAB's built-in functions that operate on vectors.

10.2 BASIC OPERATIONS

10.2.1 Defining and Accessing Vectors

There are only a few basic operations we need to know to work with vectors in MATLAB: how to define them, how to refer to an entire vector as a group, and how to access elements of a vector individually.

Defining a List of Values   To define a vector in MATLAB, enter the list of values separated either by spaces or commas, and surround the list by a pair of square brackets. The following example shows two ways of entering a vector:

```
>> [2 4 6 8 10]
ans =
     2    4    6    8   10
>> [2,4,6,8,10]
ans =
     2    4    6    8   10
```

To give a vector a name, assign it to a variable just as you would an individual value:

```
>> r = [2 4 6 8 10]
r =
     2    4    6    8   10
>> r
r =
     2    4    6    8   10
```

Note that when you define a vector, all the elements *must* be entered on a single command line. If you want to break a single command into multiple lines in a script (or in the command window), use ellipses (...). For example, the following is the correct way to break a line:

```
longvector = [1 2 3 4 5 6 ...
        7 8 9]
>> longvector =
      1     2     3     4     5     6     7     8
```

but if you don't use ellipses, you'll get an error:

```
longvector = [1 2 3 4 5 6
        7 8 9]
Error using ==> vertcat
All rows in the bracketed expression must have the same
number of columns.
```

Defining a Series   The MATLAB colon (:) operator automatically creates vectors with values regularly spaced across a range. The syntax of the colon operator is

start-value : *increment* : *end-value*

For example, the following command defines a series vector x across the interval 0 to 10 in increments of 2:

```
>> x = 0:2:10
x =
      0     2     4     6     8    10
```

The colon operator with two arguments specifies a sequence from *start-value* to *end-value* with an increment of 1:

```
>> y = 1:5
y =
      1     2     3     4     5
```

Accessing Individual Vector Elements   The elements of a vector in MATLAB are numbered sequentially, beginning with 1. *Subscripts* are used to refer to the individual elements of a vector, in which the notation $r(i)$ refers to the ith element of vector r.

```
>> r(1)
ans =
      2

>> r(3)
ans =
      6

>> r(1) + r(3)
ans =
      8
```

You can also use subscripts to change an individual value in a vector or to add elements.

```
>> r(1) = 33
r =
      33     4     6     8    10
>> r(10) = 20
r =
      33     4     6     8    10     0     0     0     0    20
```

Note that when a 10th element was added to a vector of length 5, the 6th through 9th elements were automatically assigned the value 0.

The subscript into a vector can itself be a variable. For example,

```
>> i = 2;
>> r(i)
ans =
      4
```

Many beginning programmers find this confusing. In this latest example, i is not the *value* of a vector element, rather it's the *subscript* of a vector element. Since the value of i is 2, the expression r(i) is equivalent to r(2), which is equal to 4.

Finally, to determine the length of a list, use the length command:

```
>> length(r)
ans =
      10
```

10.2.2 Element-Wise Arithmetic Operations on Vectors

MATLAB supports a suite of arithmetic operations on vectors and matrices. This suite includes the formal operations of matrix addition and multiplication from linear algebra, as well as element-wise operations. We'll discuss linear algebra later in Chapter 11, and focus for now on the element-wise operations.

Element-wise arithmetic operations perform scalar arithmetic operations between corresponding elements of two vectors of identical size on an element-by-element basis. Table 11.1 summarizes the element-wise operations. Note that the

TABLE 10.1 Element-wise vector operations

Operation	Scalar form	Array form
addition	+	+
subtraction	−	−
multiplication	*	.*
division	/	./
exponentiation	^	.^

symbols for element-wise multiplication, division, and exponentiation begin with a dot in order to distinguish the array operations from true matrix multiplication under linear algebra—an entirely different operation. Under linear algebra, matrix addition and multiplication of a matrix by a scalar *are* element-wise operations, so we don't need a dot to distinguish these. Examples of the element-wise vector operations are given below:

```
>> a = [0 1 2];
>> b = [3 4 5];
>> a + b
ans =
     3     5     7
>> 3 * a + 5
ans =
     5     8    11
>> a .* b
ans =
     0     4    10
>> a .^ b
ans =
     0     1    32
```

The trigonometric and other transcendental functions can also be applied to vectors on an element-by-element basis.

```
>> a = [0 pi/4 pi/2 3*pi/4 pi];
>> sin(a)
ans =
        0    0.7071    1.0000    0.7071    0.0000
>> b = [1 10 100 1000];
>> log10(b)
ans =
     0     1     2     3
```

10.2.3 Example: Validating Boyle's Law

In Section 5.2.1, we examined how Robert Boyle used experimental data to confirm his theory that the pressure of a gas is inversely proportional to its volume at constant temperature. In this example, we'll use MATLAB vectors to simplify the same calculations that Boyle performed by hand in his original 1660 paper. Recall that to test his theory, Boyle took a U-shaped tube with air trapped inside and slowly poured in mercury, measuring the change in the volume of air as the mercury compressed it. Table 10.2, repeated from Section 5.2.1, summarizes the results of his experiment.

Column A of the table lists the volume of air in the tube, measured in inches of tube length. We express this with a MATLAB vector as

```
% volume of trapped air, tube-inches
V = [12 10 8 6 5 4 3];
```

TABLE 10.2   Boyle's data (repeated from Section 5.2.1)

A	B	C	D	E
			P, experiment	
V	P_{Hg}	P_{atm}	$P_{Hg} + P_{atm}$	P, theory
(tube-inches)	(inches Hg)	(inches Hg)	(inches Hg)	(inches Hg)
12	0	$29\frac{1}{8}$	$29\frac{2}{16}$	$29\frac{2}{16}$
10	$6\frac{3}{16}$	$29\frac{1}{8}$	$35\frac{5}{16}$	35
8	$15\frac{1}{16}$	$29\frac{1}{8}$	$44\frac{3}{16}$	$43\frac{11}{16}$
6	$29\frac{11}{16}$	$29\frac{1}{8}$	$58\frac{3}{16}$	$58\frac{2}{8}$
5	$41\frac{9}{16}$	$29\frac{1}{8}$	$70\frac{11}{16}$	70
4	$58\frac{2}{16}$	$29\frac{1}{8}$	$87\frac{14}{16}$	$87\frac{3}{8}$
3	$88\frac{7}{16}$	$29\frac{1}{8}$	$117\frac{9}{16}$	$116\frac{4}{8}$

Column B lists the pressure exerted by the mercury, which we also express as a MATLAB vector:

```
% pressure exerted by Hg, inches Hg
P_Hg = [0 6+3/16 15+1/16 29+11/16 41+9/16 58+2/16 88+7/16];
```

The total pressure on the body of trapped air is equal to the pressure of the mercury plus the atmospheric pressure bearing down on the mercury. Column D lists the total experimental pressure as the sum of columns B and C, which we write in MATLAB as

```
% atmospheric pressure, inches Hg
P_atm = 29+1/8;

% experimental pressure, inches Hg
P_ex = P_Hg + P_atm
```

If V_i and P_i are the values in the ith rows of columns A and D of the table, then according to the theory that PV is constant,

$$P_1 V_1 = P_i V_i$$

When we solve this for P_i, we get the expected theoretical pressure for a given volume of air, which Boyle listed in column E.

$$P_i = \frac{P_1 V_1}{V_i}$$

In MATLAB, we write this as the product of the first elements of the vectors P_ex and V, divided element-wise by elements of V.

```
% theoretical pressure, inches Hg
P_th = P_ex(1)*V(1)./V
```

If we save the MATLAB commands above in a script and run it, we get the following result:

```
P_ex =
   29.1250 35.3125 44.1875 58.8125 70.6875 87.2500 117.5625
P_th =
   29.1250 34.9500 43.6875 58.2500 69.9000 87.3750 116.5000
```

Note that the values for P_th are the same that Boyle listed in Column E of the table, and are very close to the experimental values for P_ex. As a final step in the analysis, we can calculate the percentage error between the experimental and theoretical values as

```
>> pct_error = 100*(P_ex - P_th)./P_th
pct_error =
        0 1.0372 1.1445 0.9657 1.1266 -0.1431 0.9120
```

Thus, the error is approximately 1 percent or less.

10.3 SIMPLE TWO-DIMENSIONAL PLOTS AND GRAPHS

The basic MATLAB 2-D plotting command is plot. Below, we describe some of the most common variations of 2-D plots.

10.3.1 Plot Basics

The command plot(Y), in which Y is a vector, plots the elements of Y versus their subscript or index number. As an example, consider plotting a vector that contains the heights of a ball thrown straight up with an initial velocity of 25 m/s, measured at regular time intervals.

```
heights25 = [0 9.61 15.01 17.45 17.36 14.90 10.38 4.29];
plot(heights25)
```

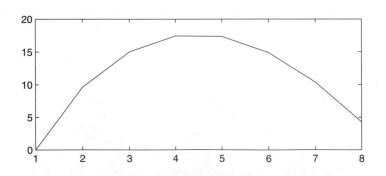

plot(X,Y), where X and Y are both vectors of the same length, plots Y with respect to corresponding values of X. For example, supposing the measurements of the ball height were taken at 0.5 s intervals, we could produce a plot of height versus time as follows:

```
heights25 = [0 9.61 15.01 17.45 17.36 14.90 10.38 4.29];
 t = 0:.5:3.5;
 plot(t, heights25)
```

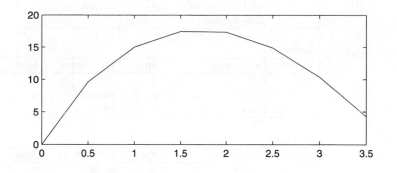

10.3.2   Adding Titles and Labels

You can add a title to a plot using the Insert > Title command from the menu on a Figure window. You can change axis properties, including adding labels and changing the range using the Edit > Axes Properties command.

   You can also add a title and labels with the commands title, xlabel, and ylabel.

```
heights25 = [0 9.61 15.01 17.45 17.36 14.90 10.38 4.29];
 t = 0:.5:3.5;
 plot(t, heights25);
 xlabel('time, s');
 ylabel('height, m');
 title('Ball Height vs. Time');
```

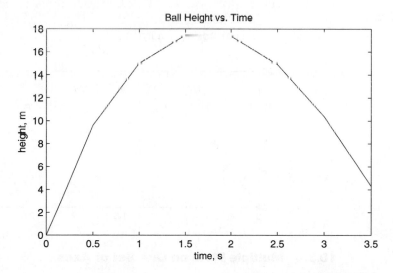

   Once you add titles and labels, you can edit them on the figure directly. First, select the Tools > Edit Plot from the figure main menu to enable plot editing. Then double-click on the text to change it.

TABLE 10.3 Line style specifiers.

Line Style

Spec	Description	Spec	Description	Spec	Description
−	solid line (default)	−−	dashed line	:	dotted line
−.	dash-dot line				

Color

Spec	Description	Spec	Description	Spec	Description
r	red	g	green	b	blue
c	cyan	m	magenta	y	yellow
k	black	w	white		

Marker

Spec	Description	Spec	Description	Spec	Description
+	plus sign	o	circle	*	asterisk
.	point	x	cross	s	square
d	diamond	^	up triangle	v	down triangle
>	right triangle	<	left triangle	p	penta-star
h	hex-star				

10.3.3 Changing Line Styles

To change the properties of a curve in a plot, including its style, color, thickness, and marker symbol, first make sure that plot editing is enabled under the `Tools` menu, and then double-click on the curve. This will display the Property Editor dialog box to make the desired changes. You can also set line styles directly as an option to the `plot` command by using the command form `plot(X,Y,LineSpec)`, in which `LineSpec` is a string of characters that denotes the style, color, and marker for a line. Table 10.3 describes the specifiers for the `LineSpec`. For example, `'-r'` specifies a red solid line with no marker, and `':kp'` specifies a black dotted line with 5-pointed star markers.

```
plot(t, heights25, ':kp')
```

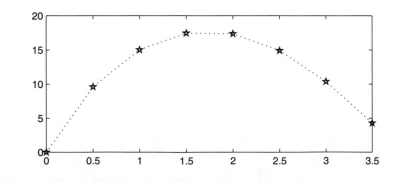

10.3.4 Multiple Plots on One Set of Axes

One particularly useful skill is plotting multiple series of data on the same set of axes, such as if we want to compare the trajectories of a ball thrown straight up with initial velocities of 25 m/s and 30 m/s. Ordinarily, when a new `plot` command is issued, it overwrites the current graph in the figure window. In order to add a new series of data to an existing graph without overwriting it, use the hold command. The command hold on holds the current axes so that the results of any subsequent plot command will appear on these axes. The command hold off releases the axes.

The complete script below defines and plots both heights25 and heights30 versus t on the same set of axes. Further, it uses a solid line for heights25 and a dashed line for heights30 and adds a legend to the plot.

```
heights25 = [0 9.6 15.0 17.4 17.4 14.9 10.4  4.3];
heights30 = [0 11.4 17.9 21.2 21.8 20.0 16.1 10.4];
plot(t, heights25, '-');
hold on;
plot(t, heights30, '--');
legend('25 m/s', '30 m/s');
xlabel('time, s')
ylabel('height, m');
hold off;
```

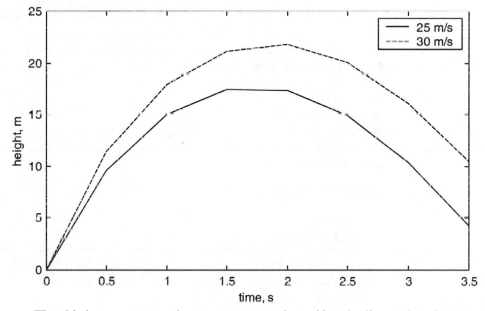

The third parameter to the plot command specifies the line style, where '-' indicates a solid line and '--' indicates a dashed line. The legend command produces a legend for the plot, pairing the text explanations with the line styles of the plots in the order in which the plots were created.

10.3.5 Multiple Sets of Axes in One Figure

The subplot function divides a single figure into multiple plotting regions or panes, organized into rows and columns. It takes three arguments, as shown below:

subplot (*rows, columns, pane*)

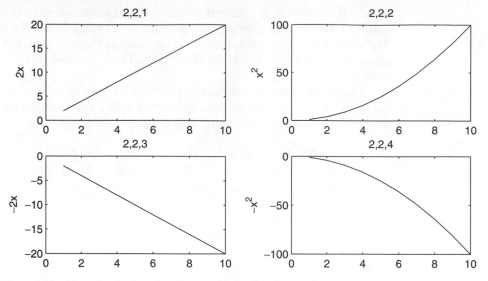

Figure 10.1 Use of `subplot` to place multiple plots in one figure.

where *rows* is the number of rows in the figure, *columns* is the number of columns, and *pane* is the pane number, counting from left to right across each row, proceeding from the top to bottom row. The following script illustrates the use of `subplot`, with the results shown in Figure 10.1.

```
x = 1:10;
subplot(2,2,1)
plot(x, 2*x)
title('2,2,1')
ylabel('2x')

subplot(2,2,2)
plot(x, x.^2)
title('2,2,2')
ylabel('x^2')

subplot(2,2,3)
plot(x, -2*x)
title('2,2,3')
ylabel('-2x')

subplot(2,2,4)
plot(x, -x.^2)
title('2,2,4')
ylabel('x^2')
```

10.3.6 Plotting Functions

One way to plot a function over a range is to define a series and then use it as an argument to an element-wise expression. The following commands generate a plot of $f(\theta) = \sin(\theta)$ over the interval $\theta = 0 \ldots 2\pi$ with 100 points.

```
theta = 0:0.01*pi:2*pi;
plot(theta, sin(theta))
```

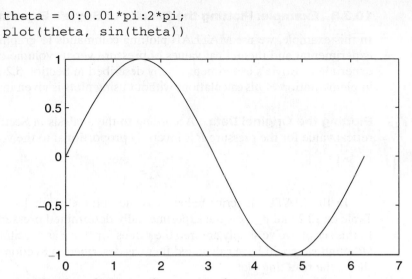

Another way is to use the command fplot(function, range), in which function is a MATLAB expression in one variable, and range is a two-element vector that specifies the upper and lower limits of the range.

```
fplot('2*sin(theta)+2', [0 2*pi]);
```

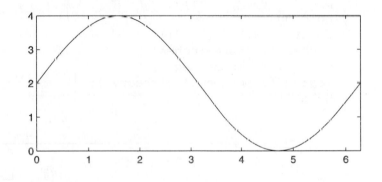

10.3.7   Specialized Plotting

MATLAB offers a very large number of specialized plots too numerous to describe here. The MATLAB online documentation describes these in complete detail, with examples that you can cut and paste into your own scripts and modify. A full listing of all the plotting functions may be found in the documentation under MATLAB > Functions by Category > Graphics. A sample of plot types includes

- area plots
- bar graphs
- pie charts
- 3-D graphs, including contour plots

Feel free to experiment on your own by following the examples in the MATLAB documentation!

10.3.8 Example: Plotting the Results of Boyle's Experiment

In this example, we use MATLAB plotting commands to graphically compare the experimental and theoretical values of pressure versus volume of air from Boyle's experiment. Boyle's experiment is fully described in Section 5.2.1 and a MATLAB implementation of his calculations without using plots is given in Section 10.2.3.

Plotting the Original Data   According to the analysis in Section 5.2.1, the theoretical value for the pressure P is inversely proportional to the volume V, where

$$P = \frac{349}{V}$$

In the MATLAB script below, V is the volume of air listed in Column A of Table or 10.2 and P_ex is the experimentally-determined pressure from Column D. In this script, we've simply entered the values for P_ex as (decimal) numeric values, although we could have calculated them as described in Section 10.2.3. Figure 10.2 shows the resulting plot.

```
V = [12 10 8 6 5 4 3];               % volume, tube-inches
P_ex = [29.1 35.3 44.2 58.8 ...
        70.7 87.3 117.6];            % pressure, inches Hg

plot(V, P_ex, 'o');
hold on
fplot('349/V', [3 12]);
xlabel('V, tube-inches');
ylabel('P, inches-Hg');
legend('experiment', 'theory');
hold off
```

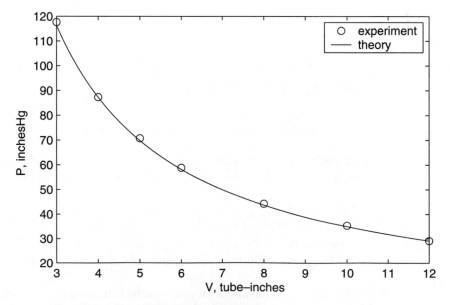

Figure 10.2   Plot of data from Boyle's experiment.

Here, we used the `plot` command to plot the experimental data as circles, and `fplot` to plot the theoretical values as a line. Alternatively, we could have calculated theoretical values for pressure at each of the points in `V` and then plotted these versus `V`:

```
P_th = 349./V;
plot(V, P_th);
```

This would work, but because there are only 7 points in `V`, the curve wouldn't be as smooth as with `fplot`.

Transforming the Data into a Straight Line   In Section 5.2.1, we learned that one of the techniques for testing how well experimental data fits a theory is to transform the data in such a way that it'll plot along a straight line if the theory is correct. The original theory for Boyle's law is

$$P(V) = \frac{k}{V},$$

which, as Figure 10.2 shows, is not a straight line. On the other hand, if we plot $1/P$ versus V, we now get

$$f(V) = \frac{1}{P(V)} = \frac{1}{k}V$$

which is the equation of a line with slope $1/k$. The following script creates this plot, where `V` and `P_ex` are the vectors defined in the previous example. The resulting plot is shown in Figure 10.3.

```
% plot using filled, black circles
plot(V, 1./P_ex, 'ok','MarkerFaceColor','k');
xlabel('V, tube-inches');
ylabel('1/P, 1/(inches-Hg)');
```

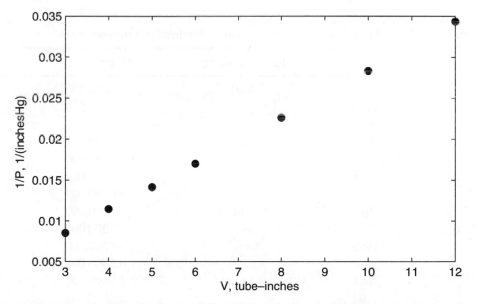

Figure 10.3   Transforming the data from Boyle's experiment so that it plots along a straight line.

This time, we plotted the data using solid black circles, determined by the following arguments to plot:

- 'ok': circles (o) that are black (k)
- 'MarkerFaceColor','k': filled with black

As we see, the data does indeed fall along a straight line. In order to determine the value of k, we could manually draw a best-fit line through the data and measure its slope.

10.3.9 Example: Moore's Law and Log Plots

In Section 5.2.2, we looked at Moore's Law as an example of a theory in which the dependent variable varies exponentially with respect to the independent variable. Recall that in 1965, Gordon Moore predicted that the number of transistors in leading-edge integrated circuits would double approximately every two years. We expressed this theory as

$$N(t) = N_0 2^{t/k}, \tag{10.1}$$

where N is the number of transistors, N_0 is the number of transistors in the starting product, t is time in years, and k is the interval between each doubling in transistor count. If the theory holds, then a plot of N versus t would clearly not be a straight line, but if we took the logarithm of both sides, the transformed function would be

$$
\begin{aligned}
\log_2 N(t) &= \log_2(N_0 2^{t/k}) \\
&= \log_2 N_0 + \log_2(2^{t/k}) \\
&= \log_2 N_0 + \tfrac{1}{k}t
\end{aligned}
$$

in which the slope of the line is $1/k$ and the intercept is $\log_2 N_0$. Table 10.4 shows the actual number of transistors over more than 30 years of Intel microprocessors.

TABLE 10.4   Transistor counts for Intel microprocessors

Year	Product Name	Transistors	$\log_2$(transistors)
1971	4004	2,300	11.2
1972		2,500	11.3
1974		4,500	12.1
1978	8086	29,000	14.8
1982		134,000	17.0
1985		275,000	18.1
1989		1,200,000	20.2
1993	Pentium	3,100,000	21.6
1997		7,500,000	22.8
1999		9,500,000	23.2
2000		42,000,000	25.3
2001	Itanium	25,000,000	24.6
2003		220,000,000	27.7
2004		592,000,000	29.1

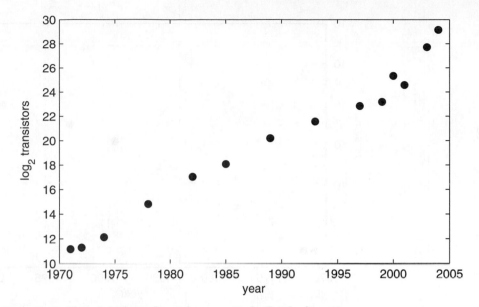

Figure 10.4 Plot of the $\log_2$ of transistor counts for Intel microprocessors.

Plotting the Log of the Data on Linear Axes The first approach to plotting exponentially changing data along a straight line is to find the log of the data and then plot it on linear axes. The following script illustrates this, with the resulting plot in Figure 10.4. Note that the y-axis label is '$\log_2$ transistors', not just the number of transistors.

```
year = [1971 1972 1974 1978 1982 ...
        1985 1989 1993 1997 1999 ...
        2000 2001 2003 2004];

transistors = [2300 2500 4500 29000 134000 ...
        275000 1200000 3100000 7500000 9500000 ...
        42000000 25000000 220000000 592000000];

% plot using filled, black circles
plot(year, log2(transistors), 'ok','MarkerFaceColor','k')
xlabel('year')
ylabel('log_2 transistors')
```

Plotting the Data on a Log Scale The other approach to plotting exponentially changing data along a straight line is to use a *semilog plot* that transforms the y-axis to a logarithmic scale, rather than transforming the data itself.

```
semilogy(year, transistors, 'ok','MarkerFaceColor','k')
xlabel('year')
ylabel('transistors')
```

The resulting plot is shown in Figure 10.5. Note that in this case, the y-axis label is the number of transistors, rather than the log of the number of transistors.

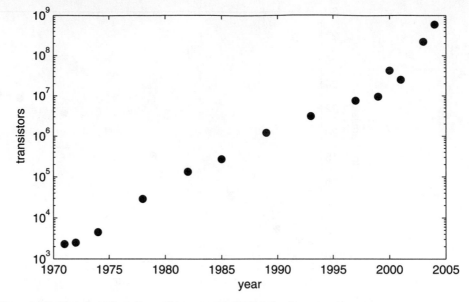

Figure 10.5 Semilog plot of transistor counts for Intel microprocessors.

10.4 STATISTICS

In mathematics terminology, a *statistic* is any quantity that's calculated from a sample of data. Examples of statistics include:

- the minimum or maximum value in the sample of data,
- the sum of the values in the sample,
- the average or mean value of the sample,
- the total number of samples,
- the number of samples whose values lie within a given range.

MATLAB includes an extensive set of functions to calculate statistics and in this section, we review some of the most common of these. As an example data set, we use the results of the slingshot experiment described in Section 5.4, in which we ran 20 trial launches of a softball with the spring pulled back a distance of 1 m, as listed in Table 5.5. We store the data values in a MATLAB vector named distance, defined below.

```
distance = [17.5 19.0 16.4 19.3 16.6 ...
        16.0 17.4 16.7 18.1 17.5 ...
        15.1 14.2 17.4 15.7 17.8 ...
        19.3 18.5 15.7 17.9 17.0];
```

Figure 5.6 shows a scatter plot of the data, generated by the following script:

```
plot(distance, '*')
xlabel('trial')
ylabel('distance, m')
set(gca,'YGrid','on') % gca, command to "get current axes"
```

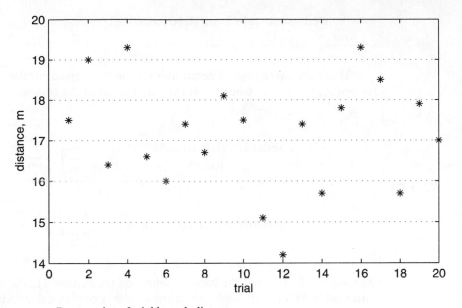

Figure 10.6 Scatter plot of trial launch distances.

10.4.1 The Basics: Minimum, Maximum, Averages, etc.

The most basic set of statistical functions in MATLAB calculate summary values such as the minimum, maximum, or mean of a vector. The following example illustrates the most common of these operations, performed on the launch distances data set.

```
>> min(distance)        % minimum of vector
ans =
   14.2000

>> max(distance)        % maximum of vector
ans =
   19.3000

>> mean(distance)       % arithmetic mean of vector
ans =
   17.1550

>> length(distance)     % number of elements in vector
ans =
    20

>> sum(distance)        % sum of elements in vector
ans =
  343.1000
```

Note that these functions can be used in combination to calculate other statistics. For example, we could alternatively calculate the mean from the sum and length:

```
>> sum(distance)/length(distance)
  ans =
    17.1550
```

There are two accepted definitions for the standard deviation of a sample of data values $x_i, i = 1 \ldots n$, where $\bar{x}$ is the mean value of the sample:

$$\text{standard deviation (definition 0):} \quad \frac{1}{n-1}\sqrt{\sum_{i=1}^{n}(x_i - \bar{x})^2}$$

$$\text{standard deviation (definition 1):} \quad \frac{1}{n}\sqrt{\sum_{i=1}^{n}(x_i - \bar{x})^2}$$

In Chapter 5, we used the second definition (definition 1). The standard deviation function in MATLAB, std, can calculate either of these, depending on the value of a parameter passed into the function:

```
>> std(distance, 0)        % n-1 in denominator
ans =
    1.3816

>> std(distance, 1)        % n in denominator
ans =
    1.3467

>> std(distance)           % defaults to n-1 in denominator
ans =
    1.3816
```

Note that the values in both cases are slightly different, and that the differences become less significant as the sample size gets bigger. Also note that the default for the std function is to use $n - 1$ in the denominator if no method is specified.

10.4.2 Counting Values in a Range

As discussed in Section 5.4.3, the estimated probability of an event, given a sample of data, is usually defined as the *relative frequency* of that event within the sample, or

$$\text{estimated probability} = \frac{\text{number of trials satisfying criterion}}{\text{total number of trials}}$$

For example, to estimate the probability that a softball would land more than 1 m beyond a target at 18 m, we would count the number of trials in which the distance was greater than or equal to 19 m and divide this by the sample size. In this section, we present some of the MATLAB functions for calculating the values in a range.

One useful first step is simply to sort the data. We can do this in MATLAB with the sort function, which returns a vector sorted in ascending order as a result.

```
>> sort(distance)

ans =
  Columns 1 through 5
    14.2000    15.1000    15.7000    15.7000    16.0000
  Columns 6 through 10
    16.4000    16.6000    16.7000    17.0000    17.4000
  Columns 11 through 15
    17.4000    17.5000    17.5000    17.8000    17.9000
  Columns 16 through 20
    18.1000    18.5000    19.0000    19.3000    19.3000
```

With the values sorted, it's easier to count the number of values in a range manually, but we'd rather let MATLAB do the work for us. To do this, we can use the MATLAB operators for "equals," "less than," "greater than," and other relations. The complete set of relational operators is listed below:

==	equal
~=	not equal
<	less than
>	greater than
<=	less than or equal
>=	greater than or equal

Relational expressions in MATLAB—that is, expressions with a relational operator in them—evaluate to either 1 or 0, where "1" means "true" and "0" means "false." For example,

```
>> value = 7
value =
     7

>> value < 5
ans =
     0

>> value > 5
ans =
     1
```

When applied to a vector, a relational operator returns a vector of 1s and 0s, indicating whether the relation with respect to the element at each position is true or false. For example,

```
>> values = [1 5 3 7]
values =
     1     5     3     7

>> values > 5
ans =
     0     0     0     1
```

We can build more complex expressions by joining two or more relational expressions with the MATLAB *logical operators* for "and," "or," and "not" listed below.

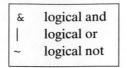

&	logical and
|	logical or
~	logical not

For example, we could express the range of values between 4 and 10 in English as "values greater than 4 and values less than 10," or write this in MATLAB as

```
>> (values > 4) & (values < 10)
ans =
     0    1    0    1
```

Note that the "and" operator (&) returns a vector with 1s for the elements in which *both* conditions are satisfied simultaneously. By contrast, the "or" operator (|) returns a 1 in which *either* condition is true, for example:

```
>> (values <= 3) | (values > 5)
ans =
     1    0    1    1
```

Given a vector of 1s and 0s, you can determine how many elements meet the test criteria simply by counting the 1s. The easiest way to do this is with the sum function. For example, the number of elements of values between 4 and 10 is:

```
>> sum((values > 4) & (values < 10))
ans =
     2
```

Given this background, we can now write a MATLAB script to perform the calculations from Example 5.1 in Chapter 5, which required us to estimate the probabilities for launches in the ranges (a) less than or equal to 17 m, (b) between 17 m and 19 m, (c) greater than 19 m.

```
n = length(distance);  % number of elements in vector
disp 'probability of launch less than or equal to 17 m'
sum(distance <= 17)/n
disp 'probability of launch between 17 m and 19 m'
sum((17 < distance) & (distance < 19))/n
disp 'probability of launch greater than or equal to 19 m'
sum(distance >= 19)/n
```

Running this script displays the result

```
probability of launch less than or equal to 17 m
ans =
    0.4500
```

```
probability of launch between 17 m and 19 m
ans =
    0.4000
probability of launch greater than or equal to 19 m
ans =
    0.1500
```

10.4.3 Bin Counts and Histograms

In the previous example, we took a rough look at the distribution of launch distances by counting the number of trials in each of three "bins": short, medium, and long distance. As discussed in 5.4.4, we can get a more detailed picture by sorting the data into finer bins. We could do this by writing a script similar to the one we wrote in the last section for Example 5.1, defining a series of ranges using relational operators, but MATLAB provides a more convenient method. The two MATLAB functions hist and histc both provide a way to specify a series of bins and count the number of vector elements in each—the difference between them is, briefly, that hist is generally simpler to use and histc provides greater control.

The hist function is the quickest way to generate a histogram plot of the bin counts. The most common use of hist is to call it with a single argument—the vector of data values—which it then divides into 10 equally spaced bins, counts the number of elements in each, and plots the results. An optional second argument lets the user specify the number of bins. The following script illustrates both options, with the results shown in Figure 10.7:

```
subplot(1,2,1)
hist(distance)        % default is 10 bins
title('hist(distance)')

subplot(1,2,2)
hist(distance, 3)  % divide into 5 bins
title('hist(distance, 3)')
```

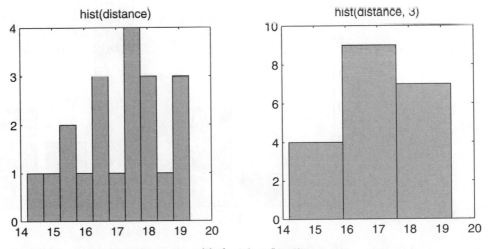

Figure 10.7 Generating histograms with the hist function.

The `histc` function lets the user specify the bin edges and returns the bin counts, but doesn't generate a plot. It takes two arguments, as shown below,

$$bin\_counts = \texttt{histc}(data,\ bin\_edges)$$

where *bin`counts* is a vector of the bin counts, *data* is a vector of the data values, and *bin`edges* is a vector of the bin edges. The following commands count the number of elements of the vector `distance` in the bins $13 \le x \langle 14,\ 14 \le x \langle 15,\ \ldots,\ 19 \le x \langle 20,\ x \rangle 20$:

```
>> bin_edges = 13:20;
>> bin_counts = histc(distance, bin_edges)
bin_counts =
     0    1    3    4    7    2    3    0
```

We can generate a bar plot of the bin counts from `histc` using the `bar` function. Like the `plot` function, `bar` takes a vector of X values and Y values as inputs, where the X values specify the position of vertical bars and the Y values specify their heights. Because we want the bars to line up between the bin edges, rather than centered on them, a trick is to shift the X values by adding 1/2 of a bin width. The following is a complete script to generate a histogram using `histc` and `bar`, with the resulting plot in Figure 10.8:

```
bin_edges = 13:20;
bin_counts = histc(distance, bin_edges);
bar(bin_edges + 0.5, bin_counts)
xlabel('distance, m')
ylabel('count')
```

Note that we could generate a histogram of the probabilities for each bin, rather than the counts, by dividing the `bin_counts` vector by the number of samples.

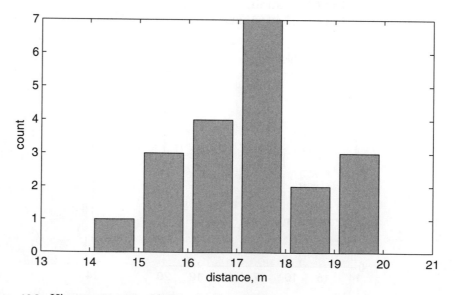

Figure 10.8 Histogram generated with `histc` and `bar`.

10.4.4 Where to Learn More

MATLAB offers a very large library of specialized statistical functions. To learn more about particular functions, see the MATLAB online documentation under `MATLAB > Functions by Category > Mathematics > Data Analysis and Fourier Transforms`.

PROBLEMS

1. **MATLAB Vector Basics**

   Suppose that MATLAB vectors x and y are defined as follows:

   ```
   >> x = [3 6 1 5 4 2];
   >> y = 0:.4:2;
   ```

   Determine the values of each of the following expressions, *without* evaluating them with MATLAB, then confirm your results with MATLAB. If an expression is invalid, indicate this as your result.

   (a) x(3)
   (b) y(2)
   (c) x(3) + y(2)
   (d) x + 3
   (e) x + 10*y
   (f) x(3) + y
   (g) length(x)
   (h) y(length(y))
   (i) x(1:3)
   (j) y(x(4))
   (k) x(y(4))

2. **Plotting Functions**

   Generate MATLAB plots of the following functions over the range indicated. Use enough points so that the function appears smooth.

   (a) $f(x) = 2x - 1, -10 \le x \le 10$
   (b) $f(x) = (x - 1)^2, 0 \le x \le 5$. Note: the MATLAB operator to raise each element of a vector to a power is ".^", not just "^".
   (c) $f(x) = \sin(10x), 0 \le x \le 4\pi$. Use at least 200 points over the range.
   (d) $f(x) = \sin(10x) + \sin(11x), 0 \le x \le 4\pi$.
   (e) $f(x) = x^{-x}\sin(10x), 0 \le x \le 2\pi$. Note: the MATLAB operator to raise each element of a vector to a power is ".*", not just "*".

3. **Plotting Temperatures**

   Write a MATLAB script that performs the following operations:

   - starts with a comment stating the name of the script and its purpose
   - creates a vector T with the values 0 to 50, incrementing by 10
   - adds 273 to each element of T (converts from centigrade to Kelvin)

- creates a vector R containing the values 27, 18, 12, 8, 6, 4
- plots $\log(R)$ versus $1/T$

4. Finding the Roots of a Polynomial

The following polynomial has three roots—values of x where it equals 0—in the interval $0 \le x \le 20$:

$$y = x^3 - 15x^2 + 54x - 43$$

(a) Using the MATLAB `fplot` function, plot the polynomial over the range `[0 20]`. Zoom in on the plot to find where it crosses the x-axis and estimate the roots.

(b) MATLAB has a built-in function that automatically finds the roots of a polynomial. Search the MATLAB online documentation to find out what this function is and learn how to use it. Use this function to find the roots of the polynomial above, and compare the results with your answer to Part (a).

5. Analyzing Force Versus Displacement for an Elastic Cord

A materials engineer is investigating how much force is needed to stretch samples of solid elastic cord by a given amount. As part of the investigation the engineer hung weights from the samples and measured the resulting displacement, which is defined as the length of the sample after the force is applied minus the original length of the sample with no applied force. The table below gives the results of this experiment.

Force (N)	0	10	20	30	40
Sample A displacement (m)	0	0.09	0.18	0.30	0.42
Sample B displacement (m)	0	0.18	0.36	0.60	0.84
Sample C displacement (m)	0	0.36	0.72	1.20	1.68

(a) Write a MATLAB script that plots the force versus displacement for sample A, where the x-axis is displacement and the y-axis is force. The script should give the plot a title, and label the axes with both the names of the variables and their units. Your solution should include both the script and the resulting plot.

(b) Modify the script to plot the force versus displacement for each of the three samples on the same axes. Choose the line and symbol styles for each of the three curves so that they can be distinguished when the graph is printed in black and white. Add a legend to the graph that indicates which curve corresponds to which sample.

(c) Suppose that each of the three samples of cord were made of the same material but were of different dimensions. The dimensions of the samples, in no particular order, were as follows:

- diameter 5 mm, length 1 m;
- diameter 5 mm, length 2 m;
- diameter 7 mm, length 1 m.

Based on the force-deflection measurements, which sets of dimensions do you think belong to samples A, B, and C? Explain your reasoning.

6. MATLAB Simulation of a Slingshot Launch

The following equation approximates the distance that a projectile will travel in a vacuum when launched from a slingshot-type launcher with horizontal and vertical settings x and y, and an experimentally determined "launcher constant" k:

$$d = kxy$$

(a) Suppose that a projectile launched with a vertical setting of 1.0 m and a horizontal setting of 1.5 m traveled a distance of 75 m. What would be the launcher constant for the launcher? Be sure to state the units.

(b) How far would a projectile travel when launched from the same launcher with a horizontal setting of 1.5 m and a vertical setting of 1.25 m?

(c) Suppose that a team using a launcher has developed a launch strategy of keeping the horizontal setting constant at 1.5 m and then adjusting the vertical setting to achieve a target distance. Assuming the launcher is known to have a constant of $k = 45\ \text{m}^{-1}$, write a MATLAB script that produces a plot of launch distance versus launcher vertical setting that the team could use in the field. Be sure that the resulting plot has a title and properly label the axes. Turn in both the script and the plot.

(d) Modify the script in part **(c)** of this problem to plot three curves on one set of axes, for $k = 45$, $k = 50$, and $k = 55$. Be sure to add a legend to the plot. Turn in both the script and the plot.

(e) Suppose that you are given four shots total to hit each of three targets at distances 50, 100, and 150 meters using a slingshot-type launcher. The precise launcher constant is unknown, but is between $k = 45\ \text{m}^{-1}$ and $k = 55\ \text{m}^{-1}$. Develop a procedure based upon the plot created in part **(d)** of this problem for conducting the launch to hit all three targets. Give step-by-step instructions that anyone who is already familiar with the launcher mechanism could follow.

7. Accuracy of a Launch System

Ten trial launches were performed with each of two slingshot-type launchers, set to fire at a target distance of 100 m. The results of the trials are:

Launcher A (m)	97	96	100	98	98	98	10	99	97	95
Launcher B (m)	96	107	95	100	108	85	116	102	90	111

(a) Write a MATLAB script that calculates the statistics listed below for the trials for each of the two launchers. Turn in both the script and the results.

- minimum distance
- maximum distance
- mean distance
- plots a histogram with bins of width 2 that spans the entire range of the sample values

(b) If you had to pick one of the launchers to use for a graded launch test, which would you choose and why? Use statistics to justify your choice.

8. Distribution of Resistor Values

A certain vendor sells carbon film resistors with a 5% tolerance for $0.05 apiece, and with a 1% tolerance for $0.10 apiece. You've just bought a lot of 1000 2.2 kΩ, 5% resistors. Curious about what the actual tolerances are of the resistors you've just purchased, you asked a college student working as a summer intern to test them all for you. The actual resistance values that the intern got are stored in the ASCII text file *resistors.dat*.

(a) What are the minimum and maximum resistor values? What is the actual tolerance range of the resistors in the shipment, expressed as plus or minus some percentage of the nominal value?

(b) What is the mean value of the resistance? What is the median value? What do you observe about these values that's somewhat odd—what might one "normally" expect of these two different averages?

(c) Plot a histogram of the resistor values. Looking at the results, what can you surmise about the process by which the vendor manufactures, sorts, and ships resistors?

9. Error Analysis of Launch Data

Consider the following data for a launcher aiming at a target 100 m downrange:

Launcher B (m)	96	107	95	100	108	85	116	102	90	111

The error for a launch i at a target distance d_{target} is defined as the actual distance d_i of the launch minus the target distance:

$$\text{err}_i = d_{\text{target}} - d_i$$

Store the trial results in a single MATLAB vector variable and then answer the questions below.

(a) Write a MATLAB vector expression to calculate the error associated with each of the trials.

(b) Write a MATLAB vector expression to calculate the average error for each of the trials. Why are the results of this calculation misleading?

(c) Write a MATLAB vector expression to calculate the average of the square of the error for each of the trials, and then takes the square root of this quantity. Why is this a better indicator of the accuracy of the launcher than the average error?

(d) Use the MATLAB `std` function to find the standard deviation of the data. How does this value compare with your results from Part **(c)**?

10. Generating Random Numbers with a Uniform Distribution

The MATLAB function `rand` generates a set of random numbers with values uniformly distributed between 0 and 1. The function returns a two-dimensional table or matrix of values; in order to generate a vector of values—which is simply a table with one row—the syntax for the function is `rand(1, N)`.

(a) Use the command

```
data = rand(1,1e6);
```

to define a variable named `data` with 1 million random numbers. (Don't forget the semicolon at the end of the line, or else MATLAB will try to echo 1 million values to the terminal—type "Ctrl-c" if you forgot and need to stop it!) Generate a histogram plot of `data`. Describe what the plot looks like. Based on your observation of the histogram, give a definition for a uniform distribution—use the words "bin" and "probability" in your definition.

(b) Write a MATLAB expression using `rand` to generate a sample of 1 million random numbers uniformly distributed between 0 and 10. Produce a histogram plot for the sample.

(c) Write a MATLAB expression using `rand` to generate a sample of 1 million random numbers uniformly distributed between 100 and 200. Produce a histogram plot for the sample.

11. Generating Random Numbers with a Normal Distribution
The MATLAB function `randn` generates a set of random numbers with values with a normal or Gaussian distribution, with a mean of 0 and a standard deviation of 1. The function returns a two-dimensional table or matrix of values; in order to generate a vector of values—which is simply a table with one row—the syntax for the function is `randn(1,N)`.

(a) Use the command

```
data = randn(1,1e6);
```

to define a variable named `data` with 1 million random numbers. (Don't forget the semicolon at the end of the line, or else MATLAB will try to echo 1 million values to the terminal—type "Ctrl-c" if you forgot and need to stop it!) Generate a histogram plot of `data`. Describe what the plot looks like. Use MATLAB functions to calculate the mean and standard deviation of the sample. Try both forms of the standard deviation, where the denominator contains either `N` or `N-1`. What effect does the form of the standard deviation have on its value?

(b) Write a short MATLAB script that calculates the fraction of samples in `data` that lie between ±1. What fraction lies between ±2 and between ±3?

(c) Write a MATLAB expression using `randn` to generate a sample of 1 million random numbers normally distributed with a mean of 0 and a standard deviation of 100.

- Calculate the mean and standard deviation of the sample.
- Plot a histogram of the sample.
- Write a short MATLAB script to calculate the fraction of the samples that lies between ±1 standard deviation from the mean. Comment on your results.

(d) Write a MATLAB expression using `randn` to generate a sample of 1 million random numbers normally distributed with a mean of 50 and a standard deviation of 25.

- Calculate the mean and standard deviation of the sample.
- Plot a histogram of the sample.
- Write a short MATLAB script to calculate the fraction of the samples that lies between ±1 standard deviation from the mean. Comment on your results.

12. Numerical Model of a Slingshot Using MATLAB

Write a MATLAB script that builds a numerical model for a softball launched from a slingshot, following the example and data from Section 5.3.3.

- Define vectors for the pullback distance X and the actual flight distance D_{exp} based on the experimental data in Table 5.3.
- Calculate the slope and intercept of the best-fit line, passing through the two (X, D) points $(0.5, 4)$ and $(1.5, 27)$.
- Calculate the vector D_{mod} of the flight distances predicted by the model at each of the values of X from Table 5.3.
- Calculate the error between the actual flight distances and those predicted by the model. Compare your results with those in Table 5.4.
- Generate a single plot with both D_{exp} and D_{mod} versus X, similar to Figure 5.10. Use the same labels and plotting styles as in the figure. (There's no need to label the two points used to evaluate the best-fit line.)
- Calculate the *root mean square* or *RMS* error for the model. RMS error is defined as the square root of the mean of the squares of the error over all of the data points. Add this value as a text comment to your plot; you may do so outside the script using tools in the MATLAB plot window.

13. Using `polyfit` to Find a Best-Fit Line

MATLAB's `polyfit` uses a technique called *linear regression* to find the coefficients of polynomial—including a simple line—that is a good fit to a set of data points. In fact, the polynomial found by `polyfit` is the best fit, in the sense that it minimizes the square of the error between the actual data points and the polynomial. In this exercise, we'll experiment with `polyfit` to see how it works.

(a) Finding a line that passes exactly through sample data points:

 i. Define two MATLAB vectors x and y that contain the *x*- and *y*-coordinates of the three points $(0, 7)$, $(1, 10)$, and $(2, 13)$. Plot the three points as small circles on a graph.

 ii. *Prove* that these points are collinear—that is, that they lie along a common line. What are the slope and *y*-intercept of this line?

 iii. Look up the MATLAB function `polyfit` in the online documentation and read the description. Use the command `polyfit(x,y,1)` to determine the coefficients of the best-fit, least-squares line through the three points. How do these coefficients compare with the slope and intercept that you found in Problem Part (**ii**)?

 iv. Using the MATLAB `hold` and `fplot` functions, plot the equation of the best-fit line, adding it to the graph from Part (**i**).

(b) Finding a best-fit line that doesn't pass through all of the data points:

 i. Define two MATLAB vectors x and y that contain the *x*- and *y*-coordinates of the three points $(0, 7)$, $(1, 11)$, and $(2, 12)$. Plot the three points as small circles on a graph.

 ii. Use `polyfit` to determine the coefficients of the best-fit, least-squares line through the three points. How do these coefficients compare with the slope and intercept that you found in Part (**ii**)?

iii. Using the MATLAB `hold` and `fplot` functions, plot the equation of the best-fit line, adding it to the graph from Part (**i**).

14. Using `polyfit` **to Build a Slingshot Model**

Write a MATLAB script that builds a numerical model for a softball launched from a slingshot, following the example and data from Section 5.3.3.

- Define vectors for the pullback distance X and the actual flight distance D_{exp} based on the experimental data in Table 5.3.
- Use MATLAB's `polyfit` function to calculate the slope and intercept of the best-fit line through the experimental data.
- Calculate the vector $D_{polyfit}$ of the flight distances predicted by the model at each of the values of X from Table 5.3.
- Generate a single plot with both D_{exp} and $D_{polyfit}$ versus X, similar to Figure 5.10. Use the same labels and plotting styles as in the figure.
- Calculate a vector with the error between the actual flight distances and those predicted by the model.
- Calculate the *root mean square* or *RMS* error for the model. RMS error is defined as the square root of the mean of the squares of the error over all of the data points. Add this value as a text comment to your plot; you may do so outside of the script using tools in the MATLAB plot window. Compare the RMS error of the model using `polyfit` with the RMS error from the model in Table 5.4.
- Can you find the equation for a line that will produce a lower RMS error than the one whose slope and intercept you calculated using `polyfit`? Explain.

15. Coin Toss

Write a MATLAB script using one of the random number generation functions to simulate 10 independent coin tosses. The script should output how many tosses were "heads" and how many were "tails."

16. Rolling Dice

Write a MATLAB script using one of the random number generation functions to simulate 100 independent rolls of a pair of six-sided dice. You may find one of the MATLAB functions `round`, `floor`, `ceil`, or `fix` helpful in converting a decimal or floating-point number to an integer. The script should output how many rolls produced each of the integer values in the range 2–12.

17. Simulating Slingshot Launches

Write a MATLAB script using one of the random number generation functions to simulate 20 launches from a slingshot with the same mean and standard deviation as the experiment shown in Table 5.5.

(a) Produce a scatter plot of the trials similar to Figure 5.12.

(b) Calculate the mean and standard deviation from your simulated trials. How do these compare with the values that you set when running the experiment?

(c) Rerun the simulation with 1000 trials and again calculate the mean and standard deviation. How do these values compare with the values that you set when running the experiment?

(d) Plot a histogram of the distances from the simulation with 1000 trials.

18. Random Number Generation

In this problem, you will use the `rand` function to produce random numbers. For each of the four exercises below, write a MATLAB statement to generate the random numbers. You will need MATLAB functions that convert a real number to an integer (do a help on: `round`, `floor`, `ceil`, and `fix`). You are encouraged to test each statement in the MATLAB console.

(a) produce random integers between 0 and 300

(b) produce random integers between two given numbers `lo` and `hi` (assume both numbers have already been assigned a value)

(c) produce random integers between two given numbers `lo` and `hi`, but making sure that the generated random numbers are not within `r` of those two numbers (assume `r` is small compared to `lo` and `hi`, and that they have all already been assigned a value)

(d) randomly produce either +1 or −1

Matrix Operations in MATLAB

LEARNING OBJECTIVES

- to use matrices as a structure for 2-dimensional arrays of data;
- to produce plots for visualizing multidimensional data, including families of curves on common axes and contour plots;
- to apply basic techniques in matrix arithmetic and linear algebra using MATLAB;
- to solve a system of linear equations using MATLAB.

11.1   BASIC OPERATIONS

11.1.1   Defining and Accessing Matrices

In mathematics, a *matrix* is a rectangular array of numbers or elements, organized in rows and columns. The *size* of a matrix is given as the number of rows in the matrix followed by the number of columns; thus, a matrix with m rows and n columns has size of m-by-n or $m \times n$. Some examples of matrices are given below.

$$\mathbf{M} = \begin{bmatrix} 16 & 5 & 9 & 4 \\ 3 & 10 & 6 & 15 \\ 2 & 11 & 7 & 14 \\ 13 & 8 & 12 & 1 \end{bmatrix} \quad \mathbf{Q} = \begin{bmatrix} 2 & 6 & 5 \\ 3 & 1 & 4 \end{bmatrix} \quad \mathbf{R} = \begin{bmatrix} 1 & 5 & 3 \end{bmatrix} \quad \mathbf{V} = \begin{bmatrix} 1 \\ 5 \\ 3 \end{bmatrix}$$

The size of matrix $\mathbf{M}$ is 4×4, $\mathbf{Q}$ is 2×3, $\mathbf{R}$ is 1×3, and $\mathbf{V}$ is 3×1. A matrix such as $\mathbf{M}$, in which the number of rows equals the number of columns, is known as a *square matrix*. Matrices that contain only one row or one column are called *vectors*. A matrix such as $\mathbf{R}$ that contains a single row is called a *row vector*, and a matrix such as $\mathbf{V}$ that contains a single column is called a *column vector*.

To define the matrix **M** as a MATLAB variable named M, enter the following at the command line:

```
>> M = [16 5 9 4
3 10 6 15
2 11 7 14
13 8 12 1]
M =
      16       5       9       4
       3      10       6      15
       2      11       7      14
      13       8      12       1
```

Alternatively, you can type

```
>> M = [16 5 9 4; 3 10 6 15; 2 11 7 14; 13 8 12 1]
M =
      16       5       9       4
       3      10       6      15
       2      11       7      14
      13       8      12       1
```

In general, the rules for entering a matrix are:

1. separate the elements in each row with either a blank or a comma,
2. separate the rows with either a new line or a semicolon,
3. surround the entire matrix with square brackets, [ ].

The following MATLAB statements define the remaining matrices from the example above.

```
>> Q = [2 6 5; 3 1 4]
Q =
       2       6       5
       3       1       4

>> R = [1 5 3]
R =
       1       5       3

>> V = [1; 5; 3]
V =
       1
       5
       3
```

Note that the row vector R is just a MATLAB "vector" as we defined it in Chapter 10.

The size command returns the dimensions of a matrix as a 2-element vector in which the first vector is the number of rows and the second element is the number of columns:

```
>> size(Q)
ans =
     2      3
>> size(R)
ans =
     1      3
>> size(V)
ans =
     3      1
```

Subscripts Just as with vectors, subscripts are used to refer to individual elements of a matrix, so that the element in the ith row and jth column of a matrix $\mathbf{A}$ is referred to as a_{ij}. In MATLAB, the numbering of subscripts begins with 1, so that's the convention we'll adopt here. Other texts and programming languages (such as C and Java) number subscripts starting with zero—so beware. The MATLAB notation A(i,j) is used to access the element in row i and column j of matrix A. The following examples refer to the matrices $\mathbf{M}$ and $\mathbf{R}$ defined above.

```
>> M(2,3)
ans =
     6
>> R(1,2)
ans =
     5
>> R(2,1)
??? Index exceeds matrix dimensions.
```

For row vectors or column vectors such as the matrices R and V above, MATLAB allows the use of a single subscript to access elements in the vector, automatically applying the subscript to the proper dimension.

```
>> R(2)
ans =
     5
>> V(2)
ans =
     5
```

When used as a subscript in MATLAB, the colon operator (:) refers to all the elements of a given row or column, and thus can be used to access an individual row or column from a matrix, as the examples below illustrate.

```
>> M(:,2)
ans =
     5
    10
    11
     8

>> M(3,:)
ans =
     2  11  7  14
```

The colon operator can also be used to refer to a range of rows or columns. The following MATLAB statements access "rows 1 through 2, columns 2 through 4" of matrix M:

```
>> M(1:2,2:4)
ans =
      5       9       4
     10       6      15
```

Transpose of a Matrix The transpose of a matrix **A**, denoted $\mathbf{A}^T$, is a matrix obtained by exchanging the rows and columns of **A**. In other words, given a matrix **A** whose elements are a_{ij}, the elements of $\mathbf{A}^T$ are a_{ji}. The transposes of each of the four example matrices from the beginning of this section are given below.

$$\mathbf{M}^T = \begin{bmatrix} 16 & 3 & 2 & 13 \\ 5 & 10 & 11 & 8 \\ 9 & 6 & 7 & 12 \\ 4 & 15 & 14 & 1 \end{bmatrix} \quad \mathbf{Q}^T = \begin{bmatrix} 2 & 3 \\ 6 & 1 \\ 5 & 4 \end{bmatrix} \quad \mathbf{R}^T = \begin{bmatrix} 1 \\ 5 \\ 3 \end{bmatrix} \quad \mathbf{V}^T = \begin{bmatrix} 1 & 5 & 3 \end{bmatrix}$$

Note that the transpose of a row vector is a column vector and vice versa. To take the transpose of a matrix in MATLAB, use the *transpose operator*, '.

```
>> M'
ans =
     16       3       2      13
      5      10      11       8
      9       6       7      12
      4      15      14       1

>> Q'
ans =
      2       3
      6       1
      5       4

>> R'
ans =
      1
      5
      3

>> V'
ans =
      1       5       3
```

The transpose operator provides a convenient way to define a long row vector in a script, which is especially useful when copying data from another source, such as a column from a spreadsheet. Rather than entering values on a single line and then breaking it with ellipses, you can instead enter each element on its own line,

defining a column vector, and then transpose the result. For example, the following
two vector definitions produce the same result.

```
row_vector_a = [1 2 3 ...    % single line continued
        4 5]

row_vector_b = [
    1
    2
    3
    4
    5]'                              % column vector transposed

>> row_vector_a
row_vector_a =
        1      2      3      4      5
>> row_vector_b
row_vector_b =
        1      2      3      4      5
```

11.1.2   Element-Wise Arithmetic Operations on Matrices

Just as with simple vectors, element-wise arithmetic operations perform scalar arithmetic operations between corresponding elements of two matrices of identical size
on an element-by-element basis. Table 11.1 summarizes the element-wise operations, which is the same list we presented in Chapter 10 for vectors. The symbols for
element-wise multiplication, division, and exponentiation begin with a dot, to distinguish them from true matrix multiplication under linear algebra. Again, under linear
algebra, matrix addition and multiplication of a matrix by a scalar *are* element-wise
operations, so a dot is not needed to distinguish these. Examples of the element-wise
matrix operations are given below:

```
>> a = [0 1 2; 3 4 5]
a =
        0      1      2
        3      4      5
>> b = [6 7 8; 9 10 11]
b =
        6      7      8
        9     10     11
>> 3 * a + 5
ans =
        5      8     11
       14     17     20
>> a .* b ans =
        0      7     16
       27     40     55
>> a .^ 2
ans =
        0      1      4
        9     16     25
```

TABLE 11.1 Element-wise vector operations.

Operation	Scalar form	Array form
addition	+	+
subtraction	−	−
multiplication	*	.*
division	/	./
exponentiation	^	.^

The trigonometric and other transcendental functions can also be applied to matrices on an element-by-element basis, as shown below.

```
>> log10(b)
ans =
    0.7782    0.8451    0.9031
    0.9542    1.0000    1.0414
```

11.2 PARAMETER SWEEPS OVER TWO VARIABLES

One of the main steps in conducting a trade study, as described in Chapter ??, is making plots or "maps" of the design space that show how system behavior varies with changes in the values of design or environmental variables. In Section 4.9.2, for example, we examined how the weight of the column of water inside the drop pipe of a pump varied with the radius of the drop pipe and the depth of the well. In this section, we consider the MATLAB tools used in creating these maps.

11.2.1 Creating Tables Using `meshgrid`

Each type of plot we used for mapping a design space in Chapter 5—3-D plots, side-view cross-section plots, and top-view contour plots—was based on data that came from sweeping the parameters of a model across a range of values, and then arranging this data in a table. In this section, we look at how to create such tables using MATLAB matrices and the `meshgrid` function.

To illustrate, let's consider the example of constructing a multiplication table such as that shown in Table 11.2. We can think of this table as a parameter sweep of the model

$$c = c(a, b) = a \times b,$$

where a varies over the range of integers 1–5 and b varies over the range of integers 1–3. Not counting the row and column headings, the multiplication table is a matrix

TABLE 11.2 Multiplication table.

$c = a \times b$	$a = 1$	$a = 2$	$a = 3$	$a = 4$	$a = 5$
$b = 1$	1	2	3	4	5
$b = 2$	2	4	6	8	10
$b = 3$	3	6	9	12	15

with 3 rows and 5 columns, in which the value of an element located at row i and column j in the matrix is equal to the product of the heading of column j and the heading of row i.

In terms of MATLAB element-wise operations, we can also think of the multiplication table as being the element-wise product of two 3×5 matrices, one that has the series of column headings in each row, and the other that has the series of row headings in each column. We call these two matrices a_grid and b_grid, respectively, and their product c_grid.

```
a_grid = [1 2 3 4 5;
          1 2 3 4 5;
          1 2 3 4 5]

b_grid = [1 1 1 1 1;
          2 2 2 2 2;
          3 3 3 3 3]

>> c_grid = a_grid .* b_grid c_grid =
    1     2     3     4     5
    2     4     6     8    10
    3     6     9    12    15
```

The tedious part of this exercise is creating the two matrices a_grid and b_grid. This is where the MATLAB meshgrid function comes in. meshgrid takes two vectors that correspond to the headings of the table as input, and produces two matrix "grids" suitable for element-wise operations. The syntax of the meshgrid function is

```
[xgrid, ygrid] = meshgrid(x,y)
```
in which

x	is an n-vector of the column headings
y	is an m-vector of the row headings
xgrid	is an $m \times n$ matrix where each of the m rows is a copy of x
ygrid	is an $m \times n$ matrix where each of the n columns is a copy of y

Thus, the following script would produce the matrices a_grid and b_grid from above:

```
a = 1:5;
b = 1:3;
[a_grid, b_grid] = meshgrid(a, b);
```

11.2.2 Example: Force on the Piston of a Pump Versus Well Depth and Cylinder Radius

In Section 4.9, we took a detailed look at the design of a pump handle assembly that a child could operate. A key part of solving the problem was making assumptions about the depth of the well and the radius of the cylinder, in order to determine the force on the load end of the pump handle. In this section, we'll explore the design further by using MATLAB to perform parameter sweeps over the cylinder radius and well depth, to see how the force varies with both of these variables. Specifically, we'll determine values for the force as the cylinder radius varies from 15 mm to 40 mm and the well depth varies from 25 m to 50 m.

Recall from Section 4.9 that the force to lift the piston is given by the function

$$W = W(r_{cylinder}, h) = \rho g(\pi r_{cylinder}^2 h) \tag{11.1}$$

in which,

W	is the load force
$r_{cylinder}$	is the radius of the cylinder, which varies 15–40 mm
h	is the depth of the well, which varies 25–50 m
ρ	is the density of water, 1000 kg/m^3

There are two steps to performing a parameter sweep across r_{pipe} and h in MAT-LAB:

1. Define grid matrices for $r_{cylinder}$ and h that sweep their values over the specified ranges.
2. Express the formula for W as element-wise matrix operations on these matrices.

The following script implements these operations:

```
rho = 1000;              % density of water, kg/m^3
g = 9.81;                % acceleration of gravity, m/s^2
r_cylinder = 15:5:40;    % cylinder radius, mm
h = 25:5:50;             % well depth, m


[r_cylinder_grid, h_grid] = meshgrid(r_cylinder, h);
W_grid = rho*g*(pi*(r_cylinder_grid/1000).^2 .* h_grid)
```

Executing the script produces the following results:

```
>> r_cylinder_grid =
    15    20    25    30    35    40
    15    20    25    30    35    40
    15    20    25    30    35    40
    15    20    25    30    35    40
    15    20    25    30    35    40
    15    20    25    30    35    40


h_grid =
    25    25    25    25    25    25
    30    30    30    30    30    30
    35    35    35    35    35    35
    40    40    40    40    40    40
    45    45    45    45    45    45
    50    50    50    50    50    50


W_grid =
  1.0e+003 *
    0.1734    0.3082    0.4815    0.6934    0.9438    1.2328
    0.2080    0.3698    0.5779    0.8321    1.1326    1.4793
    0.2427    0.4315    0.6742    0.9708    1.3214    1.7259
```

0.2774	0.4931	0.7705	1.1095	1.5101	1.9724
0.3120	0.5547	0.8668	1.2482	1.6989	2.2190
0.3467	0.6164	0.9631	1.3869	1.8877	2.4655

Note that we included the word "grid" in the names of variables that are grid matrices. This isn't necessary, but it's a convention that helps the reader—and especially the programmer—recognize that these are grid matrices and that element-wise operations are required to operate on them.

11.3   PLOTTING 3-DIMENSIONAL DATA

MATLAB provides a variety of functions for plotting 3-dimensional data that take grid matrices as input. All of the plots in the slingshot trade study in Section 5.5 were generated in this way. In this section, we illustrate the most common uses of these functions by plotting the results from the pump parameter sweep in the previous section. Each of the 3-D graphing functions has a number of options for customizing the result; the "3-D Visualization" section of the MATLAB online documentation contains an excellent explanation of these with extensive examples.

11.3.1   Mesh and Surface Plots

The MATLAB mesh function produces a "wireframe" plot of the data contained in a grid matrix. In its most common usage, mesh is called with 3 arguments:

```
mesh(Xgrid, Ygrid, Zgrid)
```

where
    Xgrid is either a vector or grid matrix that defines the x-axis of the plot
    Ygrid is either a vector or grid matrix that defines the y-axis of the plot
    Zgrid is a grid matrix that defines the height along the z-axis for each grid point

The following script illustrates the mesh function for piston force example data:

```
mesh(r_cylinder_grid, h_grid, W_grid);
xlabel('r_{cylinder}, mm');
ylabel('h, m');
zlabel('W, N');
```

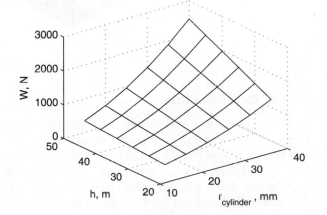

The meshz function is similar to the mesh function, but it adds a "curtain" around the plot to highlight *z*-values at the edges.

```
meshz(r_cylinder_grid, h_grid, W_grid);
xlabel('r_{cylinder}, mm');
ylabel('h, m');
zlabel('W, N');
```

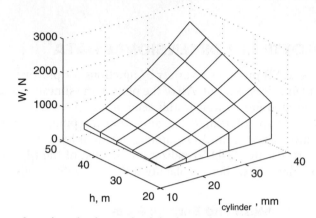

The surf function shades each of the facets in the wireframe mesh plot according to the z-value to produce a surface plot.

```
surf(r_cylinder_grid, h_grid, W_grid);
xlabel('r_{cylinder}, mm');
ylabel('h, m');
zlabel('W, N');
```

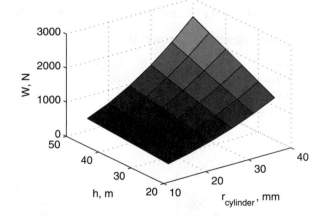

11.3.2 Contour Plots

A contour plot is a "top view" of a 3-D plot, such as a mesh or surface plot, that shows lines of equal elevation. A complete description of the construction of a contour plot was provided in Section 5.5. The MATLAB contour function generates a contour plot, and it takes the same arguments as the mesh function described above.

The contour function also returns two variables: a "handle graphics object" that contains the labels for the contours, and the "handle" to this object. It's not important for a MATLAB novice to understand much about these objects. What *is* important is that to add labels to the elevation lines in a contour plot, you need to use these objects as arguments to the `clabel` function. The following example illustrates both `contour` and `clabel`.

```
[C, hdl] = contour(r_cylinder_grid, h_grid, W_grid);
clabel(C, hdl);
xlabel('r_{cylinder}, mm');
ylabel('h, m');
```

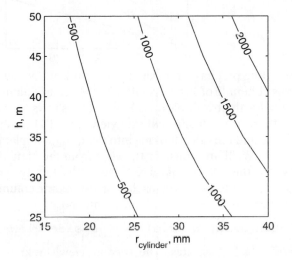

11.3.3   Side-View Cross-Section Plots

Side-view cross-section plots that include a family of "cross-section" curves are a bit trickier to generate because you need to carefully track the orientation of the data. The side-view plots we presented in Section 5.5 were produced using the MATLAB `plot` command, but with the arguments as grid matrices rather than vectors.

When the arguments of the `plot` command are matrices of the same size, it produces a family of curves that plot the columns of one matrix versus the corresponding columns of the other. For example, consider the following two matrices X and Y.

```
>> X =[0 0 0; 1 1 1; 2 2 2; 3 3 3; 4 4 4]
X =
        0        0        0
        1        1        1
        2        2        2
        3        3        3
        4        4        4

>> Y=[0 0 0; 1 2 3; 2 4 6; 3 6 9; 4 8 12]
Y =
        0        0        0
        1        2        3
        2        4        6
        3        6        9
        4        8       12
```

```
>> plot(X,Y)
```

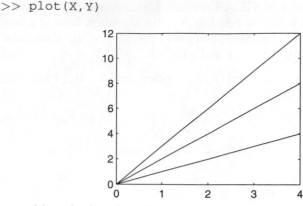

The resulting plot has three curves: column 1 of Y versus column 1 of X, column 2 of Y versus column 2 of X, and column 3 of Y versus column 3 of X.

Now for the tricky part. As Figure 11.1 shows for the water column weight example, there are two different side views of a 3-D mesh plot—one looking into the h-W plane and the other looking into the $r_{cylinder}$-W plane—that we need to consider separately. We'll first consider the view from the "left" looking into the h-W plane. In this case, the side view is an X-Y plot with a family of curves that plot the columns of W_grid on the y-axis versus the corresponding columns of h_grid on the x-axis. We can generate this view with the command

```
plot(h_grid, W_grid)        % side view from the "left"
```

as Figure 11.1 illustrates. The second view, looking from the "right" into the $r_{cylinder} - W$ plane, plots the *rows* of W_grid on the y-axis versus the *rows* of r_cylinder_grid on the x-axis. Since the plot function plots the *columns* of one matrix versus the columns of another, we need to transpose the matrices first.

```
plot(r_cylinder_grid', W_grid') % side view from the "right"
```

Labels on each of the curves in the family clarify which cross-section each curve belongs to. While there are ways of adding these labels as part of a script, in general, we add them manually, with the MATLAB plot-editing tools after generating the plot.

11.4 MATRIX ARITHMETIC

Just as simple arithmetic defines operations for adding, subtracting, and multiplying numbers, *matrix arithmetic* defines operations for adding, subtracting, and multiplying matrices. The basic rules and operations of matrix arithmetic and the corresponding MATLAB commands are summarized below.

11.4.1 Zero Matrix

A zero matrix is defined as a matrix whose elements are all equal to 0. To create an $m \times n$ zero matrix in MATLAB, use the zeros command.

```
>> zeros(2,3)
   ans =
      0      0      0
      0      0      0
```

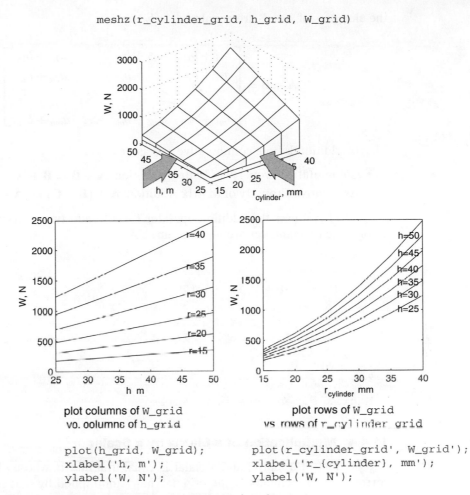

Figure 11.1   Formation of side-view cross-section plots.

11.4.2 Equality of Matrices

Two matrices are equal if and only if they are the same size and each of their corresponding elements are equal. In other words, $\mathbf{A} = \mathbf{B}$ if and only if $a_{ij} = b_{ij}$ for all i, j. From the example matrices above, $\mathbf{R}^T = \mathbf{V}$ and $\mathbf{V}^T = \mathbf{R}$, but $\mathbf{R} \neq \mathbf{V}$.

11.4.3 Matrix Addition

Two matrices may be added if and only if they are the same size. The sum of two matrices is a matrix of the same size as the original two, whose elements are the sums of the corresponding elements of the original two. More formally, given two $m \times n$ matrices $\mathbf{A}$ and $\mathbf{B}$

$$\mathbf{A} = \begin{bmatrix} a_{11} & a_{12} & \dots & a_{1n} \\ a_{21} & a_{22} & \dots & a_{2n} \\ \vdots & \vdots & \ddots & \vdots \\ a_{m1} & a_{m2} & \dots & a_{mn} \end{bmatrix}, \quad \mathbf{B} = \begin{bmatrix} b_{11} & b_{12} & \dots & b_{1n} \\ b_{21} & b_{22} & \dots & b_{2n} \\ \vdots & \vdots & \ddots & \vdots \\ b_{m1} & b_{m2} & \dots & b_{mn} \end{bmatrix},$$

the elements of $\mathbf{A} + \mathbf{B}$ are equal to $a_{ij} + b_{ij}$, or

$$\mathbf{A} + \mathbf{B} = \begin{bmatrix} a_{11} + b_{11} & a_{12} + b_{12} & \cdots & a_{1n} + b_{1n} \\ a_{21} + b_{21} & a_{22} + b_{22} & \cdots & a_{2n} + b_{2n} \\ \vdots & \vdots & \ddots & \vdots \\ a_{m1} + b_{m1} & a_{m2} + b_{m2} & \cdots & a_{mn} + b_{mn} \end{bmatrix}$$

Matrix addition is both commutative and associative.

- Commutative Property of Matrix Addition: $\mathbf{A} + \mathbf{B} = \mathbf{B} + \mathbf{A}$
- Associative Property of Matrix Addition: $\mathbf{A} + (\mathbf{B} + \mathbf{C}) = (\mathbf{A} + \mathbf{B}) + \mathbf{C}$

MATLAB uses the addition operator (+) to add two matrices. It returns an error if the two matrices are not the same size.

```
>> M+M'
ans =
    32     8    11    17
     8    20    17    23
    11    17    14    26
    17    23    26     2

>> R+V
??? Error using ==> +
Matrix dimensions must agree.
```

11.4.4 Multiplication of a Matrix by a Scalar

The product of a matrix and a scalar is equal to a matrix whose elements are the corresponding elements of the original matrix multiplied by that scalar. Thus, the elements of $\alpha\mathbf{A}$ are $\alpha \cdot a_{ij}$:

$$\alpha\mathbf{A} = \begin{bmatrix} \alpha a_{11} & \alpha a_{12} & \cdots & \alpha a_{1n} \\ \alpha a_{21} & \alpha a_{22} & \cdots & \alpha a_{2n} \\ \vdots & \vdots & \ddots & \vdots \\ \alpha a_{m1} & \alpha a_{m2} & \cdots & \alpha a_{mn} \end{bmatrix}$$

It follows that multiplication of a matrix by a scalar is commutative, $\alpha\mathbf{A} = \mathbf{A}\alpha$.

MATLAB uses the multiplication operator (*) to multiply a matrix by a scalar, as illustrated by the examples below.

```
>> 2*M ans =
    32    10    18     8
     6    20    12    30
     4    22    14    28
    26    16    24     2

>> R*2
ans =
     2    10     6
```

The arithmetic inverse of a matrix $\mathbf{A}$, $-\mathbf{A}$, is defined as $(-1) \cdot \mathbf{A}$, so that the sum of a matrix and its arithmetic inverse (negative) is a matrix of the same size containing all zeros. MATLAB uses the negative operator $(-)$ to obtain the arithmetic inverse (negative) of a matrix.

```
>> -V
ans =
    -1
    -5
    -3
```

11.4.5 Matrix Subtraction

The difference of two matrices $\mathbf{A} - \mathbf{B}$ is defined as $\mathbf{A} + (-\mathbf{B})$. As with scalar arithmetic, matrix subtraction is neither commutative nor associative. In MATLAB, matrices are subtracted using the subtraction operator $(-)$.

11.4.6 Matrix Multiplication

Two matrices $\mathbf{A}$ and $\mathbf{B}$ may be multiplied if and only if the number of columns in matrix $\mathbf{A}$ is equal to the number of rows in matrix $\mathbf{B}$. The resulting product will be a matrix with the same number of rows as $\mathbf{A}$ and the same number of columns as $\mathbf{B}$. Thus, if $\mathbf{A}$ is an $m \times p$ matrix, and $\mathbf{B}$ is a $p \times n$ matrix, then the product $\mathbf{C} - \mathbf{AB}$ will be a $m \times n$ matrix. The elements of the product $\mathbf{C}$ are defined by the following formula:

$$c_{ij} - \sum_{k=1}^{p} a_{ik} \cdot b_{kj}$$

In other words, the value of c_{ij} is obtained by taking the elements of the ith row of $\mathbf{A}$, multiplying them by the corresponding elements of the jth column of $\mathbf{B}$, and then adding the partial products together.

To get the value of $\mathbf{C}_{12}$, multiply the elements in row 1 of $\mathbf{A}$ by the corresponding elements in column 2 of $\mathbf{B}$, and add them together.

As an example, let $\mathbf{P}$ be the 3×3 matrix

$$\mathbf{P} = \begin{bmatrix} 4 & -1 & 0 \\ -1 & 6 & 2 \\ 0 & -2 & 7 \end{bmatrix}$$

The product $\mathbf{N} = \mathbf{QP}$, where $\mathbf{Q}$ is the 2×3 matrix defined above, will also be a 2×3 matrix. The value for n_{11} is

$$n_{11} = \sum_{k=1}^{3} q_{1k} \cdot p_{k1}$$
$$= (2)(4) + (6)(-1) + (5)(0)$$
$$= 2$$

Similarly,

$$n_{12} = \sum_{k=1}^{3} q_{1k} \cdot p_{k2}$$
$$= (2)(-1) + (6)(6) + (5)(-2)$$
$$= 24$$

To calculate the full matrix product using MATLAB, use the multiplication (*) operator.

```
>> P=[4 -1 0; -1 6 2; 0 -2 7]
P =
     4    -1     0
    -1     6     2
     0    -2     7

>> N = Q*P
N =
     2    24    47
    11    -5    30
```

Note that the product **PQ** is not defined because the matrix dimensions don't match.

```
>> P*Q
??? Error using ==> *
Inner matrix dimensions must agree.
```

While matrix multiplication is associative and is distributive over addition, it is *not*, however, commutative.

- Associative Property: $(\mathbf{AB})\mathbf{C} = \mathbf{A}(\mathbf{BC})$.
- Distributive Property of Multiplication over Addition: $\mathbf{A}(\mathbf{B} + \mathbf{C}) = \mathbf{AB} + \mathbf{AC}$.
- Not Commutative: $\mathbf{AB} \neq \mathbf{BA}$

Note that even if two matrices are the same size, matrix multiplication is still not commutative. For example, in general, even the product of a square matrix and its transpose cannot be commuted, as illustrated below (pay attention to the signs).

```
>> P*P'
ans =
    17   -10     2
   -10    41     2
     2     2    53

>> P'*P
ans =
    17   -10    -2
   -10    41    -2
    -2    -2    53
```

Observe that the product of a row vector and a column vector of the same length is simply a matrix with only one element, that is, a scalar value:

```
>> R*V
ans =
    35
```

Also, the product of a column vector and a row vector of the same length is a square (and symmetric) matrix.

```
>> V*R
ans =
     1            5            3
     5           25           15
     3           15            9
```

An example application of matrix multiplication is tabulating the costs of a set of purchase orders. Suppose that a company has received 4 purchase orders, each specifying a quantity of items, and wants to calculate the total cost of each individual order, as shown in the table below:

items	item 1	item 2	item 3	all
cost	1.25	7.50	4.99	total
order 1 quantities	27	42	12	?
order 2 quantities	14	0	9	?
order 3 quantities	18	6	24	?
order 4 quantities	15	21	7	?

We can express the solution of this problem as the product of an order matrix and a cost (column) vector, as illustrated by the following MATLAB script.

```
item_cost = [1.25 7.50 4.99]'
orders = [
    27 42 12
    14  0  9
    18  6 24
    15 21  7
]
order_costs = orders * item_cost
```

Running the script produces the following result:

```
item_cost =
    1.2500
    7.5000
    4.9900

orders =
    27    42    12
    14     0     9
    18     6    24
    15    21     7
order_costs =
  408.6300
   62.4100
  187.2600
  211.1800
```

11.5 SOLVING SYSTEMS OF LINEAR EQUATIONS

11.5.1 Linear Equations in Matrix Form

One of the most powerful applications of matrices is in working with systems of linear equations. A *linear equation* in the variables $x_1, x_2, \ldots, x_n$ is any equation of the form

$$a_1 x_1 + a_2 x_2 + \cdots + a_n x_n = b,$$

where $a_1, a_2, \ldots, a_n$ are constant coefficients and b is also a constant, often called simply the "right-hand side" of the equation. The following equation is a linear equation in the variables x_1, x_2, x_3.

$$-x_1 + 6x_2 - 2x_3 = 5$$

The equation below is *not* a linear equation, because it contains a product of two variables.

$$3x_1 + 2x_2 x_3 = 5$$

If we think of the coefficients of a linear equation as a row vector, and think of the variables as a column vector, then we can express the equation using a product of the two:

$$[a_1 x_1 + a_2 x_2 + \cdots + a_n x_n] = [b]$$
$$\mathbf{aX} = \mathbf{B}$$

in which

$$\mathbf{a} = \begin{bmatrix} a_1 & a_2 & \ldots & a_n \end{bmatrix} \quad \mathbf{X} = \begin{bmatrix} x_1 \\ x_2 \\ \vdots \\ x_n \end{bmatrix} \quad \mathbf{B} = [b]$$

A *system of linear equations* is a set of linear equations in the same variables $x_1, x_2, \ldots, x_n$, with different coefficients and right-hand side that are assumed to all be true simultaneously. In general, a system of linear equations has the form

$$a_{11} x_1 + a_{12} x_2 + \cdots + a_{1n} x_n = b_1$$
$$a_{21} x_1 + a_{22} x_2 + \cdots + a_{2n} x_n = b_2$$
$$\cdots$$
$$a_{n1} x_1 + a_{n2} x_2 + \cdots + a_{nn} x_n = b_n$$

We can express an entire system of linear equations as a single matrix equation of the form:

$$\begin{bmatrix} a_{11} x_1 + a_{12} x_2 + \cdots + a_{1n} x_n \\ a_{21} x_1 + a_{22} x_2 + \cdots + a_{2n} x_n \\ \cdots \\ a_{n1} x_1 + a_{n2} x_2 + \cdots + a_{nn} x_n \end{bmatrix} = \begin{bmatrix} b_1 \\ b_2 \\ \vdots \\ b_n \end{bmatrix}$$
$$\mathbf{AX} = \mathbf{B}$$

in which

$$A = \begin{bmatrix} a_{11} & a_{12} & \cdots & a_{1n} \\ a_{21} & a_{22} & \cdots & a_{2n} \\ \vdots & \vdots & \ddots & \vdots \\ a_{n1} & a_{n2} & \cdots & a_{nn} \end{bmatrix} \quad X = \begin{bmatrix} x_1 \\ x_2 \\ \vdots \\ x_n \end{bmatrix} \quad B = \begin{bmatrix} b_1 \\ b_2 \\ \vdots \\ b_n \end{bmatrix}$$

The matrix **A** is known as the coefficients matrix, the matrix **X** is called the variable vector, and the matrix **B** is called the "right-hand-side" vector.

As an example, the following system of linear equations in x_1, x_2, x_3:

$$4x_1 - x_2 \quad = 2$$
$$-x_1 + 6x_2 - 2x_3 \quad = 5$$
$$-2x_2 + 7x_3 = 17$$

can be expressed in matrix form as

$$\begin{bmatrix} 4 & -1 & 0 \\ -1 & 6 & -2 \\ 0 & -2 & 7 \end{bmatrix} \cdot \begin{bmatrix} x_1 \\ x_2 \\ x_3 \end{bmatrix} = \begin{bmatrix} 2 \\ 5 \\ 17 \end{bmatrix}$$

where in MATLAB, the **A** matrix and **B** vector are

```
>> A = [4 -1 0; -1 6 -2; 0 -2 7]
A =
        4       -1        0
       -1        6       -2
        0       -2        7

>> B = [2 5 17]'
B -
        2
        5
       17
```

11.5.2 The Identity Matrix and the Inverse of a Matrix

Before looking at the solution of a system of linear equations using matrix methods, we need to first introduce the concepts of the *identity matrix* and the *inverse* of a matrix.

Identity Matrix An identity matrix **I** is defined as a matrix of the form

$$I = \begin{bmatrix} 1 & 0 & \cdots & 0 \\ 0 & 1 & \cdots & 0 \\ \vdots & \vdots & \ddots & \vdots \\ 0 & 0 & \cdots & 1 \end{bmatrix},$$

that is, **I** is a square matrix in which the elements along the main diagonal are all equal to 1 and all other elements are equal to 0. From the definition of matrix multiplication, we can see that given any matrix **A**,

$$\mathbf{AI} = \mathbf{IA} = \mathbf{A},$$

which is why the identity matrix is so named. To create an identity matrix in MATLAB, use the `eye` command.

```
>> eye(3)
ans =
    1    0    0
    0    1    0
    0    0    1
```

Inverse of a Matrix   The inverse of a matrix **A**, denoted $\mathbf{A}^{-1}$, is that matrix such that

$$\mathbf{A} \cdot \mathbf{A}^{-1} = \mathbf{A}^{-1} \cdot \mathbf{A} = \mathbf{I}$$

To find the inverse of a matrix using MATLAB, use the `inv` command. Here, we calculate the inverse of the coefficients matrix from the system of equations at the beginning of this section.

```
>> inv(A)
ans =
    0.2621    0.0483    0.0138
    0.0483    0.1931    0.0552
    0.0138    0.0552    0.1586

>> A*inv(A)
ans =
    1.0000         0         0
    0.0000    1.0000         0
   -0.0000   -0.0000    1.0000

>> inv(A)*A ans =
    1.0000    0.0000   -0.0000
         0    1.0000   -0.0000
         0         0    1.0000
```

Note that only square matrices can have inverses, and not even all square matrices have one. A matrix that doesn't have an inverse is called a *singular* matrix. For example, the following matrix doesn't have an inverse:

$$\begin{bmatrix} 1 & 1 & 1 \\ 1 & 1 & 1 \\ 1 & 1 & 1 \end{bmatrix}$$

If we try to take the inverse of a singular matrix in MATLAB, it will produce an error.

```
>> inv([1 1 1; 1 1 1; 1 1 1])
Warning: Matrix is singular to working precision.
(Type "warning off MATLAB:singularMatrix" to sup-
press this warning.)
ans =
   Inf   Inf   Inf
   Inf   Inf   Inf
   Inf   Inf   Inf
```

11.5.3   Solving Matrix Equations Using Inversion

The goal in solving a system of linear equations is finding a set of values for the variables that causes all of the equations to be true simultaneously. In terms of matrices, given a matrix equation of the form

$$AX = B,$$

where A is an $n \times n$ matrix of coefficients, X is an $n \times 1$ column vector of variables, and B is an $n \times 1$ column vector of the right-hand-side constants, we want to find a value for the vector X that satisfies the equation.

Let's suppose for a moment the equation above were a scalar equation, rather than a matrix equation. Using the rules of algebra, we could solve for X by dividing both sides of the equation by A. Of course, we *can't* simply do this because the equation is a matrix equation and not a scalar equation, and division isn't even defined for matrices. There is a way, however, of solving for X algebraically that does follow the rules for arithmetic operations on matrices as defined in the last section. First, recall that the product of a matrix and its inverse is equal to the identity matrix, and that any matrix multiplied by the identity matrix is equal to itself. Left-multiplying both sides of the equation by A^{-1}—remember that matrix multiplication isn't commutative—we obtain the following:

$$AX = B$$
$$A^{-1}(AX) = A^{-1}B$$
$$(A^{-1}A)X = A^{-1}B$$
$$IX = A^{-1}B$$
$$X = A^{-1}B$$

Thus to obtain X, we can left-multiply B by the inverse of A, A^{-1}.

Let's try this out using MATLAB for the example from the last section, in which

$$A = \begin{bmatrix} 4 & -1 & 0 \\ -1 & 6 & -2 \\ 0 & -2 & 7 \end{bmatrix} \quad X = \begin{bmatrix} x_1 \\ x_2 \\ x_3 \end{bmatrix} \quad B = \begin{bmatrix} 2 \\ 5 \\ 17 \end{bmatrix}$$

The following script defines the matrices and solves for X using the inverse of A:

```
>> A = [4 -1 0; -1 6 -2; 0 -2 7]
A =
    4   -1    0
   -1    6   -2
    0   -2    7
```

```
>> B = [2 5 17]'
B =
      2
      5
     17

>> X = inv(A)*B
X =
    1.0000
    2.0000
    3.0000
```

Thus, **X** is equal to the column vector $\mathbf{X} = \begin{bmatrix} 1 & 2 & 3 \end{bmatrix}^T$. To verify the solution, we can calculate the product **AX** and see if it does in fact equal **B**.

```
>> A*X
ans =
    2.0000
    5.0000
   17.0000
```

It does.

11.5.4 Solving Matrix Equations Using the Backslash Operator

We now have a completely workable approach to solving systems of linear equations using MATLAB, namely, calculating the inverse of the coefficient matrix and left-multiplying the right-hand-side vector by it. From a numerical perspective, however, this approach is inefficient. It turns out that the algorithm for calculating the inverse of a matrix is rather slow, and there are faster ways available for calculating **X** that don't involve explicitly inverting a matrix. One technique for obtaining **X** is the *Gaussian elimination* algorithm. The backslash operator in MATLAB (\ ) uses the Gaussian elimination method to obtain the equivalent of $\mathbf{A}^{-1}\mathbf{B}$ more efficiently.

```
>> X = A\B
X =
    1.0000
    2.0000
    3.0000
```

For small, simple systems of equations, you probably won't notice the difference between explicitly calculating the inverse with `inv` and using Gaussian elimination via the backslash operator. With large systems of equations, however, the difference becomes clearer. In general, you should *always* use the backslash operator to solve systems of linear equations rather than explicitly inverting a matrix with `inv`.

11.5.5 Example: Analysis of a Truss

In Section 6.2.6, we performed a static analysis of the 3-bar truss pictured in Figure 11.2, solving for the external support forces as well as the internal tension or compression forces in the bars. The analysis produced 6 equations in 6 unknowns, which earlier we solved manually. In this section, we'll use MATLAB to solve these equations.

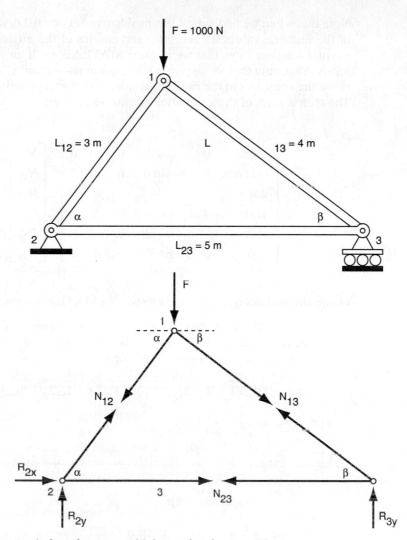

Figure 11.2   A three-bar truss, with internal and external forces.

Section 6.2.6 provides a detailed description of the problem and its solution, but in brief, the 6 unknown values are the three internal forces N_{12}, N_{23}, and N_{13}, as well as the three reaction forces R_{2x}, R_{2y}, and R_{3y}, as shown in Figure 11.2. The 6 equations express the fact that the sum of the x- and y-components of the forces at each of the three joints must equal zero. As derived in Section 6.2.6, these equations are:

x-components of forces at joint 1:	$-\cos\alpha N_{12} + \cos\beta N_{13} = 0$
y-components of forces at joint 1:	$-\sin\alpha N_{12} - \sin\beta N_{13} = 1000$
x-components of forces at joint 2:	$\cos\alpha N_{12} + N_{23} + R_{2x} = 0$
y-components of forces at joint 2:	$\sin\alpha N_{12} + R_{2y} = 0$
x-components of forces at joint 3:	$-N_{23} - \cos\beta N_{13} = 0$
y-components of forces at joint 3:	$\sin\beta N_{13} + R_{3y} = 0$

Note that when we first solved this problem in Section 6.2.6, we wrote them in terms of the numeric values for the sines and cosines of the angles, which simplified the manual solution. Now that we're using MATLAB, we'll let it do these calculations for us. Also, note that we've juggled the equations—actually, just the second one—to move the constants to the right-hand side, so that we can write them in matrix form. The matrix form of these equations is thus as follows:

$$\begin{bmatrix} -\cos\alpha & 0 & \cos\beta & 0 & 0 & 0 \\ -\sin\alpha & 0 & -\sin\beta & 0 & 0 & 0 \\ \cos\alpha & 1 & 0 & 1 & 0 & 0 \\ \sin\alpha & 0 & 0 & 0 & 1 & 0 \\ 0 & -1 & -\cos\beta & 0 & 0 & 0 \\ 0 & 0 & \sin\beta & 0 & 0 & 1 \end{bmatrix} \cdot \begin{bmatrix} N_{12} \\ N_{23} \\ N_{13} \\ R_{2x} \\ R_{2y} \\ R_{3y} \end{bmatrix} = \begin{bmatrix} 0 \\ 1000 \\ 0 \\ 0 \\ 0 \\ 0 \end{bmatrix}$$

Given the matrix equations, we now write a MATLAB script to solve them.

```
% Solution to 3-bar truss problem from Chapter 6
% Solves the system of equations AX = B, where
% X is a column vector of unknown forces, of the form
%
%    X = [N_12  N_23  N_13  R_2x  R_2y  R_3y]'

sin_a = 4/5;
cos_a = 3/5;
sin_b = 3/5;
cos_b = 4/5;

% coefficients matrix
A = [
    -cos_a  0  cos_b 0 0 0
    -sin_a  0 -sin_b 0 0 0
     cos_a  1   0    1 0 0
     sin_a  0   0    0 1 0
     0     -1 -cos_b 0 0 0
     0      0  sin_b 0 0 1
];

% right-hand side, external forces
B = [0 1000 0 0 0 0]';

% solve for the unknown forces
X = A\B;

% display the results in a nice format
disp('Solution:')
disp(' N_12 N_23 N_13 R_2x R_2y R_3y') % display labels
disp(X')                               % display transpose of X
```

Running the script produces the result:

```
>> Solution:
   N_12   N_23   N_13   R_2x   R_2y   R_3y
   -800    480   -600      0    640    360
```

These are the same results we got when we solved the problem manually, but note how much simpler—and less error prone—it is to use MATLAB! In particular, with the manual solution, we expended considerable effort in finding a clever order for solving the equations, to simplify the arithmetic. When we solve the system of equations using MATLAB, we don't have to worry about this—all we need to do is write the equations in matrix form and let the computer do the rest.

11.5.6 Example: Analysis of Electrical Circuits

In Section 7.4.3, we presented a method for solving for unknown currents and voltages in an electrical circuit. Similar to the analysis of a truss described in Section 6.2.6, circuit analysis also results in a system of linear equations. In Section 7.4.3, we analyzed a very simple circuit, shown in Figure 11.3, solving 5 equations in 5 unknowns manually. In this section, we'll first use MATLAB to solve for the currents and voltages in this earlier example, and then analyze a more complex circuit that converts a digital signal that represents a binary number into a corresponding analog voltage signal. This type of circuit, called a *digital-to-analog converter* or *DAC*, is commonly used in products such as CD or MP3 players to convert the bits in a digital music recording to an audio signal.

In brief, the analysis methodology described in detail in Section 7.4.3 has two categories of unknown variables:

- unknown currents through the circuit elements or *branches*
- unknown voltages at the circuit connection points or *nodes*

There are also two categories of equations:

- Kirchhoff's Current Law (KCL) equations, which state that the sum of the currents flowing into a node equals zero.
- Branch Constitutive Relationships (BCR), which state the basic physical relationships between current and voltage for each branch. For example, the BCR for a resistor is a statement of Ohm's Law for that element, and the BCR for a voltage source states that the voltage between the nodes at either end is a constant.

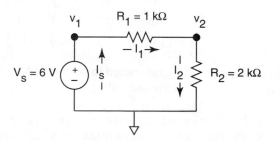

Figure 11.3 Example circuit for nodal analysis.

A Simple Example First we'll repeat the analysis from Section 7.4.3, using matrix techniques and MATLAB. The unknown currents and voltages in the simple circuit in Figure 11.3 are:

unknown node voltages: v_1 v_2
unknown branch currents: I_s I_1 I_2

The system of circuit equations is:

$$\begin{aligned}
\text{KCL at node 1:} && I_s - I_1 &= 0 \\
\text{KCL at node 2:} && I_1 - I_2 &= 0 \\
\text{BCR for } V_s: && v_1 - 0 &= V_s \\
\text{BCR for } R_1: && v_1 - v_2 &= R_1 I_1 \\
\text{BCR for } R_2: && v_2 - 0 &= R_2 I_2
\end{aligned}$$

This system of equations is actually fairly easy to solve manually—for example, we get V_s directly from the third equation. The main point here, however, is to illustrate how we can solve the equations *automatically*, without having to think much about algebra. Further, the approach we use in this example can be readily applied to more complicated circuits that would be very tedious to solve without a computer tool. In order to solve the equations in MATLAB, we need to first express them in matrix form. To prepare for this, we'll juggle the equations so that the constants are on the right-hand side, and so that the variables on the left-hand side appear in a consistent order, with the currents listed first and the voltages listed after.

$$\begin{aligned}
-I_1 + I_s &= 0 \\
I_1 - I_2 &= 0 \\
v_1 &= V_s \\
-R_1 I_1 + v_1 - v_2 &= 0 \\
-R_2 I_2 + v_2 &= 0
\end{aligned}$$

Writing this in matrix form, we get

$$\begin{bmatrix}
-1 & 0 & 1 & 0 & 0 \\
1 & -1 & 0 & 0 & 0 \\
0 & 0 & 0 & 1 & 0 \\
-R_1 & 0 & 0 & 1 & -1 \\
0 & -R_2 & 0 & 0 & 1
\end{bmatrix}
\cdot
\begin{bmatrix}
I_1 \\
I_2 \\
I_s \\
v_1 \\
v_2
\end{bmatrix}
=
\begin{bmatrix}
0 \\
0 \\
V_s \\
0 \\
0
\end{bmatrix}$$

From here, it's straightforward to write a MATLAB script to solve the equations for the unknown currents and voltages.

```
% Analysis of simple resistive circuit
% using system of equations in the form AX=B, where
% X = [I1 I2 Is v1 v2]'
```

```
Vs = 6;          % voltage source, V
R1 = 1000;       % resistance, ohms
R2 = 2000;       % resistance, ohms

% coefficients matrix
A = [
    -1 0 1 0 0
    1 -1 0 0 0
    0 0 0 1 0
    -R1 0 0 1 -1
    0 -R2 0 0 1
];

% right-hand side column vector
B = [0 0 Vs 0 0]';

% solve for unknown currents and voltages column vector
X = A\B;

% display currents and voltages in a nice format
disp('Solution:')
disp('    I1    I2    Is    v1    v2')
disp(X')
```

Running the script, we get:

```
>> Solution:
    I1        I2        Is        v1        v2
    0.0020    0.0020    0.0020    6.0000    4.0000
```

which is the same solution that we obtained in Section 7.4.3.

Digital-to-Analog Converter Circuit   Having seen how to analyze a simple circuit using matrices and MATLAB, we'll now turn to a more interesting example. Figure 11.4 illustrates a circuit known as an R2R ladder that's used to convert a digital input in the form of a binary number to a corresponding analog output voltage. In this example, we'll analyze an R2R ladder that takes a 2-bit number as input, but the circuit is readily extended to more bits, such as the 4-bit ladder, also shown.

In the 2-bit version, the inputs to the circuit are the most significant bit, MSB, and the least significant bit, LSB. These digital input signals control pairs of switches such that the node voltages V_{MSB} and V_{LSB} will either be 1 V or 0 V. The ladder circuit then processes these input voltages to produce the output voltage, v_1. One of the interesting features of this circuit is that the output voltage doesn't depend on the value of the resistance R—it'll produce the same output as long as $2R$ is twice as large as R. The unknown variables in this circuit are the 4 currents through the resistors, $I_1, I_2, I_3,$ and I_4, as well as the two node voltages v_1 and v_2. The input node voltages V_{MSB} and V_{LSB} are *not* unknowns; depending upon the switch settings, they will either be 0 V or 1 V. In our analysis of the circuit, therefore, we'll treat them as constants rather than as variables. Just as we did for the simple circuit in the previous

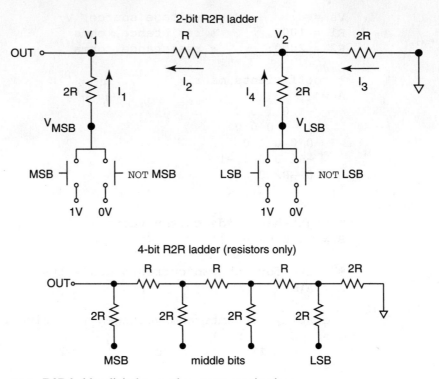

Figure 11.4 R2R ladder digital-to-analog converter circuits.

example, we can now write the circuit equations for Kirchhoff's Voltage Law (KVL) and the branch constitutive relationships (BCR):

$$
\begin{aligned}
\text{KCL at node 1:} && I_1 + I_2 &= 0 \\
\text{KCL at node 2:} && -I_2 + I_3 + I_4 &= 0 \\
\text{BCR for } R_1: && V_{\text{MSB}} - v_1 &= 2RI_1 \\
\text{BCR for } R_2: && v_2 - v_1 &= RI_2 \\
\text{BCR for } R_3: && V_{\text{LSB}} - v_2 &= 2RI_4 \\
\text{BCR for } R_4: && 0 - v_2 &= 2RI_3
\end{aligned}
$$

As before, we'll juggle the equations to move the variables to the left-hand side and the constants to the right-hand side:

$$
\begin{aligned}
-I_2 + I_3 + I_4 &= 0 \\
I_1 + I_2 &= 0 \\
-2RI_1 - v_1 &= -V_{\text{MSB}} \\
-RI_2 - v_1 + v_2 &= 0 \\
-2RI_4 - v_2 &= -V_{\text{LSB}} \\
-2RI_3 - v_2 &= 0
\end{aligned}
$$

Now, we write the matrix equations as:

$$
\begin{bmatrix}
1 & 1 & 0 & 0 & 0 & 0 \\
0 & -1 & 1 & 1 & 0 & 0 \\
-2R & 0 & 0 & 0 & -1 & 0 \\
0 & -R & 0 & 0 & -1 & 1 \\
0 & 0 & 0 & -2R & 0 & -1 \\
0 & 0 & -2R & 0 & 0 & -1
\end{bmatrix}
\cdot
\begin{bmatrix}
I_1 \\
I_2 \\
I_3 \\
I_4 \\
v_1 \\
v_2
\end{bmatrix}
=
\begin{bmatrix}
0 \\
0 \\
-V_{\text{MSB}} \\
0 \\
-V_{\text{MSB}} \\
0
\end{bmatrix}
$$

From this point, writing the MATLAB script is straightforward.

```
% Analysis of 2-bit R2R ladder circuit
% Solved as system of equations AX = B, where
% X = [I1 I2 I3 I4 V1 V2]'

% run using 4 combinations of Vmsb and Vlsb = 0 or 1

Vmsb = 1;        % voltage at most significant bit, V
Vlsb = 1;        % voltage at least significant bit, V
R = 1000;        % resistance, ohms

% coefficients matrix
A = [
   1 1 0 0 0 0
   0 -1 1 1 0 0
   -2*R 0 0 0 -1 0
   0 R 0 0 -1 1
   0 0 0 -2*R 0 -1
   0 0 -2*R 0 0 -1
];

% right hand side column vector
B = [0 0 -Vmsb 0 -Vlsb 0]';

% solve for unknown currents and voltages

X = A\B;

% display output voltage
disp(' Vmsb      Vlsb        output voltage')
disp([Vmsb Vlsb X(5)])
```

Running the script 4 times, once with each of the 4 possible combinations of V_{MSB} and V_{MSB} equal to either 0 V or 1 V, yields the following results:

Vmsb	Vlsb	output voltage
0	0	0.0000
0	1	0.2500
1	0	0.5000
1	1	0.7500

What we see is that as the inputs count from 00 to 11 in binary—or 0 to 3 in decimal—the output voltage rises in increments of 0.25 V from 0 V to 0.75 V. More generally, the voltage step size for an R2R ladder with n bits of input would be $1/2^n$. Thus, the more the bits, the more smoothly the output voltage changes with the digital input values. This is important, for example, in order to produce a high-quality audio output.

PROBLEMS

1. **Magic Square**
   Consider a 2-D MATLAB array that is a magic square of size n (n is an integer entered by the user; assume n is in the 5–10 range for this problem). In order to generate such an array, you will make use of MATLAB's magic function. Create a program that starts with the following lines of code:

   ```
   clear all
   format compact
   n = input('Enter the magic square size: ');
   arr = magic(n);
   ```

   Continue the program by writing single-line MATLAB statements to generate each of the following from the array (for each one, use as the variable name the letter of the corresponding letter bullet; also, in order to display the result, do not suppress the echo):

   (a) The magic number
   (b) The value that is in the 4th row and 5th column
   (c) The 3rd row
   (d) The 5th column, displayed as a row vector
   (e) The rectangular 2×3 array formed by the first two rows and first three columns
   (f) The largest value
   (g) The main diagonal (top left to bottom right)
   (h) The secondary diagonal (top right to bottom left)
   (i) The element-by-element product of the first and last columns
   (j) The sum of all the values whose row number and column number are both odd

2. **Matrix Operations**
   Write a MATLAB program that sets up a 5×5 array of random integers between 1 and 9 and finds each of the following in single-line statements (for each one of them, use as the variable name the letter of the corresponding letter bullet; also, in order to display the result, do not suppress the echo):

   (a) The main diagonal (top left to bottom right)
   (b) The secondary diagonal (top right to bottom left)
   (c) The element-by-element product of the first and last rows
   (d) The rectangular 4×2 array formed by the first four rows and first two columns
   (e) The sum of all the values whose row number and column number are both even

(f) The determinant of the array. (Search the MATLAB online documentation to find the function that does this.)

(g) The transpose of the array

(h) The inverse of the array

(i) The element-by-element square of the array

(j) The matrix square of the array

3. **Reading and Writing Data as Text Files**

MATLAB provides a variety of functions for reading and writing data as text files. One of the most common formats for storing data as text is a *comma-separated value* or *CSV* file. In this exercise, we'll examine some of MATLAB's facilities for working with CSV files. As an example data set, suppose that we have recorded the velocity of a moving object at regular time intervals as shown in the table below:

time (s)	velocity (m/s)
5.0	3.4
10.0	4.1
15.0	6.0
20.0	8.8
25.0	11.9
30.0	12.6
35.0	16.1
40.0	16.8
45.0	20.2

(a) Open a new MATLAB script and begin it by defining a matrix named m1 with two columns that contain the data from the table above.

(b) Look up the function csvwrite in the MATLAB online documentation. Add a line to your script that writes the matrix m1 to a CSV file named data.csv.

(c) Use a text editor to open data.csv. Describe what you see. Change the velocity at 30.0 s to be 14 m/s in the file and save.

(d) Use a spreadsheet program to open the modified version of data.csv. Describe what you see. Change the velocity at 40.0 s to be 18 m/s in the file and save.

(e) Look up the function csvread in the MATLAB online documentation. Add a line to your script that reads the values in the file data.csv into a matrix named m2 and displays the values in the matrix.

(f) Describe a scenario where it would be particularly useful to read and write MATLAB data in CSV files.

4. **Creating Tables with Matrix Operations**

Using meshgrid and the MATLAB element-wise matrix operations, create tables of the values for z versus x and y for the following functions. Use the disp command to make the rows and columns of your table line up properly.

(a) $z = x - 4y$

(b) $z = 2x^2 + y^2 - xy - 2$

(c) $z = 2x^3 - y^3 + y + 2$

5. Mesh Plots

Create a mesh plot for each of the following functions of two variables. Use a range for both x and y of -5 to $+5$, with a step size of 0.1. Make sure you use the `meshgrid` function to create the data points.

(a) $z = x - 4y$

(b) $z = 2x^2 + y^2 - xy - 2$

(c) $z = 2x^3 - y^3 + y + 2$

6. Contour Plots

Create a contour plot for each of the following functions of two variables. Use a range for both x and y of -5 to $+5$, with a step size of 0.1. Make sure you use the `meshgrid` function to create the data points. Use `clabel` so that the plot places the z values on the contour lines.

(a) $z = 5x + 4y$

(b) $z = 2x^2 + 3(y - 1)^2 + 3$

(c) $z = x^3 - 2y^2 + 2x^2y - xy + 2$

7. Finding the Minimum Value in a Mesh Plot

The following MATLAB script generates a mesh plot of the function

$$z = 0.05x + 0.03y - 0.01x^2 - 0.03y^2 + 0.8$$

```
[x,y] = meshgrid(-5:.1:8);
z =   .05*x + .03*y - .01*x.^2 - .03*y.^2 + .8;
mesh(x,y,z);
grid on
xlabel('x')
ylabel('y')
```

Run the script, and from the resulting plot, estimate the minimum value of z, as well as the (x, y) coordinates where the minimum occurs.

8. Identifying Points in a Contour Plot

Consider the following MATLAB code, which produces a contour plot for the function $z = 0.1x^2 + 0.5y$:

```
[x,y] = meshgrid(-10:.1:10);
z = (x.^2)/10 + y/2;
[C,h] = contour(x,y,z);
clabel(C,h)
xlabel('x')
ylabel('y')
grid on
```

Run the above code in MATLAB. Note the contour lines. For each of the following questions, find an estimate for the answer, and verify mathematically by plugging into the function's equation:

(a) find z if $x = 2$ and $y = 4$

(b) find y if $z = 6$ at $x = 6$

(c) find x if $z = 2$ at $y = -2$

(d) if $y = x$, find the value of x so that $z = 2$ (note here that you are basically finding the roots of a quadratic equation)

9. Determining Values in a Contour Plot

A contour plot for the function $z = x^2 + y^2 - x - y + 1$ is shown below:

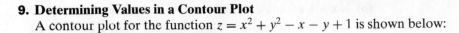

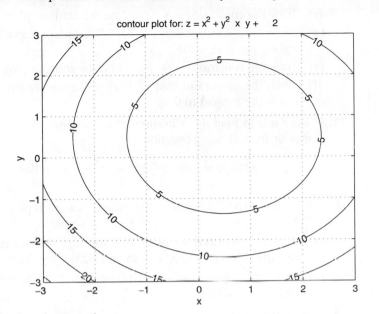

Use the plot to estimate:

(a) the value of z if $x = 2$ and $y - 2$

(b) the value of y if $x = 1$ and $z = 5$

(c) the value of x if $y - 1.5$ and $z = 10$

10. Values and Constraints in a Contour Plot

The figure below is a contour plot for the equation

$$z = 2x^2 - 3y^2 - xy + 1$$

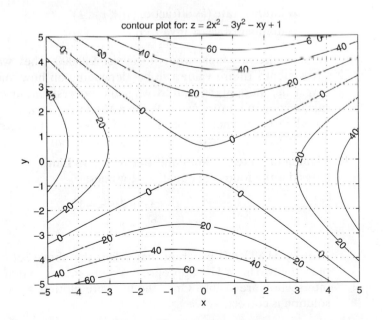

(a) If $x = 4$, estimate the values of y that makes z equal to 20.

(b) If $y = 1$, estimate the values of x that make z equal to 20.

(c) Draw the constraint $y = 2x - 2$ on the contour plot.

(d) Given the constraint that $y = 2x - 2$, estimate the value(s) of (x, y) pairs that make z equal to -20.

(e) Draw the constraint $x^2 + y^2 = 4$ on the contour plot.

(f) Given the constraint that $x^2 + y^2 = 4$, estimate the value(s) of (x, y) pairs that make z equal to 0.

11. **Using Plots to Find the Minimum of a Function**
    Consider the following function:

$$f(x, y) = -(x^3 - 25x^2 - 7y^2 + 30x + 10y - 40)/100$$

(a) Write a MATLAB script that produces a mesh plot of f versus x and y over the interval $-10 \le x \le 25$ and $-25 \le y \le 25$.

(b) Write a MATLAB script that produces a contour plot of f versus x and y over the interval $-10 \le x \le 25$ and $-25 \le y \le 25$.

(c) From the plots, estimate the minimum value of f over the interval $-10 \le x \le 25$ and $-25 \le y \le 25$.

12. **Compound Interest**
    The compound interest formula shown below is used to determine the future value of a monetary deposit given the interest rate and the number of years of investment (where interest is compounded once per year):

$$F = D(1 + r)^t$$

where

F = future value of deposit

D = initial deposit

r = interest rate (as a fraction, $0 < r < 1$)

t = number of years invested

Assume that a couple wants to invest $40,000. They want to study what the investment's future value will be, depending on how many years they plan to invest the amount, as well as what the interest rate may be, with the goal of maximizing the future value. Write a MATLAB script to plot contour lines that will show them what to expect for an interest rate between 4% and 6% (in steps of 0.1%), for a range of 5–10 years of investment.

13. **Finding the Intersection Point Between Two Lines**
    Consider the following equations for lines:

$$3x + 2y = -12$$
$$2x - y = 1$$

Write a MATLAB program that solves the above set of equations, and then plots the two lines. Use appropriate x and y ranges so that the two lines and their intersection are visible. Once the lines are plotted, zoom in to verify that your solution is correct.

14. Solving Four Equations in Four Unknowns

Consider the following set of four equations with four unknowns:

$$16a + 32b + 33c + 13d = 91$$
$$5a + 11b + 10c + 8d = 16$$
$$9a + 7b + 6c + 12d = 5$$
$$34a + 14b + 15c + d = 43$$

Write a MATLAB program that writes the above in matrix form, and then finds the solution for a, b, c, and d.

15. Solving Five Equations in Five Unknowns

Write a MATLAB M-file to solve the following system of five equations with five unknowns:

$$2.1a - 3.4b + 6.8c + 7.3d + 4.2e = 7.1$$
$$5a + 2.3b - 5.1d = -3.2$$
$$8.3a - b + 4.8c + 2.9e = 4.2$$
$$3b - 5.5d - 8.3e = -5.7$$
$$-2.7a + 2.5c - 7.1d + 4.5e = -3.1$$

16. Determining Prices

Consider the following statements about prices of fruits and vegetables:

- three apples, one banana, and four carrots together cost $3.80
- one apple and two carrots together cost two pennies less than three bananas
- two bananas and three carrots together cost as much as four apples

Write a MATLAB program that uses matrix operations to find the cost each of an apple, a banana, and a carrot.

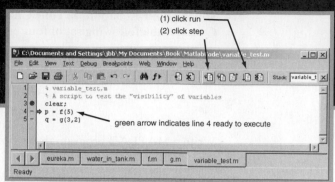

(1) click run
(2) click step

C:\Documents and Settings\jbb\My Documents\Book\Matlab code\variable_test.m

File Edit View Text Debug Breakpoints Web Window Help

```
1      % variable_test.m
2      % A script to test the "visibility" of variables
3  ●   clear;
4  -▷  p = f(5)          green arrow indicates line 4 ready to execute
5  -   q = g(3,2)
```

Stack: variable_t

eureka.m   water_in_tank.m   f.m   g.m   variable_test.m

Ready

Introduction to Algorithms and Programming In MATLAB

LEARNING OBJECTIVES

- to use flowcharts and pseudocode notation to describe simple algorithms and to give clear directions for carrying out a simple procedure using both flowcharts and pseudocode notation;
- to use MATLAB functions to break programs down into manageable-sized pieces;
- to use IF-THEN-ELSE statements to implement conditional branches;
- to write FOR and WHILE loops;
- to use the MATLAB debugger to trace the execution of a program.

12.1 ALGORITHMS, FLOW CHARTS, AND PSEUDOCODE

12.1.1 What Is an Algorithm?

In earlier chapters, we saw how we can devise complex solutions to complex problems by successively breaking the problem down into subproblems, and forming a chain of simpler solutions. This ability to figure out a process on the fly, as you perform it, is critical the first time you encounter a new problem. The next time you encounter this problem or a closely related one, however, you don't need to go through the same steps of figuring out the solution. Instead, you can simply follow the trail that you, or someone else, blazed earlier. Consider, for example, the problem of baking a loaf of bread. Someone figured out the problem of how to use a live yeast culture to add gas bubbles to the dough and hence make it lighter, as well as the baking times and temperatures, so you don't have to do that—instead you can follow a recipe. The kind of knowledge used in writing a recipe is called *algorithmic* knowledge. An *algorithm* is a detailed, ordered set of instructions for solving a problem. The term is named

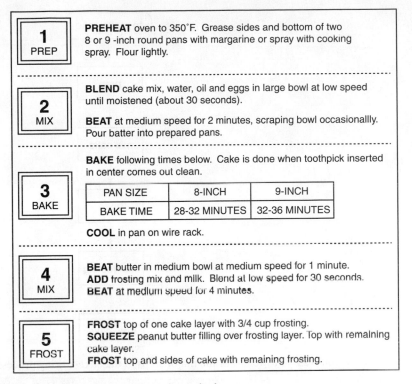

Figure 12.1   Cake baking instructions, from the box.

after Muhammad ibn Musa Al-Khawarızmı (780–850), a Persian mathematician who introduced Arabic numerals and decimal calculation to the West. Just as it takes skill and patience to give good driving directions, it also takes skill and patience to design an algorithm that's efficient, and most importantly, correct.

There are two common forms for describing algorithms. The first is a graphical form using diagrams called *flowcharts*. The second is a textual form using a precise notation called *pseudocode*. To illustrate the basic features of both approaches, we consider the example of describing a process for baking a cake. Figure 12.1 shows the instructions as they might be found right on the box of a mix, in this case a chocolate peanut butter cake mix [Dun05]. Before introducing either the notation of flowcharts or pseudocode, let's first look at some of the properties of the instructions for baking this cake.

- The process is a sequence of five major steps: PREP, MIX CAKE, BAKE, MIX FROSTING, and FROST.
- Each of these major steps consists of a sequence of minor steps.
- The BAKE step involves choosing one of two paths, depending upon the size of the pan.

12.1.2   Describing Simple Sequences of Operations

A flowchart is a directed graph in which each node corresponds to a step in a process and the edges indicate the ordering between two steps. Figure 12.2 shows a flowchart for the five major steps in baking the cake. By convention, the process steps are

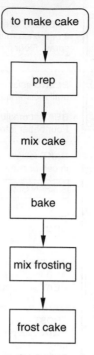

Figure 12.2 Flowchart of the major steps in the cake-baking process.

drawn using square boxes. The name of the process is specified in a rounded box at the top of the flow. By convention, we'll always begin the name with the word "to," in this case, "to make cake."

Pseudocode is a textual description of a process that uses a restricted vocabulary and precise formatting to minimize ambiguity. The name is a reference to slang for a computer program, which is sometimes called "code" by programmers. Whereas computer code is interpreted by a computer, pseudocode is meant to be interpreted by people. For a simple process such as the main flow in baking a cake, which consists of a linear sequence of steps without any choices, the pseudocode description isn't particularly interesting. Each step in the process is merely listed on a separate line, as shown below:

```
to make cake:
prep
mix cake
bake
mix frosting
frost
```

If the description of a step is too long to fit on a single line, then the convention is to indent the continuation of the description:

```
do the first step
do the second step, which is much more complicated
  than the first step
do the third step
```

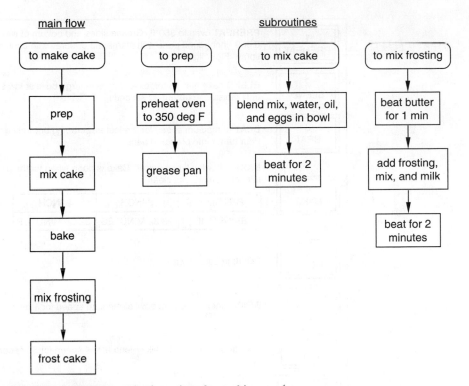

Figure 12.3   Main flow and subroutines for making a cake.

12.1.3   Subroutines

In the language of algorithms, the term for a minor sequence that is part of a major sequence is a *subroutine*. The term for jumping from the main flow into a subroutine is *calling a subroutine*. Thus each of the five steps in the main process for baking a cake calls a subroutine that implements the sequence of finer steps. Figure 12.3 shows the flowchart for the main process, along with the flowcharts for the subroutines, "to prep," "to mix cake," and "to mix frosting." The pseudocode description of the subroutines, not shown, simply lists the steps of each subroutine one per line as we did for the main process. We'll describe the subroutine "to bake," which involves making a choice, a bit later.

Subroutines don't change the steps in a task; rather they break a complex process into smaller chunks and organize them hierarchically to make the process easier to manage. Subroutines provide a mechanism for abstraction of a process that lets you ignore low-level details when you're looking at the high-level big picture. This is a concept the designers of the cake box understood very well. They could have listed a "flattened" version of the instructions, without hierarchy, as shown in Figure 12.4. In either case, the steps involved in making the cake are exactly the same, but the process seems more complicated in the flattened case, because it lists 11 top-level steps, whereas the hierarchical case in Figure 12.1 lists only 5. In terms of the box layout and the ability of a consumer to digest the information, it *is* more complicated. This would likely be an important factor to a consumer—who after all is buying a cake *mix* to simplify the process of making a cake—in choosing between brands. A long algorithm without subroutines is like a long book without chapters: unnecessarily

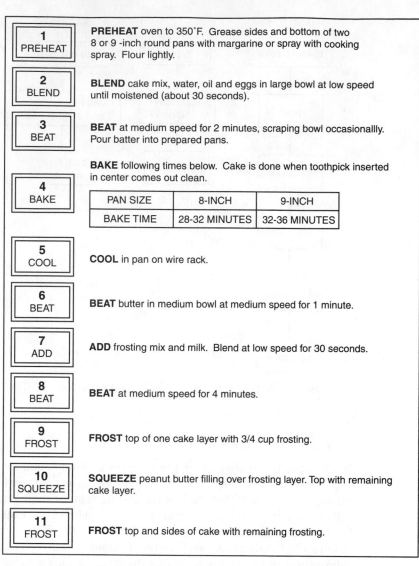

1 PREHEAT	**PREHEAT** oven to 350°F. Grease sides and bottom of two 8 or 9 -inch round pans with margarine or spray with cooking spray. Flour lightly.
2 BLEND	**BLEND** cake mix, water, oil and eggs in large bowl at low speed until moistened (about 30 seconds).
3 BEAT	**BEAT** at medium speed for 2 minutes, scraping bowl occasionallly. Pour batter into prepared pans.

BAKE following times below. Cake is done when toothpick inserted in center comes out clean.

4 BAKE	PAN SIZE	8-INCH	9-INCH
	BAKE TIME	28-32 MINUTES	32-36 MINUTES

5 COOL	**COOL** in pan on wire rack.
6 BEAT	**BEAT** butter in medium bowl at medium speed for 1 minute.
7 ADD	**ADD** frosting mix and milk. Blend at low speed for 30 seconds.
8 BEAT	**BEAT** at medium speed for 4 minutes.
9 FROST	**FROST** top of one cake layer with 3/4 cup frosting.
10 SQUEEZE	**SQUEEZE** peanut butter filling over frosting layer. Top with remaining cake layer.
11 FROST	**FROST** top and sides of cake with remaining frosting.

Figure 12.4 A "flattened" version of the cake baking instructions that doesn't describe the process hierarchically.

difficult to read and navigate. As a general rule of thumb, remember that people can only hold 5–9 concepts in short-term working memory, so if a process has many more steps, consider breaking it into subroutines. This applies to breaking the first-level subroutines into finer ones, as well.

12.1.4 Conditional Branches

In the language of algorithms, a choice between alternative courses of action is called a *conditional branch*, or simply a *branch*. In the example of making a cake, the subroutine "to bake" involves a branch to determine the cooking time, depending upon the size of the pan. Figure 12.5 illustrates a flowchart for the "to bake" subroutine, showing the branch. There are two parts to using a branch in a flowchart. The first

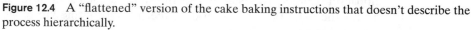

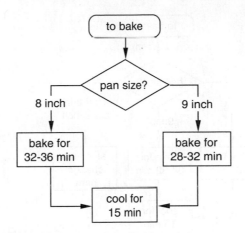

Figure 12.5 Representation of a branch in a flowchart.

part is to include a special test step in which the branch occurs. This step tests the condition that determines the course of action following the branch; in this case, the condition is the pan size. As illustrated in Figure 12.5, decision steps are conventionally drawn as diamonds in a flowchart, to distinguish them from other steps. The second part is to include edges in the flowchart leading from the decision step to each possible next step, and to label these edges with the value of the condition associated with that path. In the example, the two possible values of the condition are "8 inch" and "9 inch." After the branch, the separate paths may continue independently for some number of steps, or they may re-converge. In this case, the two branches each perform independent baking steps for different times, but then re-converge at the cooling step.

In describing a simple sequence of operations, such as the main flow for making a cake, the pseudocode representation wasn't appreciably different from standard text. In describing a branch, however, pseudocode begins to take on a personality distinct from natural language of the sort that we use in loose conversation. In natural language, there are many possible ways we could describe the branch shown in Figure 12.5. Here's one:

> Bake the cake for 32–36 minutes or 28–32 minutes, depending upon whether the pan is 8 inches or whether it is 9 inches, respectively.

It takes several readings of the above sentence to be able to reproduce the flowchart in Figure 12.5. The statement is verbose, and it's difficult to pick out the branch condition. For pseudocode to be effective, each statement must be concise and unambiguous. We achieve this by limiting the expression of flow control statements such as branches to a small set of acceptable constructs, using a restricted vocabulary of *keywords*. For branches, we'll adopt the same construct we used for productions: a condition-action pair using the keywords IF and THEN of the form:

```
IF condition THEN perform action
```

Using this construct, the pseudocode for the "to bake" subroutine becomes

```
to bake:
IF the pan size is 8 inches THEN bake for 32-36 minutes
IF the pan size is 9 inches THEN bake for 28-32 minutes
```

This form of the description is much clearer than the natural language version above.

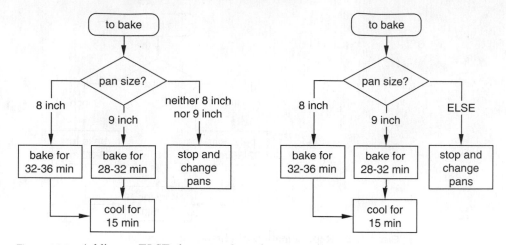

Figure 12.6 Adding an ELSE clause to a branch.

Both the flowchart and pseudocode versions of the "to bake" subroutine provide clear direction on what to do if you have either an 8 inch or 9 inch pan. But what do you do if you have a pan of any other size? The instructions on the box aren't explicit about this, but presumably the recipe wouldn't work well with a different sized pan, and the action should be to stop and switch pans. Given this new case, we now need to work it into our algorithm description, both for the flowchart and pseudocode formats, as a third branch option.

Figure 12.6 illustrates two versions of a flowchart that adds a branch for the case where none of the other branches are taken. In the version on the left, we explicitly defined the value of the condition for this branch in terms of the values of the other branches as "neither 8 inch nor 9 inch." This, however, is rather awkward. A cleaner approach is to add a new keyword, ELSE, which means "in the case where neither of the other options are satisfied explicitly." Adding the ELSE option to the conditional branch makes the subroutine definition complete and unambiguous, in that there's now a clearly specified action for any size pan.

We can also add an ELSE clause to the IF-THEN statement in the pseudocode representation, to create a new construct of the form

```
IF condition THEN perform action ELSE perform alternative-action
```

Before writing out the pseudocode for the revised "to bake" subroutine with three branches, let's first consider a simpler example with only two branches. Suppose that the instructions on the box only listed a baking time for an 8 inch pan. In this situation, you would need to stop and change pans. The pseudocode for this scenario, using the ELSE clause, would be as follows (note that we split the single IF-THEN-ELSE statement into two lines and indented the continuation):

```
to bake:
IF the pan size is 8 inches THEN bake for 32-36 minutes
   ELSE stop and change pans
```

Now suppose we want to add the third case of a 9-inch pan back into the algorithm. The question is where would you put it in the pseudocode description? The IF-THEN-ELSE statement is only set up to test *one* condition, but we need a

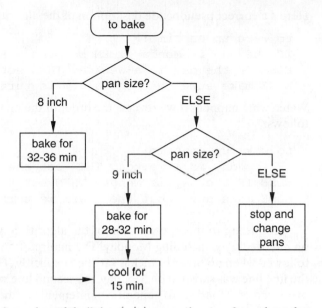

Figure 12.7 A 3-way branch built by chaining together two 2-way branches.

construct that lets us test *two* conditions. Put another way, an IF-THEN-ELSE statement only supports a 2-way branch, but we need a 3-way branch. The solution to this problem is to chain two IF-THEN-ELSE statements together, as follows:

```
to bake:
IF the pan size is 8 inches THEN bake for 32-36 minutes
  ELSE IF the pan size is 9 inches THEN bake for 28-32 minutes
    ELSE stop and change pans
```

While this construct has the exact same behavior as the three-way branch, note that a flowchart that represents it literally would have two decision steps in it, rather than one. Figure 12.7 shows the resulting flowchart. In general, however, we'll adopt the convention of representing multiway branches in flowcharts by using a single decision node with multiple branch links attached to it, which is easier to read, even though the pseudocode version chains 2-way IF-THEN-ELSE statements together. To improve the readability of multiway branches in pseudocode, the usual convention is *not* to indent them. Thus we would write the "to bake" algorithm as follows:

```
to bake:
IF the pan size is 8 inches THEN bake for 32-36 minutes
ELSE IF the pan size is 9 inches THEN bake for 28-32 minutes
ELSE stop and change pans
```

This is nice and simple.

One additional fine point we need to consider regarding branches is that order matters. To illustrate, let's consider writing pseudocode for an algorithm "to wake up on time" according to the day of the week. A natural language description of the algorithm is as follows:

On Mondays, set the alarm for 6 AM. On other weekdays, set the alarm for 7 AM. Otherwise, make sure that the alarm is turned off.

Here's a correct pseudocode description of the algorithm:

```
to wake up on time:
IF the day is Monday THEN set the alarm for 6 AM
ELSE IF the day is a weekday THEN set the alarm for 7 AM
ELSE make sure that the alarm is turned off
```

Watch what happens if we reverse the order of the first two parts of the statement as follows:

```
to wake up on time:
IF the day is a weekday THEN set the alarm for 7 AM
ELSE IF the day is Monday THEN set the alarm for 6 AM
ELSE make sure that the alarm is turned off
```

According to this specification of the algorithm, you'll set the alarm for 7 AM on *any* weekday, including Monday. The nature of an ELSE clause is that it's *only* followed when no other branch is satisfied explicitly. Thus, because the condition for the first line was satisfied on Monday, the second line will be skipped over, and you'll oversleep on Monday. To generalize, statements of the form

```
IF condition-1 THEN action-1
ELSE IF condition-2 THEN action-2
  ...
ELSE IF condition-n THEN action-n
ELSE default-action
```

define a set of $n + 1$ mutually exclusive choices, consisting of n explicit choices plus a default, in which exactly one condition will be satisfied and one action will be performed.

As a final example, let's consider what happens if we remove the ELSE keyword from the second line:

```
to wake up on time:
IF the day is a weekday THEN set the alarm for 7 AM
IF the day is Monday THEN set the alarm for 6 AM
ELSE make sure that the alarm is turned off
```

According to this version of the algorithm, you'll set the clock for 7 AM on Monday and then *reset* it for 6 AM. Your alarm will go off at the right time, but you performed unnecessary extra work in making this happen. If you find this confusing, bear in mind that this isn't a clean implementation of the algorithm. Whenever you write pseudocode, your goal should be to specify the algorithm as clearly as possible, avoiding combinations of statements that make the algorithm difficult to follow or have unexpected behaviors.

12.1.5 Loops

Many processes involve some form of repetition. A section of an algorithm in which a sequence of steps is repeated is called a *loop*. As an example of a process with a loop, consider an algorithm for counting the number of coins in your pocket. Here is a natural language description of one algorithm that'll do the job:

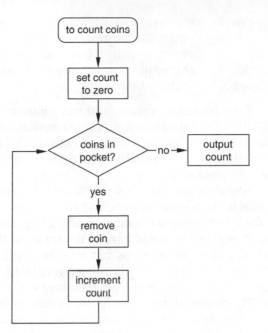

Figure 12.8   Representation of a loop in a flowchart, for counting the number of coins in your pocket.

Start with an initial count of zero. Remove coins from your pocket one at a time, and increment the count each time you remove a coin. When there are no coins left, the count will be equal to the total number of coins.

Figure 12.8 illustrates a flowchart for this algorithm. The loop in the algorithm shows up graphically as a "loop" in the flowchart, in which the arrow out of one step directs the flow back to a step "upstream" or earlier in the process. Somewhere within the body of the loop there needs to be a conditional branch that breaks you out of the loop. Otherwise, the loop would repeat forever in what's called an *infinite loop*, and the algorithm would never complete. In this case, the conditional branch is at the top of the loop, and tests whether or not there are any coins left in your pocket. If the value of the condition is "yes," then you add to the count and repeat the loop; otherwise, the current value of the count is output as the result.

In drawing a flowchart of the loop, we didn't need to introduce any new notation beyond what we already had for describing conditional branches. Technically, this is also true for pseudocode if we use a technique called *recursion*, where a subroutine can call itself. Recursion is a powerful concept, and we'll look briefly at an example of it in the next section, but many people find recursion difficult to understand. In most situations, the more intuitive approach is to describe the loop explicitly, and to do so, we need to introduce new constructs and keywords into our notation. There are several ways we can achieve this, and we will consider three options in the paragraphs below.

The first approach is to add a statement that lets us explicitly specify which step to "go to" next. This approach is commonly used in assembly manuals. The following version of the algorithm demonstrates this approach:

```
to count coins:
step 1: set the count to zero
```

```
step 2: IF there are more coins in your pocket THEN GOTO step 3
        ELSE GOTO step 5
step 3: remove a coin from your pocket
step 4: increment the count, GOTO step 2
step 5: output the count
```

For this version of the algorithm, we introduced two new pieces of notation. First we labeled the statements with step numbers. Second we added a GOTO keyword that indicates which step to "go to" next. This approach is actually a fairly literal translation of the flow chart, where the combination of GOTO statements and labels take the place of arrows.

While the approach is straightforward, it does have drawbacks. One drawback is that it can be cumbersome and error-prone to make changes to your algorithm. If for some reason we wanted to add a step somewhere in the middle of an existing algorithm, we would have to update all of the step numbers. This includes not only the labels of the steps, but also the targets of the GOTO statements. A second drawback is specific to writing algorithms that'll eventually be implemented as computer programs. Some programming languages don't have the equivalent of a GOTO statement, and even for those that do, its use is generally considered to be bad form. So next, we'll consider two ways of describing loops commonly found in many programming languages.

The first construct is the WHILE statement. A WHILE loop has the following form:

```
WHILE condition is true
   keep repeating an action
```

Using a WHILE loop, we can express the algorithm for counting coins as

```
to count coins:
set the count to zero
WHILE there are more coins in your pocket
    remove a coin from your pocket
    increment the count
output the count
```

In this example, the action part of the loop, called the *loop body*, consists of the two statements of removing a coin from your pocket and incrementing the count. These are both indented to show that they're within the loop body. Observe that the structure of a WHILE loop is very similar to the structure of the loop in the flowchart in Figure 12.8, with the condition check at the top of the loop. When the condition is true, all the steps in the loop body are executed, and then the condition is checked again. Just as with the flowchart, there needs to be some statement in the loop body that eventually sets the condition to be false, in order to avoid an infinite loop. In this case, removing a coin from your pocket serves this purpose. When the condition finally becomes false, the flow of the algorithm skips ahead to the next statement after the loop, which in this case is the statement to output the count.

The second common loop construct found in many programming languages is the FOR statement, which operates on each of the elements of a set in sequence. A FOR loop has the following format:

```
FOR each element of a set
   perform some action on the current element
```

A FOR loop version of the algorithm for counting coins is

```
to count coins:
set the count to zero
FOR each coin in your pocket
    remove the coin
    increment the count
output the count
```

One very common use of a FOR loop is when the set is a range of numbers, and the goal is to perform some operation using each of the numbers in the range, in turn. As an example, consider an algorithm for calculating the factorial of a positive integer. Recall that the factorial of a number is the product of all the numbers from 1 up to and including the number. A version of this algorithm using a FOR loop is

```
to calculate the factorial of n:
initialize the product to 1
FOR each integer i in the range from 1 to n
    multiply the product by i
output the product
```

Here, we've introduced a variable, i, to represent the current value from the set at each pass through the loop. We could also describe this algorithm using a WHILE loop as follows:

```
to calculate the factorial of n:
initialize the product to 1
initialize i to 1
WHILE i is less than or equal to n
    multiply the product by i
    increment i by 1
output the product
```

Note that the WHILE loop version of the algorithm is slightly more complicated. Specifically, we had to initialize the value of the variable i before the loop, and explicitly increment it inside the loop. These operations were handled implicitly in the FOR loop.

As a final example of a loop, we examine a loop that has a branch inside of the loop body. Consider an algorithm for the process of adding up the value of coins in your pocket. Figure 12.9 shows a flowchart for one version of the algorithm that for simplicity neglects coins other than pennies, nickels, dimes, or quarters. This flowchart contains two separate branch steps, one that determines the value to add depending upon the type of coin, and one that implements the loop that cycles through each of the coins in your pocket. A pseudocode version of the program using a FOR loop is

```
to add coins:
set total to zero
FOR each coin in your pocket
    remove the coin
    IF the coin is a penny THEN the value is 1
    ELSE IF the coin is a nickel THEN the value is 5
    ELSE IF the coin is a dime THEN the value is 10
    ELSE IF the coin is a quarter THEN the value is 25
    add the value to the total
output the total
```

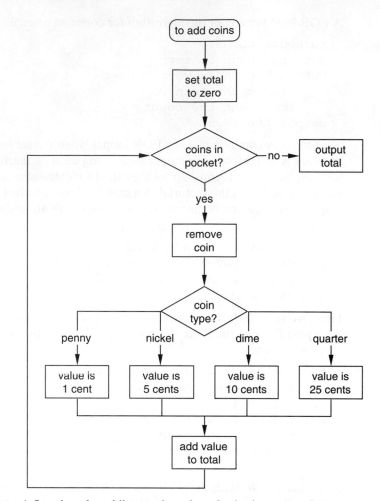

Figure 12.9   A flowchart for adding up the value of coins in your pocket.

12.2 MATLAB FUNCTIONS

12.2.1   Mathematical Functions Versus MATLAB Functions

The concept of a function in a programming language is related to, but not exactly the same as, the concept of a function in mathematics. Here, we'll just briefly review some things you already know about mathematical functions to show their relationship to MATLAB functions. In mathematics, a function is an operation that determines a unique "output" value when applied to a given ordered list of "input" values. When defining a function, we express the inputs and outputs of the function as variables. An example function f with a single input variable is given below:

$$y = f(x) = 3x + 2 \qquad (12.1)$$

Here, x is the input of the function and y is the output. We say that y is a function of x, and that the value of y for a particular value of x is obtained by substituting that value for x on the right side of the equation. Thus, the value for y when x equals 4 is

$$y = f(4) = (3 \times 4) + 2 = 14 \qquad (12.2)$$

When we define a function such as f above, the choice of which *symbol*—letter—to use for the input variable does not affect the definition. For example, if we rewrote (12.1) with a as the input variable instead of x,

$$f(a) = 3a + 2 \tag{12.3}$$

the value of $f(4)$ would still be $(3 \times 4) + 2 = 14$. The input variable simply serves as a placeholder to be substituted when the function is evaluated with a particular input value.

Like mathematical functions, functions in MATLAB also have a name and input and output variables, and express a relationship between these variables. Below is the definition of a MATLAB function named f that expresses the same relationship as (12.1):

```
function y = f(x)
% f(x)
% Calculates the value of 3x + 2
y = 3*x + 2;
```

The first line of the function definition is called the function *header*. The keyword `function` declares that this is the beginning of a function definition. In the above example, the name of this function is f, the name of the output variable is y, and the name of the input variable is x. In general, a MATLAB function header has the form

```
function output-variable = function-name(input-variables)
```

The next two lines in the function definition, beginning with percent signs, are comments; in a moment we'll see the value of these. Any lines following the comments make up the *body* of the function, which will be executed when we use or *call* the function. The body of the function *must* have a line that assigns a value to the output variable. If more than one line assigns a value to the output variable, than the last assignment will be taken as the value for the output.

Like scripts, MATLAB functions must be saved in an M-file before they can be called. Whereas scripts may have any file name, the name of a function M-file *must* be the same as the name of the function, in this case f.m. When you save a function M file, MATLAB will fill this in as the default name for the file in the Save dialog box, and you should always accept this default without making any changes.

Once the definition of the function has been saved in an M-file, we can call the function with a particular value for the input variable. To call f with an input value of 4, enter the following command at the prompt:

```
>> f(4)
```

MATLAB responds:

```
ans =
    14
```

Just as with the mathematical definition of a function, the choice of name for the input and output variable has no effect on the operation of the function. Try changing the output variable to z and the input variable to a and convince yourself that this is so.

```
function z = f(a)
% f(a)
```

```
% Calculates the value of 3a + 2
z = 3*a + 2;
```

Mathematical functions as well as MATLAB functions may have more than one input variable.[1] When a function has more than one input variable, they're specified by an ordered list. For example, the following mathematical function g has two input variables, x and y.

$$z = g(x, y) = x^2 + y \qquad (12.4)$$

Here, (x, y) is the ordered list of input variables. When the function is evaluated for particular input values, they're substituted for the variables in the order in which they're declared. Thus,

$$g(4, 5) = 4^2 + 5 = 21$$

in which 4 has been substituted for x and 5 has been substituted for y because this is the order in which the variables are listed in the function definition. As with the earlier example in (12.1) and (12.3), changing the names of input and output variables has no effect on the meaning of the function, as long as the ordering of the variables is preserved.

$$w = g(b, a) = b^2 + a \qquad (12.5)$$

The value of $g(4, 5)$ is still equal to 21. A MATLAB function that implements the function g in (12.4) is given below:

```
function z = g(x,y)
% g(x,y)
% Calculates the value of x^2 + y
z = x^2 + y;
```

After saving the function definition in a file g.m, we can call g from the command window with particular inputs as shown below:

```
>> g(4,5)
ans =
    21
```

One nice feature of how functions are implemented in MATLAB is that we can use the help facility to display the comments that immediately follow the declaration of the function interface. Thus if we forgot the order of the input parameters of g, we could type:

```
>> help g

g(x,y)
Calculates the value of x^2 + y
```

To get the most out of this feature, your function comments should always list the parameters of the function and give a brief description of what the function does.

Now, we have two mechanisms for writing and saving MATLAB programs: scripts and functions. The main difference between them is that it's possible to pass input and output variables, also called *parameters* or *arguments*, into and out of functions, while it isn't possible to pass parameters into scripts. There is another important

---

[1] Unlike mathematical functions, MATLAB functions may also have more than one output variable, but we will disregard this feature in this discussion.

difference: any variables defined in a script can be accessed via the command window after the script has finished execution, while the variables used in a function cannot. Unlike scripts, variables used in functions only have meaning inside of the function definition, and are "invisible" outside of the function. The following example illustrates this feature. Let's define a script named `variable_test` that calls functions f and g as defined above:

```
% variable_test.m
% A script to test the "visibility" of variables
clear;
p = f(5)
q = g(3,2)
```

Running the script displays the following results:

```
>> variable_test
p =
    17
q =
    11
```

Next, we check which variables are defined in the global workspace using the `who` command.

```
>> who

Your variables are:

p        q
```

Note that p and q are defined in the global workspace, but x and y aren't, even though variables with these names were used in the definitions of both f and g. When you call a function in MATLAB, all variables declared inside the function are kept in a private workspace that's invisible to the global workspace. The fact that functions run in their own private workspaces is a very important feature for writing predictable, modular software systems. One of the main implications of having private workspaces for function calls it that programmers can reuse the same names for variables in different function definitions while still having these names refer to different objects. In other words, the variable named "x" used inside function f is completely unrelated to the variable with the same name used inside function g. If this were *not* the case, then edits to function f could potentially affect the operation of function g, which is clearly not desirable.

12.2.2 Functions Calling Functions

Just as one mathematical function can call another, so can MATLAB. For example, given the mathematical expression

$$
\begin{aligned}
r &= g(f(1), f(2)) \\
&= f(1)^2 + f(2) \\
&= [3(1) + 2]^2 + [3(2) + 2] \\
&= 25 + 8 \\
&= 33
\end{aligned}
$$

the corresponding MATLAB statement is

```
>> r = g(f(1), f(2))
r =
    33
```

Functions calling functions provide a way for breaking a complex problem down into manageable pieces. As an illustration, Section 12.5.1 provides a MATLAB implementation of the cake recipe from Section 12.1.3.

12.2.3   Watching a Function Call Through the MATLAB Debugger

To get a clearer view of how functions in MATLAB work, we can examine the execution of a program in the MATLAB debugger. The debugger is a tool that lets us trace the execution of a program one step at a time and examine the values of variables at each step. A debugger is called a "debugger" because of its value in locating *bugs* or errors in a computer program, by letting the programmer compare values in the program at each step against expectations. In this example, we're not looking for bugs, but rather looking at how a correct program should execute. Specifically, we'll use the debugger to single-step through the execution of the script `variable_test` that was presented in Section 12.2.1.

Figure 12.10 shows the buttons that are used to control the debugger, which are located at the top of a MATLAB editor window. The function of each of these buttons is summarized below:

set/clear breakpoint A *breakpoint* identifies a line in the program where execution should be paused. When you set a breakpoint at a given line of a program and then run the program, the program will run up to the first breakpoint and then pause. A program may have multiple breakpoints set.

clear all breakpoints Clears all the breakpoints set in the current file.

step Execute one line of a program. If the current line is a function call, execute all of the lines of code in the function without explicitly showing this.

step in Similar to `step`, but if the current line is a function call, "step in" to the function so that single-stepping may proceed within the body of the function.

step out If the current line is inside a function, step out of the function and back to the part of the program that called the function.

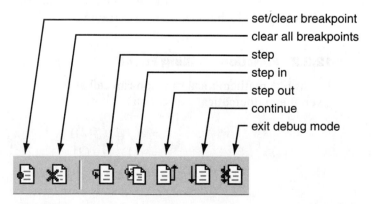

set/clear breakpoint
clear all breakpoints
step
step in
step out
continue
exit debug mode

Figure 12.10   MATLAB debugger control buttons.

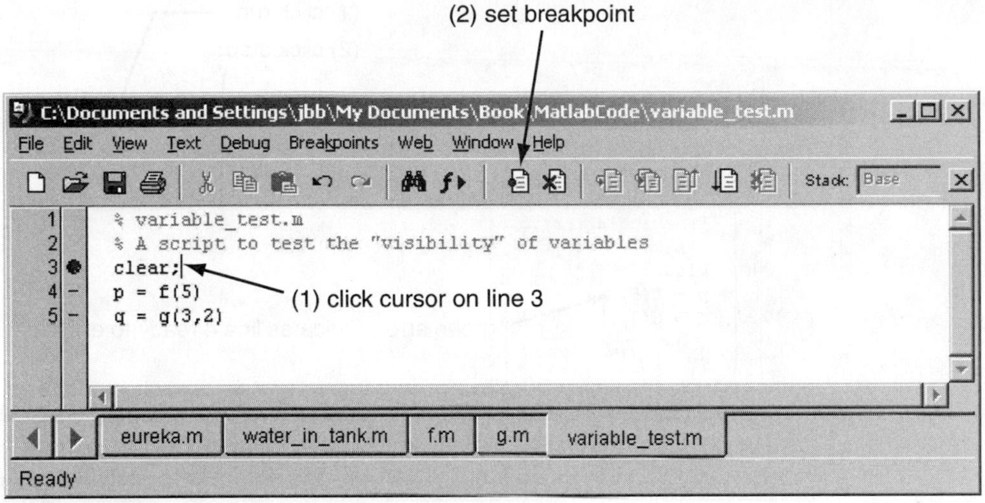

(2) set breakpoint

Figure 12.11 Setting a breakpoint with the MATLAB debugger.

continue Continue execution of the program up to the next breakpoint. If there are no remaining breakpoints, continue execution to the end of the program.

exit debug mode Quit debugging.

The first step in using the debugger is to open the program that you want to trace in a MATLAB editor window. For this example, open the file `variable_test` created in the last section. The following steps will lead you through the process of using the debugger to single-step through the execution of the program.

We want to step through the program beginning with the first command, so set a breakpoint at the line with the `clear` command by first clicking the cursor on that line and then pressing the `set/clear breakpoint` button. A red circle should appear to the left of the line to indicate the breakpoint, illustrated in Figure 12.11.

Press the `run` button. A green arrow will appear to the left of the line with the `clear` command, indicating that the debugger has paused here. The green arrow always indicates the next line to be executed—it has not been executed yet. Now press the `step` button. The green arrow will advance to the next line. At this stage the editor window should appear as shown in Figure 12.12.

The green arrow is now pointing at the first line in the script that contains a function call, specifically, a call to function `f` with an input value of 5. We have the choice of either stepping past the function call to line 5, or "stepping in" to the function `f` itself. We'll choose to step in. Press the `step in` button and the file in the editor window should change to the file `f.m`, with the green arrow pointing at the first line with a command, line 4. This is shown in Figure 12.13.

We've now left the global workspace and entered the function `f` workspace, and the only variables that will be visible are those defined within the function `f`. There are several ways we can examine our variables. The easiest is to simply place the cursor over the name of a variable. For example, place the cursor over variable `a` on line 4. A small box will pop up indicating that `a = 5`. Why is this? Because the call to the function in the script `variable_test` was `f(5)`, which caused the value 5 to be substituted for the single input parameter `a` of the function.

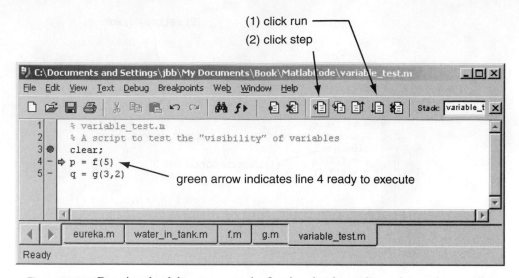

Figure 12.12 Running the debugger up to the first breakpoint and stepping to the next line.

Another way to examine the values of variables during debugging is through the command window. The first thing you'll notice when you look at the command window at this point is that the prompt has changed to K>>, indicating that MATLAB is paused in the middle of running a program. To see what variables are currently active, use the who command, then enter the name of a variable to see its value.

```
K>> who

Your variables are:

a       ans

K>> a
a =
        5
```

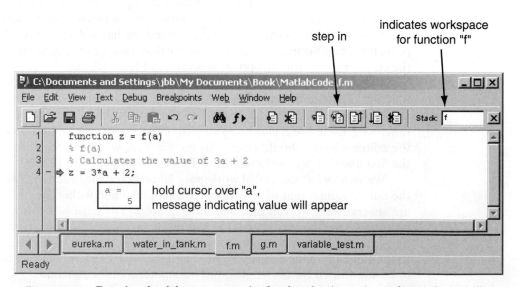

Figure 12.13 Running the debugger up to the first breakpoint and stepping to the next line.

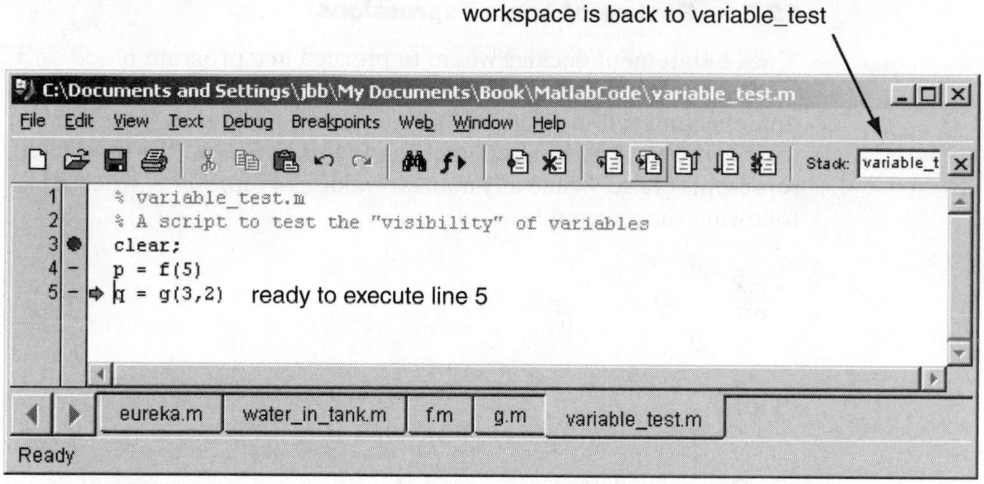

workspace is back to variable_test

Figure 12.14 Debugger returns to `variable_test.m` after execution of function call `f(5)` has completed.

The only variable that's currently defined in the current workspace is a. None of the variables from the global workspace are visible. The variable z doesn't appear because the line of the program that creates it hasn't been executed yet.

Press the step button one more time. This will execute line 4 in f.m. Now you'll be able to examine the value of variable z, either by moving the cursor over it in the editor window, or by entering its name in the command window. Note that the green arrow has changed to pointing downward, meaning that all of the lines in the function have been run, and the program is now ready to return from the function call back to the point in the program that called it.

Press the step button again. The execution of the program will jump back to the script variable_test, with the green arrow pointing at line 5, as shown in Figure 12.14. The call to function f has completed and the program is now ready to run the line with the call to function g. Go to the command window and enter the who command to see which variables are currently visible. The variables that were used inside of the call to function f are no longer defined. The only variable that is currently visible is p, because it was defined when line 4 was executed. Variable q won't be defined until after line 5 is executed.

Again we have the choice of either stepping in or stepping past line 5. This time, let's just step past it. Press the step button once. This will execute line 5 in its entirety, without single-stepping into the function g. The green arrow is now pointing downward, meaning that all the lines in the file variable_test.m have been executed. Examine the values of variables p and q; they should be what you expect.

Finally, press the exit debug mode button. You're done with this exercise.

12.3 CONDITIONAL SELECTION STATEMENTS

Conditional selection statements determine which line in a program to execute next when a decision point is reached. MATLAB offers several alternative conditional selection statements, the most common of which is the if statement. In this section, we examine the if statement and some of the optional ways of using it.

12.3.1 Review of Logic Expressions

The `if` statement decides where to proceed in a program based on a condition that may either be "true" or "false." These conditions are *logic expressions*, which we first encountered in Section 10.4.2, and will briefly review here. An example of a logic expression is the statement "six is less than five." In MATLAB, the value "0" represents "false" while any nonzero value can represent "true." This is shown in the following short example.

```
>> 6 < 5
ans =
     0
>> 5 < 6
ans =
     1
```

Operators that compare two values, such as the operator "<", are called *relational operators*. The table below summarizes the relational operators in MATLAB.

symbol	operation
==	equals
~=	does not equal
<	less than
>	greater than
<=	less than or equal to
>=	greater than or equal to

Note that the double-equals symbol "==" denotes the relational operator "equals" whereas the symbol "=" is the assignment operator. Inadvertently confusing these two symbols is a very common source of bugs in programs.

Logic expressions can be joined to form compound logical expressions using *Boolean operators*. In MATLAB, the symbols for the Boolean operators, "and", "or", and "not" are:

For example, the following expressions will evaluate to true for any value of x between 5 and 12, inclusive:

```
(x >= 5) & (x <= 12)
```

symbol	operation
&	and
|	or
~	not

Alternatively, we can express the same condition as:

```
~((x < 5) | (x > 12))
```

12.3.2 IF/ELSE Statements

The `if` statement evaluates a condition in the form of a logical expression and then executes a consequent block of statements if the condition is "true." The syntax of the `if` statement is

```
if condition
    consequent statements
end
```

The following program fragment will display the string "You pass" if `grade` is greater than 60:

```
if (grade > 60)
    disp('You pass');
end
```

The following program fragment will display the string "in range" if x is between 5 and 12, inclusive:

```
if (x >= 5) & (x <= 12)
    disp('in range');
end
```

You could accomplish the same thing by *nesting* one `if` statement inside another:

```
if (x >= 5)
    if (x <= 12)
        disp('in range');
    end
end
```

Adding an `else` clause to an `if` statement specifies an alternative course of action if the condition is "false." The syntax for an `if-else` statement is:

```
if condition
    consequent statements
else
    alternative statements
end
```

The following function, `absolute`, uses an `if-else` statement to calculate the absolute value of a number:

```
function y = absolute(x)
  if x < 0
      y = -x;
  else
      y = x;
  end
```

Finally, the `elseif` clause can be used to specify multiple consequences for multiple conditions. The syntax for an `if` statement with `elseif` clauses is:

```
if condition 1
    consequent statements 1
```

```
elseif condition 2
    consequent statements 2
. . .

elseif condition n
    consequent statements n
else
    alternative statements
end
```

The following function tests the value of variable t, and displays whether water at that temperature, in degrees Fahrenheit, would be a solid, liquid, or gas at a pressure of 1 ATM:[2]

```
function state_of_matter(t)
% state_of_matter(temperature_fahrenheit)
% Displays the state of matter for water,
% given a temperature in degrees Fahrenheit
if t <= 32
    state = 'solid';
elseif t >= 212
    state = 'gas';
else
    state = 'liquid';
end
disp(state);
```

12.3.3 Stepping Through an IF Statement in the Debugger

As we did with functions, in this section we'll use the MATLAB debugger to step through a decision statement to illustrate how these statements operate. Furthermore, we'll illustrate how the debugger can be used to actually track down a bug. The following steps will guide you through the process of tracing a working function.

1. First, save the function state_of_matter from the last section in a file and open it in a MATLAB editor/debugger window. Set a breakpoint at the beginning of the if. From the command window, enter the command state_of_matter(10) as shown in Figure 12.15.

2. Place the cursor over variable t and confirm that it has the value 10.

3. Click the Step button. The green arrow should point to the line state = 'solid'; because the condition t <= 32 is true.

4. Click the Step button again. The green arrow will jump down to the statement disp(state);. This is because in an if...elseif...else... structure, each of the conditions are evaluated sequentially, and the structure exits after a true condition is found and evaluated.

---

[2]Note that this naive example glosses over what really happens at the temperatures of 32° and 212° F (0° and 100° C). At these temperatures, the water is undergoing a phase change and will exist as combinations of solid and liquid or liquid and gas, respectively.

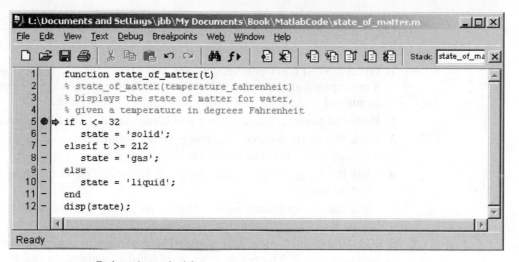

Figure 12.15 Debugging a decision statement.

5. Click the `Step` button once more. The message `solid` will display in the command window and the green arrow will point downward, indicating that the function is ready to exit.

6. Click the `Exit Debug Mode` button.

7. Repeat the above procedure two more times, with temperatures in the gas and liquid range. Carefully note which lines in the function are executed and which are skipped.

Now, let's see how the debugger helps us locate a bug. Suppose that the function to display the state of matter for water is implemented as follows:

```
function state_of_matter_bug(t)
% state_of_matter(temperature fahrenheit)
% Displays the state of matter for water,
% given a temperature in degrees Fahrenheit
if t <= 32
    state = 'solid';
end
if t <= 212
    state = 'liquid';
else
    state = 'gas';
end
disp(state);
```

Running this version of the function with the value 10 as input produces the following results:

```
>> state_of_matter_bug(10)
liquid
```

What went wrong? You might be able to locate the bug by tracing through the code by hand, but since you put the bug in there, you might not recognize it for a while. The debugger can be used to help you test your assumptions. In this program, the main assumption is that the variable `state` will have the value `'solid'` when the last line of the function is reached. The following steps will guide you through the process of locating the error in logic.

1. Open a MATLAB editor/debugger and save `state_of_matter_bug`. Set a breakpoint at the line `if t <=32`. From the command window, enter the command `state_of_matter_bug(10)`.

2. Place the cursor over variable `t` and confirm that it has the value 10.

3. Click the `Step` button. The green arrow should point to the line `state = 'solid';` because the condition `t <= 32` is true.

4. Click the `Step` button again. The green arrow will move to the `end` of the first `if` statement.

5. Click the `Step` button again. The green arrow will move to the statement `if t <= 212`, ready to test the condition of the second `if` statement.

6. Click the `Step` button again. Aha! The green arrow is now at the line `state = 'liquid';`. The previous value of `state`, namely `'solid'`, will now be overwritten with `'liquid'`. The bug was that we meant each of the three cases to be exclusive, but in using two separate `if` statements, they are not. An `elseif` clause could be used to solve the problem.

12.4 LOOPS OR REPETITION STATEMENTS

Repetition statements, also called *loops*, cause a block of statements in a program to be repeated. MATLAB has two types of iteration statements: `while` loops, which repeat as long as a condition remains "true," and `for` loops, which repeat a fixed number of times.

12.4.1 WHILE Loops

A `while` loop causes a block of statements to be repeated for as long as a condition specified by a logical expression remains "true." The syntax for a `while` loop is:

```
while condition
    statements
end
```

The following script counts from 1 to 3, displaying the values one per line:

```
i = 1;
while i <= 3
   disp(i);
   i = i + 1;
end
disp('Finished');
```

Calling the function produces the following output:

```
>> count_to_3
    1
    2
    3
Finished
```

12.4.2 FOR Loops

A `for` loop repeats a block of statements a fixed number of times. The syntax of a `for` loop is

```
for variable = a-vector
    statements
end
```

A `for` loop will repeat the statements for as many times as there are elements in *a-vector*. With each iteration of a `for` loop, the next sequential element in *a-vector*, beginning with the first, is stored in `variable`. The term `array-of-elements` can be specified by any MATLAB expression, but in practice it is usually a sequence constructed with the colon (:) operator. As an example, the following program calculates the sum of the series 0.0, 0.1, 0.2, . . . 10.0 and stores the result in the variable `total`:

```
total = 0;
for i = 0:0.1:10
   total = total + i;
end
```

Note that this example is equivalent to the array operation:

```
total = sum(0:0.1:10)
```

In both cases, the resulting value for total is 505.

Any procedure performed using a `for` loop could also be performed using a `while` loop. Here is a `while` loop that is equivalent to the `for` loop above:

```
total = 0;
i = 0;
while i <= 10
   total = total + i;
   i = i + 0.1;
end
```

The main difference in using a `while` loop over a `for` loop is that in a `while` loop, you need to remember to increment the loop counter in the loop body, whereas a `for` loop automatically uses the next element in the vector with each pass. Mostly, there isn't a universal advantage of using one type of loop over the other, as long as you're careful when you write your program.

12.4.3 Watching a Loop in the Debugger

As we did with the `if` statement in Section 12.3.3, step through the execution of this script using the MATLAB debugger. The following steps will guide you through the process.

1. Save the script `count_to_3` in a MATLAB editor/debugger window. Set a breakpoint at the first line, `i = 1;`. Run the script from the command window. The green arrow should point to the first line in the script, as shown in Figure 12.16.

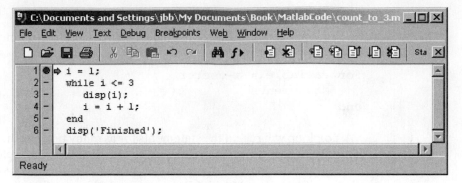

Figure 12.16 Debugging a loop.

2. Click the Step button, and the green arrow will move to the line while i <= 3, ready to test the condition. Place the cursor over variable i to confirm that its value is 1.

3. Click Step again. The green arrow will move into the loop body. Click Step again and the value 1 will display in the command window, and the green arrow will indicate that the statement i =i + 1 is ready to execute. Place the cursor over variable i and confirm that its value is still 1.

4. Click Step again. The green arrow moves to the end of the loop. Check the value of i and see that it's now 2.

5. Click Step again. The green arrow jumps back to the top of the loop to evaluate the logic statement i <= 3 again. Since i equals 2, the condition is true and the loop body will execute again.

6. Keep clicking the Step button until you read the end statement. Now the value of i should be 3. The condition i <= 3 is still true, so the loop body will execute again.

7. Keep clicking the Step button until you read the end statement again. Confirm that i now equals 4.

8. Click Step again. The green arrow jumps back to the top of the loop, but this time the condition i <= 3 is false.

9. Click Step again. The green arrow jumps over the loop body to the line disp('Finished'). Click once more to display this message in the command window and then exit the debugger.

12.4.4 Nested Loops

Just as we can "nest" one if statement inside of another, we can also nest loop statements. The following script shows a very basic example of two nested for loops.

```
for i = 1:2
    for j = 1:3
        disp([i j])   % display i and j
    end
end
```

The outer loop, which uses the variable i as a counter, executes the inner loop 2 times. The inner loop, which uses the variable j as a counter, executes the disp statement 3 times. As a result the disp statement will execute a total of i times j, or 6 times. The disp statement itself simply displays the values of i and j each time it executes. The results of running the script are shown below:

```
%    i        j
% ------------
     1        1
     1        2
     1        3
     2        1
     2        2
     2        3
```

As shown, the first three times the disp statement executes, i has a value of 1 and j counts from 1 to 3. This ends the first iteration of the outer loop, after which i increments to 2. The inner loop once again increments j from 1 to 3. At this point, the outer loop has counted to its upper limit, and the script is complete.

12.4.5 Common Loop Bugs

There are two very common bugs associated with loops. The first bug is called an *infinite loop*, and occurs when the loop condition is always true, causing the statement inside the loop to repeat forever. We could "accidentally" turn the above example into an infinite loop by forgetting to increment i with each iteration:

```
i = 1; while i <= 3
    disp(i);
    % oops, forgot to increment i
end
```

It's fairly easy to tell when a program that you're running contains an infinite loop—the command prompt never returns! To break out of an infinite loop in MATLAB, type control+C in the command window.

The second type of common bug associated with loops is called an *off-by-one error*. One form of off-by-one error occurs when you run through a loop either one too few or one too many times. Typically, this results from an improperly specified loop conditions, such as using a less-than operator in place of less-than-or-equal-to. Another form of off-by-one error relates to the order in which variables are used and reassigned within the loop body. Suppose we changed the order of the statements in the example from above:

```
% counts from 2 to 4
i = 1; while i <= 3
    i = i + 1;
    disp(i);
end
```

This version of the program will count from 2 to 4 rather than from 1 to 3. That's fine if the purpose is to count from 2 to 4, but otherwise, it's a bug. Using the MATLAB debugger to step through a problematic loop and comparing expected values of variables with the actual values during execution will usually help you track down such bugs.

12.5 EXAMPLES OF FUNCTIONS, CONDITIONALS, AND LOOPS

12.5.1 Subfunctions: The Cake Recipe

In Section 12.1.3, we saw that one way of making a complex algorithm more understandable is to divide it into subroutines. MATLAB provides a mechanism for doing this through "subfunctions," which are functions called by another function within the same program.

The following program implements the algorithm for making a cake as described in Section 12.1. The entire program below is saved in file make_cake.m. The main function in the program must be listed first and must also correspond to the name of the file. Other subfunctions may be listed in any order. Note that functions without any input parameters or that don't produce output values need not have these declared in the function definition.

```matlab
function make_cake(pan_size)
% make_cake(pan_size)
% Simulate making a cake. pan_size is a number, in inches
%
prep;
mix_cake;
bake(pan_size);
mix_frosting;
frost;

function prep
disp 'IN PREP:'
disp 'preheat oven to 350 degrees F'
disp 'grease pan'

function mix_cake
disp 'IN MIX_CAKE:'
disp 'blend mix, water, oil, and eggs in bowl'
disp 'beat for 2 minutes'

function bake(pan_size)
disp 'IN BAKE:'
if pan_size == 8
    disp 'bake for 32-36 minutes'
elseif pan_size == 9
    disp 'bake for 28-32 minutes'
else
    disp 'stop and change pan to 8 inch'
    disp 'bake for 32-36 minutes'
end

function mix_frosting
disp 'IN MIX_FROSTING:'
disp 'beat butter for 1 minute'
disp 'add frosting, mix, and milk'
disp 'beat for 2 minutes'
```

```
function frost
disp 'IN FROST:'
disp 'cover with frosting'
```

The result of running this program for a 9-inch pan is as follows:

```
>> make_cake(9)
IN PREP:
preheat oven to 350 degrees F
grease pan
IN MIX_CAKE:
blend mix, water, oil, and eggs in bowl
beat for 2 minutes
IN BAKE:
bake for 28-32 minutes
IN MIX_FROSTING:
beat butter for 1 minute
add frosting, mix, and milk
beat for 2 minutes
IN FROST:
cover with frosting
```

12.5.2 Vector and Matrix Functions

One of the most powerful features of MATLAB is its built-in support for vectors and matrices. In the early years of computing, programmers typically had to write their own procedures for operations such as vector addition or matrix multiplication, but the advent of programming environments such as MATLAB has made this obsolete for many users.[3] As an illustration of loops and conditionals, however, in this section we'll look at how one might write functions for several basic vector and matrix operations.

Vector Addition As discussed in Section 11.4.3, two vectors can be added together if and only if they are of the same size. If so, then their sum is the vector composed of the sum of their corresponding elements. The following MATLAB function first checks the lengths of the input vectors and, if they're the same, calculates their sum by marching through the vectors X, Y, and Z one element at a time in a loop.[4]

```
function Z = vector_add(X, Y)
% vector_add(X, Y)
% Returns the sum of vectors X + Y

if length(X) ~= length(Y)
    disp('Error: X and Y must be of same length')
```

---

[3]MATLAB was originally written in the 1970s on top of a set of publicly available subroutines in FORTRAN called the Linear Algebra Package or LAPACK. LAPACK itself is available free on the web at www.netlib.org.

[4]To be really correct, we'd also have to make sure that both inputs are either row vectors or column vectors, but we've left that out for the sake of simplicity.

```
        else
            n = length(X);
            Z = zeros(1,n);  % makes function run faster
            for i = 1:n
                Z(i) = X(i) + Y(i);
            end
        end
```

Running the `help` command displays the comment at the beginning of the function definition.

```
>> help vector_add

  vector_add(X, Y)
  Returns the sum of vectors X + Y
```

If the input vectors aren't of the same length, we get an error message when we call the function; otherwise, we get their sum.

```
>> a = [1 2 3];
>> b = [4 5 6];
>> c = [7 8 9 10];

>> vector_add(a,c)
Error: X and Y must be of same length

>> vector_add(a,b)
ans =
     5    7    9
```

Note that variable `i` in this example represents the *subscript* of an element of the vectors X, Y, and Z—not the *value* of a vector element. The expressions X(i), Y(i), and Z(i) are the values of the individual elements at subscript i. As mentioned in Section 10.2.1, beginning programmers often get confused by this. If you're unsure of this notation, go back and review Section 10.2.1 to clarify the difference between the subscript of a vector and the value at that subscript before proceeding to the next example.

One final observation in this example is that we initialized the values in the output vector Z to all zeros before performing the addition. The function would still work if we didn't do this, since MATLAB can append new elements to a vector on the fly, as described in Section 10.2.1. This takes time, however, and MATLAB runs significantly faster for larger examples if all of the vector elements are created at once in block, even if their values are changed later.

Matrix Addition and Nested Loops   Like vector addition, matrix addition computes the sum of two matrices on an element-by-element basis. To do this requires two loops, one inside of the other. As the pseudocode below illustrates, the outer loop sequences from row to row, while the inner loop iterates over the elements in a row:

```
to multiply matrices:
Make sure that the matrices are of the same
```

```
size FOR each row
     FOR each column in row
         Add corresponding elements
```

Translating this into a MATLAB function, we get:

```
function Z = matrix_add(X, Y)
% matrix_add(X, Y)
% Returns the sum of matrices X + Y

if size(X) ~= size(Y)
    disp('Error: X and Y must be of same size')
else
    dimensions = size(X);     % returns a 2-element vector
    nrows = dimensions(1);
    ncols = dimensions(2);
    Z = zeros(nrows,ncols);
    for i = 1:nrows
        for j - 1:ncols
            Z(i,j) = X(i,j) + Y(i,j);
        end
    end
end
```

In the vector addition example, we used a single variable i as a subscript or index into the vectors. For matrix addition, we need two subscripts to uniquely identify each element in a matrix. The variables i and j are the row and column subscripts or indices into the matrices X, Y, and Z. Thus, the statement for i = 1:nrows sequences over the rows, the statement j = 1:ncols sequences over the columns, and the expression X(i,j) denotes the element at row i and column j of the matrix X. The following commands demonstrate the function:

```
>> A = [1 2 3; 4 5 6];
>> B - [10 20 30; 40 50 60];
>> matrix_add(A,B)
ans =
    11    22    33
    44    55    66
>> matrix_add(A,B')
Error: X and Y must be of same size
```

12.6   ACCUMULATION OF CHANGE

In Chapter 8, we examined a method for modeling the behavior of a system over time as a series of discrete snapshots, in which we used laws describing the rate of change of the system—such as a population growing by a fixed percentage every year or an object falling according to the laws of gravity—to calculate its behavior from one time point to the next. We expressed the models as difference equations and used Euler's method to simulate them over time. In Chapter 8, to implement the simulations, we constructed tables where each column represented the state variables of interest and each row represented the values of these variables at different time steps. While filling in the table isn't too difficult conceptually—it only requires basic arithmetic—it can be extremely tedious. Fortunately, this is exactly the kind of task that computers do very well. One approach to automating the calculation is to use a spreadsheet. Another approach, which we discuss in this section, is to write a MATLAB script. Once you become familiar with it, you'll probably find the MATLAB approach to be both simpler and more flexible than using a spreadsheet for this task, although both approaches work well.

12.6.1 Review: Modeling Population Growth

In Section 8.4.4, we examined a model for how the size of a population changes with time. Here, we'll briefly review the model before showing how to implement it in MATLAB. In our model, we stated that the population growth rate was proportional to the size of the population, or

$$\frac{\text{change in population}}{\text{year}} = \frac{\text{percentage growth}}{\text{year}} \times \text{population}$$

or

$$\frac{\Delta P_i}{\Delta t} = gP_i$$

where P_i is the size of the population in year i and g is the *annual growth rate*, measured as a percentage of the current population. Writing this as a difference equation, we get:

$$\frac{P_{i+1} - P_i}{\Delta t} = gP_i \quad P_{i+1} = P_i + gP_i\Delta t \tag{12.6}$$

In Section 8.4.4, we implemented the model in the form of a table, where each row of the table calculates the population in successive years as a function of the previous year, according to Equation (12.6). Given that the world population was approximately 2.6 billion in 1950 with a growth rate of 0.18% per year, the first few rows of the table are

step	year	population (billions)
0	t_0 1950	P_0 2.56
1	$t_1 = t_0 + \Delta t$ $1950 + 1$ 1951	$P_1 = P_0 + gP_0\Delta t$ $2.56 + (0.018)(2.56)(1)$ 2.60
2	$t_2 = t_1 + \Delta t$ $1951 + 1$ 1952	$P_2 = P_1 + gP_1\Delta t$ $2.60 + (0.018)(2.60)(1)$ 2.65

Adding more rows to the table manually would be tedious, but is easily handled by MATLAB. In the MATLAB inplementation we can think of each column of the table—year and population—as a vector, and evaluating the population for the next year corresponds to adding new elements to the vectors. Thus if `year` is a vector of years, P is a vector of populations, i is the index of the current year, and g is the

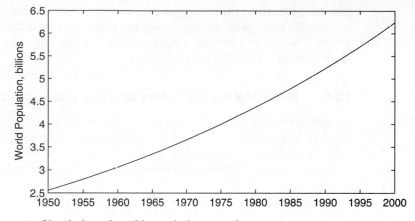

Figure 12.17 Simulation of world population growth.

growth rate, then we can add simulation results for the next year to the vectors with the MATLAB statements:

```
year(i+1) = year(i) + 1;
P(i+1) = P(i) + g*P(i);
```

Taking a step back and looking at the bigger picture, we can now write the pseudocode for the complete simulation as follows:

```
Set initial values for the year, population, and growth rate
FOR the number of years to be simulated
    Calculate and append next value to year vector
    Calculate and append next value to population vector
Plot the results
```

Writing this in MATLAB, we get the script shown below, which produces the results plotted in Figure 12.17.

```
% Simulate world population, 1950-2000
clear;              % clear all variables

year(1) = 1950;     % initial year
P(1) = 2.56e9;      % population in initial year
g = .018;           % population growth rate

for i=1:50
    year(i+1) = year(i) + 1;
    P(i+1) = P(i) + g*P(i);
end

plot(year,P/1e9);
ylabel('World Population, billions');
```

When writing scripts that append or change elements of vectors, it's important to clear all variables at the beginning of the script. The reason for this is that we don't want to leave values behind at the end of a vector from a previous run that aren't part of the current run. For example, suppose that we chose to run the simulation for only 25 years after we previously ran it for 50, with a different growth rate. The vector P would still contain values from the earlier run, and would be erroneously plotted at the end of the script.

12.6.2 Modeling the Trajectory of a Softball with Drag

As a slightly more complex example of simulating a changing system with MATLAB, in this section we present the complete script for modeling the trajectory of a softball, including the effects of drag, as presented in Section 8.3.3. Recall that the model consisted of three difference equations, one each for height y, velocity v, and acceleration a:

$$y_{i+1} = y_i + v_i \Delta t$$
$$v_{i+1} = v_i + a_i \Delta t$$
$$a_{i+1} = \frac{mg - \text{sign}(v_{i+1}) k_d v_{i+1}^2}{m}$$

in which

$$k_d = \frac{C_D}{2} \rho A$$

with the following physical constants:

r:	radius of softball	4.85 cm
m:	mass of softball	0.185 kg
C_D:	drag coefficient of sphere	0.5
ρ:	density of air	1.29 kg/m^3

The main difference between this example and the population growth model is that there are three state variables and difference equations, whereas the population model only had one. Still, the basic pattern for simulating the softball trajectory over time is very similar to that for population growth, except that for each time step we evaluate three equations instead of one. Another difference is that we want to continue the simulation until the ball returns to the ground, rather than simulate for a fixed number of steps. A pseudocode description of the script is as follows:

```
Set physical constants
Set initial conditions
WHILE the softball altitude is greater than or equal to 0
   Evaluate the differences equations for y, v, and a
Find the maximum altitude, impact time, and impact velocity
Plot altitude, velocity, and acceleration versus time
```

Translating this into MATLAB, we get the following script.

```
% A model for the trajectory of a softball launched vertically,
% including the effects of drag.
clear;
% physical constants
g = -9.81;               % acceleration, m/s^2
r = 0.0485;              % radius of softball, m;
```

```
m = 0.185;                  % mass of softball, kg;
rho = 1.29;                 % density of air, kg/m^3
Cd = 0.5;                   % drag coefficient of sphere
A = pi*r^2;                 % cross-sectional area of sphere, m^2


kd = Cd*rho*A/2;            % drag force coefficient

% set initial conditions
i = 1;
v0 = 20;                    % initial velocity, m/s
dt = .001;                  % time step size, s
y(1) = 0;                   % altitude, m
v(1) = v0;                  % velocity, m/s
t(1) = 0;                   % time, s
a(1) = (m*g - kd*v(1)^2)/m; % acceleration, m/s^2

% Euler loop, until softball hits ground
while y(i) >= 0
    t(i+1) = t(i) + dt;
    y(i+1) = y(i) + v(i)*dt;
    v(i+1) = v(i) + a(i)*dt;
    a(i+1) = (m*g - sign(v(i+1))*kd*v(i+1)^2)/m;
    i = i + 1;
end

% set final values
t_impact = t(i)
v_impact = v(i)
y_apogee = max(y)

%generate plots
subplot(3,1,1);
plot(t,y);
ylabel('y, m');
subplot(3,1,2);
plot(t,v),
ylabel('v, m/s');
subplot(3,1,3);
plot(t,a);
ylabel('a, m/s^2');
xlabel('t,s');
```

Running the script prints the following values and produces the plot shown in Figure 12.18. These are the same values that we presented in Section 8.3.3.

```
t_impact =
     3.6580
v_impact =
   -16.2043
y_apogee =
    16.3931
```

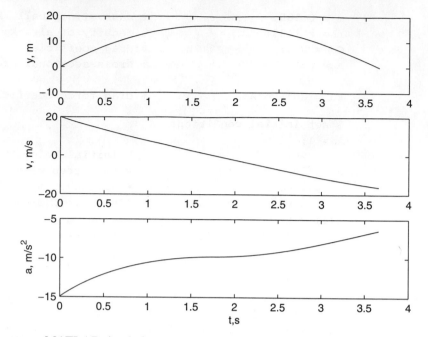

Figure 12.18   MATLAB simulation of softball trajectory.

PROBLEMS

1. Mathematical Functions in MATLAB

Write MATLAB functions that implement the following mathematical functions:

$$f(x) \quad = 2x + 7$$
$$g(x, y) = x^2 + 2y^2$$

Write MATLAB expressions that evaluate the following mathematical expressions using the functions defined above. Show the results of the evaluation from MATLAB.

(a) $f(1)$

(b) $f(2)$

(c) $g(1, 2)$

(d) $g(f(1), f(2))$

(e) $f(g(1, 2))$

2. Unit Conversion Functions

The following table lists conversions between metric units and American units:

1 meter (m)	=	39.37 inch (in)
1 kilometer (km)	=	0.6214 mile (mi)
1 liter (L)	=	0.2642 gallon (gal)
1 kilogram (kg)	=	2.205 pound-mass (lbm)
1 newton (N)	=	0.2247 pound (lb)

(a) Write a set of MATLAB functions that convert from metric units to American units and vice versa for each of the lines in the table above. Use the following convention to name your functions: `y = A2B(x)`, where A and B are abbreviations for units. For example, the function to convert from inches to meters would be `y = in2m(x)`. Test each function in MATLAB by showing that `A2B(B2A(x))` equals x.

(b) Using functions developed for the previous problem, write functions `mph2mps` and `mps2mph` that convert velocities from units of miles per hour to meters per second and vice versa. Test by converting 30 m/s to mph and 60 mph to m/s.

3. MATLAB Functions and Workspaces

Consider the following function, `seven`, that returns the value 7 for any inputs x and y:

```
function z = seven(x,y)

% always returns the value 7

x = 7; y = x; z = y;
```

If we define four variables w, x, y, z in the global workspace and then call the function seven, what will be the values of these variables after the function call returns?

```
>> w = 0;
>> x = 1;
>> y = 2;
>> z = 3;
>> w = seven(x,y)
```

Figure this out *without* running the program and explain your answer. Then type it in and run it in MATLAB and see if you were correct. If you weren't correct with your original answer, come up with the right explanation for what occurred.

4. Theoretical Model for a Launch System Using MATLAB

A system is being designed to launch a projectile at a target downrange. In this exercise, you will develop a simple model of the launch system and use it to experiment with different launcher settings to see their effect on reaching the target.

The launcher consists of a slingshot composed of a length of elastic tubing with a pouch mounted on a frame, as shown in Figure P12.1. To use the slingshot, first set the vertical height adjustment on the frame and then position the trigger mechanism at a desired distance along the base. Next, stretch the tubing, lock the pouch in place in the trigger mechanism, and place the projectile in the pouch. When the trigger is released, the projectile will take off at the angle determined by the launcher vertical and horizontal settings, with an initial velocity that depends on how much the tubing was stretched. Once the projectile is released, its trajectory is governed by Newton's Laws of Motion, and the distance that it will travel before hitting the ground depends on the forces that act on it, such as gravity and wind resistance or drag.

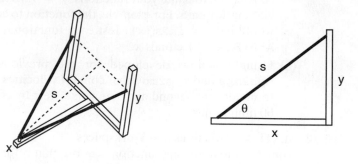

Figure P12.1 Launcher geometry

The entire launch process may be modeled by a single function

```
d=launch(x,y)
```

that calculates the distance d that the projectile will travel given horizontal and vertical launcher settings, x and y, as illustrated in Figure P12.2. This model may in turn be decomposed into two main parts: *propulsion*, which models the slingshot, and *flight*, which models the distance that the projectile travels after it is launched.

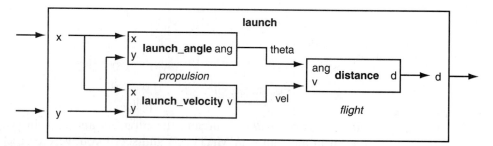

Figure P12.2 System of functions modeling the launch.

The propulsion part of the model consists of two functions,

```
ang=launch_angle(x,y)
v=launch_velocity(x,y)
```

which calculate the launch angle and initial velocity of the projectile, respectively, given the horizontal and vertical settings of the launcher. The flight portion of the model consists of a single function, d=distance(ang,v), that calculates the distance traveled by a projectile launched at a given angle and initial velocity. To complete the model for the launch process, follow the steps below.

(a) Write an outline for the function d=launch(x,y) that corresponds to the box in the figure. Make sure that the outline includes all the important parts: a statement of the purpose, a list of inputs and outputs, a summary of the steps involved in the computation, and any helpful illustrations.

(b) As above, write an outline for the function `launch_angle`. Figure out the necessary trigonometric relationship to find the angle given the launcher settings. Implement the function in MATLAB, and then test it in isolation before moving on. Describe and show the results of your test.

(c) The launch velocity of the projectile depends upon the amount that the tubing is stretched prior to launch, which may in turn be calculated from the launcher settings. If we assume that the tubing has no slack prior to stretching, then the stretch amount is s as shown in the figure. Suppose that it has been determined experimentally that the relationship between the launch velocity v in m/s and the stretch distance s in m is

$$v = 22s$$

Given this information, develop the function `launch_velocity`, starting with an outline, then writing the MATLAB code, and then testing the function.

(d) Under vacuum conditions, it may be shown that for a projectile launched with an initial velocity v in m/s at an angle θ in radians on level ground, the total horizontal distance covered before hitting the ground is, in meters:

$$d = \frac{2v^2 \sin \theta \cos \theta}{9.81}$$

Given this assumption, develop the function `launch_velocity` from outline through testing.

(e) Implement the function `launch` in MATLAB, using the functions developed in the previous parts of this problem.

(f) With the function `launch` that you developed experimentally using multiple trial launches, determine launcher settings that will send the projectile a distance of 1001 m. Describe the strategy that you used to determine the settings. How many tries did it take for you to hit the target, and what were the distances at each trial?

5. Vector Graphics

Before inexpensive *raster* displays became available with large arrays of *pixels* or "picture elements," early computer graphics programs used *vector graphics* that produced images by tracing lines between points. The early arcade video game *Battle Zone* is a classic example of vector graphics.

We can produce vector graphics images in MATLAB using the `plot` command. While this command is typically used to plot functions, it will draw lines between any series of *x-y* points in order. For example, plotting the vectors x and y as defined below will produce the diamond shape shown on the left in Figure P12.3.

```
>> x = [1 2 1 0 1];
>> y = [0 1 2 1 0];
>> plot(x, y);
```

Primitive shapes can be *scaled* by multiplying their *x*- and *y*-coordinate vectors by a constant. They can be *translated* left or right by adding a constant to the *x*-coordinate vector, and translated up or down by adding a constant to the *y*-coordinate vector. For example, the scaled and translated diamond shape on the right in Figure P12.3 was produced with the following command:

```
>> plot(2*x+1, 2*y+2);
```

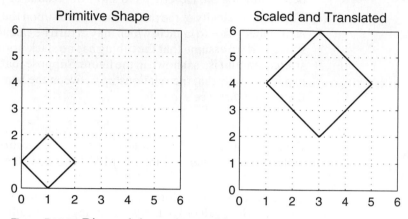

Figure P12.3 Diamond shapes drawn with `plot`.

For this problem, you are to develop a set of vector graphics functions that use the MATLAB `plot` command to draw scaled and translated squares, triangles, and circles as described below.

(a) Write a function `draw_square(s, tx, ty)`, where s is the scale factor, tx is the translation to the right, and ty is the translation up. Figure P12.4(a) shows a primitive square with a scale factor of 1 and no translation.

(b) Similarly, write a function `draw_triangle(s, tx, ty)` that draws a triangle as illustrated in Figure P12.4(b).

(c) Write a function `draw_circle(s, tx, ty)` that draws a circle as shown in Figure P12.4(c). Approximate the circle as a 100-sided polygon. To calculate the x- and y-coordinates of the vertices of this polygon, define a vector theta that is a series of angles from 0 to 2π in increments of 0.02π. The x-coordinates of the vertices will then be `cos(theta)` and the y-coordinates will be `sin(theta)`.

(d) Write a script that calls the shape-drawing function that you just wrote to draw a simple scene of your choosing, such as the one in Figure P12.5.

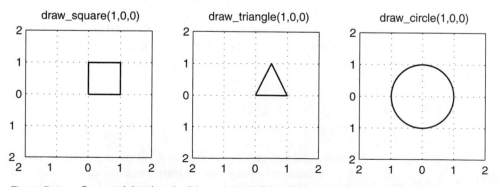

Figure P12.4 Square (a), triangle (b), and circle (c) produced with shape drawing functions.

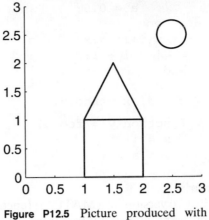

Figure P12.5 Picture produced with shape-drawing functions.

6. **Total Value of Coins**

Develop a function that determines the value, in dollars, of a collection of pennies, nickels, dimes, and quarters. (Or solve an equivalent problem in your local currency!)

(a) Write a pseudocode description of the function, clearly specifying the purpose of the function, its inputs, and its outputs.

(b) Implement the function in MATLAB.

(c) Test the function using the following sets of input data: 3 nickels, 7 quarters, 4 dimes, and 2 pennies; 3 dimes, 7 pennies, 1 quarter, and 2 nickels.

7. **MATLAB Vectors, Conditionals, and Loops**

What is the value of x at the end of execution of each of the following MATLAB scripts? Trace through each of the scripts by hand to get your answer, before checking it (if you wish) by typing it in. If you are unsure how any of these scripts work, single-step through them using the MATLAB debugger.

(a) `0:2:10`

(b) `0:2:11`

(c) `x = 0:3; 1 - 4, x(i+1) = x(i) + 2;`

(d)
```
n = 17; x = 0;
while (n > 1)
    n = n/2;
    x = x + 1;
end
```

(e)
```
x = 1;
for i = 1:5
    x = x*i;
end
```

(f)
```
x(1) = 0; i = 1;
while x(i) < 5
    if x(i) < 3
        a = 1;
    else
```

```
                    a = 0.5;
                  end
                  x(i+1) = x(i) + a;
                  i = i + 1;
              end
```

(g)
```
      x = 5; x = foo(x);
      function y = foo(a)
        x = 1;
        y = a + x;
```

8. Factorial

Write two versions of a MATLAB function called `my_factorial(n)` that computes the product of integers from 1 to n first using a `while` loop and then using a `for` loop. Test the function by comparing the results with the MATLAB library function `factorial`.

9. Vector Multiplication

Write a MATLAB function called `my_vmult(x,y)` that performs an element-wise multiplication of vectors x and y using a loop. Test the function by comparing the results with the MATLAB ".*" operator.

10. Finding the Maximum Value in a Vector the Hard Way

Write a MATLAB function called `my_max` that finds the maximum value in a vector. Your program should implement the algorithm described below:

```
Initialize the current maximum to the smallest number that
     MATLAB can represent (built-in variable realmin)
For each element of the vector
    If the current element is greater than current maximum
          Reset the current maximum to the current element
Return the current maximum
```

Test your function by comparing the results with the MATLAB library function `max`.

11. Finding the Greatest Common Divisor by Euclid's Method

The following algorithm was proposed by Euclid to find the greatest common divisor between two positive integers n1 and n2:

```
Assign n1 to variable q and assign n2 to variable p
Assign the remainder of q divided by p to variable r
While r is not zero
    Copy p into q
    Copy r into p
    Assign the remainder of q divided by p to variable r
Return p as the greatest common divisor
```

Implement this algorithm as a MATLAB function `euclid_gcd(n1,n2)`. Use either the MATLAB `mod` or `rem` library function to calculate the remainder of integer division. Test your function by comparing results with the MATLAB library function `gcd`.

12. Testing Divisibility
Write a MATLAB function called `sum_no_35` that sums only those numbers in the range 1 to *n* that are not evenly divisible by either 3 or 5. Use an `if-else` statement and the remainder function `rem`.

13. Equivalent Logic Expressions
Two logic expressions are said to be *tautologically equivalent* (symbol) if for all possible combinations of inputs, each case evaluates to the same truth value. If A and B are any valid logical expression, DeMorgan's rule states the following:

$$\sim A \ \& \ \sim B \ \Leftrightarrow \ \sim (A \ | \ B)$$
$$\sim A \ | \ \sim B \ \Leftrightarrow \ \sim (A \ \& \ B)$$

Taking advantage of DeMorgan's rule, write two versions of a function called `isinrange` that takes three numbers x, y, z as input and returns a 1 if the value of y is between x and z, inclusive, and a 0 otherwise. The two versions must use only the operators specified below:

- version 1, using only the relational operators <,> and the logical operators ~, &
- version 2, using only the relational operators <,> and the logical operators ~, |

You are not allowed to use `if` statements as part of your solution.

14. Tabulating Test Scores

(a) Write a script that uses MATLAB statistics functions that:
- sets up an array with five test score values of your choice
- computes the average (mean) test score and displays it
- computes the standard deviation, and displays it

(b) Write a MATLAB script similar to the previous one, but this time the five test scores must be read in from the keyboard. The test scores must be stored in an array, after which the mean and standard deviation must be computed and displayed.

(c) In the script that you wrote for the second part of the problem, the number of test scores was known beforehand (namely 5). You will now write a similar program, but for a situation where you lack prior knowledge of how many test scores there are: thus the program should work for any number of input values. The trick is to assume that valid test scores will always be positive. When a negative value is entered, this will serve as an "end-of-data" indicator, also known as a *sentinel*. Again, store the numbers in an array. You may assume there will always be at lease one test score. Your program must display how many numbers were entered (excluding the sentinel) before displaying the mean and standard deviation.

15. Simulating a Leaky Faucet
A leaky faucet is dripping at a rate of one 0.1 milliliter drop every 3 seconds. Write a MATLAB script that plots the amount of water lost over 1 hour.

16. A Parameter Sweep using Loops

Rewrite the script for Problem 10.6b, using two "nested loops" to automate the generation of the plot. This time, plot distance versus launcher vertical setting for five different launcher constants, rather than three. Follow the outline for the procedure given below.

```
% Outer loop
For launcher constants in the range 45 to 55, by 2
   % Inner loop
   For launcher settings in the range 0 to 2, by .1
      Include current launcher setting in a vector
       Calculate current distance and include in a vector
   Plot distance versus launcher setting
   Hold plotting to current axes
Add title, labels, and legend to plot
 Turn off holding of axes
```

17. Dropping a Computer Out of the Window

Having run into one too many inexplicable bugs, and utterly convinced of its inferiority to his trusty old UNIX workstation, a certain engineering professor tossed his new personal computer out the window of the top floor of the tallest building on campus. Estimate to within 0.2 s how long it will take for the computer to hit the ground.

(a) Use Euler's method *by hand* (that is, without using a computer other than a calculator) to obtain a first estimate of the time that it takes to hit the ground, assuming a step size of 0.5 s.

(b) Refine your estimate by writing a MATLAB program that implements Euler's method and plots both the height and the velocity of the "projectile" versus time. Use a step size of 0.01 s.

(c) Write out a top-down design and a bottom-up implementation and testing plan for your program *before* you sit down at the computer and start typing MATLAB code. Include a comment at the start of the program that describes what the program does, who wrote it, and when. Also include comments to describe variables and their units.

(d) In the discussion of your result, comment on the effects of step size. As you increase the step size, will you tend to overestimate or underestimate the time that it takes for the projectile to hit the ground? Explain.

18. Simulating a Sliding Block

The rate of change of speed, $\Delta V(t)/\Delta t$, for a block sliding down a smooth inclined plane can be written as:

$$\frac{\Delta V}{\Delta t} = g \cdot [\sin \theta - \mu \cos \theta]$$

where the following symbols represent constant parameters that influence the behavior of the sliding block:

- g is the acceleration of gravity (9.81 m/s^2 or 32.2 ft/s^2),
- θ is the angle of inclination of the plane,
- μ is the coefficient of sliding friction.

Note that the acceleration of the block—i.e., its rate of change of velocity—does not depend upon its weight, something you will study in your "statics" course.

Assuming: $\theta = 40$ and $\mu = 0.15$, use Euler's method to develop a numerical solution that will allow you to predict the speed of the block as it accelerates down the plane. Assume the block begins to slide as it is dislodged from rest and determine:

(a) How fast is the block moving after 4 seconds?

(b) How long does it take for a block to slide down a plane that is 10 feet long?

(c) How do your solutions depend upon the time step (Δt) you select for the Euler method?

19. Slowdown on the Toll Road

Suppose you are traveling westbound on I-90 and as you cross from Indiana into Illinois you are going 65 miles/hr. However, because of the approaching Chicago metropolitan area your velocity for the next 45 minutes satisfies the exponential relationship $v(t) = 65e^{-4t}$, where t is the time (in hours) since you entered Illinois and $v(t)$ is expressed in miles/hour. Using Euler's method, plot your velocity (in miles/hr) and your distance from the state line (in miles) as a function of time for the 45 minutes after you leave Indiana.

20. Simulating Pushing a Car into a Garage

You and a friend need to push a stalled car along a flat, straight driveway into a garage. The back wall of the garage is 20 m away from the front of the car and the garage is 5 m deep. The car has a mass of 1000 kg and is 3 m long. Assuming that you and your friend together can push the car with a constant force of 800 N and that the wheel bearings produce a constant friction force of 300 N, how far should you push the car before letting go so that it will roll to a stop inside the garage without hitting the wall?

While you might be able to solve this problem analytically, for this assignment you will build a numerical model for the motion of the car, and then use that model to search for a solution. Specifically, you must write a MATLAB function that uses Euler's method to calculate the total distance that the car travels for a given push distance. Once you have this model for a "trial run," you will then run trials for different push distances until you find one that is acceptable.

(a) Using a top-down design/bottom-up implementation methodology, write a MATLAB function to determine the total distance that the car travels for a given push distance. Think of a trial as having two phases that can each be modeled separately:

- *push phase*: From a dead stop, the car accelerates under a constant force $F_{net} = F_{push} - F_{friction}$ for some distance D_{push}, at which point it has a certain "launch velocity" V_{launch};
- *free-roll phase*: starting at a distance D_{push} and with an initial velocity V_{launch}, the car decelerates under a constant force $F_{friction}$ until it comes to rest at a distance D_{total}.

Your function should use Euler's method to simulate the motion of the car through both phases, generate a plot of velocity vs. time and position vs.

time, and return the total distance that the car has traveled by the end of the free-roll phase.

(b) Once you have a working version of the function, use it to search for a value of D_{push} that will park the car in the garage. You can do this simply by calling the function repeatedly from the command line with different values of D_{push}, or if you're feeling creative and ambitious (and you've already found an answer the "easy" way), you might try writing a script to automate the process. Be sure to clearly indicate your "acceptable" value for D_{push} and also include plots of velocity vs. time and position vs. time for this acceptable trial.

(c) Do your results seem reasonable? Explain.

21. Simulating a Damped Propulsion System

An automobile propulsion system can be modeled as a dynamic system with an input $u(t)$ (representing pedal position), and an output $v(t)$ (representing vehicle velocity) related by the equation

$$\frac{\Delta v(t)}{\Delta t} + p \cdot v(t) = b \cdot u(t)$$

where p is called the *damping factor* and b is a constant. This system can be simulated with the following MATLAB code:

```
dt = 0.01;
p = 3.8;
b = 27;
u(1) = 1;
v(1) = 0;
a(1) = b*u(1) - p*v(1);

for i=1:200
    u(i+1) = u(i);
    v(i+1) = v(i) + a(i)*dt;
    a(i+1) = b*u(i) - p*v(i+1);
end
```

This MATLAB script simulates 2.0 seconds of automobile operation when the vehicle begins at rest ($v(1) = 0$) and the input is a "unit step"—i.e., the accelerator is depressed 1.0 unit starting at time zero. (Note that the index value i corresponds to the time $t = (i - 1) \cdot \Delta t$, so i = 1 corresponds to time $t = 0$ and i = 201 corresponds to time $t = 2.0$ sec.) You can display the car's velocity and its acceleration as a function of time with the following MATLAB commands:

```
time = 0:0.01:2.0;
figure(1)
plot(time,v)
figure(2)
plot(time,a)}
```

(a) Execute the above MATLAB script and plot the results. Verify that the velocity starts at zero and approaches a final velocity of $b/p = 27/3.8 = 7.1$ units. Verify that the acceleration starts at 27.0 units and decreases to zero. Plot the velocity and acceleration for this case. Please label your graph with correct axes labels and a title that describes the problem number and what it is you are plotting, such as "Problem 1: Acceleration vs. Time ($p = 3.8$, $b = 27$)". Be sure to include values of p and b in the title!

(b) Change the script so that the input function $u(t) = 1.0$ for the first second of the simulation and $u(t) = 0.5$ for the second second of the simulation. Plot the velocity and acceleration for this case.

(c) Vary the damping factor p and the constant b and observe the effects on the final velocity and on how quickly the velocity changes. In particular, compare ($p = 3.8$, $b = 27$) with ($p = 38$, $b = 270$). Plot the velocity and acceleration for this case. Write out some comments on what's happening.

22. Trajectory of a Softball Without Drag

Write a MATLAB function of the form d = freeflight(v0,theta0) that determines the total horizontal distance d that a projectile will travel before hitting the ground if launched in a vacuum with an initial velocity of magnitude v0 at an angle from horizontal of theta0. The function must use Euler's method to determine the trajectory and calculate the following quantities at each time step i:

- horizontal displacement from the origin, x(i)
- vertical displacement from the origin, y(i)
- horizontal velocity, vx(i)
- vertical velocity, vy(i)
- magnitude of the total velocity, v(i)
- angle of the total velocity, theta(i)

In addition to returning total horizontal distance, the function must also produce plots (vertically aligned in a single figure) of y vs. x, v vs. x, and theta vs. x.

(a) Using freeflight, how far will the projectile travel given an initial velocity of 25 m/s and an initial angle of 45 degrees? Include plots of y vs. x, v vs. x, and theta vs. x as part of your solution.

(b) Using freeflight to search for a solution, find an initial velocity (in m/s) that will cause the projectile to travel a horizontal distance of 50 m when launched with an angle of 40 degrees. Include final plots as above.

(c) Using freeflight to search for a solution, find two different angles (in degrees) that will cause the projectile to travel a horizontal distance of 55 m when launched with an initial velocity of 25 m/s. Include final plots as above.

23. Powered Flight Trajectory

A projectile with mass 0.1 kg is equipped with a propulsion system that provides a constant thrust force $T = 5$ N for 0.8 s after launch, which always acts at an angle θ_0 relative to the horizontal throughout the powered portion of the flight. (After 0.8 s, the thrust force immediately goes to zero.)

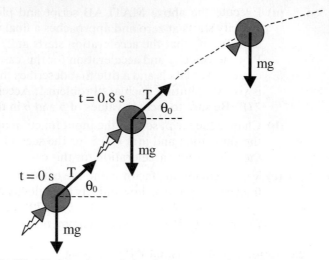

Write a MATLAB function of the form d = powerflight(theta0) that determines the total horizontal distance d that the projectile will travel before hitting the ground if launched in a vacuum at an angle of theta0. The function must use Euler's method to determine the trajectory and calculate the following quantities at each time step i:

- horizontal displacement from the origin, x(i)
- vertical displacement from the origin, y(i)
- horizontal velocity, vx(i)
- vertical velocity, vy(i)
- magnitude of the total velocity, v(i)
- angle of the total velocity, theta(i)

In addition to returning total horizontal distance, the function must also produce plots (vertically aligned in a single figure) of y vs. x, v vs. x, and theta vs. x.

(a) Using powerflight, how far will the projectile travel given a launch angle of 45 degrees? Include plots of y vs. x, v vs. x, and theta vs. x as part of your solution.

(b) Using powerflight to search for a solution, find two different angles (in degrees) that will cause the projectile to travel a horizontal distance of 100 m. Include final plots as above.

24. Trajectory of a Softball with Drag

Modify the function freeflight from the previous problem to create a new function d = dragflight(v0, theta0, m, A, Cd) that calculates the total horizontal distance covered by a projectile launched with an initial velocity v0 at an angle theta0 that includes the effects of drag. m, A, and Cd are the mass, cross-sectional area, and drag coefficient, respectively. Use MKS units throughout, and units of degrees for theta0. The density of air is 1.225 kg/m^3, and the drag coefficient of a sphere is 0.5. In addition to returning the horizontal distance, dragflight should also produce a plot of height versus distance.

Recall that under the vacuum conditions in free flight the acceleration was constant, whereas when we consider the effects of drag, the acceleration depends upon the velocity. Hence your Euler loop must calculate new values for the horizontal and vertical accelerations at each time step.

(a) Using dragflight, how far will a ball with a mass of 0.2 kg and a radius of 0.05 m travel given an initial velocity of 25 m/s and an initial angle of 45 degrees? Include a plot of y vs. x as part of your solution.

(b) Using dragflight to search for a solution, find an initial velocity (in m/s) that will cause the ball to travel a horizontal distance of 50 m when launched with an angle of 40 degrees. Include your final plot of y vs. x.

(c) Using dragflight to search for a solution, what angle will cause the ball to travel the furthest when launched with an initial velocity of 25 m/s? Include your final plot as above.

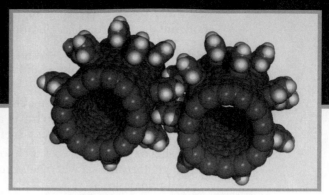

Problem Solving Process

Step 0: I Can   Have a positive attitude—view the problem as a challenge and don't give up too easily!

Step 1: Define   Be sure you fully understand the problem before you try to solve it.

- Identify the known and unknown quantities in the problem.
- Draw a diagram and define the problem in terms of variables.
- Restate the problem in clearer or more specific terms.

Step 2: Explore   A preplanning step—what is the problem actually asking you to solve, what general strategies and approaches apply, and what additional information do you need?

- Does the problem make sense?
- What are the key concepts and possible approaches?
- Do you need to make any assumptions?
- What levels of understanding are required?

Step 3: Plan   Determine the sequence of steps you will take *before* committing resources to the implementation of a solution.

- This is the toughest part of the process.
- Refer to the list of heuristics to help you get "unstuck."
- Concept maps are a convenient and dynamic way to organize a plan.

Step 4: Implement   The "do it" step, where you will often write out and solve equations.

- Don't start implementation until after planning.
- If implementation isn't going smoothly, consider going back and re-planning.

Step 5: Check   Convince yourself that the answer is acceptable.

- Sanity check—does the answer make sense?
- Test case—try plugging in a known good solution if you have one available.

Step 6: Generalize   Learn from the experience of solving this problem

- Did you learn any new specific facts?
- Could you have done something more efficiently?
- Can this solution be applied to another problem?
- Did you encounter any bugs or problems worth remembering in case you run into them again?

Step 7: Present the Results   Clearly communicate your results and the rationale to your audience.

- Show your work.
- Give good directions.
- Be neat.

Bloom's Taxonomy: Levels of Understanding

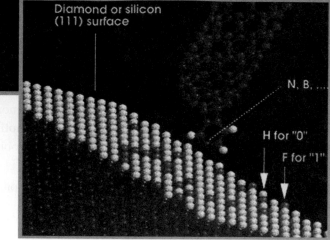

Diamond or silicon
(111) surface

N. B. ...

H for "0"

F for "1"

Levels of understanding in Bloom's Taxonomy, from most basic to most complex.

Knowledge: recalling facts from memory

Keywords: who, what, when, where, define, identify, state, select, recall

Comprehension: understanding meaning

Keywords: state in your own words, summarize, explain, paraphrase, classify, what is the meaning of, interpret, give an example of, what part doesn't fit, which is the best explanation

Application: using in new situations

Keywords: apply, use, construct, demonstrate, predict, explain how, what would happen if, show that, find

Analysis: breaking down into parts

Keywords: break down, what ideas apply, compare, contrast, what is the relationship, what conclusions, decompose, what is the justification, what might cause

Synthesis: constructing a new, integrated whole

Keywords: create, compose, design, devise, develop, solve, formulate, prove, plan

Evaluation: using judgment to make decisions

Keywords: judge, optimize, what is the best, decide, appraise, critique, defend, evaluate, what is the most appropriate

APPENDIX C

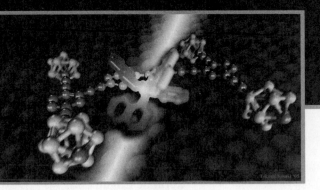

Engineering Societies and Professional Organizations

The following is a partial list of professional associations serving engineers. A broader list of professional associations, with links to their web sites is available at the Sloan Career Cornerstone Web site (http://www.careercornerstone.org/assoc.htm).

- Aerospace Engineering
  - Aerospace and Electronics Systems Society of the IEEE
  - Aerospace Industries Association
  - American Astronautical Society
  - American Institute of Aeronautics and Astronautics
  - Society of Flight Test Engineers
- Agricultural Engineering
  - American Society of Agricultural and Biological Engineers
- Architectural Engineering
  - American Society of Heating, Refrigerating and Air-Conditioning Engineers
  - Architectural Engineering Institute
  - Association for Computer Aided Design in Architecture (ACADIA)
  - International Council for Building Research Studies and Documentation
  - The Chartered Institution of Building Services Engineers
- Bioengineering
  - American Chemical Society
  - American Institute of Chemical Engineers Food, Pharmaceutical and Bio-engineering Division
  - American Society for Biochemistry and Molecular Biology

 – American Society of Agricultural and Biological Engineers
 – ASME Bioengineering Division
 – American Society For Microbiology
 – Biomedical Engineering Society
 – Biophysical Society
 – IEEE Engineering in Medicine and Biology Society
 – Institute of Biological Engineering
 – Society for Biological Engineering
 – Society of Industrial Microbiology
- Ceramic Engineering
 – The American Ceramic Society
- Chemical Engineering
 – American Chemical Society
 – American Institute of Chemical Engineers
 – Association of Consulting Chemists and Chemical Engineers
 – Electrochemical Society
 – Institution of Chemical Engineers
- Civil Engineering
 – American Society of Civil Engineers
 – American Congress on Surveying and Mapping
 – Institution of Civil Engineers–UK
- Computer Engineering and Computer Science
 – Association for Computing Machinery
 – Institute of Electrical and Electronics Engineers
 – Computer and Automated Systems Association of SME
 – IEEE Computer Society
- Electrical Engineering
 – Institute of Electrical and Electronics Engineers
 – IEEE Student Concourse
 – Institution of Electrical Engineers
 – Institution of Electrical Engineers Student, Education and Careers Area
- Environmental Engineering
 – American Academy of Environmental Engineers
 – Air and Waste Management Association
 – American Water Works Association
 – National Ground Water Association
- Geological/Geophysical Engineering
 – American Geologic Institute
 – American Geophysical Union
 – Canadian Society of Exploration Geophysicists
 – Environmental and Engineering Geophysical Society
 – Geological Society of America
 – Society of Exploration Geophysicists

- Industrial Engineering
  - Institute of Industrial Engineers
- Manufacturing Engineering
  - Society of Manufacturing Engineers
- Materials Engineering and Materials Science
  - The Minerals, Metals and Materials Society
  - The American Ceramic Society
  - Association for Iron and Steel Technology
  - ASM–The Materials Information Society
  - The Materials Research Society
- Mechanical Engineering
  - American Society of Mechanical Engineers
  - Institution of Mechanical Engineers
- Metallurgical, Mineral, and Mining Engineering
  - American Institute of Mining, Metallurgical, and Petroleum Engineers
  - The Minerals, Metals and Materials Society
  - Society for Mining, Metallurgy, and Exploration
- Naval and Marine Engineering
  - Society of Naval and Marine Engineers
- Petroleum Engineering
  - Society of Petroleum Engineers
- Plastics Engineering
  - American Institute of Chemical Engineers
  - Society of Automotive Engineers
  - Society of Manufacturing Engineers
  - Society of Petroleum Engineers
  - The Minerals, Metals and Materials Society
- Welding Engineering
  - Association of Mechanical Engineers
  - American Welding Society
  - Edison Welding Institute
  - Society of Manufacturing Engineers
  - The Minerals, Metals and Materials Society

Systems of Units

The International System of Units, SI, is the modern version of the metric system of measurement, and is widely accepted as the standard system of measurement for science and industry worldwide. The abbreviation "SI" comes from the French name for the system, *Le Système International d'Unités*, and the system is maintained by the International Bureau of Weights and Measures or *Bureau International des Poids et Mesures* (BIPM), which is based in France. As a rule, most countries today use SI units as the basis for everyday measurements; the most glaring exception to this—of course—is the United States, which uses a mix of English customary units along with some metric units, often in combination. For example, Americans purchase beverages in 2 liter bottles but 12 ounce cans. We buy gasoline by the gallon, and rate engines in horsepower, but measure cylinder displacement in liters. For this book, we've mostly used SI units, but as a concession to the American conventions, we've added some usage of English units to the mix. In this appendix, we provide a brief summary of SI, along with conversions to common English units. A complete discussion of the SI system and conventional English units may be found at the BIPM Web site (http://www.bipm.org/en/si/) as well as at the U.S. National Institute of Standards Web site (http://physics.nist.gov/cuu/Units/index.html).

D.1 | THE SI SYSTEM

The SI system is founded on 7 *base units* that are regarded as being dimensionally independent. Table D.1 lists the 7 base units. Built on top of the base units are a set of *derived units*, defined by a system of quantity equations. There are many different officially approved derived units; Table D.2 lists those derived units that are used in this text. Finally, there is another set of units that are not officially part of the SI system, but they are accepted for use by the BIPM. Table D.3 lists examples of these units.

The SI system also defines a set of *prefixes* to defines multiples or decimal parts of units. Table D.4 lists these prefixes.

TABLE D.1   SI base units

Base Quantity	Name	Symbol
length	meter (metre)	m
mass	kilogram	kg
time, duration	second	s
electric current	ampere	A
thermodynamic temperature	kelvin	K
amount of substance	mole	mol
luminous intensity	candela	cd

TABLE D.2   SI derived units

Derived Quantity	Name	Symbol	Expressed in Terms of Other SI Units	Expressed in Terms of SI Base Units
plane angle	radian	rad	1	m/m
force	newton	N		$m\ kg\ s^2$
pressure, stress	pascal	Pa	N/m^2	$m^{-1}\ kg\ s^{-2}$
energy, work, amount of heat	joule	J	N m	$m^2\ kg\ s^{-2}$
power	watt	W	J/s	$m^2\ kg\ s^{-3}$
electric charge	coulomb	C		s A
electric potential difference	volt	V	W/A	$m^2\ kg\ s^{-3}\ A^{-1}$
electric resistance	ohm	Ω	V/A	$m^2\ kg\ s^{-3}\ A^{-2}$
Celsius temperature	degree Celsius	°C		K

TABLE D.3   Non-SI units accepted for use with SI

Quantity	Name	Symbol	Value in SI Units
time	minute	min	1 min = 60 s
	hour	h	1 h = 60 min = 3,600 s
	day	d	1 d = 24 h = 86,400 s
plane angle	degree	°	$1° = (\pi/180)$ rad
	minute	′	$1' = (1/60)° = (\pi/10,800)$ rad
	second	″	$1'' = (1/60)' = (\pi/648,000)$ rad
area	hectare	ha	$1\ ha = 1\ hm^2 = 10^4\ m^2$
volume	liter (litre)	L, l	$1\ L = 1\ dm^3 = 10^3\ cm^3 = 10^{-3}\ m^3$
mass	metric ton (tonne)	t	$1\ t = 10^3\ kg$

TABLE D.4 SI prefixes

Factor	Name	Symbol	Factor	Name	Symbol
10^1	deca	da	10^{-1}	deci	d
10^2	hecto	h	10^{-2}	centi	c
10^3	kilo	k	10^{-3}	milli	m
10^6	mega	M	10^{-6}	micro	μ
10^9	giga	G	10^{-9}	nano	n
10^{12}	tera	T	10^{-12}	pico	p
10^{15}	peta	P	10^{-15}	femto	f
10^{18}	exa	E	10^{-18}	atto	a
10^{21}	zetta	Z	10^{-21}	zepto	z
10^{24}	yotta	Y	10^{-24}	yocto	y

D.2 | NON-SI UNITS AND CONVERSION FACTORS

Although unapproved by the BIPM, there are nonetheless many other units that scientists and engineers use frequently. Table D.5 lists some of the more common of these, along with factors for converting them to equivalent SI units.

TABLE D.5 Conversion of non-SI units to SI units

Quantity	To Convert From	To	Multiply By	
Acceleration:	ft/s^2	m/s^2	3.048	$\times 10^{-1}$
Area:	acre	m^2	4.046 873	$\times 10^3$
	ft^2	m^2	9.290 304	$\times 10^{-2}$
	in^2	m^2	6.4516	$\times 10^{-4}$
Energy and Work:	British thermal unit (Btu)	J	1.055 056	$\times 10^3$
	calorie (cal)	J	4.1868	$\times 10^0$
	calorie, nutritional	J	4.1868	$\times 10^3$
	electron volt (eV)	J	1.602 177	$\times 10^{-19}$
	foot pound (ft·lbf)	J	1.355 818	$\times 10^0$
	kilowatt hour (kW·h)	J	3.6	$\times 10^6$
Force:	kip (1 kip = 1000 lbf)	N	4.448 222	$\times 10^3$
	ounce-force (ozf)	N	2.780 139	$\times 10^{-1}$
	pound-force (lbf)	N	4.448 222	$\times 10^0$
	ton-force (2000 lbf)	N	8.896 443	$\times 10^3$
Fuel Consumption:	mi/gal (U.S.) (mpg)	km/L	4.251 437	$\times 10^{-1}$
	mi/gal (U.S.) (mpg)	L per 100 km	divide 235.215 by mpg	
Length:	angstrom (Å)	m	1.0	$\times 10^{-10}$
	astronomical unit (AU)	m	1.495 979	$\times 10^{11}$
	foot (ft)	m	3.048	$\times 10^{-1}$

TABLE D.5 Conversion of non-SI units to SI units (Continued)

Quantity	To Convert From	To	Multiply By	
	inch (in)	m	2.54	$\times 10^{-2}$
	light year	m	9.460 73	$\times 10^{15}$
	micron (μ)	m	1.0	$\times 10^{-6}$
	mil (0.001 in)	m	2.54	$\times 10^{-5}$
	mile (mi)	m	1.609 344	$\times 10^{3}$
	yard (yd)	m	9.144	$\times 10^{-1}$
Mass:	carat, metric	gram (g)	2.0	$\times 10^{-1}$
	ounce (oz)	gram (g)	2.834 952	$\times 10^{1}$
	pound (lb, lbm)	kg	4.535 924	$\times 10^{-1}$
	slug (slug)	kg	1.459 390	$\times 10^{1}$
	ton, short (2000 lb)	kg	9.071 847	$\times 10^{2}$
Power:	horsepower (550 ft lbf/s)	W	7.456 999	$\times 10^{2}$
Pressure or Stress:	atmosphere (atm)	Pa	1.013 25	$\times 10^{5}$
	cm of mercury (cmHg)	Pa	1.333 224	$\times 10^{3}$
	in of mercury (inHg)	Pa	3.386 389	$\times 10^{3}$
	lbf/in^2 (psi)	Pa	6.894 757	$\times 10^{3}$
Temperature:	degree Celsius ($^{\circ}$C)	K	$K = {}^{\circ}C + 273.15$	
	degree Fahrenheit ($^{\circ}$F)	$^{\circ}$C	${}^{\circ}C = ({}^{\circ}F - 32)/1.8$	
Velocity:	ft/s	m/s	3.048	$\times 10^{-1}$
	km/h	m/s	2.777 778	$\times 10^{-1}$
	knot (nautical mph)	m/s	5.144 444	$\times 10^{-1}$
	mi/h (mph)	m/s	4.4704	$\times 10^{-1}$
	mi/h (mph)	km/h	1.609 344	$\times 10^{0}$
Volume:	barrel (42 U.S. gallons) (bbl)	m^3	1.589 873	$\times 10^{-1}$
	gallon (Imperial) (gal)	m^3	4.546 09	$\times 10^{-3}$
	gallon (U.S.) (gal)	m^3	3.785 412	$\times 10^{-3}$

Bibliography

[AIS77]   Christopher Alexander, Sara Ishikawa, and Murray Silverstein. *A Pattern Language.* Oxford University Press, New York, NY, 1977.

[Ame05]   American Society of Civil Engineers. *Report Card for America's Infrastructure.* Technical report, 2005. http://www.asce.org/reportcard/2005/index.cfm.

[And80]   John R. Anderson. *Cognitive Psychology and Its Implications.* W. H. Freeman and Company, San Francisco, CA, 1980.

[And83]   John R. Anderson. *The Architecture of Cognition.* Harvard University Press, Cambridge, MA, 1983.

[ANH78]   D. P. Ausubel, J. D. Novak, and H. Hanesian. *Educational Psychology: A Cognitive View, Second Edition.* Holt, Rinehart, and Winston, New York, NY, 1978.

[Arb92]   John Arbuthnot. *Of the Laws of Chance, or a Method of Calculation of the Hazards of Game (English translation of Huygens' de Ratiociniis in Ludo Aleae).* Benjamin Motte, London, 1692.

[Ban06]   The World Bank. *World Development Indicators.* Washington, D.C., 2006. http://devdata.worldbank.org/wdi2006.

[Bar03] Albert-László Barabási. *Linked.* Plume, New York, 2003.

[Bau02]   Ray H. Baughman. Carbon Nanotubes–the Route Toward Applications. *Science,* 297:787–792, August 2 2002.

[BBC00]   John D. Bransford, Ann L. Brown, and Rodney R. Cocking, editors. *How People Learn: Brain, Mind, Experience, and School.* National Academy Press, Washington, D.C., 2000.

[BBK+03]   Gary H. Bernstein, Jay B. Brockman, Peter M. Kogge, Gregory L. Snider, and Barbara E. Walvoord. From Bits to Chips: A Multidisciplinary curriculum for microelectronics System Design Education. In *MSB' 03: Proceedings of the 2003 International Conference on Microelectronics Systems Education,* page 95, Washington, DC, USA, 2003. IEEE Computer Society.

[Bea97]   Tim Beardsley. The Machinery of Thought. *Scientific American,* 277(2):78–83, August 1997.

[BES+99]   Tom Benson, John Eigenauer, Roger Storm, Bruce Bream, and Darryl Palmer. FoilSim II. WWW document, 1999. http://www.grc.nasa.go v/WWW/K-12/airplane/foil2.html (visited June 21, 2005).

[BFB02]   Jay B. Brockman, Thomas E. Fuja, and Stephen M. Batill. A Multidisciplinary Course Sequence for First-year Engineering Students. In *2002 ASEE Annual Conference and Exposition,* Montreal, Quebec, Canada, June 2002.

[BIA+90]   J. E. Black, K. R. Isaacs, B. J. Anderson, A. A. Alcantara, and W. T. Greenough. Learning Causes Synaptogenesis, whereas Motor Activity Causes Angiogenesis, in Cerebellar Cortex of Adult Rats. *Proceedings of the National Academy of Sciences, U.S.A,* 87:5568–5572, 1990.

[Bla13]   Newton Henry Black. *Practical Physics for Secondary Schools: Fundamental Principles and Applications to Everyday Life.* Macmillan and Co., 1913. available at http://www.archive.org/details/practicalphysics00blacrich.

[Bla66]   Richard J. Blackwell. Descartes' Laws of Motion. *Isis*, 57(2):220–234, 1966.

[Bos06]   Pradip Bose. Designing Reliable Systems with Unreliable Components. *IEEE Micro*, pages 5–6, September/October 2006.

[Boy60]   Robert Boyle. *New Experiments Physico-mechanicall, Touching the Spring of the Air, and Its Effects*. H. Hall, Oxford, 1660.

[Boy61]   Robert Boyle. *The Sceptical Chymist, or, Chymico-physical Doubts & Paradoxes*. J. Caldwell, London, 1661.

[bp:06]   *BP Statistical Review of World Energy 2006*. London, UK, June 2006. Available online at www.bp.com/statisticalreview.

[BR06]   Stephen M. Batill and John E. Renaud. Ame40463: Senior Design Project, Course Handout Package. Www document, Department of Aerospace and Mechanical Engineering, University of Notre Dame, http://www.nd.edu/batill/www.ame40463f06/AME40463_handout_f06.pdf, August 2006.

[Buz95]   Tony Buzan. *The Mind Map Book*. BBC Books, London, UK, 1995.

[Car90]   Sadi Carnot. *Reflections on the Motive Power of Heat*. John Wiley and Sons, New York, 1890. Available online through the Internet Archive, www.archive.org.

[CDM96]   John C. Carneson, George Delpierre, and Ken Masters. Designing and Managing Multiple Choice Questions. WWW document, 1996. http://web.uct.ac.za/projects/cbe/mcqman/mcqman01.html (visited May 30, 2005).

[Cha96]   Allan Chapman. England's Leonardo: Robert Hooke (1635–1703) and the Art of Experiment in Restoration England. *Proceedings of the Royal Institution of Great Britain*, 67:239–275, 1996. Available online at http://www.roberthooke.org.uk/leonardo.htm (visited August 19, 2007).

[Cla79]   Rudolf Clausius. *Mechanical Theory of Heat*. Macmillan and Co., 1879.

[DA94]   Cliff I. Davidson and Susan A. Ambrose. *New Professor's Handbook: A Guide to Teaching and Research in Engineering and Science*. Anker Publishing Company, Inc., Bolton, MA, 1994.

[DB05]   M. Suzanne Donavan and John D. Bransford, editors. *How Students Learn: History, Mathematics, and Science in the Classroom*. National Academy Press, Washington, D.C., 2005.

[Des37]   René Descartes. *Discours de la Methode: Pour Bien Conduire Sa Raison, & Chercher la Verité Dans les Sciences. Plus La Dioptrique. Les Meteores. Et La Geometrie. Qui Sont des Essais de cete Methode*. Ian Maire, Leyden, 1637.

[Des91]   René Descartes. *Principles of Philosophy*. Kluwer Academic Publishers, Dordrecht, The Netherlands, 1991. Translation of the 1644 Latin edition, *Principia Philosophiae*, with additional material from the 1647 French translation, translated into English by R.P. Miller.

[Dun05]   Duncan Hines. Signature Desserts: Chocolate Peanut Butter. Cake mix, 2005.

[EDNK92]   B. H. Litwiller, E. D. Nichols, and P. A. Kennedy. *Holt Pre-Algebra*. Holt, Rinehart, and Winston, Orlando, Florida, 1992.

[Ene]   Energy Information Administration. Official Energy Statistics from the U.S. Government.

[FC86]   S. L. Friedman and R. R. Cocking. *The Brain, Cognition, and Education*, chapter Instructional Influences on Cognition and on the Brain, pages 319–343. Academic Press, Orlando, FL, 1986.

[Fey94]   Richard P. Feynman. *Six Easy Pieces: Essentials of Physics Explained by Its Most Brilliant Teacher*. Perseus Books, Reading, MA, 1994.

[FJ05]   Robert A. Freitas Jr. Microbivores: Artificial Mechanical Phagocytes Using Digest and Discharge Protocol. *Journal of Evolution and Technology*, 14, 2005. http://jetpress.org/volume14/freitas.html.

[FTG96]   Craig Pass, Brian Turtle, and Mike Ginelli. *Six Degrees of Kevin Bacon*. Plume, New York, 1996.

[Gal81]   Robert L. Galloway. *The Steam Engine and Its Inventors*. Macmillan and Co., 1881.

[Gal14]   Galileo Galilei. *Dialogues Concerning Two New Sciences*. Macmillan and Co., New York, 1914. English translation of 1638 Latin and Italian edition by H. Crew and A. De Salvio.

[GHJV95]   Erich Gamma, Richard Helm, Ralph Johnson, and John Vlissides. *Design Patterns: Elements of Reusable Object-Oriented Software*. Addison-Wesley, Reading, MA, 1995.

[Goe75]   G. Daniel Goehring. 17th Century Treatments of One dimensional Collisions. *Physics Education*, pages 457–460, September 1975.

[Hal90]   Anders Hald. *A History of Probability and Statistics and their Applications before 1750*. John Wiley and Sons, New York, 1990.

[har69]   Engraving of Joule's Apparatus for Measuring the Mechanical Equivalent of Heat. *Harper's New Monthly Magazine*, (231), August 1869. from Wikimedia Commons, public domain photo.

[HGJD97]   J. Han, A. Globus, R. Jaffe, and Glenn Deardorff. Molecular Dynamics Simulation of Carbon Nanotube Based Gears. *Nanotechnology*, 8:95–102, 1997.

[Hof05]   H.P. Hofstee. Power Efficient Processor Architecture and the Cell Processor. In *llth International Symposium on High-Performance Computer Architecture*, pages 258–262. IEEE, February 2005.

[Hol75]   John H. Holland. *Adaptation in Natural and Artificial Systems*. University of Michigan Press, Ann Arbor, MI, 1975.

[Hol84]   P. Holgate. The Influence of Huygens' Work in Dynamics on His Contribution to Probability. *International Statistical Review*, 52(2):137–140, August 1984.

[Hol92]   John H. Holland. Genetic Algorithms. *Scientific American*, pages 66–72, July 1992.

[Hoo78]   Robert Hooke. *Lectures de Potentia Restitutiva, or, Of Spring: Explaining the Power of Springing Bodies*. John Martyn, London, 1678.

[How]   HowStuffWorks.com. How the Wii Works, Nov. 12, 2007. http://electronics.howstuffworks.com/wii.htm.

[HrR04]   Pat S. Hu and Timothy R. Reuscher. Summary of Travel Trends, 2001 National Household Travel Survey. Technical report, U.S. Department of Transportation, Federal Highway Administration, Springfield, VA, December 2004.

[Hub88]   David H. Hubel. *Eye, Brain, and Vision*. Scientific American Library Series, No. 22. W. H. Freeman and Co., New York, NY, 1988.

[Hun]   Michael Hunter. Robert boyle: An Introduction. http://www.bbk.ac.uk/boyle/boyle_learn/boyle_introduction.htm (viewed August 18, 2007).

[Huy88]   Christiaan Huygens. De Motu Corporum ex Percussione. In *Oeuvres Completes de Christiaan Huygens*. M. Nijhoff, 1888.

[Ilt71]   Carolyn Iltis. Leibniz and the Vis Viva Controversy. *Isis*, 62(1):21–35, 1971.

[Ins07]   Institute and Museum of History of Science, Florence, Italy. Horror Vacui, 2007. http://www.imss.fi.it/vuoto/index.html (visited May 17, 2007).

[Jou63a]   James Prescott Joule. On Matter, Living Force, and Heat (a lecture at

St. Ann's church reading-room; and published in the Manchester Courier newspaper, May 5 and 12, 1847). In *The Scientific Works of James Prescott Joule*, pages 265–276. Dawsons of Pall Mall, 1963.

[Jou63b]   James Prescott Joule. On the Heat Evolved by Metallic Conductors of Electricity, and in the Cells of a Battery During Electrolysis (*Philosophical Magazine*, Vol. xix, p. 200). In *The Scientific Papers of James Prescott Joule*. Dawsons of Pall Mall, 1963.

[Jou63c]   James Prescott Joule. On the Mechanical Equivalent of Heat (Brit. Assoc. Rep. 1845, trans, chemical sect. p. 31.). In *The Scientific Papers of James Prescott Joule*, page 202. Dawsons of Pall Mall, London, 1963.

[Jou63d]   James Prescott Joule. On the Mechanical Equivalent of Heat (*Philosophical Transactions*, 1850, part i.). In *The Scientific Works of James Prescott Joule*, pages 298–328. Dawsons of Pall Mall, 1963.

[KGV83]   S. Kirkpatrick, C. D. Gelatt, and M. P. Vecchi. Optimization by Simulated Annealing. *Science*, 220(4598):671–680, 1983.

[KY06]   Kristopher Kubicki and Marcus Yam. The Sony Playstation 3 Dissected. *DailyTech*, November 11 2006.

[Lav89]   Antoine Lavoisier. *Traite Elementaire de Chemie*. Cruchet, Paris, 1789.

[Len00]   Craig S. Lent. Molecular Electronics: Bypassing the Transistor Paradigm. *Science*, 288(5471):1597–1599, June 2 2000.

[LIL03]   C.S. Lent, B. Isaksen, and M. Lieberman. Molecular Quantum-dot Cellular Automata. *Journal of the American Chemical Society*, (125):1056–1063, 2003.

[LL95]   Edward Lumsdaine and Monika Lumsdaine. *Creative Problem Solving: Thinking Skills for a Changing World*. McGraw-Hill, Inc., 1995.

[MA06]   J. J. Macintosh and Peter Anstey. Robert Boyle. *Stanford Encyclopedia of Philosophy*, 2006. http://plato.stanford.edu/entries/boyle/.

[Mai31]   Norman R. F. Maier. Reasoning in Humans II: The Solution of a Problem and Its Appearance in Consciousness. *Journal of Comparative Psychology*, (12):181–194, 1931.

[Mid63]   W. E. Knowles Middleton. The Place of Torricelli in the History of the Barometer. *Isis*, 54(1):11–28, March 1963.

[Mor07]   Eric Morales. A Tiger by the Tail: HPC Designs High Performance Golf Clubs. In *Supercomputing 07*, Reno, NV, November 2007.

[Moy77]   Albert E. Moyer. Robert Hooke's Ambiguous Presentation of "Hooke's Law". *Isis*, 68(242):266–275, 1977.

[MRR$^+$53]   N. Metropolis, A. Rosenbluth, M. Rosenbluth, A. Teller, and E. Teller. Equation of State Calculations by Fast Computing Machines. *Journal of Chemical Physics*, 21(6):1087–1092, 1953.

[Nat04]   National Academy of Engineering. *The Engineer of 2020: Visions of Engineering in the New Century*. National Academies Press, Washington, DC, 2004.

[Nat05]   National Academy of Engineering. *Educating the Engineer of 2020: Adapting Engineering Education to a New Century*. National Academies Press, Washington, DC, 2005.

[Nat06]   National Science Foundation. Engineering and Science Indicators 2006. Technical report, http://www.nsf.gov/statistics/seind06/, 2006.

[New29]   Sir Isaac Newton. *The Mathematical Principles of Natural Philosopy*. Benjamin Motte, London, 1729. Translated into English by Andrew Motte and John Machin.

[Newton1686]   Sir Isaac Newton. *Principia Mathematica*.

[nob]   nobelprize.org.

[Nov91]   Joseph D. Novak. Clarify with Concept Maps: A Tool for Students and Teachers Alike. *The Science Teacher*, 58(7):45–49, oct 1991.

[NS63]   Allan Newell and Herbert A. Simon. GPS, a Program that Simulates Human Thought. In Edward Feigenbaum A. and Julian Feldman, editors, *Computers and Thought*, pages 279–293. McGraw-Hill, New York, 1963.

[NS72]   Allen Newell and Herbert A. Simon. *Human Problem Solving*. Prentice-Hall, Inc., Englewood Cliffs, NJ, 1972.

[oil05]   *Oil and Gas Journal*, 103(47), December, 19 2005.

[Par]   Parametric Technology Corporation. Pro/ENGINEER. www.ptc.com.

[PB84]   A. S. Palincsar and A. L. Brown. Reciprocal Teaching of Comprehension Monitoring Activities. *Cognition and Instruction*, 1:117–175, 1984.

[Pirsig2006]   Robert M. Pirsig, *Zen and the Art of Motorcycle Maintenance*. (New York: HarperTorch, 1992) page 417.

[PMS03]   C. Pieronek, L. McWiliams, and S. Silliman. Initial Observations on Student Retention and Course Satisfaction Based on First-year Engineering Student Survey and Interviews. In *ASEE Annual Conference and Exposition*, Nashville, TN, June 2003.

[Pol45]   G. Polya. *How to Solve It*. Princeton University Press, Princeton, NJ, 1945.

[Pou05]   William Poundstone. *Fortune's Formula: The Untold Story of the Scientific Betting System That Beat the Casinos and Wall Street*. Hill and Wang Publishers, 2005.

[Qui78]   M. R. Quillian. Semantic Memory. In M. Minsky, editor, *Semantic Information Processing*. The M.I.T. Press, Cambridge, MA, 1978.

[Rey92]   John Reynolds. Handpumps: Toward a Sustainable Technology. Report 11426, World Bank Water and Sanitation Program, Washington, DC, October 1992.

[RH97]   Michael Riordan and Lillian Hoddeson. *Crystal Fire: the Invention of the Transistor and the Birth of the Information Age*. Norton, New York, 1997.

[Ric58]   R. H. Richens. Interlingual Machine Translation. *The Computer Journal*, 1(3):144–147, October 1958.

[Rub75]   Moshe F. Rubinstein. *Patterns of Problem Solving*. Prentice-Hall, Inc., Englewood Cliffs, NJ, 1975.

[Sha37]   Claude E. Shannon. A Symbolic Analysis of Relay and Switching Circuits. Master's thesis, Massachusetts Institute of Technology, August 1937.

[Sha48]   Claude E. Shannon. A Mathematical Theory of Communication. *Bell System Technical Journal*, 27:379–423, 623–656, July, October 1948.

[Sim73]   Robert Simmons. Semantic Networks : Their Computation and Use for Understanding English Sentences. In R. C. Schank and K. M. Colby, editors, *Computer Models of Thought and Language*, pages 63–113. W. H. Freeman, San Francisco, CA, 1973.

[Sim96]   Herbert A. Simon. *The Sciences of the Artificial, Third Edition*. MIT Press, Cambridge, MA, 1996.

[Sky05]   Skywest Airlines. Skywest/United Express Route Map. WWW document, 2005. http://www.skywest.com/ (visited June 11, 2005).

[SMS+06]   Yasuhiro Shirai, Jean-Franois Morin, Takashi Sasaki, Jason M. Guerrero, and James M. Tour. Recent Progress on Nanovehicles. *Chemical Society Reviews*, (35):1043–1055, September 2006.

[SOZ+05]   Y. Shirai, A. J. Osgood, Y. Zhao, K. F. Kelly, and J. M. Tour. Directional Control in Thermally Driven Single-molecule Nanocars. *Nano Letters*, 5(ll):2330–2334, 2005.

[Tay00]   Taylor Research and Consulting Group. Student and Academic Research Study: Final Quantitative Study. Technical report, http://www.aicpa.org/members/div/career/edu/taylor.htm, 2000.

[The]   The Royal Society, www.royalsoc.ac.uk.

[TLK52]   William Thomson (Lord Kelvin). On a Universal Tendency in Nature to the Dissipation of Mechanical Energy. *Proceedings of the Royal Society of Edinburgh,* April 19 1852.

[Tou]   James M. Tour. Research Group Homepage. WWW document. http://www.jmtour.com.

[U.S05]   U.S. Department of Energy, Argonne National Laboratory. Newton bbs. WWW document, 2005. http://www.newton.dep.anl.gov/ (visited June 6, 2005).

[U.S06a]   U.S. Census Bureau. *International Database.* 2006. http://www.census.gov/ipc/www/idbnew.html.

[U.S06b]   U.S. Central Intelligence Agency. *The World Factbook.* 2006. https://www.cia.gov/cia/publications/factbook/index.html.

[Vala]   Robert Valdes. How Playstation 3 Works, Nov. 12, 2007. http://electronics.howstuffworks.com/playstation-three.htm.

[Valb]   Robert Valdes. How Xbox 360 Works, Nov. 12, 2007. http://electronics.howstuffworks.com/xbox three-sixty.htm.

[VB89]   S. Vosniadou and W.F. Brewer. The Concept of the Earth's shape: A Study of Conceptual Change in Childhood. Unpublished paper, Center for the Study of Reading, University of Illinois, Champaign, Illinois, 1989.

[Wat99]   Water, Engineering, and Development Center, Loughborough University. VLOM pumps. In Rod Shaw, editor, *Running Water: More Technical Briefs on Health Water, and Sanitation*, number 41. Intermediate Technology Publications, London, UK, 1999. http://www.lboro.ac.uk/well/resources/technical-briefs/41-vlom-pumps.pdf (Visited November 3, 2005).

[WCHW79]   D. R. Woods, C. M. Crowe, T. W. Hoffman, and J. D. Wright. Major Challenges to Teaching Problem Solving Skills. *Engineering Education,* December 1979.

[Wes05]   John B. West. The Original Presentation of Boyle's Law. *Journal of Applied Physiology*, 98(1):31–39, January 2005.

[Wik]   Wikimedia Commons. First Piston Steam Engine, by Papin. 19th century encyclopedia. http://en.wikipedia.org/wiki/Image:Papinengine.jpg (accessed Nov. 15, 2007).

[WO93]   Philip Wankat and Frank Oreovicz. *Teaching Engineering.* McGraw-Hill, Inc., New York, 1993.

[Wor05a]   World Bank Water and Sanitation Program. WWW document, 2005. http://www.wsp.org.

[Wor05b]   World Health Organization/UNICEF. Water for life: Making It Happen. Report http://www.who.int/entity/water_sanitation_health/ waterforlife.pdf, WHO/UNICEF Joint Monitoring Programme for Water Supply and Sanitation, Geneva, Switzerland, 2005.

[WWH+75]   D. R. Woods, J. D. Wright, T. W. Hoffman, R. K. Swartman, and I. K. Doig. Teaching Problem Solving Skills. *Engineering Education*, December 1975.

[You07]   Thomas Young. Lecture 8. *Course of Lectures on Natural Philosophy*, 1807.

[You93]   J. M. Younse. Mirrors on a Chip. *IEEE Spectrum*, pages 27–31, November 1993.

Index

Page references followed by italic *t* indicate material in tables, while page references followed by italic *n* indicate material in footnotes. A limited number of names of famous figures in science and engineering are also included.

unnecessary constraints, 133–134
urban planning, 21
urban railway systems, tricks in design, 209
urban redevelopment, civil engineers work in, 21
U.S. National Institute of Standards, 554
U.S. Strategic Petroleum Reserve, 392
use, keyword in Bloom's Taxonomy, 93, 550

V

vacuum, horror of, 171–173
vacuum conditions, 373
vacuum tubes, 344–345
variables
    in Boolean algebra, 326–329
    change of variables, 147
    in MATLAB, 421–422
variable vector, of system of linear equations, 481
vector addition, 265–269
    MATLAB, 527–528
vector graphics, 537–538
vector quantities, 264
vectors, 433. *See also* MATLAB vector operations
VELCRO, 138
velocity, 164
video games, 14, 36
village-level operation and maintenance
    (VLOM), 75
VisiCalc, 34
vision systems, manufacturing engineers work
    with, 24
visual cortex, 46
vis viva (living force), 169–170
vocabulary, 89
voicemail, 329–330
voltage
    and energy, 338–339
    Kirchoff's Laws for determining, 348–353
    and Ohm's law, 342–343
voltage divider, 351–353, 355
voltage source, 344

W

wage and salary administration systems, industrial
    engineers work develop, 24
Wallis, John, 167
waste disposal, environmental engineers work
    with, 23
water, basic human need, 3
water irrigation systems, agricultural engineers
    work on, 19
water pollution control
    environmental engineers work on, 23
    mining engineers work on, 25
water resources engineering, 21
water tower painting, 108–109
    MATLAB application, 429–430
Watt, James, 156, 181–183
Watt's steam engine, 156, 181–183
web browsers, computer scientists work with, 22

web of innovation, 11–15
welding engineering, 26
    professional associations, 553
what, keyword in Bloom's Taxonomy, 92, 550
what conclusions, keywords in Bloom's
    Taxonomy, 94, 550
what ideas apply, keywords in Bloom's Taxonomy,
    94, 550
"what is" (declarative) knowledge, 40, 42
what is the best, keywords in Bloom's Taxonomy,
    95, 550
what is the justification, keywords in Bloom's
    Taxonomy, 94, 550
what is the meaning of, keywords in Bloom's
    Taxonomy, 93, 550
what is the most appropriate, keywords in
    Bloom's Taxonomy, 95, 550
what is the relationship between, keywords in
    Bloom's Taxonomy, 94, 550
what might cause, keywords in Bloom's
    Taxonomy, 94, 550
what part doesn't fit, keywords in Bloom's
    Taxonomy, 93, 550
what would happen if, keywords in Bloom's
    Taxonomy, 93, 550
when, keyword in Bloom's Taxonomy, 92, 550
where, keyword in Bloom's Taxonomy, 92, 550
which is the best explanation, keywords in
    Bloom's Taxonomy, 93, 550
WHILE loops, 508, 509
    syntax of MATLAB, 522, 523
WHILE statement, 508
who, keyword in Bloom's Taxonomy, 92, 550
who command, MATLAB, 422
    in debugging, 516
wildlife protection, environmental engineers work
    on, 23
wind energy, 385
    power engineers work with, 23
wind tunnel aircraft models, 9
wire, power dissipation through, 362–363
women, opportunities in engineering, 12–14
wood, 288–289
    mechanical properties, 289t
word problems, 34, 154
work, 169–170
    and heat loss in circuits, 341
work backward heuristic, 108, 109, 130–131
work forward heuristic, 108, 130–131
workspace, in MATLAB, 422
world population growth model, MATLAB
    application, 530–532
world view, 59
world wide oil reserves, 385–407
World Wide Web, 14–15

X

x-ray imaging systems, biomedical engineers work
    with, 20